Student Study G

Chemistry

Seventh Edition

Raymond Chang
Williams College

Prepared by
Ken Watkins
Colorado State University

Boston   Burr Ridge, IL   Dubuque, IA   Madison, WI   New York   San Francisco   St. Louis
Bangkok   Bogotá   Caracas   Kuala Lumpur   Lisbon   London   Madrid   Mexico City
Milan   Montreal   New Delhi   Santiago   Seoul   Singapore   Sydney   Taipei   Toronto

McGraw-Hill Higher Education

A Division of The McGraw-Hill Companies

Student Study Guide
CHEMISTRY, SEVENTH EDITION
RAYMOND CHANG

Published by McGraw-Hill Higher Education, an imprint of The McGraw-Hill Companies, Inc.,
1221 Avenue of the Americas, New York, NY 10020. Copyright © The McGraw-Hill Companies,
Inc., 2002, 1998. All rights reserved.

This book is printed on acid-free paper.

2 3 4 5 6 7 8 9 0 QPD/QPD 0 3 2

ISBN 0-07-231801-5

www.mhhe.com

CONTENTS

To Jordan, Chloe, and Chelsea; the promise of tomorrow.

PREFACE

This ***Study Guide*** is intended for use with the seventh edition of Raymond Chang's *Chemistry*. Its purpose is to help you learn chemistry. The ***Study Guide*** contains material to help you organize your studying, practice your problem-solving skills, and test yourself. Each chapter in the guide corresponds to one in the text. The chapters are arranged into four or five sections for easier presentation and comprehension. After the title of each section you will find the number of the corresponding section of the text in parentheses. Each chapter contains the following main features.

Study Objectives. Each chapter section begins with a set of performance objectives. These state in a straightforward manner what you must be able to do with the topics discussed in the section. One problem students frequently experience during a lecture is knowing what is important enough to put in notes. Previewing the objectives on specific topics before attending a lecture will help you focus on important material, listen better, and retain more from your first exposure to the material.

Reviews. The reviews provide a summary of essential material. They are meant to reinforce and enhance your understanding of the material presented in the text and in lectures. The thorough descriptions and explanations provided by the reviews should help you solve the accompanying exercises.

Example Problems. After each summary there are example problems with detailed solutions. These solutions often contain applications of important concepts. They emphasize how to initiate the solution. The steps toward the solution are discussed first and are then worked through. Almost every example contains *Method of Solution* and *Calculation* sections. The method of solution section relates the solution of the problem to the principles involved and suggests equations that are applicable. The calculation section shows how to set up equations and substitute into them to achieve a numerical answer.

Exercises. Exercises appear after each section. These questions provide a quick test of your familiarity with the terminology and principles, and provide drill problems.

Practice Tests. Each chapter contains a practice test to help you to test your knowledge of the material and to prepare for exams. Many of the questions in the practice tests have been taken from old examinations that the author has written. Some questions are more general problems that require integration of concepts from previous material.

If you cannot work a significant number of the problems in the practice test, you should go back and study the material on that subject in the textbook. Work through the practice problems by *writing down* the main steps of the solutions. Don't look at the answers first. Act as if you are taking an exam. Use the answers only to check your own work.

Conceptual Questions. These questions ask you to think through the solution to a question or problem. Understanding and applying concepts is an important part of problem solving.

Answers. Answers to all the exercises and problems are conveniently found at the end of each chapter.

Hints. For this guide to be of maximum assistance, it should be incorporated into a plan for studying general chemstry. The responsibility for organizing your study so that you learn the required material, of course, rests with you. You must organize your study and test yourself in order to be certain that you know the material. A helpful rule is "Don't get behind." Keep in mind that the content of the course builds on concepts and principles developed earlier in the course. General Chemistry will be much more rewarding if you set aside time each day to learn the material as it is presented in class. For most chemistry courses, it is necessary to study 2 to 3 hours outside of class for every hour in class.

Study should include reading the text, working lots of problems, and memorizing facts and vocabulary. Reading in science takes longer than reading in other subject areas. You cannot expect to "speed-read" a chemistry text. It is important that what you read makes sense. Cramming is not a proven or successful way to learn chemistry or any other science.

Learning chemistry, like learning anything else, takes time, effort, and concentration. In my teaching experience I have never had anyone say "I *plan to fail* your course." Yet, there are some who *fail to plan*, and end up with the same unfortunate result. My hope is that you will be a successful learner and chemistry student.

Kenneth W. Watkins

Chapter One
CHEMISTRY: THE STUDY OF CHANGE

- Introduction (1.2 – 1.3)
- Matter and Its Properties (1.4 – 1.6)
- Units of Measurement (1.7)
- Math Review and Significant Figures (1.8)
- Factor-Label Method of Problem Solving (1.9)

INTRODUCTION (1.2 – 1.3)

STUDY OBJECTIVES

1. Describe matter and its physical states.
2. Describe the four main parts of the scientific method.
3. Define chemistry.

About Chemistry. **Chemistry** is the study of matter and the changes it undergoes. **Matter** is defined as anything that has mass and occupies space. The three physical states of matter are solid, liquid, and gas. All matter exists in one or another of these three states, depending on the temperature and pressure of the surrounding environment. Chemists are concerned with developing the tools used to study matter and the concepts useful in describing the properties of matter. They direct their efforts toward the purposeful changing of given forms of matter into new and different substances, and to the discovery of the properties and uses of these new materials. Chemists usually observe matter and the changes it undergoes in the *macroscopic world*. This refers to the objects we can see and touch, and deal with everyday. However, our interpretations of matter involve atoms and molecules and their properties. Because atoms and molecules are so extremely small, we refer to them as belonging to the *microscopic world*.

Chemists, like other scientists, make use of the **scientific method** which can be broken down into four parts: observation and experiment, hypothesis, laws, and theory. The first step is to define the problem clearly. That is you must know what you are trying to find out about. During **experiments** observations are made and information is collected about the system being studied. After a large amount of data related to a certain phenomenon have been collected, sometimes the information can be summarized in a simple verbal or mathematical statement called a **law**. At this point a **hypothesis** may be formulated to provide a tentative explanation of the facts. A hypothesis is only temporary and is meant to be a working model or explanation. It is often adjusted as new information is discovered. As a hypothesis grows and successfully survives many experimental tests, it develops into a theory. A **theory** is a unifying principle that explains a large body of facts and laws. Besides providing explanations for the laws of science, its role is to aid in making predictions that lead to new knowledge. In fact theories are often tested by carefully devised experiments that confirm or disprove their predictions.

MATTER AND ITS PROPERTIES (1.4 – 1.6)

STUDY OBJECTIVES

1. Distinguish between elements, compounds, and solutions.
2. Distinguish between homogeneous and heterogeneous mixtures.
3. Distinguish between physical and chemical properties of a substance.

Elements and Compounds. A **pure substance** is a form of matter that has a definite composition and distinct properties. Examples are water, table salt, and iron. Just as each individual person has a set of characteristics, such as fingerprints and color of eyes and hair, each pure substance has characteristic properties. There are two types of pure substance: elements and compounds. An **element** is a pure substance that cannot be decomposed into simpler substances by ordinary chemical reactions. Elements are the building blocks of which all compounds are composed. Nitrogen, oxygen, and iron are examples of elements. A number of common elements and their symbols are listed in the textbook in Table 1.1.

Compounds are pure substances that are composed of two or more elements combined in definite proportions. Compounds can be broken down into the elements of which they are composed by chemical means. Water and table salt, mentioned earlier, are compounds. The action of an electric current, called *electrolysis*, is one method that can be used to decompose both water and molten table salt into their constituent elements. Pure water consists of 89 percent oxygen and 11 percent hydrogen by mass. Pure salt contains 39 percent sodium and 61 percent chlorine.

Mixtures. Pure substances can be brought together to form mixtures. **Mixtures** are combinations of two or more substances with variable composition. They can be homogeneous or heterogeneous depending on the state of subdivision of the components. Salt water is a uniform mixture of table salt (NaCl) and water. The original crystals of salt have dissolved and are dispersed evenly. On the ordinary scale of observation we cannot detect any chemical or physical differences between adjacent regions of the mixture. The particles of salt are too small to observe. A mixture that has the same composition throughout is said to be a **homogeneous mixture**. The properties of a homogeneous mixture vary since they depend on the percent composition. For example, the hardness of steel, a solid mixture of iron and carbon, depends on the percentage of carbon that is added to iron. Homogeneous mixtures are also called **solutions**.

Heterogeneous mixtures are not uniform in composition. And indeed the individual particles of their components can often be seen by the unaided eye. For example, when preparing home-made ice cream, you use a mixture of ice and rock salt. This is a heterogeneous mixture. The individual chunks (particles) of ice and salt are clearly visible, and the particles are so large, they are not evenly dispersed. The composition of this mixture varies from place to place within the mixture itself. Natural air is actually a heterogeneous mixture. It consists of nitrogen, oxygen, and argon gases, but also contains solid particles of pollen and dust. Any mixture, whether it be homogeneous or heterogeneous, can be separated into its pure components by physical means.

Properties of Matter. **Physical properties** are those properties that can be measured and observed without changing the identity or composition of the substance. Physical properties include color, hardness, solubility, density, specific heat, melting point, and boiling point. **Physical changes** are those that take place with no change in chemical composition. Changes of a substance from one state of matter to another do not change its chemical composition and are examples of physical changes. The three forms of water we call ice, liquid water, and steam are all the same substance, just different physical states.

Chemical properties most often are descriptions of reactions that a substance undergoes when brought in contact with other substances. In a chemical reaction the original substance or substances are changed into new substances. When sodium metal and chlorine gas are heated together, a white solid called sodium chloride is formed. That this is a chemical change is evident by the observation that sodium, a shiny metal, and chlorine, a pale yellow green gas, have disappeared, and in their place is a substance with a completely new set of properties. Sodium chloride is a white solid that melts at very high temperatures.

All properties of matter are either extensive or intensive properties. **Extensive properties** depend on the amount of matter being considered. Volume and mass are examples. In contrast, temperature and density are two properties that do not depend on the amount of mass present. Thus, they are **intensive properties.**

EXAMPLE 1.1 Chemical and Physical Properties

The following are properties of the element silicon; classify them as physical or chemical properties.
a. Melting point, 1410°C
b. Reacts with fluorine to form silicon tetrafluoride
c. Gray color
d. Not affected by most acids

•Method of Solution

Physical properties can be observed without a change in composition, while chemical properties describe reactions with other substances.
a. Melting involves a change in physical state but no chemical change. *Answer:* The melting point is a physical property.
b. This statement describes the change of silicon into another substance on reaction with fluorine. *Answer:* The reaction is a chemical property.
c. *Answer:* The color of a substance is a physical property. No change in composition occurs while observing the color..
d. *Answer:* The lack of reactivity with another substance or class of substances such as acids is a chemical property.

EXAMPLE 1.2 Homogeneous or Heterogeneous Mixture

Classify each of the following as a homogeneous or a heterogeneous mixture.
a. The beverage tea
b. Oil and water
c. Cow's milk
d. Wine

•Method of Solution

Recall that homogeneous mixtures are uniform throughout, while heterogeneous mixtures have components that can be physically observed to be separate.
a. The mixture called tea is uniform throughout and no particles of tea can be observed (if there are no tea leaves). *Answer:* A homogeneous mixture.
b. Since oil floats on water, the oil and water components can be observed to be separate. Samples from the top part of the mixture are different from samples taken from the bottom. *Answer:* A heterogeneous mixture.
c. Cow's milk contains fats and solids suspended in water. On standing, the cream (fat) will rise to the top and the solids will settle. *Answer:* A heterogeneous mixture.
d. Wine contains ethyl alcohol, water, flavor components, dye molecules, and many other substances. The composition is uniform throughout. No solid particles are visible in the liquid. It appears clear. *Answer:* A homogeneous mixture.

EXAMPLE 1.3   Chemical and Physical Changes

Classify the following changes as physical or chemical.
a.   Solid shortening turns to liquid when heated
b.   Gasoline burning in air

•**Method of Solution**

a.   Solid fats melt at a fairly low temperature. Melting is a physical change. Physical changes can be reversed by lowering the temperature. Removing the source of heat in this case will cause the fat to solidify. *Answer:* A physical change.
b.   Liquid gasoline undergoes combustion with oxygen (burning) to produce carbon dioxide and water. This process is accompanied by the evolution of heat. *Answer:* A chemical change.

EXERCISES

1.   Consider the following and decide whether they are homogeneous or heterogeneous mixtures.
a. Wood   b. Wine   c. Cranberry juice   d. Milk   e. Italian dressing

2.   Classify each of the following as a element, compound, or mixture.
a. Dry ice   b. Iodine   c. Beer   d. Margarine   e. Steel

3.   Identify the following as chemical or physical properties.
a. Iron melts at 1530°C   b. Wood is less dense than water   c. Wood burns in the presence of air
d. Shiny iron metal becomes rust in the presence of air and water.

4.   Identify the following as chemical or physical changes.
a. Frost forms on a automobile windshield   b. Bleach changes the color of your favorite jeans
c. Burning food on the stove produces a black solid   d. Heating water produces steam

UNITS OF MEASUREMENT (1.7)

STUDY OBJECTIVES

1.   Express base units and derived units of the SI.
2.   Learn the prefixes used in the SI.
3.   Distinguish between weight and mass.
4.   Convert temperatures between the Celsius and Fahrenheit scales.

   The International System of Units. Since 1960 a coherent system of units, known as the SI, has been in effect and is accepted among scientists and engineers. SI is the abbreviation for *Le Systeme International d'Unites.* Most of the units used in the textbook will be from the SI.

The SI has as its **base units**: the kilogram (kg) for mass, the meter (m) for length, the second (s) for time, the kelvin (K) for temperature, and the mole (mol) for amount of substance. See Table 1.1.

Table 1.1 Frequently Used Base Units

Quantity	Name of Unit	Symbol
Mass	kilogram	kg
Length	meter	m
Time	second	s
Temperature	kelvin	K
Amount of substance	mole	mol

Combinations of base units produce **derived units**. For example, velocity is distance traveled per unit of time. Therefore, velocity has units of meters per second (m/s) and is a derived unit rather than a base unit. The derived units for volume, density, and velocity are given in the Table 1.2.

Table 1.2 Some Derived Units

Quantity	Definition of Quantity	Symbol
Volume	length × length × length	m^3
Density	mass/volume	kg/m^3
Velocity	meter/second	m/s

Prefixes. A major advantage of the SI is that it uses the decimal system. A list of the SI prefixes is given in Table 1.3. These prefixes will be used throughout your study of chemistry. Note that in the SI the same prefix can be applied to any base unit or derived unit. Thus the prefix *milli* can be used to describe a unit that is 1/1000 of a gram, or 1/1000 of a meter, or 1/1000 of any SI unit.

1 milligram = 1/1000 gram
1 millimeter = 1/1000 meter
1 millisecond = 1/1000 second

Conversions from base units to units with prefixes will be illustrated in the section called the factor-label method.

Table 1.3 Prefixes Used in the SI

Prefix	Symbol	Meaning	
tera	T	1,000,000,000,000	or 10^{12}
giga	G	1,000,000,000	or 10^9
mega	M	1,000,000	or 10^6
kilo	k	1,000	or 10^3
deci	d	0.1	or 10^{-1}
centi	c	0.01	or 10^{-2}
milli	m	0.001	or 10^{-3}
micro	μ	0.000 001	or 10^{-6}
nano	n	0.000 000 001	or 10^{-9}
pico	p	0.000 000 000 001	or 10^{-12}

Mass and Weight. **Mass** is a property measuring the quantity of matter in an object. The property of mass is very obvious when comparing the resistance of two bodies to a change in their motion. The greater the mass of a body, the less its motion will be influenced by outside forces. Which is easier to stop, a truck going 50 mph, or a baseball moving at the same speed?

The terms mass and weight are often used interchangeably. Although, they actually refer to different properties. Mass is a measure of the amount of matter in an object, and is independent of the position or surroundings of the object, whereas **weight** is defined as the force exerted on an object by the pull of gravity. The mass of an object is constant, but its weight varies with its location. Your weight would change drastically if you went to the moon because the moon, being smaller than the Earth, exerts only 1/6 the attraction due to its gravity that the Earth does. If you weighed 150 lb on Earth you would weigh only 25 lbs of the moon.

Volume. Volume is a derived unit. It can be expressed in terms of length cubed because for rectangular solids volume is equal to length times width times height all of which have the base unit meters.

$$\text{volume} = \text{length} \times \text{width} \times \text{height} = \text{length}^3$$
$$\text{volume} = \quad m \quad \times \quad m \quad \times \quad m = m^3$$

The SI unit for volume is the cubic meter (m^3) which is the volume of a cube 1 m on each edge. This unit is too large and related units such as the cubic centimeter (cm^3) and cubic decimeter (dm^3) are often used.

The volume of a fluid is usually measured in liters (L). A liter is roughly the size of a quart (1.06 qt). A liter is the volume of a cube with an edge of 10 cm, and therefore is 1000 cm^3. The volume unit you are most likely to use in chemistry is the milliliter (mL). One mL is 0.001 L, and so 1000 mL equal 1 L. Since 1 L = 1000 cm^3 = 1000 mL, then one milliliter (mL) is equal to one cubic centimeter (cm^3).

$$1 \text{ mL} = 1 \text{ cm}^3$$

Density. Density is an important physical property of objects and substances. The **density** of an object is defined as the ratio of its mass to its volume.

$$\text{density} = \frac{\text{mass}}{\text{volume}}$$

or in terms of symbols:

$$d = \frac{m}{V}$$

Density has units that are derived from the base units for mass and volume which are kilograms per cubic meter. However, it is more convenient to report densities in units of grams per cubic centimeter or grams per milliliter.

Temperature Scales. Chemists use two temperature scales, the **Kelvin scale** (K) and the **Celsius scale** (°C). A third scale, the Fahrenheit scale (°F), is commonly used in the United States. The Celsius scale defines 0°C as the freezing point of water and 100°C as the boiling point of water. In the Fahrenheit scale the freezing point of water is 32°F and the boiling point is 212°F. The equation for converting from degrees Fahrenheit to degrees Celsius is:

$$°C = (°F - 32°F) \times \frac{5°C}{9°F}$$

To convert from degrees Celsius to degrees Fahrenheit, solve the equation for °F. See Example 1.6. The Kelvin scale will be discussed in Chapter 5.

EXAMPLE 1.4 Density

A flask filled to the 25.0 mL mark contained 29.97 g of a concentrated salt–water solution. What is the density of the solution?

•Method of Solution

The density (d) of an object or a solution is defined as the ratio of its mass (m) to its volume (V).

$$d = \frac{m}{V}$$

Substituting in the given quantities:

$$d = \frac{29.97 \text{ g}}{25.0 \text{ mL}}$$

$$d = 1.19 \text{ g/mL}$$

•Comment

Density problems will be worked using the factor label method of problem solving in Example 1.10.

EXAMPLE 1.5 Temperature Scale Conversion

The freezing point of a 50-50 mixture of antifreeze (ethylene glycol) and water is –36.5°C. Convert this temperature to degrees Fahrenheit.

•Method of Solution

First write the equation for temperature conversions:

$$°C = (°F - 32°F) \times \frac{5°C}{9°F}$$

Next substitute –36.5°C and solve for degrees Fahrenheit:

•Calculation

$$-36.5°C = (°F - 32°F) \times \frac{5°C}{9°F}$$

$$-65.7°C = (°F - 32°F)$$

$$°F = -33.7°F$$

EXERCISES

5. Calculate the volume of a brick that is 34 cm long by 7.0 cm wide by 14 cm high.

6. A certain metal ingot has a mass of 3951 g, and measures 10.2 cm by 8.2 cm by 4.2 cm. Calculate the density of the metal.

7. The melting point of cesium metal is 28.4°C. What is the melting point in degrees Fahrenheit?

8. On a sunny summer day the temperature inside a closed automobile can reach 128°F. Convert this to degrees Celsius.

9. Express the following amounts in exponential notation in terms of the base unit with no prefix.
   a. 7 μg
   b. 8.0 nm
   c. 0.14 ML
   d. 1.0 ks

MATH REVIEW AND SIGNIFICANT FIGURES (1.8)

STUDY OBJECTIVES

1. Express numbers and perform calculations in exponential notation.
2. Determine the number of significant figures in a given number and in a number that is the result of a mathematical calculation.
3. Round off numbers to the desired digit.

Exponential Notation. Many of the quantities (numbers) encountered in chemistry are more easily manipulated when they are written in a form known as scientific or exponential notation. Very large or very small numbers are expressed as $N \times 10^n$, where N is a number between 1 and 10 and n corresponds to the exponents 1, 2, 3, etc. Recall that the exponent tells you how many times to multiply the number 10 by itself. Therefore:

$$10^3 = 10 \times 10 \times 10 = 1000$$
$$10^2 = 10 \times 10 \quad\quad = 100$$
$$10^1 = 10 \quad\quad\quad\quad = 10$$
$$10^0 \quad\quad\quad\quad\quad = 1$$

To write the number 5000 in exponential notation, we could rewrite it as 5×1000. But 1000 can be written $10 \times 10 \times 10$, or just 10^3. Therefore, 5000 can be written as 5×10^3.

More simply, to find the exponent of 10, just count the number of places that the decimal point must be moved to the left to give the number N. For the number 5000, moving the decimal point three places to the left gives 5×10^3.

Fractional numbers can also be expressed in scientific notation, but in this case the decimal point will be moved to the right and the exponent will be a negative number (–n). For the number 0.05, the decimal point must be moved two places to the right to give N (a number between 1 and 10). In exponential notation 0.05 becomes 5×10^{-2}.

Multiplication and division of numbers that are expressed in exponential notation are accomplished by operating on the coefficients (N) and exponentials (n) separately. Recall when multiplying two exponential numbers that the exponents are added. See Example 1.8. When dividing exponential numbers, we subtract the exponent in the denominator from the exponent in the numerator, as shown in Example 1.9. These examples also show how to do calculations using a calculator.

In some calculations you must take the square or cube root of an exponential number. The general rule is:

$$(X^a)^n = X^{an}$$

where n is 1/2 for the square root, 1/3 for the cube root, and so on. Therefore, the square root of 9.0×10^{-16} is:

$$\sqrt{9.0 \times 10^{-16}} = (9.0 \times 10^{-16})^{1/2} = (9.0)^{1/2} \times 10^{-16/2}$$

Taking the square root of the coefficient and exponent separately, we get: 3.0×10^{-8}

$$\sqrt{9.0 \times 10^{-16}} = 3.0 \times 10^{-8}$$

Significant Figures. Measurements, and calculations based on measurements, must be reported so as to convey information about the number of meaningful digits. The **significant figures** are those digits in a measured number that include all certain digits plus one having uncertainty. If we measure the length of a desk top with a meterstick calibrated in millimeters and find the length to be 972.5 mm, then the digits 9, 7, and 2 mean there are 9 hundreds, 7 tens, and 2 ones, and because of the calibrations, these digits are certain. Five-tenths, however, is an estimate because it falls between the finest calibrations. Someone else might read it as 0.6 or 0.4. The result could be expressed with error limits as 972.5 ± 0.1 mm. This measurement provides three certain digits and one uncertain digit. All of these provide useful information, and we say the number 972.5 has four significant figures.

Always take care that the numbers you write reflect the proper number of meaningful digits. The process of determining the correct number of significant figures after a calculation depends on the type of calculation. First let us review the rules for expressing significant figures.

Guidelines for Writing Significant Figures. To determine the number of significant digits that are present in a *written number,* use the following rules:

- Any nonzero digit is significant.
- Zeros between nonzero digits are significant. The digits in the number 506 mean there are 5 hundreds, 0 tens, and 6 ones. The zero digit is significant because it tells us there are no tens. All three digits are significant.
- Leading zeros are those to the left of the first nonzero digit as in 0.02. They are not significant figures. These zeros serve to mark the position of the decimal point. The rule is that zeros used only to place the decimal point are *not* significant. Thus 0.6 g and 0.0006 g both have only 1 significant figure.
- Zeros written to the right of nonzero digits are called trailing zeros. They are considered to be significant only if the number contains a *decimal point*. Thus the number 1.0 mg has two significant figures, and 0.500 mg has three significant figures. The following two numbers further exemplify these rules:

0.00260740 (six significant figures)

nonsignificant zeros   significant zeros

ambiguous zeros

238,600 (four - six significant figures)

- Trailing zeros in numbers that do not contain decimal points may or may not be significant. The number 450 g has two significant figures if the number is 450 ± 10, and three significant figures if the number is 450 ± 1. The number of significant figures can be indicated unambiguously with scientific notation. The number 450 can be written as 4.5×10^2, if it has two significant figures, and as 4.50×10^2, if it has three significant figures.

- Exact numbers are those obtained from a definition. When we say 100 cm = 1 m, both the 100 and the 1 are exact numbers, and they are said to have an infinite number of significant figures. 100 cm is exactly 1 m. There is no uncertainty. On the other hand, conversion factors between unit systems usually are not exact numbers. For example, the conversion from quarts to liters is 1.06 qt = 1 L. Here the 1 is one exact liter, but the 1.06 has three significant figures.

Significant Figures and Calculations. In most cases the numbers we measure are used to calculate other quantities. Care must be exercised to report the proper number of significant figures in the calculated result. This is extremely important when electronic calculators are used because they give answers with eight and ten digits. The rule used is that the accuracy of the result is limited by the least accurate measurement. In **multiplication and division**, the number of significant figures in a calculated result is determined by the original measurement that has the *fewest* number of significant digits. Thus, if a number with three significant figures is multiplied by a number with four significant figures, the computed result will have only three significant figures. Even if a calculator computes eight digits, most of them will be meaningless in terms of their physical significance.

For example, the density of an object is its mass divided by its volume. The density of an object with a mass of 2.157 g that occupies 1.66 cm^3 can be known to only three significant figures:

$$d = \frac{2.157 \text{ g}}{1.66 \text{ cm}^3} = 1.29939759 \text{ g/cm}^3 \quad \leftarrow \text{calculator result}$$

$$d = 1.30 \text{ g/cm}^3 \qquad \leftarrow \text{rounded off}$$

Rounding Off Numbers. A number is rounded off to the desired number of significant figures by dropping one or more digits to the right. The following rules should be observed when rounding off numbers:

1. When the first digit dropped is less than 5, the last digit retained remains unchanged. Therefore, 4.2643 rounded to three significant figures becomes 4.26.
2. When the first digit dropped is equal to or greater than 5, the last digit retained is increased by 1. Thus, 2.38 and 2.35 rounded to two significant figures both become 2.4.
3. For calculations involving more than one step, carry out the calculations without rounding off the intermediate results. Round off the final answer to the correct number of significant figures.

Addition and Subtraction. In addition and subtraction, the number of significant figures in the answer depends on the original number in the calculation that has the *fewest digits to the right of the decimal point*. For example, when adding:

$$
\begin{array}{l}
102.226 \\
2.51 \\
\underline{846.} \quad \leftarrow \text{ fewest digits to the right} \\
950.736 \qquad \text{of the decimal point}
\end{array}
$$

7, 3 and 6 are nonsignificant digits

$$951 \qquad \leftarrow \text{rounded off}$$

and when subtracting:

$$
\begin{array}{l}
102.25 \\
\underline{-99.3} \quad \leftarrow \text{ fewest digits to the right} \\
2.95 \qquad \text{of the decimal point}
\end{array}
$$

5 is a nonsignificant digit

$$3.0 \quad \leftarrow \text{ rounded off}$$

It is important to remember when adding or subtracting, that the number of significant figures in the answer is not necessarily determined by the quantity having the fewest significant figures, which is true for multiplication and division.

EXAMPLE 1.6   Exponential Notation

Write the following numbers using exponential notation.
a. 7,620,000
b. 0.000495

•Method of Solution

a. 7,620,000—Count the number of places that the decimal point must be moved to the left to give a number between 1 and 10, in this case, 7.62. Since this requires six places, the exponent of 10 must be 6. *Answer:* 7.62×10^6.

b. 0.000495—Fractional numbers like this have negative exponents in scientific notation. Count the number of places that the decimal point must be moved to the right to give 4.95. Since this requires a move of four places, the exponent must be –4. *Answer:* 4.95×10^{-4}.

EXAMPLE 1.7   Math Operations with Exponential Notation

Multiply 4.0×10^4 by 3.5×10^{-6}.

•Method of Solution

$$(4.0 \times 10^4)(3.5 \times 10^{-6})$$

Regroup so that the coefficients are separated from the exponentials, and add the exponents algebraically.

•Calculation

$$(4.0 \times 3.5) \times 10^{4+(-6)} = 14 \times 10^{-2} = 1.4 \times 10^{-1}$$

•Calculator Calculation

Most people use an electronic calculator for these types of calculations. On most, but not all, calculators, you would carry out the above multiplication problem in the following way:

$$(4.0 \times 10^4)(3.5 \times 10^{-6})$$

When entering the exponential terms, the exponents (4) and (–6) are entered as [EXP] then 4, and [EXP] then [+/–] 6. Therefore, 4.0×10^4 is entered as 4.0 [EXP] 4, and 3.5×10^{-6} is entered as 3.5 [EXP] [+/–] 6. Note that the $\times$ sign within the exponential must not be entered. Some calculators have an [EE] key for exponents.

The calculation can be carried out with the following sequence of key strokes:

4.0 [EXP] 4 [×] 3.5 [EXP] [+/–] 6 [=]

The calculator will display the answer either as just 0.14, or in exponential form as 1.4^{-01}, or as 1.4   –01 depending on the brand of calculator.

•Comment

Always write out your answer in exponential notation, rather than calculator format.

EXAMPLE 1.8   Math Operations with Exponential Notation

Divide 4.2×10^{-7} by 5.0×10^{-5}.

•Method of Solution

$$\frac{4.2 \times 10^{-7}}{5.0 \times 10^{-5}}$$

Divide the coefficients, and subtract the exponent in the denominator from the exponent in the numerator.

•Calculation

$$= 0.84 \times 10^{-7-(-5)}$$
$$= 0.84 \times 10^{-7+5}$$
$$= 0.84 \times 10^{-2}$$

Because 0.84 is less than 1, the usual practice is to move the decimal point to give a coefficient between 1 and 10. Move the decimal point one place to the right, and add (−1) to the exponent.

$$= 8.4 \times 10^{-3}$$

•Calculator Calculation

On most calculators, you would carry out the above division problem:

$$\frac{4.2 \times 10^{-7}}{5.0 \times 10^{-5}}$$

by entering the numerator first, then pushing the divide key [÷], and then entering the denominator followed by the [=] key. As before, the exponential key is used to enter exponents.

The calculation, if written out, would look like this:

$$4.2 \ [EXP] \ [+/-] \ 7 \ [÷] \ 5.0 \ [EXP] \ [+/-] \ 5 \ [=]$$

The calculator will display the answer as: 8.4^{-03} or 8.4 −03; both stand for 8.4×10^{-3}.

EXAMPLE 1.9   Significant Figures

Determine the number of significant figures in each of the following numbers.
a.   6.02
b.   0.012
c. 1.23×10^7
d.   1.5400

•Method of Solution

Refer to the guidelines for writing significant figures.
a.   Recall that zeros between nonzero digits are significant. *Answer:* three significant figures.
b.   Zeros to the left of the first nonzero digit are not significant. *Answer:* two significant figures.
c.   The use of scientific notation implies that only significant figures are shown. *Answer:* three significant figures.
d.   When a number contains a decimal point and has trailing zeros, these zeros are significant. *Answer:* five significant figures.

EXAMPLE 1.10   Significant Figures

Carry out the following operations, rounding off the answer to the correct number of significant digits.
a. Subtract:

$$287.12$$
$$\underline{-95.333}$$

b. $7.25 \times 10^{12} \div 92 =$

c. $\dfrac{(6.0 + 5.21)}{4781} =$

•Method of Solution

a. *Answer:* 191.79. Only two digits to the right of the decimal are significant.
b. *Answer:* 7.9×10^{10}. Two significant figures.
c. *Answer:* 2.34×10^{-3}. Three significant figures.

EXERCISES

10. Express the following numbers in scientific notation:
    a. 24,000   b. 0.00014   c. 740,000,000   d. 0.0906

11. Write the following exponential numbers as ordinary numbers by moving the decimal point in the proper direction:
    a. 5.2×10^{-3}
    b. 1.4×10^{3}
    c. 7.5×10^{6}
    d. 7.06×10^{-5}

12. Determine the number of significant figures expressed in the following numbers:
    a. 0.609
    b. 1.0×10^{3}
    c. 0.000222
    d. 238.0
    e. 1.030×10^{-2}

13. Round off the following numbers to the number of significant figures requested:
    a. 0.60945 to three significant figures
    b. 1.012×10^{3} to two significant figures
    c. 0.00022174 to three significant figures
    d. 237.95 to four significant figures
    e. 1.303 to two significant figures

14. Carry out the following operations and express the result to the correct number of significant figures:
    a. 12×2143.1
    b. $3.09 \div 7$
    c. $(2.2 \times 10^{-3})(1.40 \times 10^{6})$
    d. $12.70 + 1.222$
    e. $595.2 \times (24.330 - 16.271)$

15. Carry out the following operations and express the result to the correct number of significant figures:

   a. 125 g + 64 g + 10.837 g =

   b. 11.2 cm + 0.0093 cm + 0.80 cm =

   c. $\dfrac{15.01 \text{ g}}{(7.13 \text{ mL} - 6.1 \text{ mL})} =$

FACTOR-LABEL METHOD OF PROBLEM SOLVING (1.9)

STUDY OBJECTIVES

1. Construct conversion factors from known equalities.
2. Solve problems using the factor-label method.
3. Solve problems involving density.

Conversion Factors. The **factor-label method** (also called dimensional analysis) is a simple but powerful technique that can be applied to a great variety of calculational problems. In this method the units (labels) of the quantities involved are used as a guide in setting up a calculation. The method is based on the use of conversion factors and on the idea that units or dimensions of various quantities can be handled algebraically in the same way as numbers are handled.

   Conversion factors are constructed from equalities such as the following:

$$2.540 \text{ cm} = 1 \text{ in}$$
$$1000 \text{ g} = 1 \text{ kg}$$
$$24 \text{ h} = 1 \text{ day}$$

From the first one, we can also write "2.54 cm per inch." Likewise, there are 1000 grams per kilogram, and 24 h per day. Since the "per" can mean "divided by," we can write the following conversion factors:

$$\frac{2.54 \text{ cm}}{1 \text{ in}}, \quad \frac{1000 \text{ g}}{1 \text{ kg}}, \quad \text{and} \quad \frac{24 \text{ h}}{1 \text{ day}}$$

These conversion factors are examples of *unit factors*. Each unit factor is equal to 1 (unity) because the numerator equals the denominator.

   Treating units in the same fashion as numbers are treated in algebra is the essential element of the factor-label method. For example, multiplying inch by inch yields inches squared, or square inches.

$$\text{in} \times \text{in} = \text{in}^2$$

One hour divided by one hour is equal to 1, and we say that the units cancel:

$$\frac{h}{h} = 1$$

The Factor-Label Method. In the factor-label method we multiply a given quantity by appropriate conversion factors in order to arrive at a desired quantity.

   desired quantity = given quantity × conversion factors

The method can be illustrated with a simple problem. How many hours are there in 5 days? First write down in algebraic form what is being asked.

   $? \text{ h} = 5 \text{ days}$

Notice that ? h is the desired quantity and 5 days is the given quantity. Hours and days are related by the equality:

24 h = 1 day

This equality can be made into two possible unit factors:

$$\frac{24 \text{ h}}{1 \text{ day}} \quad \text{"24 h per day"} \quad \text{and} \quad \frac{1 \text{ day}}{24 \text{ h}} \quad \text{"1 day per 24 h"}$$

Using the first unit factor, we obtain

$$? \text{ h} = 5 \text{ days} \times \frac{24 \text{h}}{1 \text{day}}$$

Now treating the units algebraically, the days "cancel":

$$? \text{ h} = 5 \text{ days} \times \frac{24 \text{ h}}{1 \text{ day}}$$

and taking 5 × 24 h = 120 h, we obtain:

? h = 120 h

With a little practice, the correct conversion factor can be chosen every time. In this case, we wanted the unit "days" to cancel; therefore, we put days in the *denominator*. We also wanted "hours" for the units of the answer; therefore, we put hours in the *numerator* of the conversion factor. Also, writing "? h" first serves as a guide to remind us of what units we want in the result.

Note that we *might* have picked the second of the two conversion factors. In this case we would have obtained:

$$? \text{ hours} = 5 \text{ days} \times \frac{1 \text{ day}}{24 \text{ h}} = 0.208 \ \frac{\text{day}^2}{\text{h}}$$

The units in this answer are incorrect and tell us we have chosen the wrong conversion factor. The units serve as a good check of your answer.

When several steps are necessary to carry out the conversion, a "road map" is very helpful. A football field has a length of 100.0 yd. What is its length in meters? We know that 2.540 cm = 1 in. First we convert the yards to inches, then convert the inches to centimeters, and then the centimeters to meters. Our road map, or plan, is:

yards → inches → centimeters → meters

Each arrow requires a conversion factor. The first conversion is within the U.S. system:

1 yd = 36 in

The second conversion is a U.S. to SI conversion (1 in = 2.540 cm), and the third is a conversion within the SI:

$10^{-2} \text{ m} = 1 \text{ cm}$

Writing the problem in the form desired quantity = given quantity × conversion factors, we get:

? m = 100.0 yd

Now we write the first unit factor with yards in the denominator so that the units of yards cancel:

$$? \text{ m} = 100.0 \text{ yd} \times \frac{36 \text{ in}}{1 \text{ yd}}$$

Write the second unit factor with inches in the denominator so that the units of inches cancel, and units of cm are left:

$$? \text{ m} = 100.0 \text{ yd} \times \frac{36 \text{ in}}{1 \text{ yd}} \times \frac{2.540 \text{ cm}}{1 \text{ in}}$$

Then we multiply by the third conversion factor so that the centimeter units cancel and the meter units remain:

$$? \text{ m} = 100.00 \text{ yd} \times \frac{36 \text{ in}}{1 \text{ yd}} \times \frac{2.540 \text{ cm}}{1 \text{ in}} \times \frac{1 \times 10^{-2} \text{ m}}{1 \text{ cm}}$$

Carrying out the indicated operations yields:

$$? \text{ m} = 91.44 \text{ m}$$

The key to problem solving by the factor-label method lies in knowing where to find and how to use conversion factors.

Outline of Problem Solving Steps in the Factor-Label Method

1. Identify the known (given) and unknown quantities.
2. Restate the question and identify principles and mathematical relationships appropriate to the situation.
   desired quantity = given quantity × conversion factors
3. Plan a solution (road map). Decide what steps are needed to solve the problem.
4. Setup the calculation by inserting the conversion factors. For each arrow in the road map, one conversion factor is needed.
5. Calculate the answer and cancel units. Always include the units of the answer (desired quantity).
6. Check the solution by doing the calculation again. Ask: Is the magnitude of the answer reasonable and does it have the correct units?

EXAMPLE 1.11   Conversions with SI Prefixes

Express the following amounts using prefixes from the SI.
a. Convert 827 m to kilometers.
b. Convert 257 mg to micrograms.

•Method of Solution for (a)

Conversions within the SI are based on knowing the SI prefixes. First state the problem in the format:

desired quantity = given quantity × conversion factors.

$$? \text{ km} = 827 \text{ m}$$

Our road map is: m → km

The relationship is 1×10^3 m = 1 km, which leads to two possible conversion factors:

$$\frac{1 \times 10^3 \text{ m}}{1 \text{ km}} \quad \text{and} \quad \frac{1 \text{ km}}{1 \times 10^3 \text{ m}}$$

Choose the one that allows cancellation of the unit m.

$$? \text{ km} = 827 \text{ m} \times \frac{1 \text{ km}}{1 \times 10^3 \text{ m}}$$

$$? \text{ km} = 0.827 \text{ km}$$

•Method of Solution for (b)

First state the problem:

$$? \text{ }\mu\text{m} = 275 \text{ mg}$$

Since we know how many mg are in a gram and how many μg are in a gram, we will convert mg to grams and then grams to μg.

Road map: mg → g → μm

$$? \; \mu m = 275 \; mg \times \frac{1 \times 10^{-3} \; g}{1 \; mg} \times \frac{1 \; \mu g}{1 \times 10^{-6} \; g} = 275 \times 10^3 \; \mu g$$

$$? \; \mu m = 2.75 \times 10^5 \; \mu g$$

EXAMPLE 1.12   The Factor-Label Method

The Voyager 2 mission to the outer planets of the solar system transmitted spectacular photographs of Neptune to Earth by radio signals. Radio waves, like light waves, travel at a speed of 3.00×10^8 m/s. If Neptune was 2.75 billion miles from Earth during these transmissions, how many hours were required for radio signals to travel from Neptune to Earth?

•Method of Solution

First state the problem.

$$? \; h = 2.75 \times 10^9 \; mi$$

The speed of light can be written as a conversion factor because 1 s of time is equivalent to 3.00×10^8 m of distance.

$$\frac{3.0 \times 10^8 \; m}{1 \; s} \quad or \quad \frac{1 \; s}{3.00 \times 10^8 \; m}$$

This will convert units of distance into units of time in seconds, but first the distance must be expressed in meters rather than miles. To convert from the U.S. system to the SI, we can use 1.61 km = 1.0 mile. In general, our plan will be:

distance → time

miles → km → m → s → min → h

Notice that the time in seconds must be converted to hours in order to answer the question.

•Calculation

Insert one conversion factor per arrow in the road map.

$$? \; h = 2.75 \times 10^9 \; mi \times \frac{1.61 \; km}{1.0 \; mi} \times \frac{10^3 \; m}{1 \; km} \times \frac{1 \; s}{3.00 \times 10^8 \; m} \times \frac{1 \; min}{60 \; s} \times \frac{1 \; h}{60 \; min}$$

$$= 4.099537 \; h$$

Following the rules for significant figures, we must round off this answer to three significant figures. Therefore

$$? \; h = 4.10 \; h$$

EXAMPLE 1.13   Using Density

What is the mass in grams of a lead (Pb) brick that measures 21.5 cm × 9.7 cm × 62 mm? The density of Pb is 11.4 g/cm³.

•Method of Solution

Density is defined by the formula:

$$d = \frac{m}{V}$$

If you know any two of the variables in the equation, you can solve for the remaining unknown. First rearrange the equation to solve for mass.

$$m = d \times V$$

•Calculation

The density is given but you still need the volume of the brick. For a regular solid the volume is:

$$V = length \times width \times height$$

We note that the dimensions are all in cm except the 62 mm. Convert this to centimeters.

$$62 \text{ mm} \times \frac{1 \text{ cm}}{10 \text{ mm}} = 6.2 \text{ cm}$$

The volume is

$$V = 20.5 \text{ cm} \times 9.7 \text{ cm} \times 6.2 \text{ cm} = 1233 \text{ cm}^3$$

$$V = 1.2 \times 10^3 \text{ cm}^3 \quad \text{(two significant figures)}$$

Substituting into the equation for mass

$$m = d \times V$$

$$m = \frac{11.4 \text{ g}}{1 \text{ cm}^3} \times 1.2 \times 10^3 \text{ cm}^3 = 1.4 \times 10^3 \text{ g}$$

The last step in solving the problem is checking the answer. Is it reasonable in size and units? One way to check an answer is to estimate it in your head with rounded numbers. If the density of lead is about 10 and the volume was 1200 cm^3, then the answer should be close to 12,000 g, which it is. Also the units of grams are correct for mass.

•Comment

When solving problems, based on an equation such as this example was, notice that we followed three steps.
1. Rearrange the equation so that the unknown quantity is by itself on one side.
2. Substitute in known quantities including their units.
3. Calculate the answer including its units.

EXAMPLE 1.14   A Conversion of Volume Units

How many cubic centimeters are equal to 5 m^3?

•Method of Solution

The desired unit is cm^3.

$$? \text{ cm}^3 = 5 \text{ m}^3$$

Road map: $m^3 \rightarrow cm^3$

The relationship of meters to centimeters is: $1 \text{ cm} = 10^{-2} \text{ m}$
Writing in the conversion factor

$$? \text{ cm}^3 = 5 \text{ m}^3 \times \frac{1 \text{ cm}}{1 \times 10^{-2} \text{ m}}$$

We can see that the "m" unit does not entirely cancel the "m^3" unit.

The conversion factor must also be cubed.

$$\left(\frac{1\ cm}{10^{-2}\ m}\right)^3 = \frac{1\ cm}{10^{-2}\ m} \times \frac{1\ cm}{10^{-2}\ m} \times \frac{1\ cm}{10^{-2}\ m} = \frac{1\ cm^3}{10^{-6}\ m^3}$$

Therefore, we write:

$$?\ cm^3 = 5\ m^3 \times \frac{1\ cm^3}{10^{-6}\ m^3}$$

$$= 5 \times 10^6\ m^3$$

EXERCISES

16. Make the following conversions of metric lengths.
    a. 12.5 cm = _____ m
    b. 8.0×10^{-8} m = _____ nm
    c. 445 cm = _____ km
    d. 32.5 mm = _____ μm
    e. 5.73×10^3 nm = _____ mm

17. Calculate the density of mercury given that a spherical droplet of mercury with a radius of 0.328 cm has a mass of 2.00 g. The volume of a sphere is $4/3\pi r^3$.

18. The world record for the 100 yd dash is 9.10 s. What is the average speed in miles per hour?

19. The density of a 26% salt solution is 1.199 g/mL. What is the volume occupied by 20.0 g of solution?

20. The displacement volume of a certain automobile engine is 350 in^3. How many liters is this?

CONCEPTUAL QUESTIONS

1. Why is it incorrect to say that "iron is heavier than water?"

2. Explain why intensive properties are more useful than extensive properties for comparing to substances.

3. You are given a shiny piece of metal that might be aluminum. What information would you need to prove it is Al?

4. In 1773 Ben Franklin poured a teaspoon of oil onto a pond. He observed that it spread out to form a film that covered an area of about 22,000 ft^2. If a teaspoon has a volume of about 5 cm^3 and if the film was roughly one molecule thick, then what is the average size (length) of an oil molecule in centimeters?

PRACTICE TEST

1. Which of the following describe physical properties and which describe chemical properties?
   - a. Color
   - b. Decomposition upon heating
   - c. Boiling point
   - d. Hardness
   - e. Density
   - f. A change of color on exposure to air

2. Which of the following statements describe physical changes and which describe chemical changes?
   - a. Fresh cut apples turning brown
   - b. Baking bread
   - c. Cutting a soft metal with a knife
   - d. Drying clothes
   - e. Fermentation of sugar

3. Classify each of the following as a pure substance, a heterogeneous mixture, or a homogeneous mixture:
   - a. Raw milk
   - b. Tap water
   - c. Gold
   - d. Table sugar
   - e. Concrete

4. Classify each of the following substances as an element, a compound, or a mixture.
   - a. Hot tea
   - b. Dry ice
   - c. Sulfur
   - d. Pure aspirin
   - e. Natural gas
   - f. Beer

5. Convert:
   - a. 9.8 cm to millimeters
   - b. 170 mm to centimeters
   - c. 325 mL to liters
   - d. 50 mi/h to m/s
   - e. 20 GJ to kilojoules

6. Carry out the following calculation and report the answer to the correct number of significant figures.

$$\frac{82.0715}{(32.107 - 28.224)} =$$

7. The melting point of iron is 1540°C. What is this temperature in degrees Fahrenheit?

8. To determine the density of a piece of metal, a student finds its mass to be 13.62 g. The metal is placed into a flask that can hold exactly 25.00 mL. The student finds that 19.26 g of water is required to fill the flask. Given that the density of water is 0.9970 g/mL, what is the density of the metal?

9. The density of lead is 11.4 g/cm³. What is the mass of a lead brick with dimensions 2.0 in by 2.0 cm by 80.0 mm?

10. What volume does 1.00 lb of gold occupy at 20°C? The density of gold is 19.3 g/cm³ at 20°C.

11. If gasoline costs $1.58 per gallon, compare the expense of driving exactly 10,000 miles in an SUV that gets 12 miles per gallon to that of a compact car that gets 12 km/L.

12. Calculate the density of water in units of kilograms per cubic meter. Density = 1.00 g/cm³

13. The radius of an aluminum (Al) atom is 0.125 nm. How many Al atoms would have to be lined up in a row to form a line 1 cm in length?

14. Convert atmospheric pressure from 14.7 pounds per square inch to megagrams per square centimeter.

15. Soft solder contains 70.0% tin and 30.0% lead by mass. If you had 10.0 g of tin, how many grams of solder could you make?

16. Copper (Cu) is a trace element that is essential for nutrition. Newborn infants require 80 μg of Cu per kilogram of body mass per day. The Cu content of a popular baby formula is 0.48 μg of Cu per milliliter. How many milliliters of formula should a 7.0 lb baby consume per day to obtain the minimum daily Cu requirement?

17. It takes 31,000 J of energy to make 1.0 g of aluminum (Al) metal from the ore bauxite. Burning 10.0 g of gasoline releases 4.20×10^5 J. If burning gasoline is used to supply the energy to produce aluminum from bauxite, then how many grams of gasoline are needed to make enough aluminum for one beverage can containing approximately 15 g Al?

18. What mass of carbon monoxide (CO) in kilograms is present in the lower 500 m of air above a section of a city that measures 10 km $\times$ 10 km if the concentration of CO is 10 mg/m^3?

19. A horse can run a 10 furlong race in 2 min and 5.5 s. Calculate the average speed in miles per hour. Given: 1 furlong = 1/8 mile

ANSWERS

Exercises

1. a. d. e. are heterogeneous   b. c. are homogeneous
2. a. compound   b. element   c. mixture   d. mixture   e. mixture
3. a. b. physical properties   c. d. chemical properties
4. a. d. physical change   b. c. chemical change
5. 3330 cm^3
6. 11.2 g/cm^3
7. 83°F
8. 53 3°C
9. a. 7×10^{-6} g   b. 8.0×10^{-9} m   c. 1.4×10^5 L   d. 1.0×10^3 s
10. a. 2.4×10^4 b. 1.4×10^{-4} c. 7.4×10^8 d. 9.06×10^{-2}
11. a. 0.0052   b. 1400   c. 7,050,000   d. 0.0000706
12. a. 3 b. 2 c. 3 d. 4 e. 4
13. a. 0.609   b. 1.0×10^3 c. 0.000222   d. 238.0   e. 1.3
14. a. 2.6×10^4 b. 0.4   c. 3.1×10^3 d. 13.92   e. 73.85
15. a. 199 g   b. 12.0 cm   c. 15 g/mL
16. a. 0.125 m   b. 80 nm   c. 4.45×10^{-5} km   d. 3.25×10^4 μm   e. 5.73×10^{-3} mm
17. 13.5 g/cm^3
18. 22.5 mi per hr
19. 16.7 mL
20. 5.74 L

Conceptual Questions

1. It is the density of iron that is always greater than the density of water. Not the weight. After all, a small piece of iron could be lighter than a large amount of water.
2. Extensive properties are proportional to the amount of matter present, but intensive properties are not. Many intensive properties, such as density, are characteristic of the substance.
3. Hint. Use an intensive property.
4. 2×10^{-7} cm

Practice Test

1. a. physical b. chemical c. physical d. physical e. physical f. chemical
2. a. chemical b. chemical c. physical d. physical e. chemical
3. a. heterogeneous b. homogeneous c. pure substance
   d. pure substance e. homogeneous
4. a. mixture b. compound c. element d. compound e. mixture
   f. mixture
5. a. 98 mm b. 17.0 cm c. 0.325 L d. 22 m/s e. 20,000,000 kJ
6. 21.14
7. 2804°F
8. 2.37 g/mL
9. 12 kg
10. 23.5 cm^3
11. SUV: $1317 per 10,000 miles; compact car: $562 per 10,000 miles
12. 1.0×10^3 kg/m^3
13. 8.0×10^8 atoms (800 million)
14. 1.04×10^{-3} Mg/cm^2
15. 14.3 g solder
16. 530 mL
17. 11 g gasoline
18. 5.0×10^5 kg
19. 35.86 mph

Chapter Two
ATOMS, MOLECULES, AND IONS

- The Atomic Theory (2.1)
- Subatomic Particles and the Structure of the Atom (2.2 – 2.3)
- Molecules and Ions (2.5)
- Chemical Formulas (2.6)
- Naming Inorganic Compounds (2.7)

THE ATOMIC THEORY (2.1)

STUDY OBJECTIVES

1. State the laws of definite proportions, multiple proportions, and conservation of mass.
2. List and discuss the postulates of Dalton's atomic theory.
3. Describe how Dalton's atomic theory explains these three laws.

Atoms. According to John Dalton's atomic theory, which he proposed in 1808, elements are composed of extremely small, indivisible particles called *atoms*. Dalton assumed that **atoms** were the smallest unit of an element that can enter into chemical combination. Atoms of the same element all have the same mass, and atoms of different elements have different masses. His theory explained three laws that were known at that time. The atomic theory consists of a set of postulates which can explain the law of definite proportions, the law of multiple proportions, and law of conservation of mass. The postulates are summarized below.

Dalton's Atomic Theory
1. All matter is made of extremely small, indivisible particles called atoms. An *atom* is the smallest unit of an element that can enter into chemical combination. Atoms of the same element are identical in size, mass, and chemical properties.
2. Atoms of one element are different from atoms of another element.
3. Compounds are formed when atoms from two or more elements combine. Atoms combine in the ratio of small whole numbers. For a given compound, the number of atoms of one element that combine per atom of the other element is a fixed ratio. The smallest particle that has the properties of the compound is called a molecule.
4. Atoms are the smallest units of chemical change. Chemical change involves the combination, separation, or rearrangement of atoms. Atoms are not destroyed or created in chemical processes.

The **law of definite proportions** refers to the elemental constituents of compounds and states that all purified samples of a compound contain its constituent elements combined in the same proportions by mass. For example, all samples of the compound carbon monoxide, no matter what size, show that it is 43% carbon and 57% oxygen by mass. Dalton's proposal was that the smallest particle of carbon monoxide was a molecule consisting of one carbon atom and one oxygen atom. If all carbon atoms had the same mass, and all oxygen atoms had a mass about 1.3 times greater than carbon atoms, then the composition of carbon monoxide would be exactly as given above.

While working on the atomic theory, Dalton discovered the **law of multiple proportions:** when two elements form more than one compound, the various masses of one element combining with a fixed mass of another element are related by small whole-number ratios. This law can be applied to two compounds of nitrogen and oxygen; nitrogen monoxide, and nitrogen dioxide. The mass of oxygen combining with nitrogen in nitrogen dioxide was found to be twice as great as the mass of oxygen combining with nitrogen in nitrogen monoxide. Dalton proposed that the two compounds of N and O were the result of atoms of the two elements combining in different ratios to form two different molecules. When one atom of nitrogen combined with one atom of oxygen, nitrogen monoxide was formed. But when one atom of nitrogen combined with two atoms of oxygen, then nitrogen dioxide was the compound formed.

The **law of conservation of mass** refers to an observation that has been made many times: there is no detectable gain or loss of mass during a chemical reaction. The law of conservation of mass is consistent with postulate 4. Atoms are neither created nor destroyed in a chemical reaction. Rather they are merely rearranged to form new combinations of atoms which we would recognize as new compounds. Mass is conserved in chemical reactions because atoms are conserved.

EXERCISE

1. Describe briefly in your own words the four postulates of Dalton's atomic theory.

SUBATOMIC PARTICLES AND THE STRUCTURE OF THE ATOM (2.2 – 2.3)

STUDY OBJECTIVES

1. Name and describe the three subatomic particles of most importance to chemistry and give their location within the atom.
2. Write the symbol of an isotope having been given its mass number and element name.
3. Determine the number of protons, neutrons, and electrons in an atom given its isotopic symbol.

Subatomic Particles. In the early 1900s, scientists learned that all atoms are constructed from the same three subatomic particles: the electron, proton, and neutron. Whereas atoms themselves are electrically neutral they were found to consist of electrically charged particles. The electron has a single unit of negative charge; the proton has a single unit of positive charge; and the neutron has no charge. The proton and neutron have essentially the same mass, 1.673×10^{-24} g, and 1.675×10^{-24} g, respectively. The electron mass is much smaller: 9.11×10^{-28} g which is only 1/1836 the mass of a proton.

Rutherford's Experiment. Rutherford's experiments on alpha particle scattering in 1910 led the way to the nuclear model of the atom. Alpha particles are ejected from certain radioactive elements with very high kinetic energies. Remember that alpha particles have 2 units of positive charge, and a mass 4 times that of a proton. Being smaller than the atoms of most elements they can act as probes of the interior of atoms. Rutherford's experiments lead to the following conclusions and about the structure of atoms.

Observations made by Rutherford when alpha particles were scattered by gold foil:

1. The majority of particles passed through the foil and were either undeflected or only slightly deflected (angle less than 2%).
2. A few particles were deflected (scattered) by more than a few degrees.
3. It was extremely rare, but 1 in 20,000 alpha particles were deflected back toward the direction from which they had come.

Interpretations made by Rutherford about the nature of the atom:

1. Most of the atom must be empty space.
2. The atom's positive charges are concentrated into a dense central core called the *nucleus*.
3. The diameter of a typical atom is about 10,000 times greater than the diameter of a typical nucleus.

Atomic Number and Mass Number. An atom of one element is distinguished from an atom of another element by its number of protons. The **atomic number, Z,** of an element is the number of protons in the atomic nucleus. For example, the atomic number of oxygen is 8; therefore, all oxygen atoms contain eight protons and also eight electrons.

The total mass of an atom is determined almost entirely by the number of protons and neutrons. In many cases the mass of the electrons can be neglected because it is so much smaller. The **mass number, A,** is the same as the total number of neutrons and protons present in the nucleus of an atom. The number of neutrons in an atom is $A - Z$.

Isotopes. Atoms of a given element that differ in the number of neutrons, and consequently in mass are called **isotopes.** For example, there are two atoms of the element lithium, one with a mass number of 6 and another with a mass number of 7. Isotopes can be referred to by their mass numbers as in lithium-6 (pronounced lithium six) and lithium-7. The different mass numbers are the result of different numbers of neutrons per atom. An atom of the isotope lithium-6 contains three protons, three electrons, and three neutrons, whereas an atom of lithium-7 contains three protons, three electrons and four neutrons. Both atoms contain three protons, and so both are isotopes of the element lithium. In nature, elements are found as mixtures of isotopes.

Special symbols are used to designate specific isotopes.
The general symbol for an isotope is:

The symbols for the isotopes of lithium are $^{6}_{3}Li$ and $^{7}_{3}Li$.

Isotopes of an element have similar chemical properties, and form the same types of compounds. The chemical properties of an element depend on the number of protons and electrons in an atom, not the number of neutrons.

EXAMPLE 2.1   Isotopic Symbols

The three isotopes of oxygen found in nature are oxygen-16, oxygen-17, and oxygen-18. Write their isotopic symbols.

•**Method of Solution**

The general form of the symbols is $^A_Z O$, where A is the mass number and Z is the atomic number. The atomic number of oxygen is 8, and so all oxygen atoms contain eight protons. The mass numbers are 16, 17, and 18, respectively. *Answer:* The isotopic symbols are:

$$^{16}_{8}O \qquad\qquad ^{17}_{8}O \qquad\qquad ^{18}_{8}O$$

EXAMPLE 2.2   Isotopic Symbols

How many neutrons are present in the nucleus of each of the oxygen isotopes in the previous example?

•**Method of Solution**

The number of neutrons is given by $A - Z$, the mass number minus the proton number. *Answer:*

> For ^{16}O, there are $16 - 8 = 8$ neutrons.
> For ^{17}O, there are $17 - 8 = 9$ neutrons.
> For ^{18}O, there are $18 - 8 = 10$ neutrons

EXERCISES

2. What are the three fundamental particles from which atoms are made? What are their electric charges. Which particles are in the atomic nucleus?

3. What evidence did Rutherford find that supported his theory of the atomic nucleus?

4. If the nucleus of an atom was the diameter of a nickel (about 2 cm), what would the diameter be of a such an atom?

5. What is the mass number of a sodium atom that has 13 neutrons?

6. How many neutrons are in an atom of Ag-109?

7. Write the symbol for each of the following isotopes.
    a. An atom with $Z = 30$ and 37 neutrons.
    b. An atom with $Z = 51$ and 69 neutrons.

MOLECULES AND IONS (2.5)

STUDY OBJECTIVES

1. Define the terms molecule and ion.
2. Determine the numbers of protons and electrons in given monatomic cations and anions.
3. Learn the formulas of several polyatomic ions.

Molecules. Compounds are produced by the combination of atoms of different elements. Compounds fall into two general types depending on how their atoms are held together: *molecular* and *ionic*. For a given compound, the number of atoms of one element that combine per atom of the other element is a fixed ratio. **A molecule** is a discrete particle composed of atoms that are bonded together. A molecule is the smallest particle that has the properties of the compound. Diatomic molecules contain two atoms, while polyatomic molecules contain more than two atoms. Some representations of simple molecules are given in Figure 2.12 of the text.

Ions. Atoms are electrically neutral because they have equal numbers of protons and electrons. However, an atom can acquire a charge by gaining or losing electrons. These processes do not change the identity of an atom because the atomic number (the number of protons) is unaffected. A change in the number of electrons an atom has will alter its electrical charge. Any atom or group of atoms having a net electrical charge is called an **ion**. Certain atoms (typically metal elements) always tend to lose electrons (1 or 2, some atoms lose 3 electrons). These form positive ions which are called *cations*. Others atoms (typically those of nonmetal elements) tend to gain 1 or 2 electrons forming *negative ions* called **anions**.

Take for example, the magnesium atom, which has 12 protons in the nucleus and 12 electrons outside the nucleus. It can lose two of its electrons. The result is a particle called the magnesium ion which has 12 protons and 10 electrons. Its net charge is +2 because in the ion the number of positive charges outnumber the negative charges by two. The symbol for the magnesium ion is Mg^{2+}. The superscript (2+) represents the net charge of the ion. A few elements form ions with +3 charges.

Negative ions are formed when atoms gain extra electrons. For example, the oxygen atom which has 8 protons and 8 electrons can gain two electrons. The result is an oxygen ion which has 8 protons and 10 electrons. The net charge of the ion is –2, because there are two more negative charges than positive charges. The symbol for this ion is O^{2-}. The O^{2-} ion is called the oxide ion. The names of monatomic anions end in *-ide*.

Polyatomic Ions. In addition to monatomic ions such as Mg^{2+} and O^{2-}, more complex ions are known. A **polyatomic ion** is a group of two or more atoms that carries an electrical charge. The hydroxide ion (OH^-) is a simple example. The oxygen atom and hydrogen atom are held together much as they would be in a molecule, but they have gained an extra electron. This gives the ion a –1 charge. Sometimes the net charge on these ions seems unpredicable and so the best thing to do is memorize the names, formula, and charge of the most common polyatomic ions. Table 2.1 lists some important polyatomic ions.

Table 2.1 Some Polyatomic Ions

Formula	Name
OH^-	hydroxide
CO_3^{2-}	carbonate
HCO_3^-	hydrogen carbonate
ClO_3^-	chlorate
NO_3^-	nitrate
NO_2^-	nitrite
PO_4^{3-}	phosphate
SO_4^{2-}	sulfate
SO_3^{2-}	sulfite

Be careful not to confuse a formula such as NO_2^- with NO_2. The latter formula is a molecular compound as is electrically neutral, whereas the former is a polyatomic ion. The anion does not exist by itself, and will alway be a part of an ionic compound such as $NaNO_2$.

EXAMPLE 2.3   Electrons and Protons in Ions

Determine the number of protons and electrons in the ions represented by K^+, Fe^{3+}, and I^-.

•**Method of Solution**

The net positive charge of the K^+ ion is due to the loss of an electron from a K atom.
A K atom has 19 p^+ and 19 e^-. Therefore, the K^+ ion has 19 p^+ and 18 e^-.

The net positive charge of the Fe^{3+} ion is due to the loss of three electrons from an Fe atom.
An Fe atom has 26 p^+ and 26 e^-. Therefore, the Fe^{3+} ion has 26 p^+ and 23 e^-.

The net negative charge of the I^- ion is due to a gain of one electron by an I atom.
An I atom has 53 p^+ and 53 e^-. Therefore, the I^- ion has 53 p^+ and 54 e^-.

EXERCISES

8. Identify the following as elements or compounds.
    a. ClF_3 b. HCl c. O_3 d. I_2 e. Se   f. NaI

9. Give an example of: a. a monatomic cation. b. a polyatomic anion

10. Give the number of protons and electrons in each of the following ions.
    a. Li^+ b. Sr^{2+} c. Fe^{3+} d. N^{3-} e. Se^{2-} f. Cl^-

CHEMICAL FORMULAS (2.6)

STUDY OBJECTIVES

1. Distinguish between molecular and empirical formulas, and write empirical formulas when molecular formulas are given.
2. Write the formulas of inorganic compounds from their names.

Molecular Formulas. A molecule of a compound is formed by the combination of atoms of different elements. It is the smallest unit of a compound that has properties of the compound. The **molecular formula** indicates the numbers of atoms of each element in a molecule of the compound. The formula for ethylene glycol is $C_2H_6O_2$ because there are two carbon atoms, six hydrogen atoms, and two oxygen atoms in a molecule of the compound.

All molecules are not necessarily compounds, since some elements exist as molecules. H_2, N_2, O_2, F_2, Cl_2, Br_2, and I_2 are molecular formulas just as CH_4, NH_3, and $C_2H_6O_2$ are molecular formulas. Figure 2.12 of the text shows some molecular models of several molecular compounds.

The **empirical formula** of a compound gives the simplest whole-number ratio of the various kinds of atoms of the elements making up the compound. They are written by reducing the subscripts in molecular formulas to the smallest whole-number ratio. Benzene, for example, has the molecular formula C_6H_6. Dividing 6 by 6 gives the simplest ratio. The empirical formula is just CH, the simplest ratio of C atoms to H atoms being 1 : 1. The empirical formula of many compounds is the same as the molecular formula. For ammonia (NH_3) and methane (CH_4), for example, the ratios of subscripts cannot be reduced and still remain whole numbers.

Ionic compounds consist of individual positive and negative ions. Magnesium chloride ($MgCl_2$), for example, consists of individual magnesium and chloride ions. The formula $MgCl_2$ just gives the ratio of Mg^{2+} ions to Cl^- ions in any sample of magnesium chloride. Therefore, magnesium chloride is a compound in which the ratio of Mg^{2+} ions to Cl^- ions is 1 : 2. See Figure 2.13 of the text which shows the crystal structure of sodium chloride.

The formulas of ionic compounds are always the same as their *empirical formulas*. Ionic compounds consist of positive and negative ions, rather than molecules. Their formulas already give the smallest whole-number ratio of the ions in the compound.

Formulas of Ionic Compounds. The names and formulas of the cations and anions in can be used to write formulas of compounds. To write the correct formula for an ionic compound, *you must know the ionic charges of the cations and anions.* See Table 2.3 of the textbook. *The main rule is that all compounds must be electrically neutral.* Therefore, the ratio of cations to anions must be such that the sum of the positive charges of cations and the negative charges of anions must be equal to zero. Consider the following examples:

1. *Potassium bromide.* The ions are K^+ and Br^-. Adding the charges of these ions, we get
   $(+1) + (-1) = 0$
   The ratio of K^+ to Br^- must be 1 : 1. The formula is KBr.

2. *Potassium sulfide.* The ions are K^+ and S^{2-}. When the charge on the positive ion is less than the charge on the negative ion, more positive ions will be required to balance the negative charge. If there are two K^+ ions for each S^{2-} ion, the compound will be neutral.
   2 cations$(+1)$ + 1 anion(-2) = $(+2) + (-2) = 0$.
   The formula is K_2S
   Notice that the ionic charge is written in parenthesis preceded by the number of ions.

3. Magnesium sulfate. The ions are Mg^{2+} and SO_4^{2-}. A compound with one Mg^{2+} ion and one SO_4^{2-} ion will be neutral.
   $(+2) + (-2) = 0$
   The formula is $MgSO_4$.

4. Aluminum oxide. The ions are Al^{3+} and O^{2-}. Two Al^{3+} ions have a charge of +6. It takes three O^{2-} ions to have a charge of –6.
   2 cations$(+3)$ + 3 anions$(-2) = (+6) + (-6) = 0$
   The formula is Al_2O_3.

EXAMPLE 2.4   Empirical Formulas

What are the empirical formulas of the following compounds?
a. B_2H_6 b. P_4O_{10}

•Method of Solution

To find the simplest whole-number ratio of atoms divide the smaller subscript into the larger one.
a. The simplest ratio of H to B is 6 to 2 which 3 to 1. Therefore the empirical formula is BH_3.
b. The simplest ratio of O to P is 10 to 4 which reduces to 5 to 2. Notice that you must keep a whole-number ratio. An atom ratio of 2.5 to 1 would not be physically reasonable. The empirical formula is P_2O_5.

EXAMPLE 2.5   Formulas of Ionic Compounds

Write the formulas of the following compounds:
a. magnesium chloride   b. magnesium oxide   c. magnesium phosphate

•Method of Solution

The formula must represent an electrically neutral grouping of ions. Identify the ions first.
a. The ions in magnesium chloride are Mg^{2+} and Cl^- ions. Two chloride ions are needed for each magnesium ion. The formula is $MgCl_2$. 1 cation$(+2)$ + 2 cations$(-1) = 0$

b. The ions in magnesium oxide are Mg^{2+} and O^{2-}. In this case just one cation and one anion are will give electrical neutrality. $(+2) + (-2) = 0$ The formula is MgO.

c. The ions in magnesium phosphate are Mg^{2+} and PO_4^{3-}. In this case 2 anions have a –6 charge and it will take 3 cations to have the needed +6 charge.   3 cations $(+2)$ + 2 anions$(-3) = 0$
   The formula is $Mg_3(PO_4)_2$.

EXERCISES

11. Styrene has the molecular formula C_8H_8. What is its empirical formula?

12. What are the empirical formulas of the following compounds?
    a. N_2O_4 b. C_4H_8 c. $AlCl_3$ d. Fe_2O_3 e. S_2F_{10}

13. Which of the following compounds are likely to be ionic?
    a. KCl b. CH_4 c. $AlCl_3$ d. SO_2 e. MgO f. CCl_4

14. Write formulas of the following binary compounds.
    a. barium chloride
    b. magnesium nitride
    c. iron(III) oxide
    d. iron(II) fluoride

15. Write the formulas for the following compounds.
    a. ammonium chloride
    b. sodium phosphate
    c. potassium sulfate
    d. calcium carbonate
    e. potassium hydrogen carbonate
    f. magnesium nitrite
    g. sodium nitrate
    h. ammonium perchlorate
    i. strontium hydroxide
    j. copper(II) cyanide

NAMING INORGANIC COMPOUNDS (2.7)

STUDY OBJECTIVE

1.   Write the names of several types of inorganic compounds given their formulas.

Chemical Names. **Nomenclature** refers to the naming of chemical substances. Chemical nomenclature and formulas form the basis of the language of chemistry. In the textbook, inorganic compounds are divided into four categories: *ionic compounds, molecular compounds, acids and bases,* and *hydrates*. **Binary compounds** are those with only two elements and are the easiest to name.

   Learning chemical nomenclature is like learning a new language. First, you build up a vocabulary of words, and then you learn rules for putting words together meaningfully. To learn a vocabulary requires strict memorization, and so treat the names of cations and anions as you would words in a new language. Get a package of 3×5 cards, and write the name of an ion on one side of each card and its chemical formula on the other side. Table 2.3 in the textbook has 20 cations and 26 anions. If you learn their names and formulas, you will be able to name 572 possible chemical compounds!

Ionic Compounds. In binary ionic compounds, the metal ion is positively charged and is named first. The name of a simple cation is the same as the name of the element. The negative ion name appears second, and is derived from the name of the nonmetallic element in which an *-ide* ending replaces the usual ending of the element name. The following formulas and names illustrate naming binary ionic compounds:

NaI	sodium iodide
CaF_2	calcium fluoride
SnO	tin oxide

   Some metals can form more than one kind of cation. This is true of tin, for example, where Sn^{2+} and Sn^{4+} are known. These are called tin(II) and tin(IV), respectively. SnO in the preceding list should more correctly be named tin(II) oxide. This distinguishes it from SnO_2, tin(IV) oxide, which is a different compound. The transition metals commonly exhibit a variety of ionic charges. The system in which a Roman numeral is used to designate the ionic charge of the cation was developed by Alfred Stock and is called the Stock system.

MnO_2	manganese(IV) oxide
Mn_2O_3	manganese(III) oxide

In *ternary ionic compounds*, those consisting of three elements, the metal ion is named first followed by the name of the *polyatomic anion*. Some examples are:

Na_2SO_4	sodium sulfate
$Fe(NO_2)_2$	iron(II) nitrite
$Mg(NO_3)_2$	magnesium nitrate

These polyatomic ions are examples of *oxoanions*. They contain a central atom, such as sulfur or nitrogen, that is bonded to several oxygen atoms. Some elements form several oxoanions that differ only in the number of oxygen atoms. In the case of an element that can have two oxoanions, the name of the anion with more oxygen atoms ends in *-ate*, and the name of the anion with less oxygen atoms ends in *-ite*. See nitr*ate* and nitr*ite* in the preceding examples.

Molecular Compounds. In binary molecular compounds, the more metallic element is named first, followed by the less metallic elements, whose name ends in *-ide*. When a pair of elements can form several compounds, questions about which compound is being referred to are removed by adding prefixes to denote the number of atoms of each element. The prefixes taken from Greek are given in Table 2.4 of the textbook. Some examples are:

SF_4	sulfur tetrafluoride
SF_6	sulfur hexafluoride
PCl_3	phosphorus trichloride
P_2Cl_4	diphosphorus tetrachloride

Acids and Bases. Binary acids are compounds of hydrogen and a nonmetal. Many hydrogen compounds have completely different properties when they are in the pure gaseous or liquid state than when they are dissolved in water. In the gaseous state, HCl and H_2S, for instance, are molecular compounds called hydrogen chloride and hydrogen sulfide, respectively, whereas water solutions of these compounds contain hydrogen ions (H^+) and anions and have slightly different names. **Acids** are substances that release H^+ ions when they dissolve in water.

In naming binary acids, the word *hydrogen* is replaced by *hydro-*, and the *-ide* ending of the anion name is replaced with *-ic acid*. For example:

HCl(g)	hydrogen chloride
HCl(aq)	hydrochloric acid
H_2S(g)	hydrogen sulfide
H_2S(aq)	hydrosulfuric acid

The *oxoacids* are compounds containing hydrogen and oxoanions. The acids of nitrate and sulfate are:

HNO_3	nitric acid
H_2SO_4	sulfuric acid

Note that the *-ate* ending of the anion name has been changed to *-ic acid*. For oxoanions with fewer oxygen atoms, the *-ite* ending of the anion name is changed to *-ous acid* for the oxyacid.

HNO_2	nitrous acid
H_2SO_3	sulfurous acid

Bases are substances that release hydroxide ions (OH⁻) in aqueous solution. The metal hydroxides are named as if they were binary compounds and employ the *-ide* ending.

LiOH   lithium hydroxide
Ca(OH)$_2$ calcium hydroxide

EXAMPLE 2.6   Naming Compounds

Name the following compounds according to the Stock system:
a.   CuBr
b.   CuSO$_4$

•Method of Solution

In the Stock system the symbol of the metallic element in a compound is followed by a Roman numeral derived from the charge of the cation.

a.   Assuming bromide to be –1 as in the Br⁻ ion, Cu must be a +1 ion in order to have a neutral compound. Therefore, CuBr is named copper(I) bromide.

b.   In this case copper must be a +2 ion and the compound is copper(II) sulfate.

EXERCISES

16. Name the following compounds:
    a. K$_3$N   b. Ag$_2$CO$_3$ c. Mg(OH)$_2$ d. NaCN   e. NH$_4$I
    f. Fe(NO$_3$)$_2$ g. CaSO$_4$·2H$_2$O

17. Name the following compounds:
    a. PCl$_5$ b. SO$_3$ c. P$_4$O$_{10}$ d. N$_2$O   e. NO$_2$

18. Name the following acids:
    a. HNO$_3$ b. HNO$_2$ c. HBr   d. HCN   e. H$_2$S

CONCEPTUAL QUESTIONS

1.   What is the difference between an atom and a molecule?

2.   Do the symbols O$_3$ and 3O signify the same thing? Explain.

3.   Why are the chemical formulas of ionic compounds the same as the empirical formulas?

PRACTICE TEST

1. Complete the following isotope table:

Name	Symbol	Number of Protons	Number of Electrons	Number of Neutrons	Mass Number
Sodium	^{23}Na	11	___	12	23
___	^{40}Ar	___	___	22	___
Arsenic	^{75}As	___	___	___	___
Lead	___	___	___	___	___
___	___	___	___	20	___

2. a. What is the total number of subatomic particles (protons, neutrons, and electrons) in an atom of $^{56}_{26}Fe$?

   b. What is the mass number of a copper atom that has 35 neutrons?

3. Give the number of protons, neutrons, and electrons in atoms of the following.
   a. $^{17}_{8}O$ b. $^{107}_{47}Ag$ c. $^{222}_{86}Rn$

4. Which of the following are isotopes of element X?
   $^{46}_{20}X$, $^{20}_{46}X$, $^{43}_{20}X$, $^{46}_{43}X$

5. Take the mass of an atom to be the sum of the masses of its protons, neutrons, and electrons. Determine the percentages by mass of the protons, the neutrons, and the electrons in a $^{12}_{6}C$ atom.

6. Which of the following molecules are forms of a pure element?
   P_4, He, N_2, O_3, N_2O_3

7. What is the empirical formula of each of the following compounds?
   a. $C_6H_8O_6$ b. C_2H_2 c. Hg_2Cl_2
   d. H_2O_2 e. $C_2H_2O_4$ f. $MgCl_2$

8. Write the formulas for the following compounds.
   a. calcium hypochlorite
   b. mercury(II) sulfate
   c. barium sulfite
   d. zinc oxide
   e. dinitrogen oxide
   f. sodium carbonate
   g. copper(II) sulfide
   h. lead(IV) oxide

9. Name the following compounds:
   a. Na_2HPO_4 b. HI(gas)   c. P_4O_6 d. $LiNO_3$ e. HI(solution)
   f. $Sr(NO_2)_2$ g. $NaHCO_3$ h. K_2SO_3 i. Na_3PO_4 j. $Al(OH)_3$

10. Name the following acids:
   a. H_2SO_3 b. HClO   c. $HClO_4$ d. H_3PO_4 e. HCN

ANSWERS

Exercises

2. Protons, electrons, and neutrons. Protons have one unit of positive charge, and electrons have one unit of negative charge. Neutrons have no charge. Protons and neutrons are found in the nucleus.
3. The evidence was related to the direction of scattering of the positively charged alpha particles by gold atoms. The fact that 1 in 10,000 alpha particles was deflected "backward" meant that all the positive charge of the gold atom must be concentrated into a very small part of the atom.
4. 200 m
5. 24
6. 62
7. a. $^{67}_{30}$Zn   b. $^{120}_{51}$Sb
8. a. compound   b. compound   c. element   d. element   e. element   f. compound
9. a. Fe^{3+} iron(III) ion   b. CO_3^{2-} carbonate ion
10. a. 3p, 2e   b. 38p, 36e   c. 26p, 23e   d. 7p, 10e   e. 34p, 36e   f. 17p, 18e
11. CH
12. a. NO_2 b. CH_2 c. $AlCl_3$ d. Fe_2O_3 e. SF_5
13. a, c, e.
14. a. $BaCl_2$ b. Mg_3N_2 c. Fe_2O_3 d. FeF_2
15. a. NH_4Cl b. Na_3PO_4 c. K_2SO_4 d. $CaCO_3$ e. $KHCO_3$
    f. $Mg(NO_2)_2$ g. $NaNO_3$ h. NH_4ClO_4 i. $Sr(OH)_2$ j. $Cu(CN)_2$
16. a. potassium nitride   b. silver carbonate   c. magnesium hydroxide   d. sodium cyanide
    e. ammonium iodide   f. iron(II) nitrate   g. calcium sulfate dihydrate
17. a. phosphorus pentachloride   b. sulfur trioxide   c. tetraphosphorus decoxide
    d. dinitrogen monoxide   e. nitrogen dioxide
18. a. nitric acid   b. nitrous acid   c. hydrobromic acid   d. hydrocyanic acid
    e. hydrosulfuric acid

Conceptual Questions

1. An atom is the smallest particle of an element. A molecule is a particle that is made of two or more atoms. When the atoms in an molecule are identical we have a molecular form of an element. When the atoms are different we have the smallest particle of a compound.
2. No. The symbol O_3 represents a molecule of ozone, a molecular form of oxygen containing 3 oxygen atoms. Whereas, 3O represents 3 separate oxygen atoms.
3. The empirical formula of a compound gives the simplest whole-number ratio of the atoms of the elements making up the compound. Ionic compounds consist of positive and negative ions in the ratio needed to give an electrically neutral substance. It is this ratio of positive to negative ions that is essential to the formation of a neutral ionic compound. Any crystal of the compound having the proper ratio of positive and negative ions will always be electrically neutral, no matter what its size.

Practice Test

1.

Name	Symbol	Number of Protons	Number of Electrons	Number of Neutrons	Mass Number
Sodium	^{23}Na	11	<u>11</u>	12	23
<u>Argon</u>	^{40}Ar	<u>18</u>	<u>18</u>	22	<u>40</u>
Arsenic	^{75}As	<u>33</u>	<u>33</u>	42	<u>75</u>
Lead	^{206}Pb	82	<u>82</u>	124	<u>206*</u>
<u>Calcium</u>	^{40}Ca	20	<u>20</u>	20	<u>40*</u>

*Arbitrarily chosen isotope; others are acceptable

2. a. 108   b. 64
3. a. 8p, 9n, 8e   b. 47 p, 60n, 47e   c. 86p, 136, 86e
4. $^{46}_{20}$X & $^{43}_{20}$X
5. 50% proton mass, 50% neutron mass, & 2.8×10^{-2}% electron mass
6. P_4, He, N_2, O_3
7. a. $C_3H_4O_3$ b. CH   c. HgCl
    d. HO   e CHO_2 f. $MgCl_2$

8. a. $Ca(OCl)_2$ b. $HgSO_4$ c. $BaSO_3$ d. ZnO
    e. N_2O f. Na_2CO_3 g. CuS   h. PbO_2

9. a. sodium hydrogen phosphate   b. hydrogen iodide   c. tetraphosphorus hexoxide
    d. lithium nitrate   e. hydroiodic acid   f. strontium nitrite
    g. sodium hydrogen carbonate   h. potassium sulfite   i. sodium phosphate
    j. aluminum hydroxide

10. a. sulfurous acid   b. hypochlorous acid   c. perchloric acid
     d. phosphoric acid   e. hydrocyanic acid

Chapter Three
MASS RELATIONSHIPS IN CHEMICAL REACTIONS

- Atomic and Molecular Masses (3.1)
- The Molar Mass of Elements and Compounds (3.2 – 3.3)
- Chemical Formulas and Percent Composition (3.6)
- Chemical Reactions and Equations (3.7)
- Stoichiometry: Amounts of Reactants and Products (3.8)
- Limiting Reagents and Reaction Yields (3.9 – 3.10)

ATOMIC AND MOLECULAR MASSES (3.1)

STUDY OBJECTIVES

1. Describe the atomic mass unit scale and the carbon-12 standard of the scale.
2. Determine the average atomic mass of an element given the masses and percent abundances of its isotopes.
3. Determine the molecular mass of a compound given its molecular formula.

The Atomic Mass Unit Scale. The atomic mass unit scale works like this. The masses of individual atoms cannot be measured with a balance; but *relative* masses of the atoms of different elements can be measured. For instance, it is possible to determine that an atom of ^4_2He is very close to 1/3 the mass of an atom of $^{12}_6C$. Next we assign a certain value for the mass of a carbon-12 atom. By international agreement an atom of carbon-12 has a mass of exactly 12 atomic mass units (amu). Carbon-12 is the standard of the amu scale. On this scale a helium-4 atom has a mass of 4.00 amu (1/3 of 12). Atoms heavier than carbon-12 have masses greater than 12 amu. Fluorine atoms, for instance, are 1.583 times heavier than carbon-12 atoms. Thus the mass of a fluorine atom is 19.00 amu. Other experiments showed that the lightest element is hydrogen, H atoms having about 1/12 the mass of carbon-12 atoms. In this way, the relative masses of atoms of all the elements have been established.

Modern Atomic Masses. The atomic masses that appear in the present-day periodic table are determined by taking into account the fact that atoms of the elements exist as isotopes. First the mass of each isotope of an element is measured with a mass spectrometer relative to the atomic mass of carbon-12. Then the percent abundance of each isotope is measured. This information permits calculation of the *average* atomic mass.

For example, the element lithium has two isotopes that occur in nature: ^6_3Li with 7.5 percent abundance, and ^7_3Li with 92.5 percent abundance. The atomic mass of lithium-6 is 6.01513 amu, and that of lithium-7 is 7.01601 amu. The average mass of such a mixture of Li atoms is given by:

average atomic mass = (fraction of isotope X)(mass of isotope X)
$$+ \text{(fraction of isotope Y)(mass of isotope Y)}$$

$$= (0.075)(6.01513 \text{ amu}) + (0.925)(7.0161 \text{ amu})$$
$$= 0.45 \text{ amu} + 6.49 \text{ amu}$$
$$= 6.94 \text{ amu}$$

Note that neither ^6_3Li nor ^7_3Li has an atomic mass of 6.94 amu. This is the average value for the mixture of the two Li isotopes.

Molecular Masses. Compounds result when atoms of different elements bond to each other. There are two kinds of compounds: molecular and ionic. As discussed in the previous chapter a molecule of a compound is formed by the combination of atoms of different elements. The **molecular mass** is the sum of the atomic masses of the atoms in the molecular formula. For example, the molecular masses of two nitrogen oxides NO_2 and N_2O_5 are as follows:

$$molecular\ mass\ of\ NO_2 = atomic\ mass\ of\ N + 2(atomic\ mass\ of\ O)$$
$$= 14.01\ amu + 2(16.00\ amu)$$
$$= 46.01\ amu$$

$$atomic\ mass\ of\ N_2O_5 = 2(\ atomic\ mass\ of\ N) + 5(atomic\ mass\ of\ O)$$
$$= 2(14.01\ amu) + 5(16.00\ amu)$$
$$= 108.02\ amu$$

EXAMPLE 3.1   Average Atomic Mass

The element boron (B) consists of two stable isotopes with atomic masses of 10.012938 amu and 11.009304 amu. The average atomic mass of B is 10.81 amu. Which isotope is more abundant, boron-10 or boron-11?

•Method of Solution

The atomic mass is an average of the masses of the naturally occurring isotopes of an element. This means that each isotopic mass is multiplied by its percent abundance. If both isotopes in this example had a natural abundance of 50 percent, then the atomic mass would be just the average of the two masses, or approximately 10.5 amu. *Answer:* Because the average mass of boron (10.81 amu) is between 10.5 and 11.0 amu, the abundance of the boron-11 isotope must be greater than that of the boron-10 isotope. The actual abundances of boron-10 and boron-11 are 19.6% and 80.4%, respectively.

EXAMPLE 3.2   Molecular Mass

Calculate the molecular mass of carbon tetrachloride (CCl_4).

•Method of Solution

The molecular mass is the sum of the atomic masses of all the atoms in the molecule:
$$molecular\ mass\ CCl_4 = (atomic\ mass\ of\ C) + 4(atomic\ mass\ of\ Cl)$$
$$= (12.01\ amu) + 4(35.45\ amu)$$
$$molecular\ mass\ CCl_4 = 153.81\ amu$$

EXERCISES

1. An atom of oxygen-18 is 1.49992 times heavier than and atom of carbon-12. What is the mass of an oxygen-18 atom in amu?

2. The element silver consists of two isotopes: $^{107}_{47}Ag$ with an atomic mass of 106.905 amu and a natural abundance of 51.83%, and $^{109}_{47}Ag$ with an atomic mass of 108.905 amu and natural abundance of 48.17%. Calculate the average atomic mass of silver.

THE MOLAR MASSES OF ELEMENTS AND COMPOUNDS (3.2 – 3.3)

STUDY OBJECTIVES

1. Define the mole as a base unit of the SI.
2. Calculate the number of moles in a given amount of an element or a compound.
3. Calculate the number of atoms, molecules, or formula units in a given amount of a substance.

The Mole. In the laboratory, amounts of elements and compounds are measured in units of grams. In order to measure out equal numbers of atoms of two elements, say carbon and silicon, we cannot simply weigh out equal masses of the two elements. Instead, we must measure a gram ratio of the two that is the same as the mass ratio of one Si atom to one C atom. The atomic masses of Si and C are 28.09 amu and 12.01 amu, respectively. Thus, one Si atom has a mass 2.3 times greater than that of a C atom. It follows that any amounts of Si and C that have a mass ratio of Si to C of 2.3 : 1.0, will have equal numbers of Si and C atoms.

The **molar mass** of an element is a mass equal to its atomic mass expressed in grams. A molar mass of any element will contain the same number of atoms as a molar mass of any other element. Therefore, 28.09 g Si contains the same number of atoms as 12.01 g C. The number of atoms in a molar mass, called **Avogadro's number**, is equal to 6.022×10^{23} atoms. The quantity of a substance that contains Avogadro's number of atoms or other entities is called a **mole**.

The following are examples of a mole of an element:

One mole of carbon contains 6.022×10^{23} C atoms and has a mass of 12.01 g (the molar mass).

One mole of silicon contains 6.022×10^{23} Si atoms and has a mass of 28.09 g (the molar mass).

The term *mole* can be used in relation to any kind of particle, such as atoms, ions, or molecules. For clarity the particle must always be specified. We say 1 mole of O_3 (ozone), or 1 mole of O_2 (diatomic oxygen), or 1 mole of Na^+ (sodium ions).

Other examples of molar amounts are:

1 mole Na^+ ions = 6.022×10^{23} Na^+ ions = 23.00 g Na^+
1 mole O_2 molecules = 6.022×10^{23} O_2 molecules = 32.00 g O_2
1 mole O_3 molecules = 6.022×10^{23} O_3 molecules = 48.00 g O_3

The mole is an SI unit that is defined in relation to the mass of the carbon-12 isotope. A **mole** is the amount of substance of a system that contains as many elementary entities as there are atoms in 0.012 kg of carbon-12. In 0.012 kg of carbon-12 there are 6.022×10^{23} carbon-12 atoms. The symbol for mole is *mol*.

Compounds. The **molar mass of a compound** is its molecular mass expressed in grams. The molar masses of NO_2 and N_2O_5 are 46.01 g and 108.0 g, respectively. Since 46.01 g of NO_2 contains 6.022×10^{23} molecules of NO_2, this amount is 1 mole of NO_2. On the other hand, 1 mole of N_2O_5 has a mass of 108.0 g.

1 mol NO_2 = 6.022×10^{23} NO_2 molecules = 46.01 g NO_2
1 mol N_2O_5 = 6.022×10^{23} N_2O_5 molecules = 108.0 g N_2O_5

Also note that 1 mole of NO_2 consists of 1 mole of N atoms, and 2 moles of O atoms.

1 mol NO_2 contains 1 mol N atoms = 14.01 g N
1 mol NO_2 contains 2 mol O atoms = 32.00 g O
1 mol NO_2 = 46.01 g NO_2

Just as the molecular mass is the sum of the atomic masses, the molar mass of a compound is the sum of the molar masses of the atoms in the molecule.

EXAMPLE 3.3   The Molar Mass

Naturally occurring lithium consists of 7.5 percent Li-6 atoms, and 92.5 percent Li-7 atoms. Complete the following sentences:

a.   The number of Li atoms in one mole of Li is _____.
b.   One mole of Li atoms consists of _____ Li-6 atoms and _____ Li-7 atoms and will have a mass of ____ g.

•Method of Solution

a.   One mole of Li contains 6.022×10^{23} Li atoms (a mixture of Li-6 atoms and Li-7 atoms).

b.   One mole of Li atoms contains $(0.074)(6.022 \times 10^{23})$ Li-6 atoms, which is equal to $\underline{4.45 \times 10^{22}}$ Li-6 atoms; and $(0.925)(6.022 \times 10^{23})$ Li-7 atoms, which is $\underline{5.57 \times 10^{23}}$ Li-7 atoms. Note that the total number of Li atoms is 6.022×10^{23}, and thus will have a mass of $\underline{6.94}$ g.

EXAMPLE 3.4   Mass of a Given Number of Moles

What is the mass of 2.25 moles of iron (Fe)?

•Method of Solution

To convert moles of Fe to grams of Fe, first write the relationship between moles, atoms, and mass.

$$1 \text{ mol Fe} = 6.022 \times 10^{23} \text{ Fe atoms} = 55.85 \text{ g Fe}$$

The problem asks:

$$? \text{ g Fe} = 2.25 \text{ mol Fe}$$

Since the atomic mass of iron is 55.85 amu, then 1 mol of iron has a mass of 55.85 g. The unit conversion factor is

$$\frac{55.85 \text{ g}}{1 \text{ mol Fe}} = 1$$

•Calculation

Since 1 mol Fe has a mass of 55.85 g, then 2.25 mol Fe must have a mass of:

$$? \text{ g Fe} = 2.25 \text{ mol Fe} \times \frac{55.85 \text{ g}}{1 \text{ mol Fe}} = 126 \text{ g Fe}$$

EXAMPLE 3.5   Number of Atoms in a Given Mass

How many zinc atoms are present in 20.0 g Zn?

•Method of Solution

We know that 1 mole of Zn contains 6.022×10^{23} atoms. First find how many moles of Zn atoms are in 20.0 g Zn, and then multiply by Avogadro's number. Finding the number of moles of zinc is the key to finding the number of atoms. The relationship between moles, atoms, and grams is:

$$1 \text{ mol Zn} = 6.022 \times 10^{23} \text{ Zn atoms} = 65.39 \text{ g Zn}$$

The problem asks:

$$? \text{ Zn atoms} = 20.0 \text{ g Zn}$$

Our road map is:

$$\text{g Zn} \rightarrow \text{mol Zn} \rightarrow \text{number of Zn atoms}$$

•Calculation

One mol of Zn has a mass of 65.39 g. Therefore, 20.0 g Zn corresponds to:

$$? \text{ Zn mol} = 20.0 \text{ g Zn} \times \frac{1 \text{ mol Zn}}{65.39 \text{ g Zn}} = 0.3058 \text{ mol Zn}$$

Here we will carry one more digit than the correct number of significant figures.
To convert moles of zinc to atoms of zinc, we use Avogadro's number, the number of atoms per mole, as the unit factor:

$$\frac{6.022 \times 10^{23} \text{ Zn atoms}}{1 \text{ mol Zn}} = 1$$

$$? \text{ Zn atoms} = 0.3058 \text{ mol Zn} \times \frac{6.022 \times 10^{23} \text{ Zn atoms}}{1 \text{ mol Zn}} = 1.84 \times 10^{23} \text{ Zn atoms}$$

•Comment

Instead of calculating the number of moles separately, we could have strung the conversions factors together into one calculation:

$$? \text{ Zn atoms} = 20.0 \text{ g Zn} \times \frac{1 \text{ mol Zn}}{65.39 \text{ g Zn}} \times \frac{6.022 \times 10^{23} \text{ Zn atoms}}{1 \text{ mol Zn}}$$

$$= 1.84 \times 10^{23} \text{ Zn atoms}$$

EXAMPLE 3.6   Moles of a Molecular Compound

a.   How many molecules of ethane (C_2H_6) are present in 50.3 g of ethane?
b.   How many atoms each of H and C are in this sample?

•Method of Solution (a)

The problem asks:

$$? \text{ } C_2H_6 \text{ molecules} = 50.3 \text{ g } C_2H_6$$

To convert grams to molecules, you need the molar mass.

The molecular mass of C_2H_6 is:

$$2(12.01 \text{ amu}) + 6(1.008 \text{ amu}) = 30.07 \text{ amu}$$

Therefore, we can write:

$$1 \text{ mol } C_2H_6 = 6.022 \times 10^{23} \text{ } C_2H_6 \text{ molecules} = 30.07 \text{ g } C_2H_6$$

Finding the number of moles of C_2H_6 is the key to finding the number of molecules.

$$\text{g } C_2H_6 \rightarrow \text{mol } C_2H_6 \rightarrow \text{number of } C_2H_6 \text{ molecules}$$

•**Calculation for (a)**

$$? \text{ mol } C_2H_6 = 50.3 \text{ g } C_2H_6$$

Since 1 mol C_2H_6 is 30.1 g, then 50.3 g C_2H_6 must be:

$$? \text{ mol } C_2H_6 = 50.3 \text{ g } C_2H_6 \times \frac{1 \text{ mol } C_2H_6}{30.07 \text{ g } C_2H_6} = 1.67 \text{ mol } C_2H_6$$

Since 1 mol C_2H_6 contains 6.022×10^{23} molecules, then 1.67 mol must contain:

$$? \text{ } C_2H_6 \text{ molecules} = 1.67 \text{ mol } C_2H_6 \times \frac{6.022 \times 10^{23} \text{ molecules}}{1 \text{ mol } C_2H_6}$$

$$= 1.01 \times 10^{24} \text{ } C_2H_6 \text{ molecules}$$

•**Method of Solution (b)**

The molecular formula shows the number and kinds of atoms in a molecule.

$$\frac{6 \text{ H atoms}}{C_2H_6 \text{ molecule}} \quad \text{and} \quad \frac{2 \text{ C atoms}}{C_2H_6 \text{ molecule}}$$

•**Calculation for (b)**

Multiplying the number of C_2H_6 molecules by the number of atoms of each kind per molecule gives the number of each kind of atom present.

$$1.01 \times 10^{24} \text{ molecules} \times \frac{6 \text{ H atoms}}{C_2H_6 \text{ molecule}} = 6.06 \times 10^{24} \text{ H atoms}$$

$$1.01 \times 10^{24} \text{ molecules} \times \frac{2 \text{ C atoms}}{C_2H_6 \text{ molecule}} = 2.02 \times 10^{24} \text{ C atoms}$$

EXERCISES

3. Calculate the molecular mass of the following:
   a. carbon tetrachloride, CCl_4
   b. formaldehyde, H_2CO
   c. xenon difluoride, XeF_2

4. Calculate the molar mass of the following:
   a. Na_2SO_4, sodium sulfate
   b. $FeCl_3$, iron(III) chloride
   c. $Ba(OH)_2$, barium hydroxide
   d. SO_3, sulfur trioxide

5. What is the mass of 1 mole of each of the following?
   a. Cu
   b. CuO
   c. $CuSO_4$
   d. $CuSO_4 \cdot 5H_2O$

6. a. How many moles of silver are in 5.00 g Ag?
   b. How many moles of sodium are in 5.00 g Na?

7. a. How many silver atoms are in 5.0 g Ag?
   b. How many H_2 molecules are there in 4.0 g H_2?
   c. How many molecules are in 25.0 g of methane (CH_4)?

8. a. What is the mass of 2.0 moles of H_2?
   b. What is the mass of 2.79×10^{22} atoms of Ag?
   c. What is the average mass in grams of one atom of aluminum?

9. a. How many moles of $NaNO_3$ are in 8.72 g $NaNO_3$?
   b. How many moles of O atoms are in 8.72 g $NaNO_3$?

CHEMICAL FORMULAS AND PERCENT COMPOSITION (3.6)

STUDY OBJECTIVES

1. Determine the percent composition of a compound with a known formula.
2. Determine the empirical formula of a compound given the percent composition.
3. Determine a molecular formula when the molar mass is given.

Percent Composition. The **percent composition** of a compound is the percentage by mass of each element in the compound. The percent composition can be determined by chemical analysis or it can be calculated directly if the formula of the compound is known. The percent is:

$$\text{percent of element} = \frac{\text{mass of element per mole of compound}}{\text{molar mass of compound}} \times 100\%$$

Let us take sodium chloride (NaCl) as an example. The molar mass, 58.44 g, is the sum of the mass of 1 mole of Na, 22.99 g, and the mass of 1 mole of Cl, 35.45 g. The percentage of Na by mass is

$$\%Na = \frac{22.99 \text{ g Na}}{58.44 \text{ g NaCl}} \times 100\% = 39.33\%$$

The percentage of Cl by mass is

$$\%Cl = \frac{35.45 \text{ g Cl}}{58.44 \text{ g NaCl}} \times 100\% = 60.66\%$$

Empirical and Molecular Formulas. There are two types of formulas that give information about compounds. The **empirical formula** gives the simplest whole-number ratio of the various kinds of atoms in a compound. In sulfur dioxide (SO_2) there are twice as many O atoms in the compound as S atoms. The **molecular formula** indicates the numbers of atoms of each element in a molecule of a substance. The molecular formula and the empirical formula may be the same for some substances, as in the case of SO_2, where one molecule contains two atoms of oxygen and one atom of sulfur. On the other hand, the molecular formula of benzene is C_6H_6, but the empirical formula is CH because the smallest or simplest ratio of C atoms to H atoms is 1 : 1.

Chemical formulas of compounds are determined from a knowledge of the percent composition of the elements. For example, aluminum oxide is a solid with a percent composition of 52.92% Al and 47.08% O. Its empirical formula will give the ratio of Al atoms to O atoms. A convenient starting point for determining its empirical formula is to assume you have 100 g of aluminum oxide. Then, 100 g of compound will contain 52.92 g Al, and 47.08 g O.

The number of moles of each element present is:

$$n_{Al} = 52.92 \text{ g Al} \times \frac{1 \text{ mol Al}}{26.98 \text{ g Al}} = 1.96 \text{ mol Al}$$

$$n_O = 47.08 \text{ g O} \times \frac{1 \text{ mol O}}{16.00 \text{ g O}} = 2.94 \text{ mol O}$$

Dividing each number of moles by the smaller of the two molar amounts yields the following mole ratios:

$$\frac{2.94 \text{ mol O}}{1.96 \text{ mol Al}} = 1.50 \text{ mol O to } 1.00 \text{ mol Al}$$

Which is the same as: 1.5 O atoms to 1 Al atom

This is not the simplest whole-number ratio and suggests a formula of $Al_1O_{1.5}$. Multiplying both kinds of atoms by 2 gives the simplest whole-number ratio of 3.0 to 2.0. The empirical formula of aluminum oxide is Al_2O_3.

The road map of the solution is:									
percent composition	$\rightarrow$	**# moles each element**	$\rightarrow$	**mole ratio**	$\rightarrow$	**atom ratio**	$\rightarrow$	**empirical formula**	

Molecular Formula. When the molar mass is known, the molecular formula can be deduced from the empirical formula. For example, in the case of benzene, which has an empirical formula of CH, the molar mass is known to be 78.11 g. Since the empirical mass is 13.01 g, then there are 78.11/13.01 or 6.00 CH units per molecule, and so the molecular formula is C_6H_6.

EXAMPLE 3.7   Percent Composition

Calculate the percent composition of glucose ($C_6H_{12}O_6$).

•**Method of Solution**

First the molar mass of glucose must be found from the atomic masses of each element.

$$\text{molar mass of } C_6H_{12}O_6 = \left[6 \text{ mol C} \times \frac{12.01 \text{ g}}{1 \text{ mol C}} \right] + \left[12 \text{ mol H} \times \frac{1.008 \text{ g}}{1 \text{ mol H}} \right] + \left[6 \text{ mol H} \times \frac{16.00 \text{ g}}{1 \text{ mol O}} \right]$$

$$= 72.06 \text{ g} + 12.10 \text{ g} + 96.00 \text{ g} = 180.16 \text{ g}$$

Then the percent of each element present is found by dividing the mass of each element in 1 mole of glucose by the total mass of 1 mole of glucose.

•Calculation

$$\text{mass percent of C} = \frac{\text{mass of C in 1 mol } C_6H_{12}O_6}{\text{mass of 1 mole } C_6H_{12}O_6} \times 100\%$$

$$\%C = \frac{6 \text{ mol C} \times (12.01 \text{ g/mol C})}{180.16 \text{ g } C_6H_{12}O_6} \times 100\% = \frac{72.06 \text{ g}}{180.16 \text{ g}} \times 100\% = 40.0\% \text{ C}$$

After practicing the calculation several times, you can simplify the notation:

$$\%H = \frac{12 \times 1.008 \text{ g}}{180.16 \text{ g}} \times 100\% = \frac{12.1 \text{ g}}{180.16 \text{ g}} \times 100\% = 6.72\% \text{ H}$$

$$\%O = \frac{6 \times 16.00 \text{ g}}{180.16 \text{ g}} \times 100\% = \frac{96.0 \text{ g}}{180.16 \text{ g}} \times 100\% = 53.3\% \text{ O}$$

Thus glucose ($C_6H_{12}O_6$) contains 40.0% C, 6.72% H, and 53.3% O by mass.

•Comment

A good way to check the results is to sum the percentages of the elements, because their total must add to 100%. Here we get 100.02%.

EXAMPLE 3.8   Using Percent Composition

Calculate the mass of carbon in 10.00 g of glucose ($C_6H_{12}O_6$).

•Method of Solution

In Example 3.7 we calculated that glucose contains 40.00% carbon. Therefore:

$$\text{mass C} = 40.00\% \text{ of } 10.00 \text{ g } C_6H_{12}O_6 = \frac{40 \text{ g C}}{100 \text{ g } C_6H_{12}O_6} \times 10.00 \text{ g } C_6H_{12}O_6 = 4.000 \text{ g C}$$

•Calculation

This calculation would have to be done from scratch if we did not already know the percentage of carbon. Our road map is:

mass of glucose → mol glucose → mol carbon → g carbon

$$? \text{ mass C} = 10.00 \text{ g } C_6H_{12}O_6 \times \frac{1 \text{ mol } C_6H_{12}O_6}{180.16 \text{ g } C_6H_{12}O_6} \times \frac{6 \text{ mol C}}{1 \text{ mol } C_6H_{12}O_6} \times \frac{12.01 \text{ g C}}{1 \text{ mol C}} = 4.000 \text{ g C}$$

EXAMPLE 3.9   Empirical Formula

An elemental analysis of a new compound reveals its percent composition to be: 50.7% C, 4.23% H, and 45.1% O. Determine the empirical formula.

•Method of Solution

The empirical formula simply gives the relative numbers of atoms of each element in a formula unit. To start, assume you have exactly 100 g of compound, and then determine the number of moles of each element present. Then find the ratios of the number of moles.

•**Calculation**

$$50.7 \text{ g C} \times \frac{1 \text{ mol C}}{12.01 \text{ g C}} = 4.22 \text{ mol C}$$

$$4.23 \text{ g H} \times \frac{1 \text{ mol H}}{1.008 \text{ g H}} = 4.20 \text{ mol H}$$

$$45.1 \text{ g O} \times \frac{1 \text{ mol O}}{16.00 \text{ g O}} = 2.82 \text{ mol O}$$

This gives the mole ratio for C : H : O of 4.22 : 4.20 : 2.82.
Dividing by the smallest of the molar amounts gives:

$$\frac{4.22 \text{ mol C}}{2.82 \text{ mol O}} = 1.50 \text{ C to } 1.00 \text{ O} \quad \text{and} \quad \frac{4.20 \text{ mol H}}{2.82 \text{ mol O}} = 1.50 \text{ H to } 1.00 \text{ O}$$

Therefore, $C_{1.5}H_{1.5}O_{1.0}$ is a possible formula.

This is not an acceptable formula because it does not contain all whole numbers. To convert these fractions to whole numbers, multiply each subscript by the same number, in this case a 2. The empirical formula therefore is

$C_3H_3O_2$

EXAMPLE 3.10   Molecular Formula

Mass spectrometer experiments on the compound in Example 3.9 show its molecular mass to be about 140 amu. What is the molecular formula?

•**Method of Solution**

First find how many empirical formula units there are in 1 mole of compound which is 140 g. The molar mass of the empirical formula $C_3H_3O_2$ is 71 g.

•**Calculation**

Dividing 140 g by 71 g shows that there are two empirical units per molecule.
Therefore, the molecular formula must show twice as many atoms as the empirical formula.

molecular formula = $C_6H_6O_4$

EXAMPLE 3.11   Carbon–Hydrogen Analysis

When a 0.761-g sample of a compound of carbon and hydrogen is burned in a C–H combustion apparatus (see Figure 3.5 in the text), 2.23 g CO_2 and 1.37 g H_2O are produced. Determine the percent composition of the compound.

•**Method of Solution**

Here we want the masses of carbon and hydrogen in the original compound. All the carbon is now present in 2.23 g of CO_2. All the hydrogen is now present in 1.37 g of H_2O. How many grams of C are there in 2.23 g of CO_2, and how many grams of H are there in 1.37 g of H_2O? We know there is 1 mole of C atoms (12.01 g) in every mole of CO_2, or 44.01 g.

Our road map is:

g $CO_2 \rightarrow$ mol $CO_2 \rightarrow$ g C

•**Calculation**

$$? \text{ g C} = 2.23 \text{ g CO}_2 \times \frac{1 \text{ mol CO}_2}{44.01 \text{ g CO}_2} \times \frac{12.01 \text{ g C}}{1 \text{ mol CO}_2} = 0.608 \text{ g C}$$

$$? \text{ g H} = 1.37 \text{ g H}_2\text{O} \times \frac{1 \text{ mol H}_2\text{O}}{18.0 \text{ g H}_2\text{O}} \times \frac{2.01 \text{ g H}}{1 \text{ mol H}_2\text{O}} = 0.153 \text{ g H}$$

The percentage composition is:

$$\%\text{H} = \frac{0.153 \text{ g H}}{0.761 \text{ g compound}} \times 100\% = 20.1\% \text{ H}$$

$$\%\text{C} = \frac{0.608 \text{ g C}}{0.761 \text{ g compound}} \times 100\% = 79.9\% \text{ C}$$

EXERCISES

10. Calculate the percent composition by mass of the following compounds:
    a. CO_2 b. H_3AsO_4 c. $CHCl_3$
    d. $NaNO_2$ e. H_2SO_4

11. What is the empirical formula of each of the following compounds?
    a. H_2O_2 b. CaF_2 c. $C_2H_4O_2$ d. B_2H_6

12. Determine the empirical formula of each compound from the given percent compositions.
    a. 46.7% N, 53.3% O   c. 55.3% K, 14.6% P, 30.1% O
    b. 63.6% N, 36.4% O   d. 26.6% K, 35.4% Cr, 38.1% O

13. Cyclohexane has the empirical formula CH_2. Its molecular mass is 84.16 amu. What is its molecular formula?

14. When 2.65 mg of the substance responsible for the green color on the yolk of a boiled egg is analyzed, it is found to contain 1.42 mg Fe and 1.23 mg S. What is the empirical formula of the compound?

15. Ascorbic acid is a compound consisting of three elements: C, H, and O. When a 0.214 g sample is burned in oxygen, 0.320 g CO_2 and 0.0874 g H_2O are formed. What is the empirical formula of ascorbic acid?

CHEMICAL REACTIONS AND EQUATIONS (3.7)

STUDY OBJECTIVES

1. Write a chemical equation from a written description of the reaction that includes the names of reactants and products.
2. Balance a chemical equation when given a "skeletal" equation.

Chemical Equations. A chemical equation is a symbolic way of representing a chemical reaction. The formulas of the reactants are written on the left side of the equation, and the formulas of the products on the right side. Equations are written according to the following format:

1. Reactants and products are separated by an arrow. The arrow means "produces" or "yields."
2. Plus signs are placed between reactants and between products. A plus sign between reactants implies that a mixture of two (and sometimes more) reactants is required for reaction. A plus between products implies that a mixture of two or more products is formed by the reaction. The plus sign can also be interpreted to mean "and" or just "plus."

Table 3.1 Symbols Used in Writing Chemical Equations

Symbol	Meaning
$\rightarrow$	yields, produces
$\rightleftharpoons$	reversible reaction, equilibrium
(s)	solid phase
(l)	liquid phase
(g)	gas phase
(aq)	aqueous solution
+	added to, and

3. Abbreviations are used to indicate states of matter. These symbols, given in Table 3.1, are written in parentheses and follow the elements or compounds to which they refer
4. Equations are balanced. That is, the number of atoms of a certain element appearing in the reactants must be equal to the number of atoms of that element appearing in the products. A balanced equation shows the same number of each type of atom on both sides of the equation. This means that mass is conserved in chemical reactions.

Balancing of equations is accomplished by making use of coefficients. A **coefficient** is a number placed before a chemical formula in an equation which changes the amount of the substance. The coefficient is a multiplier for the formula. For example, $2H_2O$ means two molecules of water. Two molecules of water consists of 4 hydrogen atoms and 2 oxygen atoms. Similarly, $3H_2O$ means three molecules of water, which stands for 6 H atoms and 3 O atoms. Absence of a coefficient is understood to mean one.

Balancing Equations. When sodium metal is heated in the presence of chlorine gas, a bright yellow flame is produced along with a white solid, sodium chloride. The unbalanced chemical equation representing the reaction is:

$$Na + Cl_2 \rightarrow NaCl$$

Writing a "balance sheet" of the numbers of atoms in the reactants and products is helpful. We see that the sodium atoms are balanced, but the chlorine atoms are not balanced.

Reactants	Products
Na (1)	Na (1)
Cl (2)	Cl (1)

Two Cl atoms appear in the reactants, but only one appears in the products. To balance the equation, we must adjust the coefficients of the formulas. Since there are already two Cl atoms in the reactants, we can make two Cl atoms appear in the product by placing a 2 in front of NaCl. Two formula units of NaCl contain two Na atoms and two Cl atoms. The equation is now balanced for chlorine, but sodium is no longer balanced.

Na + Cl$_2$ → 2NaCl

Reactants	Products
Na (1)	Na (2)
Cl (2)	Cl (2)

Next sodium can be balanced by placing the coefficient 2 in front of Na.

2Na + Cl$_2$ → 2NaCl

Reactants	Products
Na (2)	Na (2)
Cl (2)	Cl (2)

The equation is now balanced, and tells us that 2 Na atoms react with one chlorine molecule to yield 2 formula units of NaCl. When balancing an equation, you should *never alter the subscripts* in the formulas. This would change the substances involved. For instance, removing the 2 for Cl$_2$ in order to make the equation balance is incorrect.

Na + Cl → NaCl

The reactant is diatomic chlorine (Cl$_2$), a yellow-green gas, not atomic chlorine (Cl). These are two very different chemical species. Never change the formulas of the reactants or products. Adjust only their coefficients. In the examples that follow a few hints will be given that will make balancing easier.

EXAMPLE 3.12   Balancing Chemical Equations

Balance the following displacement reaction. In this reaction zinc displaces H$_2$ from chemical combination with chlorine.

Zn(s) + HCl(aq) → ZnCl$_2$(aq) + H$_2$(g)

•Method of Solution

Step 1: As a general way to begin, *identify elements that occur in only two compounds in the equation*. In this case, each of the three elements Zn, H, and Cl occurs in only two compounds.
Step 2: Of the elements that occur in only two compounds, *identify and balance first that element with the largest subscript*. Both Cl and H have the subscript 2, so balance either one of them first. Let's balance H by placing the coefficient 2 before HCl:

Zn + 2HCl → ZnCl$_2$ + H$_2$

Taking stock:

Reactants	Products
H (2)	H (2)
Cl (2)	Cl (2)
Zn (1)	Zn (1)

At this point both H and Cl are balanced. Further inspection reveals that Zn is also balanced. Thus the equation is balanced!

EXAMPLE 3.13   Balancing Chemical Equations

The reaction of a hydrocarbon with oxygen is an example of a combustion reaction. Balance the equation for the reaction of pentane with oxygen gas:

$$C_5H_{12} + O_2 \rightarrow CO_2 + H_2O$$

•Method of Solution

Following step 1 given in the previous example, we note that both H and C appear in only two compounds. According to step 2, we note that of these, H has the largest subscript. Balance H first by placing the coefficient 6 in front of H_2O. Next balance the carbon atoms by placing a 5 in front of CO_2.

$$C_5H_{12} + O_2 \rightarrow 5CO_2 + 6H_2O$$

Taking stock:

Reactants	Products
C (5)	C (5)
H (12)	H (12)
O (2)	O (16)

Finally the oxygen atoms can be balanced. The products already have their coefficients determined, and so the reactants must be adjusted to supply 16 O atoms. Eight molecules of O_2 contain 16 O atoms, and so write the coefficient 8 in front of O_2.

$$C_5H_{12} + 8O_2 \rightarrow 5CO_2 + 6H_2O$$

Reactants	Products
C (5)	C (5)
H (12)	H (12)
O (16)	O (16)

EXAMPLE 3.14   Balancing Chemical Equations

Balance the following decomposition reaction:

$$NaClO_3(s) \rightarrow NaCl(s) + O_2(g)$$

•Method of Solution

According to steps 1 and 2, we should balance O first. There are 3 O atoms on the reactant side and 2 on the product side. A third step is helpful in this situation.

Step 3: For an element that occurs in only two substances in the equation, note whether *the number of atoms of that element is even on one side and odd on the other side of the equation*. If this is the case, *use two coefficients* which increase the number of atoms on each side of the equation to the *least common multiple*.

$$NaClO_3 \rightarrow NaCl + O_2$$

Reactants	Products
O (3)	O (2)

The least common multiple for balancing O is $3 \times 2 = 6$. The two coefficients are 2 and 3.

$$2NaClO_3 \rightarrow NaCl + 3O_2$$

Reactants	Products
O (6)	O (6)

Balancing the Na and Cl atoms we get:

$$2NaClO_3 \rightarrow 2NaCl + 3O_2$$

EXAMPLE 3.15   Balancing Chemical Equations

Balance the following precipitation reaction in which sulfuric acid reacts with barium chloride in aqueous solution to yield hydrochloric acid and a precipitate of barium sulfate.

$$H_2SO_4(aq) + BaCl_2(aq) \rightarrow HCl(aq) + BaSO_4(s)$$

•Method of Solution

Here it helps to recognize certain groups of atoms that maintain their identity during the reaction. In this equation SO_4^{2-}, the sulfate ion, appears in the reactants and products as a unit.

$$H_2SO_4 + BaCl_2 \rightarrow HCl + BaSO_4$$

Reactants	Products
H (2)	H (1)
SO_4^{2-} (1)	SO_4^{2-} (1)
Ba (1)	Ba (1)
Cl (2)	Cl (1)

Either H or Cl could be balanced first. Balance the chlorine by placing the coefficient 2 in front of HCl:

$$H_2SO_4 + BaCl_2 \rightarrow BaSO_4 + 2HCl$$

A check of the H atoms shows they are now balanced along with the other elements.

EXERCISES

16. Rust (Fe_2O_3) forms readily when iron is exposed to air. Write a balanced chemical equation for the formation of rust.

17. Write a balanced chemical equation for the reaction between hydrogen gas and carbon monoxide to yield methanol (CH_3OH).

18. Balance the following equations:
    a. __ CH_4 + __ $H_2O \rightarrow$ __ CO_2 + __ H_2
    b. __ H_2SO_4 + __ $NaOH \rightarrow$ __ Na_2SO_4 + __ H_2O
    c. __ NH_3 + __ $O_2 \rightarrow$ __ NO + __ H_2O

STOICHIOMETRY: AMOUNTS OF REACTANTS AND PRODUCTS (3.8)

STUDY OBJECTIVES

1. Write mole ratios relating the amounts of reactants and products in a chemical reaction.
2. Calculate the mass of a substance that is produced or consumed in a chemical reaction, given the mass of any other species involved in the reaction.

Mass Relationships in Reactions. Chemical equations contain information about the relative masses of reactants involved in the reaction. The balanced equation can be used to decide how many grams of one substance will be needed to react with a given mass of another, or how many grams of product can be produced by the reaction of a specific mass of a reactant. For instance, the balanced equation

$$2Al \ + \ Fe_2O_3 \ \rightarrow \ Al_2O_3 \ + \ 2Fe$$

2 mol	1 mol	1 mol	2 mol
53.96 g	159.70 g	101.96 g	111.70 g

states that 2 moles of Al will react completely with 1 mole of Fe_2O_3, and will yield 1 mole of Al_2O_3 and 2 moles of Fe. Converting these molar amounts to grams, we find that 53.96 g Al reacts with 159.70 g Fe_2O_3, producing 101.96 g Al_2O_3 and 111.70 g Fe.

The mass relationships as discussed in the above reaction can be determined only from the balanced equation. The equation, however, is written in terms of moles, not grams. The quantitative relationships between elements and compounds in chemical reactions is a part of chemistry called **stoichiometry**. In all stoichiometry problems the balanced chemical equation provides the "bridge" that allows us to relate the amount of a reactant to a given amount of another reactant, or to find the amounts required to yield a given amount of a product.

Mole Ratios. To calculate the amount of a specific product that is formed when a given amount of one reactant is consumed, we need quantitative relationships. In the preceding equation, note that 2 moles of Al is consumed whenever 1 mole of Fe_2O_3 reacts; and that whenever 1 mole of Al_2O_3 is formed, 2 moles of Al must react. We can write the following equivalences:

$$2 \text{ mol Al} \ \rightleftharpoons \ 1 \text{ mol Fe}_2O_3 \qquad 1 \text{ mol Al}_2O_3 \ \rightleftharpoons \ 2 \text{ mol Al}$$

The symbol $\rightleftharpoons$ means "is stoichiometrically equivalent to." This symbol is used because the equivalency cannot be universally applied to every reaction. These amounts are equivalent only in the context of this specific reaction. We can write conversion factors called *mole ratios* relating any two quantities that are equivalent.

$$\frac{2 \text{ mol Al}}{1 \text{ mol Fe}_2O_3}, \quad \frac{1 \text{ mol Al}_2O_3}{2 \text{ mol Al}}, \quad \text{and} \quad \frac{2 \text{ mol Fe}}{1 \text{ mol Fe}_2O_3}$$

One of these mole ratios will be used in the following example.

How many grams of iron are produced when 25.0 g Fe_2O_3 reacts with aluminum?

$$2Al + Fe_2O_3 \rightarrow Al_2O_3 + 2Fe$$

Three steps are required. Since the balanced equation relates the number of moles of Fe_2O_3 to moles of Fe, (1) convert the grams of Fe_2O_3 to moles of Fe_2O_3 using its molar mass. (2) Use a conversion factor (2 mol Fe formed per 1 mol Fe_2O_3 consumed) to find the number of moles of Fe produced. (3) Convert the moles of Fe to grams using its molar mass.

$$2Al \ + \ Fe_2O_3 \ \rightarrow \ Al_2O_3 \ + \ 2Fe$$

The road map outlined above is:

$$\underset{\text{step 1}}{\text{g Fe}_2\text{O}_3} \rightarrow \underset{\text{step 2}}{\text{mol Fe}_2\text{O}_3} \rightarrow \underset{\text{step 3}}{\text{mol Fe}} \rightarrow \text{g Fe produced}$$

The three steps are:

1. $25.0 \text{ g Fe}_2\text{O}_3 \times \dfrac{1 \text{ mol Fe}_2\text{O}_3}{159.7 \text{ g Fe}_2\text{O}_3} = 0.156 \text{ mol Fe}_2\text{O}_3$

2. $0.156 \text{ mol Fe}_2\text{O}_3 \times \dfrac{2 \text{ mol Fe}}{1 \text{ mol Fe}_2\text{O}_3} = 0.312 \text{ mol Fe}$

3. $0.312 \text{ mol Fe} \times \dfrac{55.58 \text{ g Fe}_2\text{O}_3}{1 \text{ mol Fe}} = 17.3 \text{ g Fe}$

This calculation can be set up and carried out in one equation.

$$25.0 \text{ g Fe}_2\text{O}_3 \times \dfrac{1 \text{ mol Fe}_2\text{O}_3}{159.7 \text{ g Fe}_2\text{O}_3} \times \dfrac{2 \text{ mol Fe}}{1 \text{ mol Fe}_2\text{O}_3} \times \dfrac{55.58 \text{ g Fe}_2\text{O}_3}{1 \text{ mol Fe}} = 17.3 \text{ g Fe}$$

The amount of product predicted by these calculations is called the **theoretical yield**.

Summary. Once you have the balanced chemical equation, you can work all stoichiometry problems using the three steps of the mole method.
Step 1. Convert the amounts of given substances into moles if necessary.
Step 2. Use the coefficients in the balanced equation to calculate the number of moles of the sought or unknown substance in the problem from the appropriate mole ratio.
Step 3. Convert the number of moles of the needed substance into mass units if required.

Always check your answer for reasonableness.

EXAMPLE 3.16 The Mole Method

Sulfur dioxide can be removed from stack gases by reaction with quicklime, CaO.

$$SO_2(g) + CaO(s) \rightarrow CaSO_3(s)$$

If 975 g of SO_2 is to be removed from stack gases by the above reaction, what mass of CaO is required to completely react with it?

•Method of Solution

First be certain that the equation is correctly balanced. Then follow the three steps as outlined in the above section.

$$SO_2 + CaO \rightarrow CaSO_3$$

$$\underset{\text{step 1}}{\text{g SO}_2} \rightarrow \underset{\text{step 2}}{\text{mol SO}_2} \rightarrow \underset{\text{step 3}}{\text{mol CaO}} \rightarrow \text{g CaO}$$

1. Convert the 975 g of SO_2 to moles using its molar mass.
2. Use the chemical equation to find the number of moles of CaO that reacts per mole of SO_2.
   From the balanced equation we can write: 1 mol CaO $\rightleftharpoons$ 1 mol SO_2
3. Convert moles of CaO to grams of CaO using the molar mass of CaO.

•Calculation

Write out an equation which restates the problem:

$$? \text{ g CaO} = 9.75 \times 10^2 \text{ g SO}_2$$

String the conversion factors from the three steps one after the other.

$$? \text{ g CaO} = 9.75 \times 10^2 \text{ g SO}_2 \times \frac{1 \text{ mol SO}_2}{64.07 \text{ g SO}_2} \times \frac{1 \text{ mol CaO}}{1 \text{ mol SO}_2} \times \frac{56.08 \text{ g CaO}}{1 \text{ mol CaO}}$$

$$= 8.53 \times 10^2 \text{ g CaO}$$

EXERCISES

19. Silicon tetrachloride ($SiCl_4$) can be prepared by heating silicon in the presence of chlorine gas:

    $$Si(s) + 2Cl_2(g) \rightarrow SiCl_4(l)$$

    a. How many moles of $SiCl_4$ are formed when 4.24 moles of Cl_2 gas reacts?
    b. How many grams of $SiCl_4$ are produced when 4.24 moles of Cl_2 gas reacts?

20. The following reaction can be used to prepare hydrogen gas:

    $$CaH_2 + 2H_2O \rightarrow Ca(OH)_2 + 2H_2$$

    How many grams of H_2 will result from the reaction of 100 g CaH_2 with an excess of H_2O?

21. Oxygen gas can be produced by the decomposition of mercury(II) oxide, HgO. How many grams of O_2 will be produced by the reaction of 24.2 g of the oxide?

    $$2HgO(s) \rightarrow 2Hg(l) + O_2$$

22. Carbon dioxide in the air of a spacecraft can be removed by its reaction with lithium hydroxide.

    $$CO_2(g) + 2LiOH(aq) \rightarrow Li_2CO_3(aq) + H_2O(l)$$
    On average, a person will exhale about one kg of CO_2 per day. How many kg of LiOH are required to react with 1.0 kg of CO_2?

LIMITING REAGENTS AND REACTION YIELDS (3.9 – 3.10)

STUDY OBJECTIVES

1. Determine which reactant is the limiting reagent in a chemical reaction.
2. Calculate the percent yield of a product.

Limiting Reagents. Often the two reactants are not present in the exact ratio prescribed by the balanced chemical equation. In this situation, one reactant will be completely consumed before the other runs out. When this occurs, the reaction will stop and no more product will be made. The reactant that is consumed first is called the **limiting reagent** or **limiting reactant** because determines, or limits, the amount of product formed. The reactant that is not completely consumed is called the **excess reagent**.

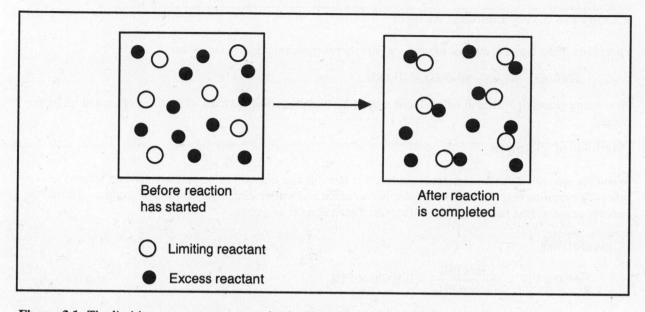

Figure 3.1 The limiting reagent reacts completely. Unreacted portions of the excess reagent remain.

Take for example, the reaction:

$$2Al + Fe_2O_3 \rightarrow Al_2O_3 + 2Fe$$

If 1.2 mol Al and 1 mol Fe_2O_3 are mixed together, the reactant that is used up first is Al. We can tell this because the 1.0 mol Fe_2O_3 given requires 2.0 mol Al for complete reaction. The 1.2 mol Al given will react with only 0.60 mol Fe_2O_3. When this reaction ceases, 0.4 mol Fe_2O_3 will be left unreacted. Fe_2O_3 is present in *excess*. In this situation aluminum is consumed first. Therefore, aluminum is the *limiting reagent*. In any stoichiometry problem, it is important to determine which reactant is the limiting one, in order to correctly predict the yield of products.

Note, in this case, that the limiting reactant (1.2 mol Al) was not the reactant in lesser molar amount (1.0 mol Fe_2O_3). It is necessary to make use of the mole ratio of the reactants given by the balanced equation (the stoichiometric ratio). The amount of Fe_2O_3 required to react with 1.2 mol Al is:

$$1.2 \text{ mol Al} \times \frac{1 \text{ mol Fe}_2\text{O}_3}{2 \text{ mol Al}} = 0.60 \text{ mol Fe}_2\text{O}_3$$

Since 1.0 mol Fe_2O_3 is given, there is more Fe_2O_3 than needed to react completely with the given amount of Al. Fe_2O_3 is the excess reagent. Therefore, Al is the limiting reagent.

Percent Yield. In practice it is not unusual for the actual yield to be less than the calculated, or "theoretical" yield. The theoretical yield is calculated by assuming that all of the limiting reagent reacts according to the balanced equation. The percent yield is defined as

$$\% \text{ yield} = \frac{\text{actual yield}}{\text{theoretical yield}} \times 100\%$$

Limiting reagents and theoretical yields are illustrated in Examples 3.17 and 3.18.

EXAMPLE 3.17   Limiting Reagent

Phosphine (PH_3) burns in oxygen to produce phosphorus pentoxide (P_2O_5) and water.

$$2PH_3(g) + 4O_2(g) \rightarrow P_2O_5(s) + 3H_2O(l)$$

How many grams of P_2O_5 will be produced when 17.0 g of phosphine is mixed with 16.0 g of O_2 and reaction occurs?

•Method of Solution

When the amounts of both reactants are given, it is possible that one will be completely used up before the other. The limiting reagent will be completely consumed and will determine the amount of products. To find the limiting reagent, first find the number of moles of each reagent available.

•Calculation

$$17.0 \text{ g } PH_3 \times \frac{1 \text{ mol } PH_3}{33.99 \text{ g } PH_3} = 0.500 \text{ mol } PH_3$$

$$16.0 \text{ g } O_2 \times \frac{1 \text{ mol } O_2}{32.00 \text{ g } O_2} = 0.500 \text{ mol } O_2$$

The balanced equation gives the stoichiometric ratio of moles of O_2 required to react per mole of PH_3.

$$\text{stoichiometric ratio} = \frac{2 \text{ mol } O_2}{1 \text{ mol } PH_3}$$

The amount of O_2 required for complete reaction of the given amount of PH_3 is:

$$0.500 \text{ mol } PH_3 \times \frac{2 \text{ mol } O_2}{1 \text{ mol } PH_3} = 1.00 \text{ mol } O_2$$

Since only 0.50 mol O_2 is given, then O_2 is the limiting reagent. When all of the O_2 is consumed, some PH_3 will be left unreacted (the excess reagent). The theoretical yield of P_2O_5 is calculated by assuming complete reaction of the limiting reagent. The road map is:

$$g \; O_2 \rightarrow mol \; O_2 \rightarrow mol \; P_2O_5 \rightarrow g \; P_2O_5$$

$$? \text{ g } P_2O_5 = 0.500 \text{ mol } O_2 \times \frac{1 \text{ mol } P_2O_5}{4 \text{ mol } O_2} \times \frac{141.9 \text{ g } P_2O_5}{1 \text{ mol } P_2O_5}$$

$$= 17.7 \text{ g } P_2O_5$$

EXAMPLE 3.18   Percent Yield

In the reaction of 4.0 moles of N_2 with 6.0 moles of H_2, a chemist obtained 1.6 moles of NH_3. What is the percent yield of NH_3?

$$3H_2 + N_2 \rightarrow 2NH_3$$

•Method of Solution

To calculate the percent yield, you must first calculate the theoretical yield. The theoretical yield of NH_3 is determined by the limiting reagent. From the balanced equation we see that the stoichiometric ratio is 3 mol H_2

to 1 mol N_2. The 4.0 mol N_2 given require 12.0 mol H_2 for complete reaction. Since only 6.0 mol H_2 are given, then H_2 is the limiting reactant. In this case, 6.0 mol H_2 will react with 2.0 mol N_2 yielding 4.0 mol NH_3.

$$\text{theoretical yield} = 6.0 \text{ mol } H_2 \times \frac{2 \text{ mol } NH_3}{3 \text{ mol } H_2} = 4.0 \text{ mol } NH_3$$

The percent yield is found by dividing the actual yield by the theoretical yield.

$$\% \text{ yield } NH_3 = \frac{\text{actual yield } NH_3}{\text{theoretical yield } NH_3} \times 100\%$$

$$= \frac{1.6 \text{ mol } NH_3}{4.0 \text{ mol } NH_3} \times 100\%$$

$$\% \text{ yield } NH_3 = 40\%$$

EXERCISES

23. Given 5.00 mol KOH and 2.00 mol H_3PO_4, how many moles of K_3PO_4 can be prepared?

    $H_3PO_4 + 3KOH \rightarrow K_3PO_4 + 3H_2O$

24. The reaction of iron ore with carbon follows the equation:

    $2Fe_2O_3 + 3C \rightarrow 4Fe + 3CO_2$

    a. How many grams of Fe can be produced from a mixture of 200 g Fe_2O_3 and 300 g C?
    b. If the actual yield of Fe is 110 g, what is the percent yield of iron?

25. Hydrofluoric acid (HF) can be prepared according to the reaction:

    $CaF_2 + H_2SO_4 \rightarrow 2HF + CaSO_4$

    In one experiment 42.0 g CaF_2 was treated with excess H_2SO_4 and a yield of 14.2 g of HF was obtained.
    a. What is the theoretical yield of HF?
    b. Calculate the percent yield of HF.

CONCEPTUAL QUESTIONS

1. Describe why the atomic mass unit scale is referred to as a relative scale.

2. A mole is the amount of a substance that contains as many elementary entities as there are atoms in exactly 12 g of the carbon-12 isotope. What is an elementary entity of a substance?

3. Sulfur atoms, on average, have twice the mass of O oxygen atoms. Why is it important to stress the wording "on average"?

PRACTICE TEST

1. If the standard of the atomic mass unit scale had been defined such that a fluorine atom was 10.0 amu, what would the atomic masses of oxygen and argon be?

2. How many times heavier are bromine atoms on the average than neon atoms?

3. A sample of magnesium could one day be brought back from a planet circling a distant star. What is the atomic mass of this magnesium if it has two isotopes with the following masses and abundances?
   magnesium-24   23.985 amu; 18.05%
   magnesium-26   25.982 amu; 81.95%

4. a. What is the mass in grams of 1.5×10^{22} molecules of PCl_5?
   b. What is the mass in grams of 10^{12} carbon atoms?

5. What is the mass of a single atom of fluorine in grams?

6. How many O atoms are in each of the following amounts?
   a. 1 mol ozone, O_3
   b. 1 mol CuO
   c. 20 g $CuSO_4 \cdot 5H_2O$

7. What mass of $Ca_3(PO_4)_2$ would contain 0.500 mol Ca?

8. a. How many moles of $CaSO_4$ are there in 600 g $CaSO_4$?
   b. How many moles of oxygen are there in 600 g $CaSO_4$?
   c. How many oxygen atoms are in 600 g $CaSO_4$?

9. How many antimony atoms are there in 5.00 mol Sb_2O_5?

10. Answer the following questions concerning sulfur trioxide.
    a. What is the mass of 17.5 mol of SO_3?
    b. What is the mass of 7.5×10^{20} SO_3 molecules?

11. What is the empirical formula of each of the following compounds?
    a. $C_6H_8O_6$
    b. P_4O_{10}
    c. $MgCl_2$

12. Calculate the percent composition by mass of the following compounds:
    a. Al_2O_3 b. HNO_3 c. CCl_2F_2

13. Calculate the percentage of fluorine by mass in the refrigerant HFC-134a. Its formula is $C_2H_2F_4$?

14. The empirical formula of styrene is CH, and its molar mass is 104 g/mol. What is its molecular formula?

15. Chromic oxide is a compound of chromium and oxygen. Determine the empirical formula of chromic oxide given that 2.00 g of the compound contains 1.04 g Cr.

16. The compound glycerol contains three elements: C, H, and O. When a 0.740 g sample is burned in oxygen, 1.06 g CO_2 and 0.580 g H_2O are formed. What is the empirical formula of glycerol?

17. Balance the following equations:
    a. $P_4O_{10} + H_2O \rightarrow H_3PO_4$
    b. $Ga + H_2SO_4 \rightarrow Ga_2(SO_4)_3 + H_2$
    c. $C_4H_{10} + O_2 \rightarrow CO_2 + H_2O$

18. Complete lines a, b, and c of the following table for the reaction:

    $2SO_2 + O_2 \rightarrow 2SO_3$

moles SO_2	grams O_2	moles SO_3	grams SO_3
a. 1.50	_____	_____	_____
b._____	20.0	_____	_____
c._____	_____	5.21	_____

19. Sodium thiosulfate, the "hypo" used in photographic developing, can be made by the reaction:

    $Na_2CO_3 + 2Na_2S + 4SO_2 \rightarrow 3Na_2S_2O_3 + CO_2$

    a. How many grams of Na_2CO_3 (sodium carbonate) are required to produce 321 g of sodium thiosulfate, $Na_2S_2O_3$?
    b. How many grams of Na_2S are required to react with 25.0 g Na_2CO_3 if SO_2 is present in excess?

20. Hydrofluoric acid (HF) can be prepared according to the following equation:

    $CaF_2 + H_2SO_4 \rightarrow 2HF + CaSO_4$

    a. How many grams of HF can be prepared from 75.0 g H_2SO_4 and 63.0 g CaF_2?
    b. How many grams of the excess reagent will remain after the reaction ceases?
    c. If the actual yield of HF is 26.2 g, what is the percent yield?

General Problems

21. The density of silver is 10.5 g/cm^3. How many Ag atoms are present in a silver bar that measures 0.10 m × 0.05 m × 0.01 m?

22. The pesticide malathion has the chemical formula $C_{10}H_{19}O_6PS_2$.
    a. What is the molecular mass?
    b. The dose that is lethal to 50 percent of a human population is about 1.25 g per kilogram of body mass. How many molecules are in a dose lethal to an adult male weighing 70 kg? To an adult female of 58 kg?

23. An unknown element combines with oxygen to form a compound with a formula MO_2. If 25.0 g of the element combines with 4.50 g of oxygen, what is the atomic mass of M?

ANSWERS

Exercises

1. 17.999 amu
2. 107.87 amu
3. a. 153.81 amu   b. 30.03 amu   c. 169.3 amu
4. a. 142.04 g   b. 162.21 g   c. 171.3 g   d. 80.08 g
5. a. 63.546 g   b. 79.545 g   c. 159.6 g   d. 249.7 g
6. a. 0.0463 mol Ag   b. 0.217 mol Na
7. a. 2.79×10^{22} atoms   b. 1.2×10^{24} atoms   c. 9.41×10^{23} molecules
8. a. 4.0 g   b. 5.0 g   c. 4.48×10^{-23} g
9. a. 0.103 mol $NaNO_3$ b. 0.308 mol O
10. a. 27.29% C, 72.71% O
    b. 2.40% H, 59.5% As, 38.1% O
    c. 10.06% C, 0.844% H, 89.09% Cl
    d. 33.32% Na, 20.30% N, 46.38% O
    e. 2.06% H, 32.70% S, 65.24% O
11. a. HO   b. CaF_2 c. CH_2O d. BH_3
12. a. NO   b. N_2O c. K_3PO_4 d. $K_2Cr_2O_7$
13. C_6H_{12}
14. Fe_2S_3
15. $C_3H_4O_3$
16. $4Fe + 3O_2 \rightarrow 2Fe_2O_3$
17. $2H_2 + CO \rightarrow CH_3OH$
18. a. $CH_4 + 2H_2O \rightarrow CO_2 + 4H_2$
    b. $H_2SO_4 + 2NaOH \rightarrow Na_2SO_4 + 2H_2O$
    c. $4NH_3 + 5O_2 \rightarrow 4NO + 6H_2O$
19. a. 2.12 mol   b. 360 g
20. 9.58 g H_2
21. 1.79 g
22. 1.09 kg
23. 1.67 mol
24. a. Fe_2O_3 is the limiting reactant; 1.40 g Fe   b. 78.6% yield
25. a. 21.5 g   b. 66%

Conceptual Questions

1. Atoms cannot be weighed directly. However, how many times heavier an atom of one element is compared to an atom of another element can be measured. These relative masses allow one to order the elements by their increasing atomic mass. Assigning a mass to one of the atoms makes it the standard for the scale. The individual masses of atoms of all the elements are assigned by comparison to the standard. The present standard of the atomic mass unit scale is the carbon-12 atom which is assigned a mass of exactly 12 amu.
2. The elementary entity in a mole depends on the nature of the substance we are referring to. For metallic iron the elementary entity is an iron atom, for water the elementary entity is an H_2O molecule. For sodium ions it would be a Na^+ ion. The elementary entity is the species that is repeated many times to make up the given mass of the substance.

3. The average atomic mass of sulfur is 32.07 amu and that of oxygen is 16.00 amu. Because of the presence of isotopes all sulfur atoms do not weigh 32.07 amu, and all oxygen atoms do not weigh 16.00 amu. So only on average do S atoms weigh 2.0 times as much as the average O atom.

Practice Test

1. Oxygen would be 8.4, and argon would be 21.0
2. 3.96 times
3. 25.62 amu
4. a. 5.2 g   b. 2.0×10^{-11} g
5. 3.15×10^{-23} g
6. a. 1.81×10^{24} atoms   b. 6.022×10^{23} atoms   c. 4.3×10^{23} atoms
7. 51.69 g
8. a. 4.41 mol $CaSO_4$ b. 17.6 mol O   c. 1.06×10^{25} O atoms
9. 6.022×10^{24} Sb atoms
10. a. 1.40×10^3 g   b. 0.0997 g
11. a. $C_3H_4O_3$ b. P_2O_5 c. $MgCl_2$
12. a. 52.91% Al, 47.08% O   b. 1.60% H, 22.23% N, 76.17% O
    c. 9.93% C, 58.63% Cl, 31.43% F
13. 74.5% F
14. C_8H_8
15. CrO_3
16. $C_3H_8O_3$
17. a. $P_4O_{10} + 6H_2O \rightarrow 4H_3PO_4$
    b. $2Ga + 3H_2SO_4 \rightarrow Ga_2(SO_4)_3 + 3H_2$
    c. $2C_4H_{10} + 13O_2 \rightarrow 8CO_2 + 10H_2O$

18.

mol SO_2	g O_2	mol SO_3	g SO_3
a. 1.50	24.0	1.50	120
b. 1.25	20.0	1.25	100
c. 5.21	83.4	5.21	417

19. a. 71.8 g Na_2CO_3 b. 36.8 g Na_2S
20. a. H_2SO_4 is the limiting reactant; 30.6 g HF
    b. 3.27 g CaF_2 c. 85.6% yield
21. 2.93×10^{24} Ag atoms
22. a. 330 amu   b. male, 1.6×10^{23} molecules; female, 1.3×10^{23} molecules.
23. 178 amu

Chapter Four
REACTIONS IN AQUEOUS SOLUTION

GENERAL PROPERTIES OF AQUEOUS SOLUTIONS (4.1)

STUDY OBJECTIVES

1. Define solute, solvent, and solution.
2. Distinguish between nonelectrolytes, weak electrolytes, and strong electrolytes.

Solutions. Many chemical reactions occur in aqueous solutions. **Solutions** are homogeneous mixtures of two or more substances. Solutions have two components, the **solute** which is the substance present in the smaller amount, and the **solvent** which is the substance present in the larger amount. A solute can be a solid, liquid, or gaseous substance. In aqueous solutions, water is the solvent.

Electrolytes and Nonelectrolytes. Many aqueous solutions of ionic compounds conduct electricity whereas water itself essentially does not conduct. These solutions are called electrolyte solutions. An **electrolyte** is a substance that, when dissolved in water, yields an electrically conducting solution. Electrolytes produce ions in solution that are capable of carrying an electrical current through the solution.

For example, salts such as KCl(s), which already consist of ions, separate into individual ions when dissolved in water. Salts dissociate in water.

$$KCl(aq) \rightarrow K^+(aq) + Cl^-(aq)$$

Acids can ionize in water. Hydrochloric acid even though it is a molecule will completely ionize in water.

$$HCl(aq) \rightarrow H^+(aq) + Cl^-(aq)$$

Thus a solution of HCl, for instance, contains H^+ ions and Cl^- ions, but <u>no</u> HCl molecules! *Strong* electrolytes are 100% dissociated into ions in solution.

A **nonelectrolyte** is a substance that does not ionize in aqueous solution. Such a solution does not conduct an electrical current. Sugar is an example of a nonelectrolyte. In solution, sugar exists as $C_6H_{12}O_6$

molecules. These are neutral (no charge) and cannot conduct an electrical current. All nonelectrolytes exist as molecules in aqueous solution.

Between electrolytes and nonelectrolytes are the weak electrolytes. These substances ionize only *partially* to yield a few ions capable of conducting a weak electrical current. Compounds that are incompletely dissociated into ions are called **weak electrolytes**. Acetic acid (CH_3COOH) is an example of a weak electrolyte. You will see in Chapter 16 that in a solution of CH_3COOH about 99% of the CH_3COOH molecules are not ionized. Only 1% are ionized into H^+ and CH_3COO^- ions as shown below.

$$CH_3COOH(aq) \rightleftharpoons CH_3COO^-(aq) + H^+(aq)$$

Recall that a double arrow means the reaction is reversible. This means that the reaction can occur in the reverse direction as well as the forward direction. The reverse reaction removes ions from the solution and leads to incomplete ionization. Table 4.2 of the text lists examples of substances that are strong electrolytes, weak electrolytes, and nonelectrolytes.

EXAMPLE 4.1   Identifying Electrolytes and Nonelectrolytes

Identify the following substances as strong, weak, or nonelectrolytes:
a. CH_3OH (methyl alcohol)
b. NH_3
c. Na_2CO_3

•Method of Solution

Refer to Table 4.1 in the text.
a. Various sugars and alcohols dissolve as neutral molecules. *Answer:* nonelectrolyte.
b. Ammonia is a weak base. See Section 4.3 of the text. *Answer:* weak electrolyte.
c. Sodium carbonate is an ionic compound. All ionic compounds dissociate completely in aqueous solution. *Answer:* strong electrolyte.

EXERCISES

1. Which of the following are ionic compounds and therefore are strong electrolytes?
    a. CaF_2 b. $C_6H_{12}O_6$ c. KBr   d. $LiNO_3$ e. $CuSO_4$

2. What ionic species are present in aqueous solutions of the following?
    a. $Mg(NO_3)_2$ b. KOH   c. CaF_2

rules a short?

SOLUBILITY RULES AND PRECIPITATION REACTIONS (4.2)

STUDY OBJECTIVES

1. Write molecular, ionic, and net ionic equations for precipitation reactions.
2. Predict which product will precipitate when two electrolyte solutions are mixed.

Solubility Rules. The **solubility** of a solute is the amount that can be dissolved in a given quantity of solvent at a given temperature. For example the solubility of $Pb(NO_3)_2$ is 56 g per 100 mL H_2O at 20°C. The solubilities of ionic solids in water vary over a wide range of values. For convenience, we divide compounds into three categories called *soluble*, *slightly soluble*, and *insoluble*. Compounds are classified as soluble if a good deal of solid visibly when added to water. On the other hand, if only a small amount of solid dissolves, the solid is slightly soluble.. If no solid can be observed to dissolve, the substance is described as insoluble. Be aware that insoluble is a relative term and does not mean that no solute dissolves. Keep in mind also that an ionic compound can have a low solubility, but still be a strong electrolyte. The solubilities of various compounds in water at 25°C are summarized below and in Table 4.2.

Solubility Rules

1. All alkali metal (Group 1A) salts and all ammonium (NH_4^+) salts are soluble.
2. All salts containing nitrate (NO_3^-), chlorate (ClO_3^-), and bicarbonate (HCO_3^-) are soluble.
3. Most hydroxides (OH^-) are insoluble, except those of the alkali metals and barium hydroxide $Ba(OH)_2$. Calcium hydroxide $Ca(OH)_2$, is slightly soluble.
4. Most compounds containing chlorides, bromides, and iodides are soluble, except those containing Ag^+, Hg_2^{2+}, and Pb^{2+} ions.
5. All carbonates (CO_3^{2-}), phosphates (PO_4^{3-}), and sulfides (S^{2-}) are insoluble, except those covered by rule 1.
6. Most sulfates (SO_4^{2-}) are soluble. Both Ag_2SO_4 and $CaSO_4$ are slightly soluble. Ba^{2+}, Sr^{2+}, and Pb^{2+} sulfates are insoluble.

Precipitation Reactions. When certain solutions of electrolytes are mixed, the cation of one solute may combine with the anion of the other solute to form an insoluble compound. This leads to the formation of a precipitate. A **precipitate** is an insoluble solid that settles from a solution. For instance, when a solution of $Pb(NO_3)_2$ is mixed with a solution of HCl, a precipitate of insoluble $PbCl_2(s)$ forms. This occurs because combining the solutions results in a new solution containing more $PbCl_2(aq)$ than is soluble. A condition referred to as supersaturation. In this case, solid $PbCl_2$ must settle out of the solution. Chemical equations for precipitation reactions can be written in several ways.

In the **molecular equation** the formulas of the compounds are written as their usual chemical formulas.

$$Pb(NO_3)_2(aq) + 2HCl(aq) \rightarrow PbCl_2(s) + 2HNO_3(aq)$$

The precipitation reaction is more accurately represented by the ionic equation. The **ionic equation** shows the compounds as being dissociated in solution.

$$Pb^{2+}(aq) + 2NO_3^-(aq) + 2H^+(aq) + 2Cl^-(aq) \rightarrow PbCl_2(s) + 2H^+(aq) + 2NO_3^-(aq)$$

If you examine the ionic equation, you will notice that the H^+ ions and NO_3^- ions are not involved in the formation of the precipitate. We refer to the ions which are not involved in the reaction as **spectator ions**. Identical species on both sides of the equation can be omitted from the equation. The **net ionic equation** shows only the species that actually undergo a chemical change.

$$Pb^{2+}(aq) + 2Cl^-(aq) \rightarrow PbCl_2(s)$$

In order to predict whether a precipitate will form when two solutions are mixed, you need to know the solubility rules for ionic compounds.

EXAMPLE 4.2   Solubility Rules

According to the solubility rules which of the following compounds is soluble in water?
a. $MgCO_3$ b. $AgNO_3$ c. $MgCl_2$ d. Na_3PO_4 e. KOH

•Method of Solution

a. According to rule 5, $MgCO_3$ is insoluble.
b. According to rule 2, $AgNO_3$ is soluble.
c. According to rule 4, $MgCl_2$ is soluble.
d. According to rules 1 and 5, Na_3PO_4 is soluble.
e. According to rules 1 and 3, KOH is soluble.

EXAMPLE 4.3   Writing Ionic and Net Ionic Equations

Write balanced molecular, ionic, and net ionic equations for the reaction that occurs when a $CaCl_2$ solution is mixed with a Na_2CO_3 solution.

•Method of Solution

The combined solution will contain Na^+, CO_3^{2-}, Ca^{2+}, and Cl^- ions. The two new possible ionic compounds are NaCl and $CaCO_3$. If one (or both) is insoluble, it (they) will precipitate. According to solubility rules 1 and 4, NaCl is soluble, and from rule 5, $CaCO_3$ is insoluble. The molecular equation should show $CaCO_3(s)$ as a product along with NaCl(aq).

Answer: The molecular equation is:

$$CaCl_2(aq) + Na_2CO_3(aq) \rightarrow CaCO_3(s) + 2NaCl(aq)$$

The ionic equation:

$$Ca^{2+}(aq) + 2Cl^-(aq) + 2Na^+(aq) + CO_3^{2-}(aq) \rightarrow CaCO_3(s) + 2Na^+(aq) + 2Cl^-(aq)$$

The net ionic equation omits spectator ions:

$$Ca^{2+}(aq) + CO_3^{2-}(aq) \rightarrow CaCO_3(s)$$

EXERCISES

3. Which of the following compounds are soluble in water?
    a. $BaCl_2$ b. $PbSO_4$ c. $Ni(OH)_2$ d. $Ca_3(PO_4)_2$ e. NH_4NO_3

4. According to the solubility rules, which of the following compounds are insoluble in water?
    a. NH_4CO_3 b. AgBr   c. $CaCO_3$ d. $FeCl_3$ e. ZnS

5. Which of the following compounds are strong electrolytes?
    a. KOH   b. Sucrose   c. MgI_2 d. Methyl alcohol   e. HBr   f. AgCl

6. Predict whether a precipitate will form when aqueous solutions of the following are mixed. Write the formulas for any precipitates.
    a. $MgBr_2$ and NaOH
    b. NaI and $AgNO_3$

7. Identify the spectator ions in parts (a) and (b) above.

8. Write net ionic equations for the reactions that occur when solutions of the following compounds are mixed.
    a. $MgBr_2$ and $Pb(NO_3)_2$
    b. NaBr and $AgNO_3$

ACID-BASE REACTIONS (4.3)

STUDY OBJECTIVES

1. Write equations that show the ionization of acids and bases in water.
2. Define acids and bases according to Brønsted's definition.
3. Describe monoprotic, diprotic, and triprotic acids.
4. Write molecular, ionic, and net ionic equations for neutralization reactions.

  The Arrhenius Acid-Base Theory. Acids and bases are general classifications of compounds that are based on their properties. In general, solutions of acids have a sour taste, change blue litmus to red, dissolve certain metals, and react with carbonates to produce carbon dioxide gas. On the other hand, bases are compounds that form solutions that have a bitter taste, feel slippery, and change red litmus to blue. The properties of an acid are neutralized by the addition of a base, and vice versa.

  An acid-base theory or definition is an attempt to explain what happens to molecules of acidic and basic substances in solution that gives rise to their characteristic properties. Svante Arrhenius proposed that an **acid** is a substance that produces hydrogen ions (H^+, also called protons) in aqueous solution, and a **base** is a substance that produces hydroxide ions (OH^-) in aqueous solution. Nitric acid is one example:

$$HNO_3(aq) \rightarrow H^+(aq) + NO_3^-(aq)$$

Calcium hydroxide is an example of a base:

$$Ca(OH)_2(aq) \rightarrow Ca^{2+}(aq) + 2OH^-(aq)$$

We can conclude that all acids and bases are electrolytes undergoing dissociation in aqueous solution.
  Arrhenius explained neutralization as the combination of H^+ ions and OH^- ions to form water.

$$H^+(aq) + OH^-(aq) \rightarrow H_2O(l)$$

Notice that the $H^+(aq)$ is called the hydrogen ion or a proton. To understand the name proton remember that a hydrogen atom consists of a proton as the nucleus and an electron which travels around the nucleus. The positive ion results from the loss of the electron. Thus the hydrogen ion is just a proton.

  The Brønsted Theory. J. N. Brønsted proposed a more general theory of acid-base reactions than that of Arrhenius. According to Brønsted's theory, an **acid** is defined as a substance that is able to donate a proton, and a **base** is a substance that accepts a proton, in a chemical reaction. For example, when HCl gas dissolves in water, it can be viewed as donating a proton to molecules of the solvent. Hydrochloric acid is a Brønsted acid.

$$HCl(aq) + H_2O(l) \rightarrow H_3O^+(aq) + Cl^-(aq)$$
  acid   base

The water molecule is the proton acceptor, and so water is a Brønsted base in this reaction. All acid-base reactions, according to Brønsted, involve the transfer of a proton. In the above reaction H_3O^+ is called the **hydronium ion**. It is a close representation of the proton in solution. All protons are hydrated, which means that they are surrounded with water molecules.
  When ammonia dissolves in water, it accepts a proton and so is classified as a Brønsted base. In this reaction water is the acid and ammonia is the base.

$$NH_3(aq) + H_2O(l) \rightleftharpoons NH_4^+(aq) + OH^-(aq)$$

Ammonia is a weak electrolyte because only a small fraction of dissolved NH_3 molecules are actually ionized at any one time.

Monoprotic, Diprotic, and Triprotic Acids. Some acids are capable of donating more than one proton to a base. Acids are referred to as monoprotic, diprotic, and triprotic depending on whether they can donate one, two, or three protons. Sulfuric acid (H_2SO_4) and carbonic acid (H_2CO_3) are diprotic acids. Phosphoric acid (H_3PO_4) is a triprotic acid.

Writing Equations for Acid-Base Neutralization Reactions. Acids and bases react with each other in a reaction called neutralization. A **neutralization reaction** is a reaction between an acid and a base that produces water and a salt.

$$acid + base \rightarrow water + a\ salt$$

In a neutralization reaction the hydrogen ion from the acid and the hydroxide ion from the base combine to form water. The metal ion from the base and the nonmetal ion from the acid constitute a solution of a salt. The neutralization reaction of hydrochloric acid and magnesium hydroxide, for example, is represented by the following molecular equation.

$$2HCl(aq) + Mg(OH)_2(aq) \rightarrow 2H_2O(l) + MgCl_2(aq)$$
acid   base   water   a salt

Note that one mole of base contains 2 moles of OH^- ions. This means that two moles of HCl are needed to neutralize one mole of base.

Since the acid and the base are both completely ionized in solution we can write the ionic equation:

$$2H^+(aq) + 2Cl^-(aq) + Mg^{2+}(aq) + 2OH^-(aq) \rightarrow 2H_2O(l) + Mg^{2+}(aq) + 2Cl^-(aq)$$

Notice that Mg^{2+} ion and Cl^- ion are spectator ions. The net ionic equation is:

$$2H^+(aq) + 2OH^-(aq) \rightarrow 2H_2O(l) \quad or\ just\ \ H^+(aq) + OH^-(aq) \rightarrow H_2O(l)$$

The coefficients can be cancelled. The neutralization equation for any strong acid and strong base is:

$$H^+(aq) + OH^-(aq) \rightarrow H_2O(l)$$

If we had carried out this reaction with two moles of HCl for each mole of base, there would be no left over acid or base, and the solution would be described as neutral.

EXAMPLE 4.4   Neutralization Reactions

Write balanced molecular, ionic, and net ionic equations for the neutralization reaction that occurs when nitric acid and barium hydroxide are mixed.

•Method of Solution

First write the formulas of the reactants

$$HNO_3(aq) + Ba(OH)_2(aq) \rightarrow$$

Neutralization reactions of acids and metal hydroxides yield water and a salt. The cation of the salt produced comes from the base and the anion of the salt comes from the acid. In order to write proper equations, you need to be able to write the formulas of salts. The salt is $Ba(NO_3)_2$. It will take two moles of the HNO_3 to neutralize one mole of $Ba(OH)_2$.

The molecular equation is:

$$2HNO_3(aq) + Ba(OH)_2(aq) \rightarrow 2H_2O(l) + Ba(NO_3)_2(aq)$$

To write the ionic equation recall that the acid and the base are 100% ionized in solution.

$$HNO_3(aq) \rightarrow H^+(aq) + NO_3^-(aq)$$

$$Ba(OH)_2(aq) \rightarrow Ba^{2+}(aq) + 2OH^-(aq)$$

The ionic equation is:

$$2H^+(aq) + 2NO_3^-(aq) + Ba^{2+}(aq) + 2OH^-(aq) \rightarrow 2H_2O(l) + Ba^{2+}(aq) + 2NO_3^-(aq)$$

The net ionic equation is the equation left after any spectator ions are cancelled:

$$H^+(aq) + OH^-(aq) \rightarrow H_2O(l)$$

EXERCISES

9. Write the equations for the ionization of the following acids in aqueous solution.
   a. HBr   b. $HClO_4$ c. H_2SO_4

10. Write the equations for the dissociation of the following bases in aqueous solution.

11. Write balanced molecular equations for the reaction of NaOH(aq) with each of the following.
    a. $HClO_4(aq)$ b. $H_2SO_4(aq)$

12. Complete the following molecular equations
    a. $HNO_3(aq) + LiOH(aq) \rightarrow$
    b. $Ca(OH)_2(aq) + HCl(aq) \rightarrow$
    c. $KOH(aq) + H_3PO_4(aq) \rightarrow$

REDOX REACTIONS AND OXIDATION NUMBERS (4.4)

STUDY OBJECTIVES

1. Define oxidation, reduction, half-reaction, oxidizing agent, and reducing agent
2. Assign oxidation numbers to the atoms in molecules, simple ions, and polyatomic ions.

Oxidation and Reduction. In this section we review an important type of chemical process called the oxidation-reduction reaction, or just redox reaction for short. In redox reactions electrons are transferred from one atom to another. **Oxidation** is a chemical change in which an atom, ion, or molecule loses electrons. **Reduction** is a chemical change in which an atom, ion, or molecule gains electrons. In the reaction of metallic sodium with chlorine gas, an electron is transferred from the sodium atom to a chlorine atom.

$$2Na + Cl_2 \rightarrow 2Na^+ + 2Cl^-$$

Sodium is oxidized and chlorine is reduced. Oxidation and reduction always occur simultaneously.

In a redox reaction the substance oxidized is called the *reducing agent* because it supplies electrons which cause reduction in another substance. The substance being reduced is called the *oxidizing agent* because it acquires electrons from another substance. In the sodium-chlorine reaction, sodium is the reducing agent and chlorine is the oxidizing agent.

Redox reactions are conveniently separated into two *half-reactions*: the oxidation half-reaction and the reduction half-reaction. For the sodium-chlorine reaction they are:

$$\text{oxidation} \qquad 2Na \rightarrow 2Na^+ + 2e^-$$
$$\text{reduction} \qquad 2e^- + Cl_2 \rightarrow 2Cl^-$$

Summation of the two half-reactions yields the overall redox reaction shown above. Half-reactions are used for convenience, but you must not forget that oxidation and reduction occur simultaneously.

The concepts of oxidation and reduction have been extended from cases of complete electron gain or loss to reactions involving varying degrees of electron transfer. When tin ore is reduced to metallic tin by heating tin oxide with charcoal, tin ions are reduced and carbon is oxidized.

$$SnO_2(s) + C(s) \rightarrow Sn(l) + CO_2(g)$$

The reduction of tin is quite obvious, but in what sense is carbon oxidized? In the context of the original definition of oxidation, carbon is oxidized because it combines with oxygen. Carbon loses its electrons in the following sense. In charcoal, each carbon atom shares its four valence electrons equally with other carbon atoms. But in CO_2 the bonding electrons are shared unequally, lying closer to the oxygen atoms. Even though a C^{4+} is not formed in CO_2, the C atoms do relinquish their electrons to some extent to the O atoms. Thus carbon is oxidized in the reaction. In this reaction oxygen is reduced.

Oxidation Numbers. The oxidation number or oxidation state is used to help keep account of electrons in chemical reactions. The **oxidation number** is the charge an atom would have in a molecule or an ionic compound, if electrons were completely transferred.

The oxidation number of an atom designates its positive or negative character. A positive oxidation number indicates that an atom has lost electrons compared to an uncombined atom. A negative oxidation number indicates that an atom has gained electrons. Oxidation numbers are useful in *writing formulas*, in *recognizing redox reactions*, and in *balancing redox reactions*. An element is oxidized in a reaction whenever its oxidation number *increases* as a result of the reaction. When the oxidation number of an element *decreases* in a reaction, it is reduced.

The following is a set of rules for assigning oxidation numbers:

1. *The oxidation number of an uncombined element is zero.* The atoms in free elements are neutral. Thus the atoms in Fe, Ne, O_2, H_2, and P_4, for example, have oxidation numbers of zero.
2. *The oxidation number of an element in a simple monatomic ion is the same as the charge of the ion.* The atoms in the monatomic ions Na^+, Mg^{2+}, and Al^{3+} have oxidation numbers of +1, +2, and +3, respectively. The oxidation number of chlorine in Cl^- is −1, and of sulfur in S^{2-} is −2.
3. *Certain elements have the same oxidation number in all (or almost all) of their compounds.*
   a. The oxidation number of oxygen is −2 in almost all of its compounds, except when it is in a peroxide, where it is −1.
   b. The oxidation number of hydrogen is +1, except when it is bonded to a metal in binary compounds. In metal hydrides, hydrogen has an oxidation number of −1. An example is AlH_3.
   c. Fluorine has an oxidation number of −1 in all of its compounds.
   d. The oxidation numbers of Group 1A and Group 2A metals in compounds are always +1 and +2, respectively. Aluminum is always +3.
4. *In a neutral compound, the algebraic sum of the oxidation numbers of all the atoms is zero.* For example, in sodium oxide, Na_2O

$$2(\text{ox. no. Na}) + (\text{ox no. O}) = 0$$
$$2(+1) + (-2) = 0$$

5. *The sum of the oxidation numbers of all atoms in a polyatomic ion is equal to the net charge of the ion.* For example, in nitrate ion, (NO_3^-) which has a net charge of -1:

$$\text{(ox. no. N)} + 3\text{(ox. no. O)} = -1$$
$$(+5) + 3(-2) = -1$$

Recognizing Redox Reactions. It is easy to spot the equation of a redox reaction when you see one. First assign oxidation numbers to atoms of the elements in the reactants and products. Then look to see whether any of the elements have changed their oxidation state as they changed from reactants to products. In the reaction of tin oxide above, the oxidation number of tin in SnO_2 is $+4$, and in Sn it is 0. Therefore, Sn is reduced in the reaction. Carbon on the other hand must be oxidized. Inspection indicates that the oxidation number of C as the element is 0, and that of C in CO_2 is $+4$. Carbon is oxidized as we thought.

Four types of redox reactions are described in the text. They are listed in Table 4.1.

Table 4.1. Types of Redox Reactions.

1. Combination Reactions
2. Decomposition Reactions
3. Displacement Reactions
    a. Hydrogen displacement
    b. Metal displacement
    c. Halogen displacement
4. Disproportionation Reactions

EXAMPLE 4.5   Determining Oxidation Numbers

Assign oxidation numbers to all the atoms in the following compounds and ions:

a. Na_2SO_4
b. $CuCl$
c. SO_3^{2-}

•Method of Solution

a. In Na_2SO_4 we can assign oxidation numbers to Na and O because their oxidation numbers are almost always the same in all compounds (rule 3). Na is an alkali metal and is $+1$. Oxygen is -2. The oxidation number of sulfur varies from compound to compound. Apply rule 4 that the sum of the oxidation numbers of all atoms in the compound is zero.

$$2\text{(ox. no. Na)} + \text{(ox. no. S)} + 4\text{(ox. no. O)} = 0$$
$$2(+1) + \text{(ox. no. S)} + 4(-2) = 0$$
$$+ \text{(ox. no. S)} = 8 - 2 = +6$$

b. In $CuCl$, the chlorine ion has an oxidation number of -1. Thus Cu must be $+1$ by rule 4.

c. In SO_3^{2-} we assume the oxidation number of oxygen is -2, and solve for the oxidation number of sulfur using rule 5.

$$\text{(ox. no. S)} + 3\text{(ox. no. O)} = -2$$
$$\text{(ox. no. S)} + 3(-2) = -2$$
$$\text{(ox. no. S)} = -2 + 6 = +4$$

EXAMPLE 4.6   Identifying Oxidation and Reduction

Determine whether iron in the reactant species is oxidized or reduced during the following reactions?

a. $2Fe + \frac{3}{2}O_2 + 2H_2O \rightarrow Fe_2O_3 \cdot 2H_2O$ (rust)

b. $FeO + CO \rightarrow Fe + CO_2$

c. $2Hg + 2Fe^{3+} + 2Cl^- \rightarrow Hg_2Cl_2 + 2Fe^{2+}$

•Method of Solution

a.   There are two ways to tell. First assign oxidation numbers to iron in Fe and in Fe_2O_3 (ignore the waters of hydration). The oxidation number changes form zero in Fe to +3 in Fe_2O_3. Since Fe becomes more positive, it must have lost electrons. Thus iron is oxidized. In the second way, when a metal combines with oxygen, its electrons are always transferred to the oxygen atom. All elements except fluorine are oxidized when they combine with O_2, and oxygen is reduced in these reactions.

b.   Using the oxidation number change as a guide, the oxidation number of iron changes from +2 in FeO to zero in Fe. Therefore, iron is gaining electrons in this reaction, and so it is reduced.

c.   The oxidation number of iron is more negative in the product, and so iron must gain an electron in this reaction. Fe is reduced.

EXERCISES

13. Assign oxidation numbers to the underlined elements in the following molecules and ions.
   a. $\underline{C}O_2$ b. $K_2\underline{Cr}O_4$ c. $H_2\underline{S}$ d. $\underline{H}_2S$ e. $\underline{H}_2$
   f. $Ca\underline{H}_2$ g. $\underline{C}O_3^{2-}$ h. $\underline{Fe}Br_3$ i. $\underline{Cr}O_4^{2-}$ j. $K\underline{Cl}O_4$

14. In the following reactions identify the elements oxidized and the elements reduced.
   a. $3Ag + 4HNO_3 \rightarrow 3AgNO_3 + NO + 2H_2O$
   b. $6Fe^{2+} + Cr_2O_7^{2-} + 14H^+ \rightarrow 6Fe^{3+} + 2Cr^{3+} + 7H_2O$
   c. $Cu^{2+} + Zn \rightarrow Cu + Zn^{2+}$

15. In the following reactions identify the oxidizing agents and reducing agents.
   a. $3Ag + 4HNO_3 \rightarrow 3AgNO_3 + NO + 2H_2O$
   b. $6Fe^{2+} + Cr_2O_7^{2-} + 14H^+ \rightarrow 6Fe^{3+} + 2Cr^{3+} + 7H_2O$
   c. $Cu^{2+} + Zn \rightarrow Cu + Zn^{2+}$

16. Identify which of the following reactions are oxidation-reduction reactions.
   a. $Ca(s) + HCl(aq) \rightarrow CaCl_2(aq) + H_2(g)$
   b. $HNO_3(aq) + NaOH(aq) \rightarrow NaNO_3(aq) + H_2O(l)$
   c. $H_2O_2 + 2HI \rightarrow I_2 + 2H_2O$
   d. $Pb^{2+}(aq) + S^{2-} \rightarrow PbS(s)$

17. What is the oxidation number of Mn before and after the reaction?

   $5H_3AsO_3 + 2MnO_4^- + 6H^+ \rightarrow 2Mn^{2+} + 5H_3AsO_4 + 3H_2O$

SOLUTION CONCENTRATION (4.5)

STUDY OBJECTIVES

1. Define molarity and calculate the molarity of a solute in solution.
2. Determine the amount of a solute needed to prepare a solution of specific concentration.
3. Describe the steps involved in the dilution of a solution of known concentration, and calculate the concentration of the diluted solution.

Solution Concentration. Solutions are characterized by their concentration, that is the amount of solute dissolved in a given amount of solvent. The most common unit of concentration used for aqueous solutions is molarity. The **molarity** is the number of moles of solute per liter of the solution.

$$\text{molarity} = \frac{\text{moles of solute}}{\text{liters of solution}}$$

To prepare a 0.50 molar (0.50 M) solution of KCl, for instance, we first measure out 0.50 mole of solid KCl, that is 37.3 g. The KCl is then added to a 1.0 L volumetric flask. This is a flask with a calibrated ring etched around the neck to mark the point at which the flask contains 1.0 liter of liquid. When water is added up to this mark, the flask will contain 1.0 L of the KCl solution, and will contain 0.50 mole of KCl.

The number of moles of KCl in given amounts of the above solution is easy to find. For instance, 0.10 L of 0.50 M KCl will contain

$$0.10 \text{ L soln} \times \frac{0.50 \text{ mol KCl}}{1 \text{ L soln}} = 0.050 \text{ mol KCl}$$

Here molarity is being used as a conversion factor (mol per L). In general, *solution volume (V) × molarity (M) = moles solute.*

Dilution. Dilution is a procedure used to prepare a less concentrated solution from a more concentrated solution. Stock solutions of laboratory acids and ammonia, in particular, are purchased in gallon jugs as highly concentrated aqueous solutions. The key to understanding dilution is to remember that adding more water to a given amount of solution does not change the number of moles of solute present in solution. See Figure 4.19 in the text.

Moles of solute before dilution = moles of solute after dilution.

Since the moles of solute equals the solution volume (V) times the molarity (M) we have

$$\text{moles}_i = \text{moles}_f$$
$$M_i V_i = M_f V_f$$

where the subscripts i and f, stand for initial solution, and final solution, respectively. The concentration after dilution is

$$M_f = M_i \times \frac{V_i}{V_f}$$

Note that V_f is always larger than V_i.

Electrolytes and Nonelectrolytes. It is important to keep in mind that molarity refers only to the amount of solute originally dissolved in water, and does not reflect any subsequent processes such as the dissociation of a salt or the ionization of an acid. The symbols 1.0 M HCl and 1.0 M $Ca(OH)_2$ mean that 1.0

mole of the solutes HCl and Ca(OH)$_2$ were dissolved to make 1 liter of solution. It, however, does not say anything about the actual chemical species in the solution. Since HCl and Ca(OH)$_2$ are strong electrolytes, there are no molecules of HCl or Ca(OH)$_2$ in these solutions. The HCl solution has 1.0 M H$^+$ ion and 1 M Cl$^-$ ion. While the Ca(OH)$_2$ solution has 1.0 M Ca^{2+} ion and 2 M OH$^-$ ion!

EXAMPLE 4.7   Calculation of Molarity

What is the molarity of a solution made from 32.1 g KNO$_3$ dissolved in enough water to make 500 mL of solution?

•Method of Solution

First write the equation that defines molarity.

$$\text{molarity} = \frac{\text{moles KNO}_3}{\text{liters of solution}}$$

Then substitute into the equation the number of moles of KNO$_3$ in 32.1 g, and the volume of solution.

•Calculation

$$\text{moles KNO}_3 = 32.1 \text{ g KNO}_3 \times \frac{1 \text{ mol KNO}_3}{101.1 \text{ g KNO}_3}$$

$$= 0.317 \text{ mol KNO}_3$$

Convert 500 mL to 0.500 L

$$\text{molarity} = \frac{0.317 \text{ mol KNO}_3}{0.500 \text{ L}} = 0.635 \text{ M}$$

This is normally written 0.635 M KNO$_3$.

EXAMPLE 4.8   Use of Molarity

How many moles of solute are in 250 mL of 0.100 M KCl?

•Method of Solution

$$? \text{ mol KCl} = 250 \text{ mL}$$

To obtain the number of moles, we will use the concentration as a conversion factor.

$$0.100 \text{ M KCl means } \frac{0.100 \text{ mol KCl}}{1 \text{ L soln}}$$

When the volume (in liters) is multiplied by the concentration (mol/L), we obtain moles of solute. The volume of 250 mL must be expressed in liters, or else the units will not cancell..

$$V \times M = \text{moles solute}$$

•Calculation

$$? \text{ mol KCl} = 0.250 \text{ L} \times \frac{0.100 \text{ mol KCl}}{1 \text{ L}} = 0.025 \text{ mol KCl}$$

EXAMPLE 4.9   Dilution

What volume of a concentrated (6.0 M) hydrochloric acid stock solution is required to prepare 725 mL of 0.12 M HCl?

•Method of Solution

The number of moles of HCl in the 800 mL solution must come from the stock solution. Therefore:

$$mol\ HCl_{initial} = mol\ HCl_{final}$$

And since moles = MV, we can write

$$M_i V_i = M_f V_f$$

Rearranging to solve for the initial volume, we get:

$$V_i = V_f \frac{M_f}{M_i}$$

•Calculation

$$V_i = ?$$
$$V_f = 725\ mL$$
$$M_f = 0.12\ M$$
$$M_i = 6.0\ M$$

$$V_i = 725\ mL \times \frac{0.120\ M}{6.0\ M} = 14.5\ mL$$

•Comment

Note that 0.725 L of 0.12 M HCl contains 0.087 mol HCl. The exact amount supplied by 0.0145 L of 6.0 M HCl. Also we were able to use milliliter units in the calculation because the units of molarity cancelled.

EXERCISES

18. What is the molarity of a solution consisting of 11.8 g of NaOH dissolved in enough water to make 300 mL of solution?

19. How many moles of H_2SO_4 are contained in 225 mL of 0.750 M H_2SO_4?

20. How many grams of HCl are contained in 250 mL of 0.500 M HCl?

21. When 147 mL of 4.25 M NH_3 solution is mixed with 353 mL of H_2O. What is the concentration of the final solution?

22. Describe how you would prepare 2.50×10^2 mL of 0.666 M KOH solution starting with a 5.0 M KOH solution.

23. Consider 125 mL of 0.550 M $MgCl_2$ solution.
    a. How many moles of $MgCl_2$ are contained in 125 mL of 0.110 M $MgCl_2$?
    b. What is the concentration of Mg^{2+} ion?
    c. What is the concentration of Cl^- ion?
    d. What is the concentration of $MgCl_2$ molecules?

GRAVIMETRIC ANALYSIS (4.6)

STUDY OBJECTIVES

1. Use the balanced chemical equation to predict the yield of a precipitate given information about the amounts of the reactants.
2. Use data about the mass of a precipitate to calculate the concentration of an unknown.

Gravimetric Analysis. Quantitative analysis is the determination of the amount or concentration of a substance in a sample. Precipitation reactions form the basis of a type of chemical analysis called gravimetric analysis. A **gravimetric analysis** experiment involves the formation of a precipitate, and measurement of its mass. By knowing the mass and chemical formula of the precipitate, we can calculate the mass (or concentration) of the cation or anion component in the original sample.

Stoichiometry of a Precipitation Reaction. First, however we will look at a precipitation reaction. When solutions of $AgNO_3$ and NaCl are mixed, a precipitate of AgCl forms because silver chloride is insoluble in water. The equation is:

$$AgNO_3(aq) + NaCl(aq) \rightarrow AgCl(s) + NaNO_3(aq)$$

If 0.275 L of 0.100 M $AgNO_3$ are mixed with 0.135 L of 1.00 M NaCl, how many grams of AgCl would be formed? This example is just like the limiting reagent problems you worked earlier except that now the reaction occurs in solution. Therefore, let's find the number of moles of each reactant first from $M \times V$.

$$\text{mol } AgNO_3 = 0.275 \text{ L} \times \frac{0.100 \text{ mol}}{L} = 0.0275 \text{ mol } AgNO_3$$

$$\text{mol NaCl} = 0.135 \text{ L} \times \frac{1.00 \text{ mol}}{L} = 0.135 \text{ mol NaCl}$$

Since 0.0275 mol $AgNO_3$ will react with only 0.0275 mol NaCl, NaCl must be present in excess. The reaction will stop when all the $AgNO_3$ (specifically the Ag^+) has reacted. Since 1 mole of AgCl forms for each mole of $AgNO_3$, then the yield of AgCl will be 0.0275 mol. The amount of AgCl(s) formed is:

$$? \text{ g AgCl} = 0.0275 \text{ mol AgCl} \times \frac{143 \text{ g AgCl}}{1 \text{ mol AgCl}} = 3.93 \text{ g AgCl}$$

In gravimetric analysis the goal is to turn this procedure around. That is to isolate and weigh the precipitate from a precipitation reaction. This makes the mass of precipitate a known quantity. Then we calculate the number of moles and concentration of the limiting reagent.

Given the amount of limiting reagent, calculate the yield of a precipitate,
or
Given the yield of a precipitate, calculate the amount of limiting reagent.

EXAMPLE 4.10   Solution Stoichiometry

Calculate the number of moles of NaCl that must be added to 0.752 L of 0.150 M $AgNO_3$ solution in order to precipitate all the Ag^+ ions as AgCl.

•Method of Solution

First we need to determine the chemical equation for the precipitation reaction. Recall that NaCl is a strong electrolyte and so dissociates entirely into Na^+ and Cl^- ions in solution. $AgNO_3$ is also a strong electrolyte yielding Ag^+ and NO_3^- ions. At first the mixed solution contains Ag^+, NO_3^-, Na^+, and Cl^- ions. Since AgCl is insoluble, the net ionic reaction is:

$$Ag^+(aq) + Cl^-(aq) \rightarrow AgCl(s)$$

The ionic equation is:

$$Ag^+(aq) + NO_3^-(aq) + Na^+(aq) + Cl^-(aq) \rightarrow AgCl(s) + Na^+(aq) + NO_3^-(aq)$$

And the molecular equation is:

$$AgNO_3(aq) + NaCl(aq) \rightarrow AgCl(s) + NaNO_3(aq)$$

•Calculation

From the molecular equation we see that 1 mol NaCl $\rightleftharpoons$ 1 mol $AgNO_3$. Therefore, if we find the number of moles of $AgNO_3$ then we will also know the number of moles of NaCl required to react with it. From the volume and concentration of $AgNO_3$, we know that :

$$mol\ AgNO_3 = V \times M$$

$$= 0.752\ L \times \frac{0.150\ mol}{1\ L} = 0.113\ mol\ AgNO_3$$

$$mol\ NaCl = mol\ AgNO_3 = 0.113\ mol\ NaCl$$

•Comment

In this case 0.113 mol NaCl supplies enough Cl^- ion to react with the given amount of Ag^+ ion.

EXAMPLE 4.11   Gravimetric Analysis

Calculate the molar concentration of Ba^{2+} ions in a 500.0 mL sample of an unknown aqueous solution if 2.47 g $BaSO_4$ is formed upon the addition of excess Na_2SO_4?

•Method of Solution

(1) First write the chemical equation for the precipitation reaction. (2) Then use the mass of the precipitate (the product) to calculate the number of moles of Ba^{2+} ions originally present. (3) Finally divide by the volume of the original Ba^{2+} solution to determine the concentration of Ba^{2+}.
Barium sulfate being insoluble is formed in the following reaction.

$$Ba^{2+}(aq) + SO_4^{2-}(aq) \rightarrow BaSO_4(s)$$

Therefore, the number of moles of Ba^{2+} originally equals the number of moles of $BaSO_4$ precipitate.

•Calculation

$$?\ mol\ BaSO_4 = 2.47\ g\ BaSO_4 \times \frac{1\ mol\ BaSO_4}{233.4\ g\ BaSO_4} = 0.0106\ mol\ BaSO_4$$

$$mol\ Ba^{2+} = mol\ BaSO_4 = 0.0106\ mol\ Ba^{2+}\ ion$$

To find the original concentration of Ba^{2+} ions, substitute into the equation for molarity:

$$\text{molarity of } Ba^{2+} = \frac{\text{mol } Ba^{2+}}{\text{L soln}} = \frac{0.0106 \text{ mol}}{0.500 \text{ L soln}}$$

$$= 0.0212 \text{ M } Ba^{2+} \text{ ions}$$

EXERCISES

24. Calculate the number of moles of NaBr that must be added to 450 mL of 0.250 M $AgNO_3$ solution in order to precipitate all the Ag^+ ions as AgBr.

25. In order to precipitate AgCl, excess $AgNO_3$ was added to 10.0 mL of a solution containing Cl^- ion. If 0.339 g of AgCl was formed, what was the concentration of Cl^- in the original solution?

26. A 0.198 g sample of an ionic compound containing the bromide ion was dissolved in water and treated with excess $AgNO_3$. If the mass of silver bromide precipitate that forms was 0.0964 g, what is the percent of Br^- ion by mass in the original compound?

ACID-BASE TITRATIONS (4.7)

STUDY OBJECTIVES

1. Calculate the amount of a base needed to neutralize a given amount of acid.
2. Use titration data to calculate the concentration of an unknown acid or base solution.

Acid-Base Titrations. A titration is a procedure for determining the concentration of an unknown acid (or base) solution using a known (standardized) concentration of a base (or acid) solution. In the titration of an acid solution of unknown concentration, a known volume of a standardized base solution is added to the acid until the acid is just neutralized. The point at which exactly enough base has been added to neutralize the acid is called the **equivalence point**. Because the concentration (M) and the required volume (V) of the base are known, the number of moles of base needed to neutralize the acid can be calculated. The number of moles of acid that were neutralized is readily calculated using the mole ratios from the balanced neutralization reaction. If the volume of acid solution was measured then the concentration of acid is:

$$\text{acid concentration} = \frac{\text{moles of acid neutralized}}{\text{liters of acid solution}}$$

EXAMPLE 4.12   Stoichiometry of an Acid-Base Reaction

What volume of 0.900 M HCl is required to completely neutralize 25.0 g of $Ca(OH)_2$?

•**Method of Solution**

Write and balance the neutralization equation

$$2HCl + Ca(OH)_2 \rightarrow CaCl_2 + 2H_2O$$

Since the balanced equation relates moles of one reactant to moles of another reactant, first determine the number of moles of $Ca(OH)_2$ in 25.0 g.

•Calculation

$$g \ Ca(OH)_2 \rightarrow mol \ Ca(OH)_2 \rightarrow mol \ HCl \rightarrow volume \ HCl$$

1. Determine the number of moles of calcium hydroxide. The formula mass of $Ca(OH)_2$ is 74.1 g/mol.

$$25.0 \ g \ Ca(OH)_2 \times \frac{1 \ mol \ Ca(OH)_2}{74.1 \ g \ Ca(OH)_2} = 0.337 \ mol \ Ca(OH)_2$$

2. Now determine the number of moles of HCl required to neutralize the base. One mole of $Ca(OH)_2$ is neutralized by 2 mol HCl.

$$0.337 \ mol \ Ca(OH)_2 \times \frac{2 \ mol \ HCl}{1 \ mol \ Ca(OH)_2} = 0.674 \ mol \ HCl$$

3. Finally determine the volume of acid that contains the required number of moles of HCl.

$$0.674 \ mol \ HCl \times \frac{1 \ L \ soln}{0.900 \ mol \ HCl} = 0.749 \ L \ soln$$

EXAMPLE 4.13 Acid-Base Titration

A volume of 128 mL of 0.650 M $Ba(OH)_2$ was required to completely neutralize 50.0 mL of nitric acid solution (HNO_3). What was the concentration of the acid solution?

•Method of Solution

This is similar to the other problems. Our goal is to fill in the equation for molarity:

$$molarity \ of \ nitric \ acid = \frac{mol \ HNO_3}{L \ acid \ soln}$$

First we need the balanced chemical equation to give us the important mole-ratio. Then we construct a roadmap:

$$Ba(OH)_2(aq) + 2HNO_3(aq) \rightarrow Ba(NO_3)_2(aq) + 2H_2O(l)$$

$$(vol \times conc) \rightarrow mol \ Ba(OH)_2 \rightarrow mol \ HNO_3 \rightarrow conc \ HNO_3$$

•Calculation

1. The number of moles of $Ba(OH)_2$ required to neutralize the HNO_3 was

$$0.128 \ L \ soln \times \frac{0.650 \ mol \ Ba(OH)_2}{1 \ L \ soln} = 0.0832 \ mol \ Ba(OH)_2$$

2. The number of moles of HNO_3 that reacted was

$$0.0832 \ mol \ Ba(OH)_2 \times \frac{2 \ mol \ HNO_3}{1 \ mol \ Ba(OH)_2} = 0.166 \ mol \ HNO_3$$

3. The concentration of the acid was

$$molarity = \frac{mol \ HNO_3}{1 \ L \ soln} = \frac{0.166 \ mol}{0.050 \ L} = 3.33 \ M \ HNO_3$$

EXERCISES

27. How many mL of 1.00 M H_2SO_4 solution are required to neutralize 2.10 g KOH?

28. What volume of 0.210 M H_2SO_4 solution is needed to exactly neutralize 50.0 mL of 0.082 M NaOH?

29. What volume of 0.0824 M NaOH solution is needed to titrate 9.8 mL of 0.210 M H_2SO_4 to the equivalence point.?

30. What is the molarity of an oxalic acid ($H_2C_2O_4$) solution, if 22.50 mL of this solution requires 35.72 mL of 0.198 M NaOH for complete neutralization?

REDOX TITRATIONS (4.8)

STUDY OBJECTIVE

1. Use the results of a redox titration to calculate the concentration of an unknown solution.

Redox Titrations. In redox titrations the oxidizing agent and reducing agent react by a known reaction with exact stoichiometry. If the amount of one substance reacting is known, the amount of the other substance that reacts can be calculated. When the concentration of the oxidizing agent solution is known, this solution is placed in a buret. A known volume of the solution of unknown concentration of reducing agent is placed in the titration flask. For example, consider the titration of arsenic(III) of unknown concentration with standard potassium permanganate ($KMnO_4$):

$$5H_3AsO_3 + 2\,MnO_4^- + 6H^+ \rightarrow 2Mn^{2+} + 5H_3AsO_4 + 3H_2O$$

$KMnO_4$ solution containing the purple permanganate ion is placed in a buret. The solution is added drop by drop to a known volume of H_3AsO_3 solution in a flask. The resulting mixture remains colorless as long as MnO_4^- is consumed by the reaction. When enough permanganate has been added to oxidize all the H_3AsO_3, the next slight excess of $KMnO_4$ gives a pale pink color to the solution and signals the endpoint. At that point the number of moles of $KMnO_4$ reacted is equal to its concentration times the volume added from the buret.

$$\text{moles of } MnO_4^- = \text{volume} \times \text{molarity}$$

In this case the number of moles of H_3AsO_3 originally present can be calculated because we know the mole ratio from the balanced equation.

$$\text{moles of } MnO_4^- \times \frac{5 \text{ mol } H_3AsO_3}{2 \text{ mol } MnO_4^-} = \text{moles of } H_3AsO_3$$

The concentration of H_3AsO_3 is:

$$\text{molarity} = \frac{\text{mol } H_3AsO_3}{\text{liters soln}}$$

EXAMPLE 4.14   A Redox Titration

In aqueous solution 31.0 mL of 0.150 M $KMnO_4$ was required to completely oxidize 50.0 mL of H_3AsO_3. What is the concentration of the H_3AsO_3 solution?

•**Method of Solution**

The balanced equation was given previously.

$$5H_3AsO_3 + 2\,MnO_4^- + 6H^+ \rightarrow 2Mn^{2+} + 5H_3AsO_4 + 3H_2O$$

We need to find the number of moles of H_3AsO_3 that reacted and then divide by the original volume.

•**Calculation**

Three steps are required.
1. The number of moles of $KMnO_4$ required to oxidize all the H_3AsO_3 to H_3AsO_4.

$$31.0\text{ mL} \times \frac{10^{-3}\text{ L}}{1\text{ mL}} \times \frac{0.15\text{ mol}}{1\text{ L}} = 4.65 \times 10^{-3}\text{ mol }KMnO_4$$

$$= 4.65 \times 10^{-3}\text{ mol }MnO_4^-$$

2. The number of moles of H_3AsO_3 oxidized:

$$4.65 \times 10^{-3}\text{ mol }MnO_4^- \times \frac{5\text{ mol }H_3AsO_3}{2\text{ mol }MnO_4^-} = 1.16 \times 10^{-2}\text{ mol }H_3AsO_3$$

3. The concentration of H_3AsO_3 is:

$$\text{molarity} = \frac{\text{mol }H_3AsO_3}{\text{liters soln}}$$

$$= \frac{1.16 \times 10^{-2}\text{ mol}}{50.0\text{ mL}} \times \frac{1\text{ mL}}{10^{-3}\text{ L}} = 0.232\text{ M}$$

EXERCISES

31. What volume of 0.200 M $K_2Cr_2O_7$ will be required to oxidize 4.0 g of H_3AsO_3?

$$14H^+ + Cr_2O_7^{2-} + 3H_3AsO_3 \rightarrow 2Cr^{3+} + 3H_3AsO_4$$

32. In an experiment, 52.0 mL of $K_2Cr_2O_7$ solution was required to oxidize 4.0 g of H_3AsO_3. What was the molarity of the dichromate solution?

$$14H^+ + Cr_2O_7^{2-} + 3H_3AsO_3 \rightarrow 2Cr^{3+} + 3H_3AsO_4$$

CONCEPTUAL QUESTIONS

1. Explain how an ionic compound can have a low solubility, but be a strong electrolyte.

2. What makes molarity a convenient concentration unit for use in chemistry.

3. Suggest a precipitation reaction and a redox reaction by which baruim sulfate can be prepared.

PRACTICE TEST

1. Identify each of the following substances as a strong electrolyte, a weak electrolyte or a nonelectrolyte.
   a. NaBr   b. HCl   c. CH_3COOH d. NH_3 e. glucose

2. Write balanced *net ionic equations* for the reactions between:
   a. HCl(aq) and $Mg(OH)_2$(aq)
   b. CH_3COOH(aq) and NaOH(aq)
   c. $Ca(OH)_2$(aq) and H_2SO_4(aq)

3. Write balanced net ionic equations for the reactions that occur when solutions of the following solutes are mixed:
   a. $Ba(NO_3)_2$ and Na_2CO_3
   b. RbCl and $AgNO_3$
   c. $Pb(NO_3)_2$ and K_2S

4. Characterize the following compounds as soluble or insoluble in water.
   a. $Mg(OH)_2$ b. AgCl   c. $BaSO_4$
   d. $CaCO_3$ e. $Pb(NO_3)_2$ f. Na_2CO_3

5. Write formulas for the acid and base whose reactions produce the following salts.
   a. $CuSO_4$(aq)   b. KBr(aq)   c. $Ca_3(PO_4)_2$(s)

6. Write ionic and net ionic equations for the reactions that occur when solutions of the following compounds are mixed.
   a. NaBr and $AgNO_3$
   b. $MgBr_2$ and $Pb(NO_3)_2$

7. Assign oxidation numbers to the underlined elements in the following molecules and ions:
   a. $\underline{N}O_2^-$ b. $\underline{N}H_3$ c. $H\underline{Cl}O$ d. $H\underline{Cl}O_3$
   e. $\underline{Cl}O_4^-$ f. $H_2\underline{S}O_3$ g. $K\underline{Mn}O_4$ h. $\underline{C}H_4$
   i. $Na_2\underline{S}_2O_3$ j. $\underline{Hg}_2Cl_2$ k. $\underline{S}_4O_6^{2-}$

8. Identify which of the following reactions are oxidation-reduction reactions.
   a. $Mg(s) + HCl(aq) \rightarrow MgCl_2(aq) + H_2(g)$
   b. $HCl(aq) + NH_3(aq) \rightarrow NH_4Cl(aq)$
   c. $Mg(s) + CO_2(g) \rightarrow MgO(s) + C(s)$
   d. $Pb^{2+}(aq) + S^{2-} \rightarrow PbS(s)$

9. Identify the oxidizing agents and reducing agents in the following reactions:
   a. $S + O_2 \rightarrow SO_2$
   b. $BrO_3^- + 6I^- + 6H^+ \rightarrow 3I_2 + 3H_2O + Br^-$
   c. $As + H^+ + NO_3^- + H_2O \rightarrow H_3AsO_3 + NO$

10. What is the molarity of a solution consisting of 11.8 g of NaOH dissolved in enough water to make exactly 300 mL of solution?

11. How many grams of NaCl are present in 45.0 mL of 1.25 M NaCl?

12. How many liters of 0.50 M glucose, $C_6H_{12}O_6$, solution will contain exactly 100 g glucose?

13. If 30 mL of 0.80 M KCl is mixed with water to make a total volume of 0.400 L, what is the final concentration of KCl?

14. When aqueous solutions of $Pb(NO_3)_2$ and Na_2SO_4 are mixed a precipitate of $PbSO_4$ is formed. Calculate the mass of $PbSO_4$ formed when 655 mL of 0.150 M $Pb(NO_3)_2$ and 525 ml of 0.0751 M Na_2SO_4 are mixed.

15. Calculate the molar concentration of Pb^{2+} ions in 500.0 mL unknown aqueous solution if 1.07 g $PbSO_4$ is formed upon the addition of excess Na_2SO_4?

16. How many mL of 0.10 M H_2SO_4 would be required to neutralize 2.5 mL of 1.0 M NaOH?

17. In a titration experiment, a student finds that 23.6 mL of 0.755 M H_2SO_4 solution is required to completely neutralize 30.0 mL of NaOH solution. Determine the concentration of the NaOH solution.

18. If 10 mL of 1.0 M HCl are required to neutralize 50 mL of a NaOH solution, how many mL of 1.0 M H_2SO_4 will neutralize another 50 mL of the same NaOH solution?

19. Calculate the volume of 0.0300 M Ce^{4+} required to reach the equivalence point when titrating 41.0 mL of 0.0200 M $C_2O_4^{2-}$ (oxalate ion).

$$Ce^{4+} + C_2O_4^{2-} \rightarrow Ce^{3+} + CO_2 \text{ (acidic solution)}$$

20. A sample of iron ore weighing 1.824 g is analyzed by converting the Fe to Fe^{2+} and then titrating with standard potassium dichromate?

$$6Fe^{2+} + Cr_2O_7^{2-} + 14H^+ \rightarrow 6Fe^{3+} + 2Cr^{3+} + 7H_2O$$

If 37.21 mL of 0.0213 M $K_2Cr_2O_7$ was required to reach the equivalence point, what was the percent of iron in the ore.

21. $KMnO_4$ solution can be standardized against As_2O_3. A 0.2661 g sample of As_2O_3 was dissolved in acidic solution.

$$As_2O_3 + 2H_2O \rightarrow 2H_3AsO_3$$

If 38.22 mL of $KMnO_4$ solution was required to react with this amount of As_2O_3, what is the molarity of the $KMnO_4$ solution?

$$2MnO_4^- + 6H^+ + 5H_3AsO_3 \rightarrow 2Mn^{2+} + 5H_3AsO_4 + 3H_2O$$

ANSWERS

Exercises

1. a, c, d, and e are ionic compounds.
2. a. Mg^{2+} & NO_3^- b. K^+ & OH^- c. Ca^{2+} & F^-
3. a., & e.
4. b., c., & e.
5. a., c., e., & f.
6. a. $Mg(OH)_2$ b. AgI
7. a. Na^+ & Br^- b. Na^+ & NO_3^-
8. a. $Pb^{2+}(aq) + 2Br^-(aq) \rightarrow PbBr_2(s)$
   b. $Ag^+(aq) + Br^-(aq) \rightarrow AgBr(s)$
9. a. $HBr \rightarrow H^+ + Br^-$ b. $HClO_4 \rightarrow H^+ + ClO_4^-$ c. $H_2SO_4 \rightarrow 2H^+ + SO_4^{2-}$
10. a. $NaOH \rightarrow Na^+ + OH^-$ b. $Mg(OH)_2 \rightarrow Mg^{2+} + 2OH^-$ c. $Al(OH)_3 \rightarrow Al^{3+} + 3OH^-$

11. a. $NaOH(aq) + HClO_4(aq) \rightarrow H_2O + NaClO_4(aq)$

    b. $2NaOH(aq) + H_2SO_4(aq) \rightarrow 2H_2O + Na_2SO_4(aq)$

12. a. $HNO_3(aq) + LiOH(aq) \rightarrow H_2O + LiNO_3(aq)$

    b. $Ca(OH)_2(aq) + HCl(aq) \rightarrow H_2O + CaCl_2(aq)$

    c. $KOH(aq) + H_3PO_4(aq) \rightarrow H_2O + K_3PO_4(aq)$

13. a. +4   b. +6   c. –2   d. +1   e. 0   f. –1   g. +4   h. +3   i. +6   j. +7

14. a. Ag is oxidized & N is reduced   b. Fe^{2+} is oxidized & Cr is reduced

    c. Zn is oxidized & Cu^{2+} is reduced

15. a. Ag is the reducing agent; HNO_3 is the oxidizing agent.

    b. Fe^{2+} is the reducing agent; $Cr_2O_7^{2-}$ is the oxidizing agent.

    c. Zn is the reducing agent; Cu^{2+} is the oxidizing agent.

16. a. & c.

17. $+7 \rightarrow +2$

18. 0.983 M

19. 0.169 mol

20. 4.56 g

21. 1.25 M

22. Take 33.3 mL of 5.0 M KOH and dilute to a final volume of 250 mL.

23. a. 0.0138 mol   b. 0.550 M Mg^{2+} c. 1.1 M Cl^- ion

24. 0.113 mol

25. 0.236 M

26. 20.7%

27. 18.7 mL

28. 9.8 mL

29. 50 mL

30. 0.157 M

31. 52.9 mL

32. 0.200 M

Conceptual Questions

1. A strong electrolyte is a compound that is 100% dissociated in solution. Its solubility may be low or its solubility may be high. The term strong does not relate to the solubility. It only relates to the degree of dissociation. So a low solubilty compound that is 100% dissociated is still a strong electrolyte.

2. Molarity is the moles of solute per liter of solution, and is useful because for a solution of known molarity and volume the product of M x V is equal to the number of moles of solute. Measuring liters of solution is a way to count moles of solute.

3. $Ba(NO_3)_2(aq) + Na_2SO_4(aq) \rightarrow BaSO_4(s) + 2NaNO_3(aq)$

    $Ba(s) + H_2SO_4(aq) \rightarrow BaSO_4(s) + H_2(g)$

Practice Test

1. a. & b. are strong electrolytes   c. & d. are weak electrolytes   e. is a nonelectrolyte

2. a. $2HCl(aq) + Mg(OH)_2(aq) \rightarrow MgCl_2(aq) + H_2O(l)$

   b. $CH_3COOH(aq) + NaOH(aq) \rightarrow CH_3COONa(aq) + H_2O(l)$

   c. $Ca(OH)_2(aq)$ and $H_2SO_4(aq) \rightarrow CaSO_4(aq) + 2H_2O(l)$

3. a. $Ba(NO_3)_2 + Na_2CO_3 \rightarrow BaCO_3(s) + 2NaNO_3(aq)$

   b. $RbCl + AgNO_3 \rightarrow AgCl(s) + RbNO_3(aq)$

   c. $Pb(NO_3)_2 + K_2S \rightarrow PbS(s) + 2KNO_3(aq)$

4. a. b. c. & d. are insoluble   e. & f. are soluble

5. a. H_2SO_4 & $Cu(OH)_2$ b. HBr & KOH   c. H_3PO_4 & $Ca(OH)_2$

6. a. $Na^+(aq) + Br^-(aq) + Ag^+(aq) + NO_3^-(aq) \rightarrow AgBr(s) + Na^+(aq) + NO_3^-(aq)$

   $Ag^+(aq) + Br^-(aq) \rightarrow AgBr(s)$

   b. $Mg^{2+}(aq) + 2Br^-(aq) + Pb^{2+}(aq) + 2NO_3^-(aq) \rightarrow PbBr_2(s) + Mg^{2+}(aq) + 2NO_3^-(aq)$

   $Pb^{2+}(aq) + 2Br^-(aq) \rightarrow PbBr_2(s)$

7. a. +3   b. −3   c. +1   d. +5

   e. +7   f. +4   g. +7   h. −4

   i. +2   j. +1   k. +2.5

8. a, & c.

9. a. O is the oxidizng agent and S is the reducing agent.

   b. BrO_3^- is the oxidizng agent and I^- is the reducing agent.

   c. NO_3^- is the oxidizng agent and As is the reducing agent.

10. 0.983 M NaOH

11. 3.29 g NaCl

12. 1.1 L

13. 0.060 M KCl

14. 12.0 g $PbSO_4$

15. 7.06×10^{-3} M Pb^{2+}

16. 12 mL of 0.10 M H_2SO_4

17. 1.19 M NaOH

18. 5.0 mL

19. $2Ce^{4+} + C_2O_4^{2-} \rightarrow 2Ce^{3+} + 2CO_2$ 54.6 mL Ce^{4+} soln

20. 14.56% Fe

21. 0.0282 M $KMnO_4$

Chapter Five
GASES

- Properties of Gases (5.1 – 5.2)
- The Gas Laws (5.3)
- The Ideal Gas Law (5.4 & 5.6)
- Gas Stoichiometry (5.5)
- The Kinetic Molecular Theory of Gases (5.7)
- Nonideal Gases (5.8)

PROPERTIES OF GASES (5.1 – 5.2)

STUDY OBJECTIVES

1. List the properties that are characteristic of gases.
2. Interconvert pressure readings among the common units used to express pressure.
3. Calculate the pressure of a gas trapped in an open-tube manometer.

Characteristics of Gases. All matter exists as either a solid, a liquid, or a gas, depending on the temperature and the pressure. Of the three states, the simplest to describe is the gaseous state. Gases have certain general characteristics.

1. *Expansion.* Gases expand indefinitely to fill the space available to them.
2. *Indefinite shape.* Gases fill all parts of a container evenly and so have no definite shape of their own.
3. *Compressibility.* Gases are the most compressible of the states of matter.
4. *Mixing.* Two or more gases will mix evenly and completely when confined to the same container.
5. *Low density.* Gases have much lower densities than liquids and solids. Typically, densities of gases are about 1/1000 those of liquids and solids.

Gas Pressure. One of the most obvious properties of a gas, to someone who has inflated bicycle or automobile tires, is its pressure. **Pressure** is force per unit area. Any exertion of force applied to an area can be expressed as pressure.

$$\text{Pressure} = \frac{\text{force}}{\text{area}}$$

Some bicycle tires must be inflated to 60 psi. That's 60 pounds per square inch of tire surface. Recall that *weight* is the force with which an object is pulled vertically downward by gravity. Thus, a pound is a measure of force (whether due to gravity or not), and pounds per square inch is a unit of pressure.

A large number of units are used to measure pressure. The **standard atmosphere** is the average pressure exerted by Earth's atmosphere at sea level. In common units this is 14.7 psi. This pressure is enough to support a column of mercury 760 mm high at 0°C. The pressure that supports one mm of Hg is called a *torr*. In the SI, pressure is measured in a unit called the *pascal* (Pa), which is defined as the pressure from a force of one newton exerted on one square meter of area.

1 Pa = 1 N/m²

It takes 1.013×10^5 Pa to equal 1 atm. Since pascals are so small, the kilopascal is sometimes more convenient to use.

1 kPa = 10^3 Pa

Table 5.1 Pressure Units in Terms of the Standard Atmosphere.

1 atmosphere	= 760.0 mm Hg = 760.0 torr
	= 1.013×10^5 Pa = 101.3 kPa
	= 33.9 ft H_2O
	= 14.7 pounds per square inch (psi)

Pressure Measurement with an Open-Tube Manometer. A manometer is a device used to measure the pressure of gases, rather than that of the atmosphere. In an open-tube manometer (Example 5.2) the pressure of the trapped gas is the sum of the pressure of a mercury column of height h, plus the pressure of the atmosphere. Example 5.1 illustrates pressure measurement with a manometer.

EXAMPLE 5.1   Pressure Measurement

What is the pressure of the gas trapped in the J-tube shown below if the atmospheric pressure is 0.803 atm?

•**Method of Solution**

The pressure of the gas is sufficient to support a column of mercury 42 mm high and hold back the pressure of the atmosphere as well. Therefore the pressure of the gas is:

$$P_{gas} = P_h + P_{atm}$$

The pressure from the mercury column is just the difference in heights of the two mercury surfaces, 42 mm Hg. The atmospheric pressure is 610 mm Hg.

$$P_{gas} = 42 \text{ mm Hg} + 610 \text{ mm Hg}$$
$$P_{gas} = 652 \text{ mm Hg}$$

EXAMPLE 5.2   Converting Pressure Units

The atmospheric pressure on Mars is about 0.22 inches of Hg as compared to about 30 inches of Hg on Earth. Express the Martian pressure in mm Hg, atmospheres, and pascals.

•Method of Solution

Treat this like any other factor label problem. First state the problem:

$$? \text{ mm Hg} = 0.22 \text{ in Hg}$$

Recall that there are 2.54 cm/in, therefore there are 25.4 mm/in.
The unit factor needed is

$$\frac{25.4 \text{ mm}}{1 \text{ in}}$$

Converting inches to millimeters gives:

$$? \text{ mm Hg} = 0.22 \text{ in Hg} \times \frac{25.4 \text{ mm}}{1 \text{ in}}$$

$$= 5.6 \text{ mm Hg}$$

The number of atmospheres is

$$? \text{ atm} = 5.6 \text{ mm Hg} \times \frac{1 \text{ atm}}{760 \text{ mm Hg}}$$

$$= 7.4 \times 10^{-3} \text{ atm}$$

The pressure in pascals is

$$? \text{ pascals} = 7.4 \times 10^{-3} \text{ atm} \times \frac{1.013 \times 10^5 \text{ Pa}}{1 \text{ atm}}$$

$$= 750 \text{ Pa}$$

EXERCISES

1. Convert a pressure of 645 mm Hg into its value in
    a. atmospheres   b. kilopascals

2. Do the following unit conversions.
    a. 125 mm Hg to torr
    b. 725 mm Hg to kilopascals
    c. 1500 psi to atmospheres

THE GAS LAWS (5.3)

STUDY OBJECTIVES

1. State Boyle's, Charles's, and Avogadro's laws and know what variables are held constant for each law. Solve problems involving these laws.
2. Solve problems involving the combined gas law.
3. Solve problems using Dalton's law of partial pressures.

The Pressure-Volume Relationship. Experiments carried out in the 17th and 18th centuries led to three gas laws the provide a basis for our understanding of the behavior of gases. It is helpful when discussing the volume changes of a gas to visualize that we are considering a gas trapped in a container with a movable barrier so that its volume is not fixed. See Figures 5.4 and 5.5 of the text, and the open-tube manometer above.

Boyle's law states that the volume of a given amount of gas at a constant temperature is inversely proportional to the pressure of the gas. In other words, as the pressure applied to a sample of a gas increases its volume decreases. Mathematically we can write this in two ways:

$$V \alpha \frac{1}{P} \quad \text{or} \quad V = \frac{k_1}{P}$$

where k is a proportionality constant.

Rearranging yields a mathematical statement of Boyle's law. The product of $P \times V$ is a constant for a fixed amount of gas at a constant temperature.

$$PV = k_1 \text{ (Boyle's Law)}$$

For changes in pressure from P_1 to P_2, we can apply Boyle's law. Since the product PV is a constant, its value at the initial pressure P_1 and inital volume V_1 will equal PV at the final pressure P_2 and final volume V_2.

$$P_1V_1 = P_2V_2 \quad \text{(n and T are constant)}$$

The Temperature-Volume Relationship. **Charles's law** states that at constant pressure the volume of a gas is directly proportional to its absolute temperature. Mathematically we can write this in two ways:

$$V \alpha T \quad \text{or} \quad V = k_2T$$

where k is a constant. This equation shows that V is proportional to the *absolute* temperature T. Rearranging we get another form of Charles's law:

$$\frac{V}{T} = k_2$$

Since V/T is equal to a constant, then V_1/T_1 at an initial temperature T_1 and volume V_1 will equal the quotient V_2/T_2 at a final temperature T_2 and volume V_2.

(nandParecons tan t)
$$\frac{V_1}{T_1} = \frac{V_2}{T_2} \quad \text{(n and P are constant)}$$

The Volume-Amount Relationship. **Avogadro's law** says that at the same temperature and pressure, equal volumes of different gases contain the same number of molecules (or moles). This is another way of saying that the volume of a gas is proportional to the number of moles present.

$$V = k_3n \text{ (P and T are constant)}$$

A particularly useful quantity is *the volume occupied by one mole of ideal gas* called the **molar volume**. Since the volume of a specific amount of gas varies with the temperature and the pressure, a set of standard conditions have been defined. **Standard temperature and pressure** (abbreviated STP) *are 0°C and 1 atm pressure.* One mole of a gas at STP will occupy 22.4 L.

In addition to these gas laws, a useful relationship for changes in pressure, volume, and temperature of *a fixed number of moles* of gas is called **the combined gas law.**

$$\frac{P_1V_1}{T_1} = \frac{P_2V_2}{T_2}$$

Dalton's Law of Partial Pressures. Each component gas of a mixture of gases uniformly fills the containing vessel. Each component exerts the same pressure as it would if it occupied that volume alone. Thus, the total pressure of a mixture is the sum of the individual pressures, called the **partial pressures**, of each component:

$$P_t = P_A + P_B + P_C + \cdots$$

Also, the total number of moles n_t is the sum of the number of moles of each component:

$$n_t = n_A + n_B + n_C + \cdots$$

The ratio n_A/n_t, the ratio of the number of moles of component A to the total number of moles of all components, is called the mole fraction of component A. The symbol X_A is used for the mole fraction of component A. According to Dalton's law, the partial pressure of a gas is proportional to its mole fraction in the mixture.

$$P_A = \frac{n_A}{n_t}P_t = X_AP_t$$

Whenever gases are collected by displacement of water, the total gas pressure is the sum of the partial pressure of the collected gas and the partial pressure of water vapor.

$$P_t = P_{gas} + P_{H_2O}$$

Consequently, the partial pressure of the collected gas is $P_{gas} = P_t - P_{H_2O}$, where the partial pressure of water depends on the temperature and corresponds to the vapor pressure of water.

EXAMPLE 5.3   Gas Law Calculation

When 2.0 L of chlorine gas (Cl_2) at STP is warmed at constant pressure to 100°C, what is the new volume?

•**Method of Solution**

The volume of a fixed amount of a gas at constant pressure is proportional to the absolute temperature according to Charles's law, where $T_1 = 273$ K and $T_2 = (100°C + 273°C)$ K $= 373$ K.

$$\frac{V_1}{T_1} = \frac{V_2}{T_2}$$

•Calculation

Rearranging gives:

$$V_2 = V_1 \times \frac{T_2}{T_1}, \text{ where } V_1, \text{ the initial volume, is 2.0 L.}$$

Substituting, we get:

$$2.0 \text{ L} \times \frac{373 \text{ K}}{273 \text{ K}} = 2.7 \text{ L}$$

•Comment

Note that we expect $V_2 > V_1$ because of the temperature increase. This provides us with a simple check of our answer. If we calculated that $V_2 < V_1$, we would know our answer was wrong.

EXAMPLE 5.4   Combined Gas Law Calculation

Given 10.0 L of neon gas at 5°C and 630 mm Hg, calculate the new volume at 400°C and 2.5 atm.

•Method of Solution

First note that both the pressure and the temperature of the gas are changed, but that the number of moles is constant. Write the ideal gas equation with all the constant terms on one side.

$$\frac{PV}{T} = nR = \text{a constant}$$

We see that PV/T is a constant. Therefore

$$\frac{P_1 V_1}{T_1} = \frac{P_2 V_2}{T_2}$$

where the subscript 2 refers to the final state, and the subscript 1 refers to the initial state.

•Calculation

Rearranging to solve for V_2 gives

$$V_2 = V_1 \times \frac{P_1}{P_2} \times \frac{T_2}{T_1}$$

where $P_1 = 630 \text{ mm Hg} \times \dfrac{1 \text{ atm}}{760 \text{ mm Hg}} = 0.829 \text{ atm}$

$$P_2 = 2.5 \text{ atm}$$
$$T_1 = (273°C + 5°C) \text{ K} = 278 \text{ K}$$
$$T_2 = (273°C + 400°C) \text{ K} = 673 \text{ K}$$

Substituting

$$V_2 = 10.0 \text{ L} \times \frac{0.829 \text{ atm}}{2.5 \text{ atm}} \times \frac{673 \text{ K}}{278 \text{ K}}$$

$$= 8.0 \text{ L}$$

•Comment

Note that both P_1 and P_2 must be expressed in the same units and both temperatures in kelvin.

EXAMPLE 5.5   Partial Pressures

When oxygen gas is collected over water at 30°C and the total pressure is 645 mm Hg:
a.   What is the partial pressure of oxygen? Given the vapor pressure of water at 30°C is 31.8 mm Hg.
b.   What are the mole fractions of oxygen and water vapor?

•**Method of Solution for (a)**

Mixtures of gases obey Dalton's law of partial pressures which says that the total pressure is the sum of the partial pressures of oxygen and water vapor.

$$P_t = P_{O_2} + P_{H_2O}$$

•**Calculation for (a)**

$$P_{O_2} = P_t - P_{H_2O} = 645 \text{ mm Hg} - 31.8 \text{ mm Hg}$$
$$P_{O_2} = 613 \text{ mm Hg}$$

•**Method of Solution for (b)**

Recall that the partial pressures of O_2 and H_2O are related to their mole fractions,

$$P_{O_2} = X_{O_2}P_t \qquad P_{H_2O} = X_{H_2O}P_t$$

•**Calculation for (b)**

Therefore, on rearranging, we get:

$$X_{O_2} = \frac{P_{O_2}}{P_t} \qquad X_{H_2O} = \frac{P_{H_2O}}{P_t}$$

$$X_{O_2} = 613/645 = 0.950, \text{ and } X_{H_2O} = 31.8/645 = 0.049$$

•**Comment**

Note also that the sum of the mole fractions is 1.0, within the number of significant figures, given:

$$X_{O_2} + X_{H_2O} = 0.950 + 0.049 = 0.999$$

EXERCISES

3.   A sample of gas has a volume of 200 cm^3 at 25°C and 700 mm Hg. If the pressure is reduced to 280 mm Hg, what volume would the gas occupy at the same temperature?

4.   Two liters of oxygen gas at –15°C are heated and the volume expands. At what temperature will the volume reach 2.31 L?

5.   A 20.0 mL sample of a gas is enclosed in a gas-tight syringe at 50.0°C. What will the volume of the gas be at the same pressure after the syringe has been immersed in ice water?

6.   If 2.00 L of oxygen at –15°C are allowed to warm to 25°C at constant pressure, what is the new volume of oxygen gas?

7. The gas pressure in an aerosol can is 1.5 atm at 25°C. What pressure would develop in the can if it were heated to 450°C?

8. A sample of gas occupies 155 mL at 21.5°C, and at 305 mm Hg. What is the pressure of the gas sample when it is placed in a 5.00×10^2 mL flask at a temperature of −10°C?

9. Hydrogen and helium are mixed in a 20.0 L flask at room temperature (20°C). The partial pressure of hydrogen is 250 mm Hg and that of helium is 75 mm Hg. How many grams of H_2 and He are present?

THE IDEAL GAS LAW (5.4 & 5.6)

STUDY OBJECTIVES

1. Apply the ideal gas equation to calculate one of its variables.
2. Determine the density of an ideal gas of known molecular mass.
3. Determine the molecular mass of an ideal gas of given density.

The Ideal Gas Equation. The properties of a gas that we are most often concerned with are pressure (P), volume (V), absolute temperature (T), and amount (number of moles, n). The experimentally observed relationship between these properties is called the **ideal gas law**:

$$PV = nRT$$

and R is the ideal gas constant, R = 0.0821 L·atm/K·mol. The dots appearing in the units are to remind us that both L and atm are in the numerator, and K and mol are in the denominator. The units of R may also be written, R = 0.0821 L·atm·K^{-1}·mol^{-1}.

The ideal gas law has great importance in the study of gases. It does not contain information that is characteristic of any particular gas. Rather, it is a generalization applicable to most gases, at pressures up to about 10 atm, and at temperatures above 0°C. An **ideal gas** is one whose behavior agrees with that predicted by the ideal gas law. A calculation using the ideal gas law is shown in Example 5.6.

Gas Density. The density of a gaseous compound is d = m/V, where m is the mass of the gas in grams. Recall the number of moles n of gas is n = m/M, where *M* is the molar mass. Substituting for n in the ideal gas equation gives

$$PV = \frac{m}{M}RT$$

Rearranging gives

$$\frac{m}{V} = \frac{PM}{RT}$$

and so the density of a gas that follows the ideal gas law is:

$$d = \frac{PM}{RT}$$

Note that the density is directly proportional to its molar mass, as we noted earlier. Note also that the density increases as gas pressure increases, and density decreases as temperature increases.

The Molar Mass of a Gas. Since the density of a gas is directly proportional to its molar mass, then a measurement of the density of a gas allows the calculation of its molar mass. By rearranging the above equation, we get

$$M = \frac{dRT}{P}$$

The molar mass of an unknown gaseous substance can be calculated, which will help in its identification.

EXAMPLE 5.6 Ideal Gas Law

What volume will be occupied by 0.833 mole of fluorine (F_2) at 645 mm Hg and 15.0°C?

•**Method of Solution**

The ideal gas law relates the volume of a gas to its temperature, pressure and number of moles.

•**Calculation**

$$V = \frac{nRT}{P}$$

First convert the pressure into atmospheres:

$$P = 645 \text{ mm Hg} \times \frac{1 \text{ atm}}{760 \text{ mm Hg}} = 0.849 \text{ atm}$$

$$= \frac{(0.833 \text{ mol})(0.0821 \text{ L} \cdot \text{atm/K} \cdot \text{mol})(288 \text{ K})}{0.849 \text{ atm}} = 23.2 \text{ L}$$

•**Comment**

Even though it takes longer to write out the units for each term in the equation, doing so is very helpful because when the units cancel and leave you with the desired units (in this case liters) you are more likely to get the right answer.

EXAMPLE 5.7 Molar Mass of a Gas

A gaseous compound has a density of 1.69 g/L at 25°C and 714 torr. What is its molar mass?

•**Method of Solution**

The density of an ideal gas is directly proportional to its molecular mass. We use the equation

$$d = \frac{PM}{RT}$$

•**Calculation**

Convert pressure into units of atmospheres.

$$P = 714 \text{ torr} \times \frac{1 \text{ atm}}{760 \text{ torr}} = 0.939 \text{ atm}$$

Rearranging and substituting, we get:

$$M = \frac{dRT}{P} = \frac{(1.69 \text{ g/L})(0.0821 \text{ L} \cdot \text{atm/K} \cdot \text{mol})(298 \text{ K})}{0.939 \text{ atm}} = 44.0 \text{ g/mol}$$

EXERCISES

10. What is the pressure in atmospheres of 1.20×10^4 moles of methane, CH_4, when stored at 22°C in a 3.00×10^3 L tank?

11. 1.75 g sample of CO_2 is contained in a 7.50×10^2 mL flask at 35°C. What is the pressure of the gas?

12. How many moles of CO_2 gas are required to fill a 5.00 L balloon to a pressure of 1.05 atm at 5.0°C?

13. How many grams of helium are required to fill a 10.0 L balloon to a pressure of 1250 torr at 25°C?

14. How many O_2 molecules occupy a 1.00 L flask at 75°C and 777 mm Hg?

15. What is the density of $H_2(g)$ at 35°C and 650 torr?

16. Which one of the following gases will have the greatest density when they are all compared at the same temperature and pressure?

    O_2 CO_2 NO_3 CF_4

17. When 2.96 g of mercuric chloride is vaporized in a 1.00 liter bulb at 680 K, the pressure is 450 mm. What is the molar mass and molecular formula of mercuric chloride?

18. Determine the molar mass of Freon-11 gas if a sample weighing 0.597 g occupies 1.00×10^2 mL at 95°C, and 1000 mm Hg.

GAS STOICHIOMETRY (5.5)

STUDY OBJECTIVE

1. Calculate the volume of any gaseous reactant consumed or of any gaseous product generated in a chemical reaction, given the mass of any species involved in the reaction.

Mole Relationships. Earlier we saw that stoichiometry involves relationships between masses and moles of the reactants and the products of a chemical reaction. The ideal gas equation furnishes a way to convert between moles and the volume of a gas. The reaction of a known number of moles of a given reactant will yield a specific volume of a gaseous product. When the balanced equation is known, the volume of the product can be calculated by the following path:

mass of a given reactant	$\rightarrow$	moles of reactant	$\rightarrow$	moles of gaseous product	$\rightarrow$	volume of the product gas at P and T

EXAMPLE 5.8   Volume of a Gaseous Product

Calculate the volume of ammonia gas, measured at 645 torr and 21°C, that is produced by the complete reaction of 25.0 g quicklime, CaO, with excess ammonium chloride, NH_4Cl, solution.

•Method of Solution

First write the balanced chemical equation.

$$CaO(s) + 2NH_4Cl(aq) \rightarrow 2NH_3(g) + CaCl_2(aq) + H_2O(l)$$

Here we must determine the number of moles of NH_3 formed in the reaction and convert this into the volume of an ideal gas. Three steps are required: (1) convert g CaO to moles CaO, (2) convert moles CaO to moles NH_3 produced, and (3) use the ideal gas equation to calculate the volume of NH_3.

$$\text{grams CaO} \xrightarrow{1} \text{moles CaO} \xrightarrow{2} \text{moles } NH_3 \xrightarrow{3} \text{volume } NH_3$$

We already know how to find conversion factors 1 and 2; therefore let's first find the number of moles of NH_3 formed .

•Calculation

Stating the problem:

$$? \text{ mol } NH_3 = 25.0 \text{ g CaO}$$

$$? \text{ mol } NH_3 = 25.0 \text{ g CaO} \times \frac{1 \text{ mol CaO}}{56.1 \text{ g CaO}} \times \frac{2 \text{ mol } NH_3}{1 \text{ mol CaO}} = 0.891 \text{ mol } NH_3$$

The volume of 0.891 mol NH_3 can be calculated from the ideal gas equation:

$$V = \frac{nRT}{P}$$

Convert pressure into units of atm before substituting into the ideal gas equation.

$$P = 645 \text{ torr} \times \frac{1 \text{ atm}}{760 \text{ torr}} = 0.849 \text{ atm}$$

$$V = \frac{0.891 \text{ mol} (0.0821 \text{ L} \cdot \text{atm} / \text{K} \cdot \text{mol})(294K)}{0.849 \text{ atm}} = 25.3 \text{ L}$$

EXERCISES

19. The discovery of oxygen resulted from the decomposition of mercury(II) oxide.

    $$2HgO \rightarrow 2Hg + O_2(g)$$

    a. What volume of oxygen will be produced by the decomposition of 25.2 grams of the oxide, if the gas is measured at STP?
    b. How many grams of mercury(II) oxide must be decomposed to yield 10.8 L of O_2 gas at 1 atm and 298 K.

20. Sodium azide decomposes according to the equation.

    $$2NaN_3(s) \rightarrow 2Na(s) + 3N_2(g)$$

    What volume of N_2 at 1.1 atm and 50.0°C will be produced by the decomposition of 5.0 g NaN_3?

21. Consider the reaction of 20.0 g calcium oxide with carbon dioxide.

    $$CaO(s) + CO_2(g) \rightarrow CaCO_3(s)$$

    If you have 5.5 L of CO_2 at 7.50 atm and 22°C, will you have enough carbon dioxide to react with all the CaO?

KINETIC MOLECULAR THEORY OF GASES (5.7)

STUDY OBJECTIVES

1. Describe the assumptions on which the kinetic molecular theory is based.
2. Calculate the root-mean-square speeds of given molecular gases.

Postulates of the Kinetic Molecular Theory. Up to this point in the chapter we have emphasized the laws of gas behavior. Theories are developed to explain observed behavior and properties, and are used to make predictions of behavior under new circumstances. The theory that has been very successful in accounting for the properties of gases is called the **kinetic molecular theory of gases**. The main assumptions that explain the behavior of an ideal gas areas follows:

1. Gases consist of molecular particles. These molecules are much smaller than the distances between them. The volume occupied by a molecule is considered negligible.
2. Gas molecules are in constant motion. They experience frequent collisions with each other and with the container walls.
3. Although energy may be transferred from one molecule to another in a collision, the total energy of all molecules remains constant.
4. The average kinetic energy of a molecule in a gas is:

$$\overline{KE} = \tfrac{1}{2}m\overline{u^2}$$

where m is the mass of the molecule and u is its speed. The quantity $\overline{u^2}$ is the mean square speed. A line drawn over a variable means an average value. The average kinetic energy is proportional to the absolute temperature:

$$\overline{KE} \propto T \qquad \text{or} \qquad \overline{KE} = CT \qquad \text{where C is a proportionality constant.}$$

5. Gas molecules behave as independent particles; all attractive and repulsive intermolecular forces are negligible.

The manner in which the kinetic theory explains many of the properties of gases and the gas laws is discussed in Section 5.7 of the text. According to the kinetic theory, gas molecules are in constant, random motion. The pressure exerted by a gas on the container walls is assumed to arise from impacts on the walls by gas molecules. During impacts, force is exerted on the wall area. The gas pressure depends on the volume, the temperature, and the amount of gas. The kinetic molecular theory explains these observations in the following ways:

1. *Boyle's law* $V = k_1/P$: As the volume of the container is increased, there are fewer impacts of gas molecules with the wall per second; thus, the gas pressure is lowered.
2. *Charles's law* $V = k_2T$: As the temperature of a gas increases, the average kinetic energy and speed of the gas molecules increase. Thus, the collision frequency and the force of impact with the wall increase. If one of the walls is a movable piston, then the volume of the container will expand until balanced by the external force.
3. *Gay-Lussac's law* $P = kT$: When a gas is heated in a container with a fixed volume, the gas molecules will impact more forcefully with the wall, thus increasing the pressure.
4. *Avogadro's law* $V = k_3n$: As more gas molecules are added to the container, the number of impacts per second with the wall increases, and so the pressure increases correspondingly. If the volume of the container is not fixed, then the volume will increase.

Speeds of Gas Molecules. Experiments show that the molecules of a gas do not all move at the same speed. Instead, the speeds are distributed over a range. The distribution, called the **Maxwell speed distribution** (Figure 5.15 in the textbook), always has the same general shape. The highest point on the curve marks the most probable speed. Some molecules move with speeds much less than the most probable speed, while others move with speeds much greater than this speed. According to Maxwell, the **root-mean-square speed, u_{rms}**, of gas molecules with molar mass M at absolute temperature T is given by the equation:

$$u_{rms} = \sqrt{\frac{3RT}{M}}$$

where R = 8.314 J/K·mol, and the molar mass M must be expressed in kg/mole in order for the units to come out in m/s. This equation reveals that gas molecules, on the average, travel at very high speeds. For instance, at 25°C the root mean square speed of He atoms is 1.36×10^3 m/s, and for N_2 molecules it is 5.15×10^2 m/s. These values correspond to 3040 miles per hour and 1150 mph, respectively. The speed of sound in a gas is about the same as the root mean square speed of the gas molecules. The speed of sound through He gas is much greater than through air.

The kinetic molecular theory predicts that the average kinetic energy of the molecules of a gas is determined *only* by the temperature. The average kinetic energy is:

$$\overline{KE} = \tfrac{1}{2}\overline{mu^2} = CT$$

and has the same value for all gases at the same temperature. This means that gas molecules with higher molar masses will have slower average speeds, while lighter molecules will have higher average speeds.

EXAMPLE 5.9   Root-Mean-Square Speed

Calculate the root-mean-square speed of gaseous argon atoms at STP.

•**Method of Solution**

The root-mean-square speed is given by the equation

$$u_{rms} = \sqrt{\frac{3RT}{M}}$$

where R = 8.314 J/K·mol. T is the absolute temperature 273 K; and M is the molar mass of Ar in kilograms, 0.0399 kg/mol. Note that standard pressure is irrelevant here because the average kinetic energy depends only on the temperature. Before substituting into the equation, we recall that 1 J = 1 kg m²/s².

•**Calculation**

$$u_{rms} = \sqrt{\frac{3(8.314 \text{ J/K} \cdot \text{mol})(273 \text{ K})}{(39.9 \times 10^{-3} \text{ kg/mol})} \times \frac{1 \text{ kg m}^2/\text{s}^2}{1 \text{ J}}}$$

$$u_{rms} = (17.1 \times 10^4 \text{ m}^2/\text{s}^2)^{1/2} = 413 \text{ m/s}$$

EXERCISES

22. Calculate the root-mean-square speed of ozone molecules (O_3) in the stratosphere where the temperature is −83°C.

23. If NO_2 molecules have a root-mean-square speed of 290 m/s. What is the temperature corresponding to this average speed?

24. If the root-mean-square speed of an O_2 molecule is 4.2×10^2 m/s at 25°C, what is the average speed of a Cl_2 molecule at the same temperature?

NONIDEAL GASES (5.8)

STUDY OBJECTIVES

1. List two molecular properties responsible for deviations from ideal gas behavior. And identify the conditions under which gases will be most likely to behave in a nonideal manner.
2. Calculate the pressure of a real gas as predicted by the van der Waals equation.

Nonideal Gas Behavior. According to the kinetic theory, molecules of a gas are quite independent of one another and do not exert attractive or repulsive forces on one another. Such molecules would obey the ideal gas law. For an ideal gas, a plot of PV/nRT versus P at constant temperature should be a horizontal line, with PV/nRT = 1 at all values of P. As Figure 5.19 of the text shows, this is true for real gases only at low pressures (less than about 10 atm).

Another way to observe nonideality of gases is to lower the temperature. Cooling a gas decreases the average kinetic energy of its molecules. All gases when cooled enough will condense to the liquid. This suggests that intermolecular forces of attraction exist between molecules of real gases. Condensation occurs when the average kinetic energy is not great enough for molecules to break away from intermolecular attractive forces. The same forces account for nonideal behavior in gases.

Van der Waals Equation. Van der Waals modified the ideal gas law to account for two molecular properties that affect gas behavior: intermolecular forces and finite molecular volume. According to van der Waals, the pressure of a real gas will be lower than that of an ideal gas because attraction to neighboring molecules tends to lessen the impact a real molecule makes with the wall. We can express this in mathematical terms:

$$P_{real} = P_{ideal} - \frac{an^2}{V^2}$$

where an^2/V is the appropriate correction. Both n and V are defined as before. The proportionality constant *a* is different for each gas. The magnitude of *a* is a measure of the strength of intermolecular attractive forces, and increases with increasing molecular mass and increasing complexity of molecular structure (Table 5.4 in the textbook).

The other correction concerns the volume of the gas. According to the assumption of the kinetic theory, molecules are volumeless points. Since every real molecule does occupy a small, though finite, volume, a certain volume within a container of volume V is excluded, or not available, to gas molecules. The effective volume is (V − nb), where *b* is a constant that represents the volume of a molecule. The term nb is a correction for the finite volume of a mole of molecules.

The van der Waals equation is written in the form of the ideal gas equation.

$$P_{ideal}V_{effective} = nRT$$

$$\left(P_{real} + \frac{an^2}{V^2}\right)(Vnb) = nRT$$

corrected	corrected
pressure	volume

EXAMPLE 5.10   Van der Waals Equation

The molar volume of isopentane, C_5H_{12}, is 1.0 L at 503 K and 30.0 atm:

a.   Does isopentane behave as an ideal gas?

b.   Given that $a = 17.0$ L^2 atm/mol^2 and $b = 0.136$ L/mol, calculate the pressure of isopentane as predicted by the van der Waals equation.

•Method of Solution

a.   According to the ideal gas equation the pressure of 1 mole of a gas at 503 K that occupies 1.0 L is:

$$P = \frac{nRT}{V} = \frac{(1.0 \text{ mol})(0.082 \text{ L} \cdot \text{atm} / \text{K} \cdot \text{mol})(503 \text{ K})}{1.0 \text{ L}}$$

P = 41.3 atm

This calculated result differs considerably from the observed pressure of 30 atm. In fact, the percent error which is the difference between the two values divided by the actual pressure is:

$$\% \text{ error} = \frac{41.3 \text{ atm} - 30.0 \text{ atm}}{30.0 \text{ atm}} \times 100\%$$

% error = 37.7%

We conclude that under these conditions C_5H_{12} behaves in a nonideal manner.

b.   In this case, write the van der Waals equation

$$\left(P_{real} + \frac{an^2}{V^2}\right)(V - nb) = nRT$$

and substitute into it, but first calculate the correction terms.

$$\frac{an^2}{V^2} = \frac{(17.0 \text{ L}^2 \cdot \text{atm/mol}^2)(1.0 \text{ mol})^2}{1.0 \text{ L}^2} = 17.0 \text{ atm}$$

nb = 1.0 mol (0.136 L/mol) = 0.136 L

nRT = (1.0 mol)(0.0821 L·atm/K·mol)(503 K) = 41.3 L atm

Now substitute using the van der Waals equation.

(P + 17.0 atm)(1.0 L – 0.136 L) = 41.3 L atm

(P + 17.0 atm)(0.864) = 41.3 atm

P + 17.0 atm = 47.8 atm

P = 30.8 atm

Thus, the pressure calculated by the van der Waals equation is much closer to the actual value of 30.0 atm. The percent error is only 2.6 percent.

EXERCISES

25. 0.50 mol CCl_4 gas is introduced into a 10.0 L flask. What fraction of the total volume of the flask is occupied by CCl_4 molecules? See table 5.4 in the text for the value of the van der Waals constant b.

26. Which gas should have the higher strength of attractive forces between its molecules: ammonia or carbon tetrachloride?

CONCEPTUAL QUESTIONS

1. Why is the density of a gas much lower than that of a solid or liqiud?

2. Refer to the figure in Example 5.1 of the gas trapped in the end of the J-tube. If water was in the tube instead of mercury what height h of water would produce the same pressure as 42 mm Hg?

3. Which has more molecules: a. 1.0 L of O_2 gas at 20°C and 2.0 atm or b. 1.0 L of SF_4 gas at 20°C and 2.0 atm? Which one has more mass?

4. How does kinetic molecular theory explain Charles's law?

PRACTICE TEST

1. A barometer reads 695 mm Hg. Calculate the pressure in units of:
   a. atm   b. torr   c. Pa   d. psi

2. The pressure of H_2 gas in a 0.50 L cylinder is 1775 psi at 70°F. What volume would the gas occupy at 1 atm and the same temperature?

3. If 30.0 L of oxygen is cooled at constant pressure from 200°C to 0°C, what would the new volume of oxygen?

4. What is the pressure of the gas trapped in the apparatus shown below when the atmospheric pressure is 0.950 atm?

open to the air

68 mm

Gas

Hg

5. A balloon has a volume of 1500 L of He at 0.925 atm and 23°C. At an altitude of 20 km the temperature is –50°C and the pressure is 0.25 bar. What is the volume of this balloon at 20 km?

6. Given that the Martian atmosphere is mostly CO_2, at a pressure of 5.5 mm Hg at a temperature of –31.4°C. What is the density of the atmosphere?

7. How many grams of chlorine (Cl_2) occupy a 0.716 L cylinder when the pressure is 10.9 atm at 30°C?

8. Fill in the blank spaces in the table.

	P	V	n	T
a.	7.25 atm	40.0 L	10.5 mol	_____
b.	451 torr	150 mL	2.50×10^{-3} mol	_____
c.	14.2 atm	12.0 L	_____	325 K
d.	152 kPa	120 mL	_____	25°C
e.	2.50 atm	_____	12.0 mol	501 K
f.	2280 torr	_____	2.00×10^2 mol	450°C
g.	_____	22.4 L	1.25 mol	301 K
h.	_____	10.0 mL	0.0625 mol	25°C

9. Calculate the volume occupied by 15.2 g of CO_2 at 0.74 atm and 24°C.

10. What is the density of uranium hexafluoride gas, UF_6, at STP?

11. When 1.48 g of mercuric chloride is vaporized in a 1.00 L bulb at 680 K, the pressure is 225 mm Hg. What is the molar mass and molecular formula of mercuric chloride vapor?

12. Determine the molecular mass of chloroform gas if a sample weighing 0.495 g is collected as a vapor (gas) in a flask of volume 127 cm^3 at 98°C. The pressure of the chloroform vapor at this temperature in the flask was determined to be 754 mm Hg.

13. A 150 mL sample of O_2 gas is collected over water at 20°C and 758 torr. What volume will the same sample of oxygen occupy at STP when it is dry?

14. A sample of nitrogen gas was bubbled through liquid water at 25°C and then collected in a volume of 750 cm^3. The total pressure of the gas, which was saturated with water vapor, was found to be 740 mm at 25°C. The vapor pressure of water at this temperature is 24 mm. How many moles of nitrogen were in the sample?

15. The volume of carbon monoxide gas (CO) collected over water at 25°C was 680 cm^3 with a total pressure of 752 mm Hg. The vapor pressure of water at 25°C is 23.8 mm Hg. Determine the partial pressure and mole fraction of CO in the container.

16. The partial pressures of N_2, O_2, and Ar in dry air are 570, 153, and 6 torr, respectively. What are the mole fractions of these three gases?

17. A mixture of 40.0 g of O_2 and 40.0 g of He has a total pressure of 0.900 atm. What are the partial pressures of O_2 and He in the mixture?

18. a. What volume of CO_2 at 1 atm and 225°C would be produced by the reaction of 12.0 g $NaHCO_3$?

    $$2NaHCO_3 + H_2SO_4 \rightarrow Na_2SO_4 + 2CO_2 + 2H_2O$$

    b. On cooling to 20°C, what volume would the CO_2 occupy?

19. How many liters of ammonia at 10 atm and 500°C can be produced by the reaction of 6.0 g of hydrogen with excess N_2?

    $$3H_2(g) + N_2(g) \rightarrow 4NO(g) + 6H_2O(l)$$

20. In the oxidation of ammonia

    $$4NH_3(g) + 5O_2(g) \rightarrow 4NO(g) + 6H_2O(l)$$

    how many liters of O_2, measured at 18°C and 1.10 atm, must be used to produce 50 liters of NO at the same condition?

21. How many liters of oxygen, measured at STP, are required for the complete combustion of 72.0 grams of hexane (C_6H_{14}), a component of gasoline?

22. If the root-mean-square speed of an N_2 molecule is 475 m/s at 25°C, what is the root-mean-square speed of a He molecule at 25°C?

23. In each of the following pairs, which gas would you expect to deviate more than the value $PV/nRT = 1$ expected for an ideal gas?
    a. N_2 or SF_6 b. He or O_2 c. CO_2 or SO_2

24. Calculate the pressure of 200 moles of NH_3 in a 10.0 L container at 500°C using (a) the ideal gas law, and (b) the Van der Waals equation.

General Problems

25. A 0.356 g sample of $XH_2(s)$ reacts with water according to the following equation:

    $$XH_2(s) + 2H_2O(l) \rightarrow X(OH)_2(s) + 2H_2(s)$$

    The hydrogen evolved is collected over water at 23°C and occupies a volume of 431 mL at 746 mm Hg total pressure. Find the number of moles of H_2 produced and the atomic mass of X.
    Vapor pressure of H_2O = 21 mm Hg.

26. A H_2 gas thermometer has a volume of 100.0 cm^3 when immersed in an ice-water bath at 0°C. When immersed in boiling liquid Cl_2, the volume of the H_2 at the same pressure was 87.2 cm^3. Determine the temperature of the boiling point of Cl_2 in K and °C.

27. a. The buoyant force on a balloon is proportional to the mass of air it displaces. What mass of air is displaced by a weather balloon at an altitude where the pressure is 212 mm Hg and the temperature is −35°C? Initially the balloon contained 10 moles of He.

    b. The lift of the balloon is the difference between its mass and that of the displaced air. What is the lift of the above balloon?

28. A certain noble gas compound contains 68.8 percent Kr and 31.2 percent F. Its density at STP is 5.44 g/L. What is the molecular formula of the compound?

ANSWERS

Exercises

1. a. 125 torr   b. 86.0 kPa
2. a. 0.849 atm   b. 96.5 kPa   c. 102 psi
3. 500 cm^3
4. 25.0°C
5. 16.9 mL
6. 2.31 L
7. 3.39 atm
8. 84.4 mL
9. 0.55 g H_2 and 0.33 g He
10. 96.9 atm
11. 1.34 atm
12. 0.230 mol
13. 2.69 g
14. 2.15×10^{22} molecules
15. 0.0677 g/L
16. CF_4
17. 279 g/mol, $HgCl_2$
18. 137 g/mol
19. a. 1.30 L   b. 191 g
20. 2.78 L
21. yes
22. 314 m/s
23. 155 K (−118°C)
24. 2.8×10^2 m/s
25. 0.0138
26. carbon tetrachloride according to van der Waals constant a.

Conceptual Questions

1. Because of the large distances between molecules in the gaseous state as compared to the liquid state.
2. Since Hg is 13.6 times more dense than water, it would take a column of water 13.6 times higher than Hg to exert the same pressure. Using the definition of 1 atm from Table 5.1 we see that 760 mm Hg = 33.9 ft of H_2O. Therefore 42 mm Hg = 1.87 ft H_2O. That's 22.4 inches H_2O.
3. Equal volumes of two gases at the same temperature and pressure contain the same number of molecules. There are more grams of SF_4 because SF_4 molecules have a higher molecular mass than that of O_2 molecules.
4. Hint. See Application to the Gas Laws in Section 5.7 of the text.

Practice Test

1. a. 0.914 atm   b. 695 torr   c. 9.26×10^4 Pa   d. 13.4 psi
2. 60 L
3. 17.3 L
4. 790 mm Hg
5. 5240 L
6. 0.016 g/L
7. 22.3 g
8. a. 336 K   b. 434 K   c. 6.39 mol   d. 7.36×10^{-3} mol
   e. 197 L   f. 3960 L   g. 1.38 atm   h. 153 atm
9. 11.4 L
10. 15.7 g/L
11. 279 g/mol, $HgCl_2$
12. 120 g/mol
13. 0.136 L
14. 0.0289 mol N_2
15. CO: 728 mm Hg; X = 0.968
16. 0.782 N_2, 0.210 O_2, 0.008 Ar
17. P_{O_2} = 0.1 atm P_{He} = 0.8 atm
18. a. 5.84 L b. 3.44 L
19. 12.6 L
20. 62.5 L
21. C_6H_{14} + 19/2$O_2 \rightarrow$ 6CO_2 + 7H_2O; 178 L O_2
22. 1360 m/s
23. a. SF_6 b. O_2 c. SO_2
24. a. 1270 atm   b. 3250 atm
25. 0.0169 mol H_2, 40.1 g/mol
26. 238 K and $-35°C$
27. a. V = 700 L, thus 290 g air.   b. 290 g air – 40 g He = 250 g lift.
38. KrF_2

Chapter Six
THERMOCHEMISTRY

- Energy Changes in Chemical Reactions (6.1 – 6.2)
- Enthalpy (6.3)
- Calorimetry (6.4)
- Standard Enthalpies of Formation and Reaction (6.5)
- Heats of Solution and Dilution (6.6)
- The First Law of Thermodynamics (6.7)

ENERGY CHANGES IN CHEMICAL REACTIONS (6.1 – 6.2)

STUDY OBJECTIVES

1. Define heat, work, and energy.
2. Distinguish between exothermic and endothermic reactions.

Thermochemistry. Chemical reactions occur with either the absorption or release of energy. When a system does work or releases heat, its energy decreases. In this chapter we are concerned with the heat absorbed or released in chemical reactions. This topic is called **thermochemistry**.

Energy can be assigned to one of two classes: kinetic or potential. **Kinetic energy** is energy that an object has because of its motion. **Thermal energy** comes under this class because it is the energy associated with the random motion of atoms and molecules. According to the kinetic-molecular theory, all matter has thermal energy, because atoms and molecules are in constant motion. They vibrate, rotate, and move from one point to another. The more energy of molecular motion the higher the temperature. **Potential energy** is energy that results from an object's position. Potential energy is stored energy, and can only be released when a process occurs that changes the position. Chemical energy is stored within molecules as a consequence of attractions between electrons and atomic nuclei. When substances undergo chemical reactions, the positions of atoms change, and energy is released or absorbed.

Endothermic and Exothermic Reactions. In the study of thermochemistry, certain terms are defined very carefully and are used in a precise way. A **thermodynamic system** is the part of the universe that is selected for study. It may be a beaker containing a mixture of reactants, a bacteria colony, or the entire Earth. The boundary of the system must be carefully defined. The **surroundings** are the part of the universe that interacts with the system. Systems can interact with the surroundings by the exchange of heat, work, and matter.

Heat is energy that is transferred because of a temperature difference between the system and the surroundings. Heat always flows from a hotter to a colder body. **Work** is energy lost or gained by a system by mechanical means, rather than by heat conduction. Systems do not "contain" work or heat. What they do contain is energy. The SI unit for energy is the joule (J). Since both work and heat are energy in transition, they also have units of joules. A non-SI unit for heat is the calorie. One calorie (cal) is the amount of heat needed to raise the temperature of 1 gram of water by 1°C from 14.5°C to 15.5°C. Heat or energy in calories can be converted to joules:

$$1 \text{ cal} = 4.184 \text{ J} \quad \text{and} \quad 1 \text{ kcal} = 4.184 \text{ kJ}$$

Chemical reactions that release heat to the surroundings are called **exothermic reactions**. Combustion reactions involving wood, coal, and hydrogen are examples of exothermic reactions. During an exothermic reaction the surroundings warm up. When heat is absorbed from the surroundings by a reaction system, the reaction is referred to as **endothermic**. The melting of ice and the boiling of water are endothermic processes. During an endothermic reaction the surroundings cool down.

The heat released during a chemical reaction comes from the decrease in chemical potential energy as reactants are converted to products. In an exothermic reaction the total chemical energy of the products is less than the total chemical energy of the reactants [Figure 6.3(a) in the text].

In an endothermic reaction the products have greater chemical energy than the reactants [Figure 6.3(b) in the text]. Here, the heat absorbed brings about an increase in the potential energy of the system. The difference in energy between the reactants and products is equal to the heat absorbed by the system.

ENTHALPY (6.3)

STUDY OBJECTIVES

1. Use the thermochemical equation to calculate the heat released or absorbed when a given amount of reactant is converted to products.
2. Relate the sign of ΔH to whether the reaction is endothermic or exothermic.

Enthalpy Changes. Most reactions that you carry out in the general chemistry laboratory occur at constant pressure because the atmospheric pressure does not change appreciably in three or four hours. The heat evolved or absorbed by a reaction carried out at a constant pressure is equal to a quantity called the enthalpy change of the system, also called the **enthalpy of reaction**, ΔH.

ΔH = heat transferred to or from a system at constant pressure

The enthalpy of reaction is the difference between the enthalpies of the products and the enthalpies of the reactants.

ΔH = enthalpy of the system after the reaction – enthalpy of the system before the reaction
ΔH = H(products) – H(reactants)

Changes in enthalpy can be positive or negative. For an endothermic reaction H(products) > H(reactants) and ΔH is positive. The heat that is added is converted into potential energy of the system. For an exothermic reaction, H(reactants) > H(products) and ΔH is negative. The loss of chemical potential energy accounts for the amount of heat released.

Enthalpy and chemical reactions:			
exothermic	H(reactants) > H(products)	$\Delta H < 0$	ΔH is negative
endothermic	H(products) > H(reactants)	$\Delta H > 0$	ΔH is positive

A **thermochemical equation** is a balanced chemical equation with a term added to show the enthalpy change, ΔH. Two examples are:

$$CaCO_3(s) \rightarrow CaO(s) + CO_2(g) \qquad \Delta H° = 178 \text{ kJ}$$
$$2SO_2(g) + O_2(g) \rightarrow 2SO_3(g) \qquad \Delta H° = -198 \text{ kJ}$$

Summary of Thermochemistry Principles

There are several points to keep in mind when interpreting thermochemical equations.

1. The enthalpy change in kilojoules applies to the reaction of the amount of substance shown in the balanced equation. As shown in the above equation, when 2 mol SO_2 react with 1 mol O_2, 198 kJ of heat will be released. The value of ΔH must always relate to the way the equation is balanced. If the above equation was divided by 2 and written:

$$SO_2(g) + \frac{1}{2}O_2(g) \rightarrow SO_3(g) \qquad \Delta H° = -99 \text{ kJ}$$

    then the enthalpy change (−198 kJ) must be divide by 2 also. Remember that the heat evolved or absorbed by a chemical reaction is an *extensive* property: that is, it depends on the amount of reactants.

2. If we write the reverse of a chemical reaction, the magnitude of ΔH is the same, but its sign changes. The following reaction absorbs heat:

$$CaCO_3(s) \rightarrow CaO(s) + CO_2(g) \qquad \Delta H° = 178 \text{ kJ}$$

    But the reverse reaction must release the same amount of heat:

$$CaO(s) + CO_2(g) \rightarrow CaCO_3(s) \qquad \Delta H° = -178 \text{ kJ}$$

3. We must always specify the physical states of all reactants and products, because they help determine the magnitude of the enthalpy change. For instance, in the combustion of methane, water is a product. The enthalpy change is different when H_2O is formed as a gas, than when H_2O is a liquid.

$$CH_4(g) + 2O_2(g) \rightarrow CO_2(g) + 2H_2O(g) \qquad \Delta H° = -802 \text{ kJ}$$
$$CH_4(g) + 2O_2(g) \rightarrow CO_2(g) + 2H_2O(l) \qquad \Delta H° = -890 \text{ kJ}$$

EXAMPLE 6.1   Thermochemical Equations

The thermochemical equation for the combustion of propane is:

$$C_3H_8(g) + 5O_2(g) \rightarrow 3CO_2(g) + 4H_2O \qquad \Delta H°_{rxn} = -2220 \text{ kJ}$$

a. How many kJ of heat are released when 0.50 mole of propane reacts?
b. How much heat is released when 88.2 g of propane reacts?

•Method of Solution

Let q = the heat absorbed or released by the reaction. The heat released by an exothermic reaction is an extensive property. This means that q depends on the amount of propane consumed. The equation indicates that 2220 kJ of heat is released per mole of propane burned. The conversion factor is:

$$\frac{-2220 \text{ kJ}}{1 \text{ mol } C_3H_8}$$

•Calculation for (a)

$$q = \text{\# mol } C_3H_8 \times \frac{\text{\# kJ}}{1 \text{ mol } C_3H_8}$$

$$q = 0.50 \text{ mol } C_3H_8 \times \frac{-2220 \text{ kJ}}{1 \text{ mol } C_3H_8}$$

$$q = -1110 \text{ kJ}$$

•Calculation for (b)

Since we know the heat of reaction per mole, we should first convert the number of grams of C_3H_8 to moles.

$$q = 88.2 \text{ g } C_3H_8 \times \frac{1 \text{ mol } C_3H_8}{44.1 \text{ g } C_3H_8} \times \frac{-2220 \text{ kJ}}{1 \text{ mol } C_3H_8}$$

$$q = -4440 \text{ kJ}$$

EXERCISES

1. Given the thermochemical equation:

$$SO_2(g) + \frac{1}{2}O_2(g) \rightarrow SO_3(g) \qquad \Delta H^\circ_{rxn} = -99.0 \text{ kJ}$$

   how much heat is evolved when 75 g SO_2 undergoes combustion?

2. How many grams of SO_2 must be burned to yield 251 kJ of heat? Given:

$$2SO_2(g) + O_2(g) \rightarrow 2SO_3(g) \qquad \Delta H^\circ_{rxn} = -198.0 \text{ kJ}$$

3. The reaction that occurs when a typical fat, glyceryl trioleate, is metabolized by the body is

$$C_{57}H_{104}O_6(s) + 80O_2(g) \rightarrow 57CO_2(g) + 52H_2O(l) \qquad \Delta H^\circ_{rxn} = -3.35 \times 10^4 \text{ kJ}.$$

   How much heat is evolved when 1 gram of this fat is completely oxidized?

4. Given $2Al(s) + \frac{3}{2}O_2(g) \rightarrow Al_2O_3(s) \qquad \Delta H^\circ = -1670 \text{ kJ}$. What is ΔH for the reaction?

$$2Al_2O_3(s) \rightarrow 4Al(s) + 3O_2(g) \qquad \Delta H^\circ = ?$$

CALORIMETRY (6.4)

STUDY OBJECTIVES

1. Calculate the heat evolved or absorbed when an object or substance undergoes a known temperature change.
2. Determine the enthalpy of reaction using data from both constant pressure and constant volume calorimeter experiments.

Heat Capacity and Specific Heat. Measurement of the heat released or absorbed by a reaction is called **calorimetry**. The device used to measure the amount of heat evolved or absorbed in a reaction is called a *calorimeter*. All calorimetry is based on measuring a temperature change (Δt) of a medium such as water. The **heat capacity** (C) of a substance is the amount of heat (q) required to raise the temperature of a given mass of substance by 1 °C.

$$C = \frac{\text{heat absorbed}}{\text{temperature rise}} = \frac{q}{\Delta t}$$

The amount of heat flowing in or out of a substance and is equal to:

$$q = C \Delta t$$

Δt is the temperature change given by $t_f - t_i$, where t_f is the final temperature and t_i is the initial temperature.

$$q = C (t_f - t_i)$$

A related term is the **specific heat** (s) which is the amount of heat required to raise the temperature of one gram of a substance by 1 °C. In terms of the specific heat (s), the heat absorbed is:

$$q = ms \Delta t$$

where m is the mass in grams of the substance absorbing the heat and the heat capacity is $C = ms$.

With this equation we can calculate the amount of heat absorbed by a given amount of water or by an object of given mass. See Example 6.2.

Constant-Volume Calorimetry. Bomb calorimeters, as shown in Figure 6.6 in the textbook, are used to measure the heat evolved in combustion reactions of foods and fuels. The stainless steel "bomb" is loaded with a small amount of a combustible substance and O_2 at 30 atm of pressure, and is immersed in a known amount of water. The heat evolved during combustion is absorbed by the water and the calorimeter assembly (bomb and contents, can, thermometer, etc.). Measurements of Δt of the water and calorimeter are related to the heat evolved by the reaction q and to ΔH.

We saw previously that the enthalpy change of a system is equal to the heat evolved at constant pressure, q.

$$\Delta H = q$$

Reactions in a "bomb" occur under constant volume conditions rather than constant pressure conditions, and so the heat released is not quite equal to ΔH:

$$\Delta H \cong q$$

However, for most reactions the difference is small and for our purpose can be neglected. For instance, for the combustion of 1 mole of pentane, the difference is only 7 kJ out of 3500 kJ. Therefore, we can often use

$$q \cong \Delta H$$

without significant error.

To measure q_{rxn}, the heat evolved by the reaction, first measure the total heat absorbed by the bomb calorimeter *and* the surrounding water. The negative sign below is necessary because the heat gained by the calorimeter and water represents heat released by the reaction system.

$$q \text{ of reaction} = - (q \text{ of bomb calorimeter} + q \text{ of water})$$
$$q_{rxn} = - (q_{bomb} + q_{water})$$

Where $q_{bomb} = C_{bomb} \Delta t$ and $q_{water} = m_{water} s \Delta t$.

Here C_{bomb} is the heat capacity of the bomb, s is the specific heat of water, and m is the mass of water surrounding the bomb.

The heat capacity of both the bomb calorimeter and the surrounding water must be known. Substituting gives:

$$q_{rxn} = -[C_{bomb} \Delta t + m_{water} s \Delta t]$$

Examples 6.3 and 6.4 illustrate the use of a bomb calorimeter.

Constant-Pressure Calorimeter. A constant-pressure calorimeter is open to the atmosphere, and the measurement of heat given off by a reaction is a direct measure of the enthalpy change.

$$q_{rxn} = \Delta H$$

Such a device is shown in Figure 6.7 (text). A reaction involving solutes occurs in the solution and liberates heat that is absorbed by the solution (which is mostly water), the stirrer, the thermometer, and the container walls. These later three parts are considered the calorimeter, and so

$$q_{rxn} = -(q_{calorimeter} + q_{water})$$

EXAMPLE 6.2   Specific Heat

a.   What is the heat capacity of a block of lead if the temperature of a 425 g block increases 2.31°C when it absorbs 492 J of heat?
b.   What is the specific heat of lead?

•Method of Solution

a. The heat capacity of the block of lead is the heat absorbed divided by the temperature rise:

$$C = \frac{q}{\Delta t} = \frac{492 \text{ J}}{(2.31°C)} = 213 \text{ J/°C}$$

b. In terms of the specific heat (s), the amount of heat absorbed when an object of mass m is heated from T_i to T_f is:

$$q = ms \, \Delta t$$

Substituting into the equation the given quantities for q, m, and Δt gives:

$$492 \text{ J} = (425 \text{ g}) \, s \, (2.31°C)$$

Rearranging to solve for the specific heat s:

$$s = \frac{492 \text{ J}}{(425 \text{ g})(2.31°C)} = 0.501 \text{ J/g·°C}$$

•Comment

Since C = ms we could have solved part b from s = C/m using our answer from part a.

EXAMPLE 6.3   Determining The Calorimeter Constant

The combustion of benzoic acid is often used as a standard source of heat for calibrating combustion bomb calorimeters. The heat of combustion of benzoic acid has been accurately determined to be 26.42 kJ/g. When 0.8000 g of benzoic acid was burned in a calorimeter containing 950 g of water, a temperature rise of 4.08°C was observed. What is the heat capacity of the bomb calorimeter (the calorimeter constant)?

•Method of Solution

The combustion of 0.8000 g of benzoic acid produces a known amount of heat.

$$q_{rxn} = 0.8000 \text{ g} \times \frac{26.42 \text{ kJ}}{1 \text{ g}} = -21.14 \text{ kJ} = -21.14 \times 10^3 \text{ J}$$

And since Δt and the amount of water are known, C_{bomb}, the heat capacity of the calorimeter, can be calculated. All the heat from the combustion reaction is absorbed by the bomb calorimeter and water.

$$q_{rxn} = -(q_{bomb} + q_{water})$$

$$q_{rxn} = -(C_{bomb} \, \Delta t + ms \, \Delta t)$$

The heat absorbed by the water was q_{water} = $ms \Delta t$, where

$$m = 950 \text{ g water}$$
$$s = 4.184 \text{ J/g·°C}$$
$$\Delta t = 4.08\text{°C}$$

•**Calculation**

$$q_{water} = 950 \text{ g } (4.184 \text{ J/g·°C})(4.08\text{°C}) = 16.20 \times 10^3 \text{ J}$$

The bomb calorimeter must have absorbed the difference between the 21.14×10^3 J released, and the 16.20×10^3 J absorbed by the water.

$$q_{rxn} = -(q_{bomb} + q_{water})$$

$$q_{bomb} = -q_{rxn} - q_{water} = -(-21.14 \times 10^3 \text{ J}) - 16.20 \times 10^3 \text{ J}$$

$$q_{bomb} = 4.94 \times 10^3 \text{ J}$$

The heat capacity of the calorimeter is

$$C_{bomb} = \frac{q_{bomb}}{\Delta t} = \frac{4.94 \times 10^3 \text{ J}}{4.08\text{°C}}$$

$$C_{bomb} = 1210 \text{ J/°C}$$

EXAMPLE 6.4　Determining the Heat of Combustion

The thermochemical equation for the combustion of pentane (C_5H_{12}) is:

$$C_5H_{12}(l) + 8O_2(g) \rightarrow 5CO_2(g) + 6H_2O(l) \qquad \Delta H° = -3509 \text{ kJ}$$

From the following information calculate the heat of combustion, $\Delta H°$, and compare your result to the value given in the above equation. 0.5521 g of C_5H_{12} was burned in the presence of excess O_2 in a bomb calorimeter. The heat capacity of the calorimeter was 1.800×10^3 J/°C, and the temperature of the calorimeter and 1.000×10^3 g of water rose from 21.22°C to 25.70°C.

•**Method of Solution**

The heat evolved by the combustion reaction is absorbed by the water and the calorimeter assembly.

$$q_{rxn} = -[q_{bomb} + q_{water}]$$

where

$$q_{rxn} = -(C_{bomb}\Delta t + ms\Delta t)$$

•**Calculation**

$$\Delta t = 25.70\text{°C} - 21.22\text{°C} = 4.48\text{°C}$$

Substituting:

$$q_{bomb} = -\frac{1.800 \times 10^3 \text{ J}}{1\text{°C}} \times 4.48\text{°C} = 8.06 \times 10^3 \text{ J}$$

$$q_{water} = 1.000 \times 10^3 \text{ g} \times \frac{4.184 \text{ J}}{\text{g·°C}} \times 4.48\text{°C} = 1.87 \times 10^3 \text{ J}$$

$q_{rxn} = -[8,060 \text{ J} + 18,700 \text{ J}]$

$q_{rxn} = -26,800 \text{ J} = -26.8 \text{ kJ}$ (three significant figures)

$\Delta H°$ refers to the reaction of 1 mole of pentane. The 26,800 J was evolved by 0.5521 g C_5H_{12}. The molar mass of pentane is 72.15 g.

$$\Delta H° = q$$

$$= \frac{26.8 \text{ kJ}}{0.5521 \text{ g}} \times \frac{72.15 \text{ g } C_5H_{12}}{1 \text{ mol}}$$

$$\Delta H° = -3500 \text{ kJ/mol} = -3.50 \times 10^3 \text{ kJ/mol}$$

•**Comment**

To three significant figures the result compares well with the value given in the thermochemical equation. Small differences will arise because the heat evolved at constant volume is not quite the same as the heat evolved at constant pressure (which is equal to ΔE). Only the heat evolved at constant pressure exactly equals ΔH, the enthalpy change.

EXERCISES

5. A piece of iron initially at a temperature of 25.2°C absorbs 16.9 kJ of heat. If its mass is 82.0 g, calculate the final temperature. Given the specific heat of iron is 0.444 J/g·°C.

6. How much heat is absorbed by 52.0 g of iron when its temperature is raised from 25°C to 275°C? The specific heat of iron is 0.444 J/g·°C.

7. A 25.0 g piece of aluminum at 4.0°C is dropped into a beaker of water. The temperature of water drops from 75.0°C to 55.0°C. What amount of heat did the aluminum absorb? The specific heat of Al is 0.902 J/g·°C.

8. Benzoic acid ($C_6H_5CO_2H$) was used to calibrate a bomb calorimeter. Its enthalpy of combustion is accurately known to be −3226.7 kJ/mol. When 1.0236 g of benzoic acid was burned in a bomb calorimeter, the temperature of the calorimeter and the 1.000×10^3 g of water surrounding it rose from 20.66°C to 24.47°C. What is the heat capacity of the calorimeter?

9. To determine the heat capacity of a bomb calorimeter, a student added 150 g of water at 50.0°C to the calorimeter. The calorimeter initially was at 20.0°C. The final temperature of the water and calorimeter was 32.0°C. What is the heat capacity of the calorimeter in J/°C?

10. Octane, C_8H_{18}, a constituent of gasoline burns according to the equation:

$$C_8H_{18}(l) + \frac{25}{2} O_2(g) \rightarrow 8CO_2(g) + 9H_2O(l)$$

0.1111 g of C_8H_{18} was burned in the presence of excess O_2 in a bomb calorimeter. The heat capacity of the calorimeter was 1.726×10^3 J/°C. The temperature of the calorimeter and 1.200×10^3 g of water rose from 21.22°C to 23.05°C. Calculate the heat of combustion per gram of octane.

11. Magnesium metal reacts with hydrochloric acid according to the following equation.

$$Mg(s) + 2HCl(aq) \rightarrow MgCl_2(aq) + H_2(g)$$

When 0.425 g of magnesium was added to 150.0 mL of 1.00 M HCl in a coffee-cup calorimeter the temperature of the solution increased from 24.5°C to 35.3°C. Given that the heat capacity of the calorimeter is 125 J/°C and that the density of the HCl solution is 1.00 g/mL, calculate the heat released per mole of magnesium.

STANDARD ENTHALPIES OF FORMATION AND REACTION (6.5)

STUDY OBJECTIVES

1. Define standard enthalpy of formation.
2. Calculate ΔH°_{rxn} for any reaction with the aid of a table of enthalpies of formation.
3. Use Hess's law to calculate ΔH° for a reaction that is the sum of several given chemical reactions.

Formation Reactions. In a formation reaction 1 mole of a compound is formed from the elements in their stable forms. For methane the formation reaction is:

$$C(s) + 2H_2(g) \rightarrow CH_4(g)$$

The **standard enthalpy of formation** ΔH°_f is the heat released or absorbed when 1 mole of a compound is formed from the elements in their standard states at a pressure of 1 atm. The enthalpy of formation of methane is –74.7 kJ. This is the enthalpy change for the reaction of carbon as graphite and hydrogen gas at 1 atm pressure that forms methane gas at 1 atm pressure at 25°C.

In some cases the enthalpies of formation can be measured directly by calorimetry. For example, both H_2O and P_4O_{10} can be formed by direct reactions in a calorimeter.

$$H_2(g) + \frac{1}{2}O_2(g) \rightarrow H_2O(l) \qquad \Delta H^{\circ}_f (H_2O(l)) = \Delta H_{comb}$$

$$P_4(white) + 5O_2(g) \rightarrow P_4O_{10}(s) \qquad \Delta H^{\circ}_f (P_4O_{10}(s)) = \Delta H_{comb}$$

The Indirect Method. In many cases, enthalpies of formation cannot be determined by a direct calorimetry experiment. For example, the formation reaction of hydrogen sulfide (H_2S) will not go cleanly in a calorimeter, and many other compounds besides H_2S are formed when sulfur and hydrogen react. In cases like these, a calculational method can used to determine ΔH°_f.

$$S(s) + H_2(g) \rightarrow H_2S(g) \qquad \Delta H^{\circ}_f (H_2S(g)) = ?$$

According to **Hess's law of heat summation** if a reaction is the sum of several reaction steps, then the enthalpy change for the overall reaction is equal to the sum of the enthalpy changes of the intermediate steps. For example, let's say that we want to determine the enthalpy change for the above reaction. The reaction enthalpies of the following reactions involving sulfur and hydrogen are known from combustion bomb calorimetry:

$$S(s) + O_2(g) \rightarrow SO_2(g) \qquad\qquad \Delta H^{\circ}_{rxn} = -296.1 \text{ kJ}$$

$$H_2(g) + \frac{1}{2}O_2(g) \rightarrow H_2O(l) \qquad\qquad \Delta H^{\circ}_{rxn} = -285.8 \text{ kJ}$$

$$H_2S(g) + \frac{3}{2}O_2(g) \rightarrow SO_2(g) + H_2O(l) \qquad \Delta H^{\circ}_{rxn} = -561.7 \text{ kJ}$$

These equations can be arranged so that their sum is the formation reaction of hydrogen sulfide.

The first one is unchanged: $S(s) + O_2(g) \rightarrow SO_2(g)$

The second one is unchanged: $H_2(g) + \frac{1}{2}O_2(g) \rightarrow H_2O(l)$

The third one is reversed: $SO_2(g) + H_2O(l) \rightarrow H_2S(g) + \frac{3}{2}O_2(g)$

Their sum is the overall equation: $S(s) + H_2(g) \rightarrow H_2S(g)$

We must reverse the sign of the enthalpy change of the third reaction before summing the enthalpy changes.

$S(s) + O_2(g) \rightarrow SO_2(g)$ $\qquad \Delta H^o_{rxn} = -296.1 \text{ kJ}$

$H_2(g) + \frac{1}{2}O_2(g) \rightarrow H_2O(l)$ $\qquad \Delta H^o_{rxn} = -285.8 \text{ kJ}$

$SO_2(g) + H_2O(l) \rightarrow H_2S(g) + \frac{3}{2}O_2(g)$ $\qquad \Delta H^o_{rxn} = 561.7 \text{ kJ}$

$S(s) + H_2(g) \rightarrow H_2S(g)$ $\qquad \Delta H^o_f = -20.2 \text{ kJ}$

Summation of the three reaction steps yields the net reaction, and summation of the $\Delta H^°$ values yields the enthalpy change of the net reaction. Example 6.5 in the text provides another example of the indirect method. The value of ΔH^o_f for a compound depends on the temperature, pressure, and physical state of the compound. Standard enthalpies of formation at 25°C are tabulated in Appendix 3 of the text.

Using Enthalpies of Formation. Rather than measuring a standard enthalpy change ΔH^o_{rxn} for every chemical reaction, chemists have devised a way to use the enthalpy of formation values of compounds involved in a reaction. Enthalpies of formation can be used to calculate the standard enthalpy change for any reaction for which ΔH^o_f is known for all reactants and products. The relationship is

$$\Delta H^o_{rxn} = \sum n \ \Delta H^o_f (\text{products}) - \sum m \ \Delta H^o_f (\text{reactants})$$

where $\sum$ (sigma) means "the sum of" and n and m are the stoichiometric coefficients. So here you simply add together all the ΔH^o_f of the products, and add together all the ΔH^o_f of the reactants, and subtract the reactant sum from the product sum. Example 6.5 illustrates the use of this equation and shows how the stoichiometric coefficients are entered into the equation.

EXAMPLE 6.5 Using Enthalpies of Formation

Using ΔH^o_f values in Appendix 3 of the textbook calculate the standard enthalpy change for the incomplete combustion of ethane (C_2H_6).

$$C_2H_6(g) + \frac{5}{2}O_2(g) \rightarrow 2CO(g) + 3H_2O(l)$$

•Method of Solution

The enthalpy change for this chemical reaction in terms of enthalpies of formation is:

$$\Delta H^o_{rxn} = \sum n \ \Delta H^o_f (\text{products}) - \sum m \ \Delta H^o_f (\text{reactants})$$

$$\Delta H^o_{rxn} = [2 \text{ mol} \times \Delta H^o_f (CO) + 3 \text{ mol} \times \Delta H^o_f (H_2O)] - [1 \text{ mol} \times \Delta H^o_f (C_2H_6) - \frac{5}{2} \text{ mol} \times \Delta H^o_f (O_2)]$$

To avoid cumbersome notation, the physical states of the reactants and products were omitted from this equation. Be careful to obtain from Appendix 2 the appropriate value of ΔH^o_f. Note that the coefficients from the chemical equation are equal to the number of moles of each substance and that all the terms involving ΔH_f^o of a *reactant* are preceded by a negative sign. Next, look up and insert values.

•**Calculation**

$$\Delta H_{rxn}^o = [2 \text{ mol}(-110.5 \text{ kJ/mol}) + 3 \text{ mol}(-285.8 \text{ kJ/mol})] - [1 \text{ mol}(-84.68 \text{ kJ}) + 5/2(0)]$$

$$= (-221.0 \text{ kJ}) + (-857.4 \text{ kJ}) - (-84.68 \text{ kJ}) = -221.0 \text{ kJ} - 857.4 \text{ kJ} + 84.68 \text{ kJ}$$

$$\Delta H_{rxn}^o = -993.7 \text{ kJ}$$

EXAMPLE 6.6   Hess's Law

The standard enthalpy change for the combustion of 1 mole of ethanol is:

$$C_2H_5OH(l) + 3O_2(g) \rightarrow 2CO_2(g) + 3H_2O(l) \qquad \Delta H_{rxn}^o = -1367 \text{ kJ}$$

What is ΔH_{rxn}^o for the following reaction in which H_2O is formed as a gas, rather than as a liquid?

$$C_2H_5OH(l) + 3O_2(g) \rightarrow 2CO_2(g) + 3H_2O(g)$$

Given the heat of vaporization of water

$$H_2O(l) \rightarrow H_2O(g) \qquad \Delta H_{vap} = 44 \text{ kJ}$$

•**Method of Solution**

We can imagine a two-step path for this reaction. In the first step, ethanol undergoes combustion to form liquid H_2O, followed by the second step in which $H_2O(l)$ is vaporized. The sum of the two steps gives the desired overall reaction, and according to Hess's law the sum of the two ΔH's gives the overall ΔH.

•**Calculation**

$C_2H_5OH(l) + 3O_2(g) \rightarrow 2CO_2(g) + 3H_2O(l)$	$\Delta H_{rxn}^o = -1367 \text{ kJ}$
$3H_2O(l) \rightarrow 3H_2O(g)$	$\Delta H_{rxn}^o = 132 \text{ kJ}$
$C_2H_5OH(l) + 3O_2(g) \rightarrow 2CO_2(g) + 3H_2O(g)$	$\Delta H_{rxn}^o = -1235 \text{ kJ}$

EXERCISES

12. Using data from Appendix 3 in the text, calculate the standard enthalpy of formation of octane, C_8H_{18}, given the following combustion reaction.

    $$2C_8H_{18}(l) + 25O_2(g) \rightarrow 16CO_2(g) + 18H_2O(l) \qquad \Delta H_{rxn}^o = -11,020 \text{ kJ}$$

13. From the following heats of combustion with fluorine, calculate the enthalpy of formation of CH_4.

$CH_4(g) + 4F_2(g) \rightarrow CF_4(g) + 4HF(g)$	$\Delta H_{rxn}^o = -1942 \text{ kJ}$
$C(graphite) + F_2(g) \rightarrow CF_4(g)$	$\Delta H_{rxn}^o = -933 \text{ kJ}$
$H_2(g) + F_2(g) \rightarrow 2HF(g)$	$\Delta H_{rxn}^o = -542 \text{ kJ}$

14. A reaction used for rocket engines is

$$N_2H_4(l) + 2H_2O_2(l) \rightarrow N_2(g) + 4H_2O(l)$$

What is the enthalpy of reaction in kJ? Given the following enthalpies of formation:

$\Delta H_f^o (N_2H_4) = 95.1$ kJ

$\Delta H_f^o (H_2O_2) = -187.8$ kJ

$\Delta H_f^o (H_2O) = -285.8$ kJ

HEATS OF SOLUTION AND DILUTION (6.6)

STUDY OBJECTIVES

1. Define the heat of solution, and the heat of dilution.
2. Describe the energy changes that occur during the solution process, and apply Hess's law to calculate the enthalpy of solution.

Heat of Solution. The *heat of solution* of a solute is the enthalpy change associated with the dissolution of a certain amount of solute in a given amount of solvent. The solution process can be represented by a thermochemical equation.

$$HCl(g) \xrightarrow{H_2O} HCl(aq) \qquad \Delta H_{soln}^o = -75.1 \text{ kJ}$$

Values of the heat of solution ΔH_{soln} of several solutes in H_2O are given in Table 6.1.

Table 6.1 Heats of solution.

Solute	ΔH_{soln}^o kJ/mol
NaOH	−44.5
Ca(OH)$_2$	−11.7
KF	+2.0
NaCl	+4.0
NH$_4$NO$_3$	+26.2

The energy changes that occur when ionic solids dissolve in water can be understood by imagining the solution process to take place in two steps. We realize that when a compound such as KF dissolves in water the K^+ and F^- ions must break away from the crystal lattice and enter the aqueous phase where they are surrounded by water molecules. First, we imagine a step in which the lattice is broken up and the ions enter the gas phase.

$$KF(s) \rightarrow K^+(g) + F^-(g)$$

The energy change for this process is called the **lattice energy**, U. All ionic crystals have a characteristic lattice energy. This is the energy required to completely separate all the ions in 1 mole of solid into ions in the gas phase. For KF(s), U = 821 kJ.

Second, the gaseous K^+ and F^- ions enter the aqueous phase, where they are hydrated. During hydration, K^+ and F^- ions are surrounded by H_2O molecules. Ions in water are attracted to the polar water molecules, and a quantity of energy is released that is called the **hydration energy**, ΔH_{hydr}. For K^+ and F^- ions, $\Delta H_{hydr} = -819$ kJ.

$$K^+(g) + F^-(g) \xrightarrow{H_2O} K^+(aq) + F^-(aq)$$

According to Hess's law, the overall energy change for the two steps will be the heat of solution.

where $\Delta H_{soln} = U + \Delta H_{hydr} = 2$ kJ.

Heat of Dilution. Dilution of concentrated laboratory reagents such as hydrochloric and sulfuric acids is accompanied by sufficient evolution of heat to make the procedure hazardous. The **heat of dilution** is *the heat released or absorbed when additional solvent is added to a solution.* If heat is released during the preparation of a certain solution, heat will also be released if that solution is diluted. It follows that when heat is absorbed during a solution process, then heat is also absorbed during dilution.

EXAMPLE 6.7 Heat of Hydration

The heat of solution of NaBr is -1.0 kJ/mol. The lattice energy of NaBr is 735 kJ/mol. Determine the heat of hydration of NaBr and write the equation for the hydration reaction.

•**Method of Solution**

Recall that the heat of solution is the sum of the lattice energy (U) and the heat of hydration (ΔH_{hydr}).

$$\Delta H_{soln} = U + \Delta H_{hydr}$$

•**Calculation**

Rearrranging so as to solve for the heat of hydration:

$$\begin{aligned} \Delta H_{hydr} &= \Delta H_{soln} - U \\ &= -1.0 \text{ kJ/mol} - 735 \text{ kJ/mol} \\ &= -736 \text{ kJ/mol} \end{aligned}$$

ΔH_{hydr} refers to the heat liberated or absorbed when the gas phase ions are dissolved in water: The equation is:

$$Na^+(g) + Br^-(g) \xrightarrow{H_2O} Na^+(aq) + Br^-(aq)$$

•Comment

For NaBr, U and ΔH_{hydr} have essentially the same values, but are opposite in sign. Therefore, when NaBr dissolves in water, the energy U required to break up the crystal lattice is "paid back" by the hydration of the Na^+ and Br^- ions. Also, NaBr does not dissolve in nonpolar solvents. Molecules of nonpolar solvents can only interact weakly with ions in the crystal lattice, and so "solvation" is not sufficient to compensate for the lattice energy.

EXERCISES

15. Given that the lattice energy of LiCl is 828 kJ/mol and the heat of hydration of Li^+ and Cl^- ions is –865 kJ/mol, calculate the heat of solution of LiCl(s).

16. Given that the heat of solution of LiF is +32 kJ/mol and that the lattice energy (U) for LiF is 1006 kJ/mol, calculate the heat of hydration (ΔH_{hydr}) of Li^+ and F^- ions.

THE FIRST LAW OF THERMODYNAMICS (6.7)

STUDY OBJECTIVES

1. Describe a state function and give several examples.
2. Calculate the work done when a gas expands.
3. Write the equation for the first law of thermodynamics and give the sign conventions for q and w.
4. Relate ΔE to ΔH.

State Functions. Internal energy (E), is an example of a **state function.** This means that the internal energy of the system depends only on the conditions (the state) of the system, and not on how it got that way. Changes in internal energy (ΔE) during a chemical change depend only on the state and identity of the products, and the state and identity of the reactants. Energy change (ΔE) is independent of the path of the reaction. Temperature, pressure, volume, and enthalpy are also state functions. On the other hand, *heat is not a state function* because the heat released or absorbed for a given change depends on how the process is carried out (the path). Heat is not a property of the system like temperature and energy are.

Work of Expansion. Energy can be transferred from one system to another by means of either heat flow or by work. Calculations involving the heat liberated by chemical reactions were performed previously in this chapter. There are many kinds of work, but in this chapter we are mainly concerned with the work done by expanding gases.

In the reaction:

$$2HgO(s) \rightarrow 2Hg(l) + O_2(g)$$

the products occupy more volume than the reactants by virtue of the fact that one mole of O_2 gas is formed. The oxygen "expanding" from an open reaction vessel must do work in pushing back the atmosphere.

The work done by expansion through a volume change (ΔV) is:

$$w = -P\,\Delta V$$

where P is the pressure opposing the expansion and ΔV is the volume change. In order for expansion to take place the pressure of the gas within the cylinder (internal pressure) must be greater than the external pressure. In this equation, P is the opposing pressure (or external pressure), not the internal pressure.

The First Law. A system contains a certain quantity of energy (E) called its internal energy. The **first law of thermodynamics** is the statement that energy is conserved during any process. Energy can be converted into many forms but the total amount of energy is constant. A useful form of the first law is :

$$\Delta E = q + w$$

where ΔE is a change in internal energy of the system, w is the work done on the system, and q is the heat absorbed by the system. A change in a system's internal energy may also be written as:

$$\Delta E = E_f - E_i$$

Here ΔE is the change in a state function (E) and depends only on the initial and final states, E_i and E_f. The change in energy of a system depends only on how much heat flows into ($+q$), or out of ($-q$) the system; and on how much energy is gained when work is done on the system ($+w$), or expended when the system does work on the surroundings ($-w$). These sign conventions are summarized below.

The Sign Conventions. Thermodynamic quantities always have two parts. The number gives the amount of change, and the sign indicates the direction of flow. Here we use the convention that the sign reflects the system's point of view, rather than the surrounding's point of view. The symbol q denotes the heat transferred to the system. *When q is positive,* it signifies that heat has been transferred to the system (endothermic), and that it contributes to an increase in the internal energy of the system. *When q is negative,* it signifies that heat has flowed out of the system (exothermic) with a corresponding decrease in internal energy. This heat is absorbed by the surroundings.

The symbol w denotes the work *done on* the system by the surroundings. *When w is positive,* it signifies that work has been done *on* the system and has contributed to an increase in the internal energy. When the volume of a system is compressed by the surroundings, work is done on the system. *When w is negative,* it signifies that work has been done *by* the system on the surroundings. If the system does work on the surroundings, its internal energy must decrease. When the volume of a system expands, the system pushes back the surroundings and work is done *by* the system. These conventions are the result of taking the system's point of view. Energy entering the system is positive. Energy leaving the system is negative.

Enthalpy. Two important conditions under which chemical reactions can be carried out are constant volume and constant pressure. The energy change is:

$$\Delta E = q + w = q - P\,\Delta V$$

where $w = -P\,\Delta V$ and is the work done on the system. Under conditions of *constant volume,* $\Delta V = 0$, and the change in energy of the system is equal to the heat absorbed or released.

$$\Delta E = q_v$$

Where the subscript "v" refers to a constant volume process.

Under conditions of *constant pressure,* when the system expands against the atmosphere:

$$\Delta E = q_p + w = q_p - P\,\Delta V$$

$$q_p = \Delta E + P\,\Delta V$$

Where the subscript "p" refers to a constant pressure process.

Since the enthalpy of reaction (ΔH) is the heat evolved by a reaction at constant pressure, then the enthalpy change ΔH is equal to q_p.

$$q_p = \Delta H$$

Therefore

$$\Delta H = \Delta E + P\,\Delta V$$

ΔH is equal to the difference between the initial and final enthalpies of the system:

$$q_p = \Delta H = H_2 - H_1$$

Figure 6.1 shows that the enthalpy change (ΔH) for the combustion of sulfur dioxide with oxygen.

$$SO_2(g) + \tfrac{1}{2}O_2(g) \rightarrow SO_3(g) \qquad \Delta H^\circ = -99.1 \text{ kJ}$$

The heat released at constant pressure (q_p) is equal to the enthalpy change of the system. The heat released by the reaction comes from the decrease in enthalpy of the system. In an exothermic reaction like this $H_1 > H_2$. In an endothermic reaction $H_2 > H_1$.

Figure 6.1. An enthalpy diagram showing the change in enthalpy of the system. Enthalpy lost by the system is transferred to the surroundings as heat.

It will be helpful if you learn to sketch enthalpy diagrams for exothermic and endothermic reactions as shown in Figures 6.4 and 6.5 in the text.

EXAMPLE 6.7   Work of Expansion

A gas, initially at a pressure of 10.0 atm and having a volume of 5.0 L, is allowed to expand at constant temperature against a constant external pressure of 4.0 atm until the new volume is 12.5 L. Calculate the work done by the gas on the surroundings.

•Method of Solution

In this problem the system does work on the surroundings as it expands, and by convention (Table 6.5 in the text) the sign is negative:

$$w = -P\Delta V = -P(V_2 - V_1)$$

where P is the pressure opposing the expansion, and ΔV is the change in volume of the system.

•Calculation

$$w = -4.0 \text{ atm } (12.5 \text{ L} - 5.0 \text{ L})$$
$$= -30 \text{ L·atm}$$

This quantity can be expressed in units of joules.

$$w = -30 \text{ L·atm} \times \frac{101.3 \text{ J}}{1 \text{ L·atm}} = -3.0 \times 10^3 \text{ J}$$

EXAMPLE 6.8   Work Done on the System

Calculate the work done on the system when 6.0 L of a gas is compressed to 1.0 L by a constant external pressure of 2.0 atm.

•Method of Solution

The work done is:

$$w = -P \Delta V = -P(V_2 - V_1)$$
$$= -2.0 \text{ atm } (1.0 \text{ L} - 6.0 \text{ L}) = + 10 \text{ L·atm}$$

As in the above example, the answer can be converted to joules.
$$w = 1.01 \times 10^3 \text{ J}$$

•Comment

Obtaining a positive value for work means, that in a compression, work is done *on* the system by the surroundings. The positive value for work means that the system gains energy.

EXAMPLE 6.9   The First Law of Thermodynamics

A gas is allowed to expand at constant temperature from a volume of 10.0 L to 20.0 L against an external pressure of 1.0 atm. If the gas also absorbs 250 J of heat from the surroundings, what are the values of q, w, and ΔE?

•Method of Solution
The work done by the system is

$$w = -P \Delta V = -P(V_2 - V_1)$$
$$= -1.0 \text{ atm } (20.0 \text{ L} - 10.0 \text{ L}) = - 10 \text{ L·atm}$$

$$w = - 10 \text{ L·atm} \times \frac{101.3 \text{ J}}{1 \text{ L·atm}} = -1.0 \times 10^3 \text{ J}$$

The amount of heat absorbed was 250 J, and so q = 250 J.

Substituting into the first law of thermodynamics gives the energy change, ΔE.

$$\Delta E = q + w$$
$$= 250 \text{ J} - 1000 \text{ J}$$
$$\Delta E = -750 \text{ J}$$

•Comment

In this example, the system did more work than the energy absorbed as heat, therefore the internal energy E decreased.

EXERCISES

17. Calculate the work done on a gas when 22.4 L of the gas is compressed to 2.24 L under a constant external pressure of 10.0 atm.

18. A system does 975 J of work on its surroundings while at the same time it absorbs 625 J of heat. What is the change in energy ΔE for the system?

19. A system does 975 J of work on the surroundings. How much heat does the system absorb at the same time if its energy charge is −350 J?

CONCEPTUAL QUESTIONS

1. Hess's law of heat summation works because enthalpy is a state function. Explain.

2. Why is the lattice energy always a positive quantity? Why is the hydration energy always negative?

3. Define energy, work, and heat. Consider a 9V "alkaline" battery. Which of these three does it contain? Name a device that uses a battery to do work. Name a device that uses a battery to generate heat.

PRACTICE TEST

1. What is a state function? Name three state functions used in thermochemistry.

2. Given the thermochemical equation:

$$SO_2(g) + \frac{1}{2}O_2(g) \rightarrow SO_3(g) \quad \Delta H^o_{rxn} = -99 \text{ kJ}$$

How much heat is liberated when
a. 0.50 mole of SO_2 reacts
b. 2 moles of SO_2 reacts

3. Given the thermochemical equation

    $$H_2 + I_2 \rightarrow 2HI \qquad \Delta H^o_{rxn} = 52 \text{ kJ}$$

    What is ΔH for the reaction

    $$HI \rightarrow \frac{1}{2}H_2 + \frac{1}{2}I_2 \qquad \Delta H^o_{rxn} = ?$$

4. Given the following reactions and their associated enthalpy changes:

    $$2H(g) \rightarrow H_2(g) \qquad \Delta H^o_{rxn} = -436 \text{ kJ}$$
    $$Br_2(g) \rightarrow 2Br(g) \qquad \Delta H^o_{rxn} = +224 \text{ kJ}$$
    $$H_2(g) + Br_2(g) \rightarrow 2HBr(g) \qquad \Delta H^o_{rxn} = -72 \text{ kJ}$$

    Calculate ΔH^o_{rxn} in kJ for the reaction:

    $$HBr(g) \rightarrow H(g) + Br(g)$$

5. Nitrogen and oxygen react according to the following thermochemical equation:

    $$N_2(g) + O_2(g) \rightarrow 2NO(g) \qquad \Delta H^o_{rxn} = 180 \text{ kJ}$$

    How many kJ of heat are absorbed when 50.0 g of N_2 reacts with excess O_2 to produce 107 g of NO?

6. How much energy is required to raise the temperature of 180 g of graphite from 25°C to 500°C? The specific heat of graphite is 0.720 J/g·°C.

7. 0.500 g of ethanol, $C_2H_5OH(l)$, was burned in a bomb calorimeter containing 2.000×10^3 g of water. The heat capacity of the bomb calorimeter was 950 J/°C, and the temperature rise was found to be 1.60°C.
    a. Write a balanced equation for the combustion of ethanol.
    b. Calculate the amount of heat transferred to the calorimeter.
    c. Calculate ΔH for the reaction as written in part (a).

8. The enthalpy of combustion of sulfur is:

    $$S(\text{rhombic}) + O_2(g) \rightarrow SO_2(g) \qquad \Delta H^o_{rxn} = -296 \text{ kJ}$$

    What is the enthalpy of formation of SO_2?

9. The combustion of methane occurs according to the equation

    $$CH_4(g) + 2O_2(g) \rightarrow CO_2(g) + 2H_2O(l) \qquad \Delta H^o_{rxn} = -890 \text{ kJ}$$

    Use enthalpies of formation for CO_2 and H_2O to determine the enthalpy of formation of methane.

10. During expansion of its volume from 1.0 L to 10.0 L against a constant external pressure of 2.0 atm, a gas absorbs 200 J of energy as heat. Calculate the change in internal energy of the gas.

11. How much work must be done on or by the system in a process in which the internal energy remains constant and 322 J of heat is transferred from the system.

ANSWERS

Exercises

1. 116 kJ
2. 162 g
3. 37.9 kJ
4. 3340 kJ
5. 498°C
6. 5.77 kJ
7. 1150 J
8. 2.91 kJ/°C
9. 941 J/°C
10. 48.2 kJ/g
11. 464 kJ/mol
12. $\Delta H_f^\circ = -250$ kJ.
13. $\Delta H_f^\circ = -75.0$ kJ
14. -862.7 kJ
15. -37 kJ/mol
16. $\Delta H_{hydr} = 1038$ kJ/mol
17. w = +202 L·atm
18. $\Delta E = -350$ J
19. q = 625 J

Conceptual Questions

1. According to Hess's law of heat summation, if a reaction is the sum of several reaction steps, then the enthalpy change for the overall reaction is equal to the sum of the enthalpy changes of the intermediate steps. The law works because enthalpy is a state function. Starting with the reactants and forming the products will give the same enthalpy change whether we go by a set of intermediate steps or by a direct path. Only the difference in initial and final states determines the change in a state function such as enthalpy.

2. The lattice energy is the energy required to completely separate one mole of a solid ionic compound into gaseous ions. This energy depends on the strength of attraction between the cations and anions. The lattice energy is a positive quantity because oppositely charged particles attract one another, therefore an input of energy is required to separate them. The hydration energy is a negative quantity because there are strong attractive forces between ions and polar water molecules that lead to ion hydration.

3. The battery contains energy which is the capacity to do work. Work is energy lost or gained by mechanical means, and heat is energy transferred because of a temperature difference. A toy truck or car can be moved (displaced) when energy in the battery is used to do work. A flashlight uses energy in the battery to produce a hot glowing filament in the bulb which transfers heat to the surroundings.

Practice Test

1. A state function is a property of a system that depends only on its present condition (state) and not on how it got that way. Energy, enthalpy, and temperature are state functions.

2. a. 5.0×10^1 kJ   b. 198 kJ

3. $\Delta H^o_{rxn} = -26$ kJ

4. $\Delta H^o_{rxn} = 366$ kJ

5. 321 kJ (q = 321 kJ)

6. 61.6 kJ

7. a. $C_2H_5OH + 3O_2 \rightarrow 2CO_2 + 2H_2O$

   b. $q = 2000(4.18)(1.6) + 950(1.6) = 14,900$ J

   c. $\Delta H^o_{rxn} = -1370$ kJ

8. $\Delta H^o_f(SO_2) = -296$ kJ

9. $\Delta H^o_f(CH_4) = -75$ kJ

10. $\Delta E = -1618$ J

11. $w = +322$ J

Chapter Seven

QUANTUM THEORY AND THE ELECTRONIC STRUCTURE OF ATOMS

ELECTROMAGNETIC RADIATION (7.1 – 7.2)

STUDY OBJECTIVES

1. Use the equations relating wavelength, frequency, and the speed of light.
2. Describe a quantum and be able to calculate the energy of a quantum emitted by an atom.
3. Explain how Einstein accounted for the frequency dependence of the photoelectric effect.

Electromagnetic Radiation. To understand some of the experiments that illuminated the behavior of electrons in atoms requires some appreciation of the nature of electromagnetic radiation. Radiation is energy that is transmitted through space in the form of waves. Electromagnetic radiation consists of an electric field component and a magnetic field component that are perpendicular to each other and to their direction of propagation (see Figure 7.3 of the text). The distance between identical points on successive waves, such as the wave crests, is called the **wavelength**, λ (lambda). All electromagnetic waves travel at the same speed through a vacuum. This speed, known as the *speed of light*, is 3.00×10^8 m/s and has the symbol c. The **frequency** of a wave is the number of waves that pass a given point per second. Its symbol is ν (nu), and the unit of frequency is cycles per second which is written as /s or s^{-1}. The SI unit for one cycle per second is called a hertz (Hz). These three quantities, c, λ, and ν, are related by the equation:

$$c = \lambda\nu$$

Figure 7.4 in the textbook shows that the wavelengths of electromagnetic waves have been observed to vary from as short as 5×10^{-4} nm for some γ rays to more than 100 m for radio waves. The visible spectrum ranges from 400 to 700 nm and is only a tiny fraction of the complete span of electromagnetic waves.

According to this equation, the frequency increases as the wavelength decreases. Therefore, radio waves with their long wavelengths have low frequencies, and γ rays with their short wavelengths have extremely high frequencies. The radio waves of an AM station broadcasting at a frequency of 1000 kilohertz (1000 kHz), or $10^6 \, s^{-1}$, emits electromagnetic waves with a wavelength of 300 m.

Quantum Theory. In 1900 Max Planck solved a long-standing riddle when he provided an explanation for the energy radiated by an object at a certain temperature. Heated bodies emit radiation usually in the infrared. However, a body will emit visible radiation if it is hot enough. According to classical physics, an object can acquire or lose any amount of energy no matter how small. But all attempts by scientists to explain the wavelengths of electromagnetic radiation emitted by heated bodies using classical theory have failed. In Planck's quantum theory, however, there is a lower limit to the smallest increment of energy (a quantum) that an atom can gain or lose. In addition, the total amount of energy gained or lost by the atom is some multiple of this increment of energy. A **quantum** is the smallest quantity of energy that can be emitted by an atom in the form of electromagnetic radiation. The energy of a quantum (E) is given by Planck's equation:

$$E = h\nu$$

where h is Planck's constant, 6.63×10^{-34} J·s (joule·seconds) and ν is the frequency of the radiation. With this equation, Planck was able to explain the entire spectrum of radiation emitted by an object at a certain temperature.

The Photoelectric Effect. In 1905 Albert Einstein used Planck's quantum hypothesis to explain the photoelectric effect. When ultraviolet light is directed onto a clean metal electrode it causes electrons to be ejected from the metal surface. If the frequency of the light is varied, the significant finding is that below a certain frequency no photoelectrons are observed. However, above this frequency, called the *threshold frequency*, the number of photoelectrons and their kinetic energy increases with increasing frequency of the radiation. Therefore, the energy absorbed from a light beam is proportional to its frequency, $E \propto \nu$. To Einstein, the significance of the light frequency was that it was a measure of the energy of particle-like entities of light that are now called photons. The energy of a photon is given by Planck's equation.

$$E_{photons} = h\nu$$

Einstein's interpretation of the existence of the threshold frequency was that a photon of light would interact with one electron in the surface of the metal. This electron is held to the metal by a force called the binding energy of the metal (BE). Below the threshold frequency, a photon does not have enough kinetic energy to overcome the binding energy and eject an electron from the metal surface, but photons associated with light above the threshold frequency do. As the frequency of light is increased, the energy of a photon increases and more energy can be transferred to the electrons. Since the binding energy is constant, then as the photon energy increases, more of its energy is available to become kinetic energy (KE) of the photoelectrons.

$$KE = h\nu - BE$$

Planck's and Einstein's use of the quantum concept demonstrated that light, in addition to having properties of a wave, has properties of particles as well. This behavior is known as the *dual nature of light*.

EXAMPLE 7.1 Wavelength and Frequency

Domestic microwave ovens generate microwaves with a frequency of 2.450 GHz. What is the wavelength of this microwave radiation?

•Method of Solution

The equation relating the wavelength to the frequency is

$$c = \lambda\nu$$

where c is the speed of light.

•Calculation

Rearranging and substituting, we get:

$$\lambda = \frac{c}{\nu} = \frac{3.0 \times 10^8 \text{ m/s}}{2.450 \text{ GHz}} \times \frac{1 \text{ GHz}}{10^9 /s}$$

$$- 0.122 \text{ m}$$

EXAMPLE 7.2   Planck's Equation

The orange light given off by a sodium vapor lamp has a wavelength of 589 nm. What is the energy of a single photon of this radiation?

•Method of Solution

The energy of a quantum is proportional to its frequency. Since wavelength is given here, then substitute c/λ into Planck's equation for ν.

$$E_{photon} = h\nu = \frac{hc}{\lambda}$$

•Calculation

Introduce values for Planck's constant and the speed of light into the equation. The units will not cancel unless the wavelength is converted from nanometers into meters.

$$E_{photon} = \frac{(6.63 \times 10^{34} \text{ J s})(3.0 \times 10^8 \text{ m/s})}{(589 \times 10^9 \text{ m})}$$

$$E_{photon} = 3.38 \times 10^{-19} \text{ J}$$

•Comment

We can compare this energy to that of a mole of substance. The energy of a mole of photons is the product of $E_{photon} \times N_A$, where N_A is Avogadro's number.

$$E = (3.38 \times 10^{-19} \text{ J})(6.02 \times 10^{23}/\text{mol})$$

$$= 203,000 \text{ J/mol} = 203 \text{ kJ/mol}$$

EXERCISES

1. What is the wavelength of electromagnetic radiation having a frequency of 5.0×10^{14} s^{-1}?

2. What is the frequency of light that has a wavelength of 750 nm?

3. What is the energy of a photon of radiation having a frequency of 6.2×10^{14} s^{-1} ?

4. The red line in the spectrum of lithium occurs at 670.8 nm. What is the energy of a photon of this light? What is the energy of 1 mole of these photons?

5. A photon has an energy of 2.05×10^{-19} J. Calculate the wavelength of the radiation.

6. The binding energy for an electron on a surface of metallic sodium is 3.9×10^{-19} J. Find the kinetic energy of the photoelectrons when sodium is illuminated with UV light of 305 nm.

BOHR'S MODEL OF THE HYDROGEN ATOM (7.3)

STUDY OBJECTIVES

1. Distinguish between a line spectrum and a continuous spectrum.
2. List the two assumptions made in Bohr's model of the hydrogen atom.
3. Calculate the wavelength of radiation emitted by a specific electron transition in the hydrogen atom.

Emission Spectra. Elements in the gaseous state that are excited by an electrical current give off electromagnetic radiation in a process called emission. By passing emitted light through a prism, an emission spectrum consisting of the wavelengths of emitted electromagnetic radiation is produced. A **continuous** spectrum contains all the wavelengths between the longest and shortest emitted by that substance. "White" light results from the emission of all wavelengths in the visible part of the spectrum.

The light emitted from excited atoms in the gaseous state produces a *line spectrum* rather than continuous spectrum. A line spectrum consists of several discrete wavelengths. The observed wavelengths are characteristic of the element and can serve to identify the presence of the element in a sample.

Emission Spectrum of Hydrogen. A portion of the line spectrum of hydrogen falls in the visible region, and is called the Balmer series. Figure 7.6(b) of the text shows that the prominent visible lines are at 656 nm, 486 nm, and 434 nm. In all, five spectral series are known for hydrogen. The Lyman series is in the ultraviolet region, and the Paschen, Brackett, and Pfund series are in the infrared region

Bohr's Model. In 1913 Niels Bohr produced his famous model of the hydrogen atom. The great achievement of this model was that it explained the source and the observed wavelengths of lines in the spectrum of the H atom. Bohr postulated a "solar system" model for the H atom in which the electron traveled in circular orbits about the proton. The basic assumptions of Bohr's model were:

1. The electron moves in a circular orbit about the nucleus. Of the infinite number of possible orbits only certain orbits with distinct radii are allowed. As long as the electron remains in an orbit, its energy remains constant.
2. The energy of the electron increases the farther its orbit lies from the nucleus. When energy is absorbed by the atom, the electron must jump to a higher energy orbit. When an electron transition occurs from a higher to a lower energy orbit, radiation is emitted by the atom.

Bohr was able to calculate the radii of the allowed orbits and their energies. The energies of the H atom are given by:

$$E_n = -R_H \left(\frac{1}{n^2} \right)$$

where R_H is the Rydberg constant, which in units of joules has the value 2.18×10^{-18} J. The minus sign in the equation simply means that the energy of the hydrogen atom is lower than that of a completely separated proton and electron for which the force of attraction is zero. It does not signify negative energy. The quantity n is called the principal quantum number, and is an integer; $n = 1, 2, 3, \cdots$. Inserting values for n into the energy equation gives the energies of all allowed orbits shown in Figure 7.11 of the text.

The radius of each electron orbit in the hydrogen atom is proportional to n^2. As n increases, the orbit radius increases rapidly. The farther the electron is from the nucleus, the higher its energy.

Bohr explained that the emission of light, when an electrical current passes through a gas, is due to the formation of excited atoms. He proposed that in the emission process an electron drops from a higher to a lower orbit. During this transition the atom emits radiation. For energy to be conserved, the energy lost as radiation must equal the difference in energy between the initial energy level E_i and the final energy level E_f. For the emission process, n_i represents the higher energy level and n_f lower level. Then the change in energy of the atom ΔE is

$$\Delta E = E_f - E_i$$

where, according to Bohr's equation, the energy change for the atom is

$$\Delta E_{atom} = R_H \left(\frac{1}{n_i^2} - \frac{1}{n_f^2} \right)$$

Bohr's stroke of genius was to equate ΔE_{atom} to the energy of a photon of emitted light.

$$E_{photon} = \Delta E_{atom}$$

Since, $E_{photon} = h\nu = \dfrac{hc}{\lambda}$

Then, $h\nu = \Delta E_{atom}$

and the frequency of light emitted should be given by

$$\nu = \frac{\Delta E_{atom}}{h}$$

The wavelengths of emitted light can be calculated from

$$\lambda = \frac{c}{\nu} = \frac{hc}{\Delta E_{atom}}$$

With this model of the hydrogen atom, Bohr was able to predict the wavelengths of all the literally hundreds of observed lines in the spectrum of hydrogen!

EXAMPLE 7.3   The Hydrogen Atom

a.   What amount of energy in joules is lost by a hydrogen atom when an electron transition from $n = 3$ to $n = 2$ occurs in the hydrogen atom?

b.   What is the wavelength of the light emitted when the electron transition $n = 3 \rightarrow n = 2$ occurs?

•**Method of Solution for (a)**

The energy of the atom is quantized and depends on the orbit of the electron. Each orbit is assigned a principal quantum number n:

$$E_n = -R_H \left(\frac{1}{n^2} \right)$$

For this transition $n_i = 3$ and $n_f = 2$:

$$\Delta E = E_f - E_i = E_2 - E_3 = \left(\frac{-R_H}{2^2} \right) - \left(\frac{-R_H}{3^2} \right)$$

•**Calculation for (a)**

Substituting for R_H gives:

$$\Delta E = E_2 - E_3 = \left(\frac{-2.18 \times 10^{-18} \text{ J}}{4} \right) - \left(\frac{-2.18 \times 10^{-18} \text{ J}}{9} \right)$$

$$\Delta E = -0.545 \times 10^{-18} \text{ J} + 0.242 \times 10^{-18} \text{ J}$$

$$\Delta E = -0.303 \times 10^{-18} \text{ J} = -3.03 \times 10^{-19} \text{ J}$$

•**Method of Solution for (b)**

The energy lost by the atom appears as a photon of radiation with its respective frequency and wavelength. Since the energy of the photon is positive, we will drop the negative sign from the calculation of the wavelength.

$$\Delta E_{atom} = E_{photon} = \frac{hc}{\lambda}$$

$$\lambda = \frac{hc}{\Delta E_{atom}}$$

•**Calculation for (b)**

$$\lambda = \frac{hc}{\Delta E_{atom}} = \frac{hc}{3.03 \times 10^{-19} \, J}$$

$$\lambda = \frac{(6.63 \times 10^{-34} \, J \cdot s)(3.00 \times 10^{8} \, ms^{-1})}{3.03 \times 10^{-19} \, J}$$

$$\lambda = 656 \times 10^{-9} \, m$$
$$= 656 \, nm$$

EXERCISES

7. Consider the following energy levels for the hydrogen atom.

   _____ n = 4
   _____ n = 3
   _____ n = 2
   _____ n = 1

   Considering only these levels:
   a. How many emission lines are possible?
   b. Which transition produces photons of the greatest energy?
   c. Which transition for the H atom produces the emission line with the longest wavelength?

8. What wavelength of radiation will be emitted when an electron in a hydrogen atom jumps from the n = 5 to the n = 1 principal energy level? Name the region of the electromagnetic spectrum corresponding to this wavelength.

9. A hydrogen emission line in the ultraviolet region of the spectrum at 95.2 nm corresponds to a transition from a higher energy level n to the n = 1 level. What is the value of n for the higher energy level?

10. The second line in the Balmer series occurs at 486.1 nm. What is the energy difference between the initial and final energy levels involved in the electron transition?

THE DUAL NATURE OF THE ELECTRON (7.4)

STUDY OBJECTIVE

1. Calculate the wavelength of a particle, given its momentum.

Matter Waves. In 1925 Louis de Broglie had the novel idea that since light had been found to have a dual nature with properties of both waves and particles, then particles of *matter* should have a dual nature and exhibit wave properties as well. He deduced that the wavelength λ of a particle depends on its mass m and velocity u:

$$\lambda = \frac{h}{mu}$$

Experimental evidence of the wave properties of small particles was provided in 1927 by diffraction experiments with electron beams. Diffraction is a phenomenon that can be explained only by wave motion. Since electrons were observed to produce diffraction patterns, then electrons must behave as waves. According to de Broglie's equation, the wavelength of a particle increases as its mass decreases, so wave properties (diffraction) are most prominent for the electron, and are the basic principle behind the electron microscope. Diffraction patterns of neutron beams have also been observed and serve as a means to learn about the structures of molecules. As the mass of a particle reaches macroscopic size (say, > 10^{-12} g) its wavelength becomes extremely short and wave properties cannot be observed.

EXAMPLE 7.4   De Broglie Wavelength

When an atom of Th-232 undergoes radioactive decay, an alpha particle which has a mass of 4.0 amu is ejected from the Th nucleus with a velocity of 1.4×10^7 m/s. What is the de Broglie wavelength of the alpha particle?

•**Method of Solution**

The de Broglie wavelength depends on the mass and velocity of the particle. The equation is

$$\lambda = \frac{h}{mu}$$

Because Planck's constant has units of J·s, and 1 J = 1 kg·m^2·s^{-2}, the mass of the alpha particle must be expressed in kg. Since 1 mole of alpha particles has a mass of 4.0 g, then the mass of one particle is:

$$\frac{?\ kg}{particle} = \frac{4.00 \times 10^{-3}\ kg/mol}{6.022 \times 10^{23}\ particles/mol} = 6.64 \times 10^{-27}\ kg/particle$$

•**Calculation**

The wavelength of this alpha particle is:

$$\lambda = \frac{h}{mu} = \frac{(6.63 \times 10^{-34}\ J \cdot s) \times \left(\dfrac{1\ kg \cdot m^2/s^2}{1\ J} \right)}{(6.64 \times 10^{-27}\ kg)(1.4 \times 10^7\ m/s)}$$

Notice that all the units cancel except meters. The wavelength is:

$$\lambda = 7.1 \times 10^{-15}\ m$$

•**Comment**

The wavelength is smaller than the diameter of the thorium nucleus, which is about 2×10^{-14} m.

EXERCISE

11. Calculate the wavelength in nanometers of a proton (mass = 1.6725×10^{-27} kg) that is moving at 10% of the speed of light.

QUANTUM MECHANICS AND QUANTUM NUMBERS (7.5 – 7.7)

STUDY OBJECTIVES

1. Describe the use of the terms probability, electron density, and orbital in the wave mechanical model of the hydrogen atom.
2. Describe the significance of the three quantum numbers that correspond to an orbital in an atom.
3. Describe the shapes of s, p, and d orbitals.
4. State the Pauli exclusion principle.

The Uncertainty Principle. The dual nature of the electron is further described by the **Heisenberg uncertainty principle** which states that it is impossible to know simultaneously both the momentum and the position of an electron with certainty. The principle can be interpreted as follows: If we want to know the energy of an electron in an atom with only a small uncertainty, then we will have a large uncertainty in knowing the position of the electron in some region of the atom. Since we know the energy levels precisely, then we can only know the probability of finding the electron in a certain region of the atom..

Quantum Mechanics. In 1926 Erwin Schrödinger incorporated all the wave and particle properties of the atom into a form of mechanics called *wave mechanics* or *quantum mechanics.* He formulated the now famous Schrödinger wave equation and obtained a set of mathematical functions called *wave functions,* ψ. For any set of coordinates within the atom, ψ has a numerical value. The square of the wave function ψ^2 is related to the probability that an electron will be found in a small region (volume) in space. Where ψ^2 is large, the probability of finding the electron is high. Where ψ^2 is small, the probability of finding the electron is low.

The term *electron density* (Figure 7.15 in the text) is used to represent the probability concept. **Electron density** refers to the probability of finding the electron in a certain region of the atom. The electron density must not be interpreted to mean that one electron occupies the whole of the space covered by the probability distribution *at one time*. Rather, the electron is considered a point charge, and the electron density refers to the probability of finding the electron in a specific volume of the atom.

As a result of the probability concept, we no longer speak of distinct Bohr orbits. In the wave mechanical model of an atom, the Bohr orbit is replaced with an *orbital* or *atomic orbital.* Orbitals are regions in an atom within which electrons have a high probability, usually 90 percent, of being found. Each atomic orbital has a characteristic shape and energy.

Atomic Orbitals. An **orbital** is a visual representation of the electron probability around the nucleus. Think of a cloud surrounding the nucleus with the density of the cloud representing the probability of finding an electron at each point (see Figure 7.18 text). Orbitals are often drawn using equal-probability contours. If we go far enough from the nucleus in a particular direction, we come to a point at which 90 percent of the time the electron, when it is in that direction, will be inside that point. Connecting all such points produces an equal-probability boundary surface such that 90 percent of the time the electron will be inside that surface no matter what direction from the nucleus we consider. This graphical representation gives the shape of the orbital.

The different types of orbitals can be recognized by their shapes. While there are more than three types of orbitals only the characteristics of s, p, and d orbitals are given in Table 7.1. The shapes of s, p, and d orbitals are shown in Figures 7.19 to 7.21 of the text.

Table 7.1 Symbols and Shapes of Orbitals

Orbital	Designation	Number of orbitals per shell	Shape of 90 percent boundary surface
s	ns	1	spherical
p	np	3	dumbbell (2 lobes)
d	nd	5	4 lobes

Quantum Numbers. Both the energy and the probability distribution are described by a set of numbers called quantum numbers. The **principal quantum number n** determines the energy of a hydrogen atom according to the formula $E_n = -R_H (1/n^2)$. This number can have any integral value, $n = 1, 2, 3, \cdots$. The larger the value of n, the greater the energy, and the farther the electron is from the nucleus on the average. When $n = 1$, the electron in the hydrogen atom is said to be in the *ground state*. All other values of n correspond to excited states. When $n = \infty$, the highest possible energy, the atom has lost the electron and exists as a H^+ ion.

The **angular momentum quantum number l** is related to the shape of the orbital. Within each principal energy level one or more orbital types will exist. Thus any one principle quantum level may have more than one value for l. For a given n, l has integral values from 0 to $(n - 1)$; $l = 1, 2, 3,\ldots, (n - 1)$. Each value of l defines a group of orbitals composing a specific sublevel or subshell.

The shape of atomic orbitals will be discussed in the next section. For now each orbital is specified by a letter designation. Orbitals for which $l = 0$ are also referred to as s orbitals. Electrons in s orbitals have zero units of angular momentum. The letters used to designate orbitals with different l values are

l	0	1	2	3	4 $\cdots$
orbital	s	p	d	f	g $\cdots$

The **magnetic quantum number m_l** describes the orientation of an orbital in space. The value of m_l may have integral values from $-l$ to $+l$, including zero.

$$m_l = -l, (-l + 1), \ldots, 0, \ldots, (l - 2), (l - 1), +l$$

The number of m_l values is very important because it designates the number of orbitals within a subshell. For example, when $l = 2$, then $m_l = -2, -1, 0, 1, 2$. We see there are five m_l values. d orbitals always occur in groups of five orbitals. These five d orbitals all have the same energy, but have five different orientations in space. They make up the d subshell. A **subshell** is a group of orbitals having the same l value within a shell. Table 7.2 shows the number of orbitals in the most common subshells.

Table 7.2 The Number of Orbitals in Various Subshells

Type of Orbital	l Value	m_l Values	Number of Orbitals
s	0	0	1
p	1	−1, 0, 1	3
d	2	−2, −1, 0, 1, 2	5
f	3	−3, −2, −1, 0, 1, 2, 3	7

The **electron spin quantum number**, m_s, designates one of two possible spin directions for electrons. The two possible values of m_s are $+1/2$ and $-1/2$. The **Pauli exclusion principle** states that no two electrons in an atom can have all four quantum numbers the same. Accordingly, this principle limits the numbers of electrons each atomic orbital can hold to no more than two electrons, one electron will have $m_s = +1/2$ and the other $m_s = -1/2$.

Let's take an electron in a particular orbital, say, a $3p_x$ orbital. The three quantum numbers associated with this orbital are $n = 3$, $l = 1$, and $m_l = -1$. Any electrons in this orbital must have these identical n, l, and m_l quantum numbers. The two sets of quantum numbers for electrons in this orbital differ in their m_s values:

electron 1: $n = 3$, $l = 1$, $m_l = -1$, $m_s = +1/2$
electron 2: $n = 3$, $l = 1$, $m_l = -1$, $m_s = -1/2$

The result of this qualitative treatment of wave mechanics gives several essential pieces of information about electrons in atoms:
1. Quantum numbers (n, l, m) and their relationships to each other.
2. Electron probability density and atomic orbitals.
3. The maximum number of electrons per orbital.
4. An electron shell is a collection of orbitals with the same value of n.
5. A subshell is a set of orbitals with the same values of n and l.

EXAMPLE 7.5   Quantum Numbers

If the principal quantum number of an electron is n = 2, what values could the following have?
a.   Its l quantum number
b.   Its m_l quantum number
c.   Its m_s quantum number

•Method of Solution

a.   Recall that the angular momentum quantum number l has values that depend on the value of the principal quantum number n. $l = 0, 1, 2$, on up to the highest value (n – 1). *Answer:* When n = 2, then $l = 0, 1$.
b.   The magnetic quantum number m_l depends on the value of l where $m_l = -l, -l +1, ..., 0, ..., l + 1, l$. *Answer:* There are two l values and so two sets of m_l values. When $l = 1$, then $m_l = -1, 0, 1$; and when $l = 0$, then $m_l = 0$.
c.   *Answer:* The m_s quantum number for a single electron could be either +1/2 or –1/2.

EXAMPLE 7.6   Quantum Numbers and Orbitals
            ~~subshell~~
List all the possible types of orbitals associated with the principal energy level n = 5.

•Method of Solution

The type of orbital is given by the angular momentum quantum number. When n = 5; $l = 0, 1, 2, 3$, and 4. *Answer:* These correspond to subshells consising of s, p, d, and f orbitals.

EXAMPLE 7.7   Quantum Numbers

For the following sets of quantum numbers for electrons, indicate which quantum numbers — n, l, m_l — could not occur and state why.
a. 3, 2, 2   b. 2, 2, 2   c. 2, 0, –1

•Method of Solution

a. When n = 3, then l = 0, 1, 2. When l = 2; then m_l = −2, −1, 0, 1, 2. *Answer:* Thus set (a) can occur.
b. When n = 2, then l = 0, 1. *Answer:* The set in (b) cannot occur because $l \neq 2$ when n = 2. This imeans that the n = 2 energy level cannot have d orbitals.
c. When n = 2, l = 0, 1. When l = 0, m_l = 0. *Answer:* The set in (c) cannot occur because $m_l \neq -1$ when l = 0.

EXAMPLE 7.8   Quantum Numbers and Orbitals

What values can m_l take for
a. a 3d orbital?
b. a 2s orbital?

•Method of Solution

The value of m_l depends only on l and so first convert each orbital designation to the corresponding value of l
a. For a d orbital l = 2; therefore, possible m_l values are −2, −1, 0, 1, and 2.
b. For an s orbital l = 0; therefore, m_l = 0.

EXERCISES

12. Which of the following sets of quantum numbers are not allowed for describing an electron in an orbital?

	n	l	m_l	m_s
a.	3	2	−3	1/2
b.	2	3	0	−1/2
c.	2	1	0	−1/2

13. Which choice is a possible set of quantum numbers for the last electron added to make up an atom of gallium (Ga) in its ground state?

	n	l	m_l	m_s
a.	4	2	0	−1/2
b.	4	1	0	1/2
c.	4	2	−2	−1/2
d.	3	1	+1	1/2
e.	3	0	0	−1/2

14. Which of the following are incorrect designations for an atomic orbital?
     a. 3f   b. 4s   c. 2d   d. 4f

15. How many orbitals in an atom can have the following designations?
     a. 2s   b. 3d   c. 4p   d. n = 3

16. For each of the following give the subshell designation, the m_l values, and the number of possible orbitals.
     a. n = 3 l = 2
     b. n = 4 l = 3
     c. n = 5 l = 1

17. For each of the following subshells give the n and l values, and the number of possible orbitals.
     a. 2s   b. 3p   c. 4f

ELECTRON CONFIGURATIONS AND THE AUFBAU PRINCIPLE (7.8–7.9)

STUDY OBJECTIVES

1. Compare the order of orbital energies in the hydrogen atom to the order in a many-electron atom.
2. Use the Aufbau principle to write the electron configuration of an element.
3. Use Hund's rule to construct orbital diagrams for the electron configurations of elements.

The Energies of Orbitals. The way electrons are distributed among the orbitals of an atom is its **electronic structure** or **electron configuration**. Each of the principal energy levels (called shells) of an atom can be divided into subshells, which are sets of orbitals. The s subshell has one s orbital, the p subshell has three p orbitals, the d subshell has five d orbitals, and so on. In addition each subshell has an electron capacity based on a maximum of two electrons per orbital. The orbitals available to a shell and the subshell capacities are summarized in Table 7.3.

Table 7.3   Allowed Numbers of Orbitals per Energy Level

Principal Quantum Number	Orbital Shape l	Orbital Name	Number of Orbitals per Subshell	Electron Capacity per Subshell
1	0	1s	1	2
2	0	2s	1	2
2	1	2p	3	6
3	0	3s	1	2
3	1	3p	3	6
3	2	3d	5	10
4	0	4s	1	2
4	1	4p	3	6
4	2	4d	5	10
4	3	4f	7	14

In a hydrogen atom all of the orbitals in a particular energy level (same n value) have the same energy. In a hydrogen atom the electron resides in the 1s orbital because it has the lowest possible energy in that orbital. The ground state electron configuration of H is $1s^1$, as shown below.

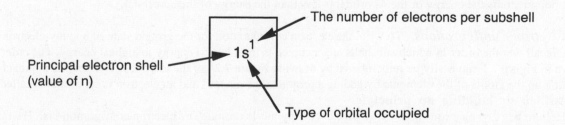

Atoms of elements other than hyrogen have more than one electron and are called **many-electron atoms**. In these atoms each principal energy level is split into several energies. These energies correspond to the orbitals

of one type having a slightly different energy than orbitals of another type. This splitting is shown in Figure 7.1. The energy of an orbital in a many-electron atom depends on both *n and l*. For a given value of n, the energy of orbitals increases as *l* increases. Therefore, for atoms other than hydrogen atoms, the 2p orbital has a higher energy than the 2s orbital.

Figure 7.1 The orbital energies in a many-electron atom. Within a shell the orbital energies increase as the *l* quantum number increases. The energy of an orbital depends on its n and *l* values. Here the 4s subshell is shown to be lower than the 3d subshell.

The order of increasing energy for orbitals within a given principal energy level is

$$s < p < d < f < \ldots$$

This trend results from the fact that s orbitals have a greater electron density near the nucleus than p orbitals, and p orbitals have greater electron density near the nucleus than d orbitals. In a many-electron atom, the energies of the orbitals increase as follows (Figure 7.1):

$$1s < 2s < 2p < 3s < 3p < 4s < 3d < 4p < 5s \cdots$$

Note also that the higher principal energy levels (n values) have more subshells, and as these split in energy some of these subshells can actually be lower in energy than subshells from a lower principal energy level. Thus the energy of a 4s orbital is very close to the energy of a 3d orbital in a many-electron atom. In potassium and calcium atoms the energy of the 4s orbital is less than the energy of the 3d orbital.

Electron Configurations. To write the electron configuration for the ground state of a many-electron atom, recall that the order in which subshells are occupied is from lowest energy to highest energy. The order shown in Figure 7.1 can easily be remembered by drawing Figure 7.24 of the textbook. The general approach of building up the atoms of the elements by adding a proton to the nucleus and an electron to an orbital is called the **Aufbau** or **building up principle**.

Helium has two electrons both of which can occupy the 1s orbital. The electron configuration is: He $1s^2$. A lithium atom has three electrons. According to the Pauli principle, we can place only two electrons in the 1s orbital; the third electron must enter the next higher orbital, the 2s orbital. Its electron configuration is: Li $1s^2 2s^1$. The electron configurations of the second period elements are:

Be $1s^2 2s^2$

With Be the 1s and 2s orbitals are filled, and so additional electrons must enter the 2p orbitals. The 2p subshell is being filled over the next six elements.

$$B \quad 1s^2 2s^2 2p^1$$
$$C \quad 1s^2 2s^2 2p^2$$
$$N \quad 1s^2 2s^2 2p^3$$
$$O \quad 1s^2 2s^2 2p^4$$
$$F \quad 1s^2 2s^2 2p^5$$
$$Ne \quad 1s^2 2s^2 2p^6$$

Orbital Diagrams. The above electron configurations tend to hide information about the number of electrons in any one outer orbital. The 2p subshell consists of three 2p orbitals: $2p_x$, $2p_y$, and $2p_z$. Therefore the capacity of the subshell is 6 electrons. If we use an orbital diagram, a choice arises about where to place the electrons. In carbon, for example, are the 2 electrons in the p subshell both in the $2p_x$ orbital, or are they distributed so that the $2p_x$ and the $2p_y$ each have one electron? To make the choice first we need to describe an orbital diagram.

An orbital diagram uses boxes to designate individual orbitals, and groups of boxes to designate subshells.

An arrow pointing up ↑ stands for an electron spinning in one direction, and an arrow pointing down ↓ stands for an electron spinning in the opposite direction.

The orbital diagrams for carbon and nitrogen are drawn as follows. Electrons are placed into orbitals according to **Hund's rule**, which states that electrons entering a subshell containing more than one orbital will have the most stable arrangement when the electrons occupy the orbitals singly, rather than in pairs. Thus carbon has one electron in the $2p_x$ and one in the $2p_y$, rather than two electrons in the $2p_x$.

According to Hund's rule, nitrogen atoms have the electrons in the 2p subshell distributed with one each in the $2p_x$, $2p_y$, and $2p_z$.

The orbital diagram indicates the number of unpaired electrons in an atom. The presence, or absence, of unpaired electrons is detected experimentally by the behavior of a element when placed in a magnetic field. **Paramagnetism** is the property of attraction to a magnetic field. **Diamagnetism** is the property of repulsion by a magnetic field. The atoms of paramagnetic substances contain unpaired electrons. Those of diamagnetic substance contain only paired electrons. Using the three principles that have been given (Aufbau, Pauli, and Hund's rule) the electron configurations of the ground states of most atoms can be estimated.

Noble Gas Core Abbreviations. The electron configurations of the noble gas elements can be used as abbreviations when writing electron configurations. Neon has the configuration Ne $1s^2 2s^2 2p^6$. All elements beyond neon have this configuration for the first 10 electrons. We use the symbol [Ne] to represent the configuration of the 10 electrons in neon and call it a neon core. Similarly 18 electrons arranged in the ground state of an argon atom ($1s^2 2s^2 2p^6 3s^2 3p^6$) are called an argon core, written [Ar]. [Kr] and [Xe] cores can also be used. In practice we select the noble gas that most nearly precedes the element being considered. Therefore, calcium and strontium could be written in full or abbreviated as follows:

$$Ca \quad 1s^2 2s^2 2p^6 3s^2 3p^6 4s^2 \qquad or \qquad [Ar]4s^2$$
$$Sr \quad 1s^2 2s^2 2p^6 3s^2 3p^6 4s^2 3d^{10} 4p^6 5s^2 \quad or \quad [Kr]5s^2$$

EXAMPLE 7.9   Electron Configurations

Which of the following electron configurations would correspond to ground states and which to excited states?
a. $1s^2 2s^2 2p^1$ b. $1s^2 2p^1$ c. $1s^2 2s^2 2p^1 3s^1$

•Method of Solution

a. In the above configuration electrons occupy the lowest possible energy states. *Answer:* It corresponds to a ground state.
b. In this configuration electrons do not occupy the lowest possible energy states. The 2s orbital which lies lower than the 2p is vacant, and the last electron is in the 2p orbital. *Answer:* This is an excited state.
c. In this case the last electron is in the 3s orbital while the lower energy 2p subshell is not completely filled. *Answer:* This is an excited state.

EXAMPLE 7.10   Electron Configurations

Write the electron configuration for a potassium atom.

•Method of Solution

Potassium has 19 electrons. The first 10 of these will occupy the same orbitals as neon which is given above: $1s^2 2s^2 2p^6$.

This leaves 9 electrons to account for. Keeping in mind the order of increasing orbital energies, the next 2 electrons fill the 3s orbital. Then 6 electrons can be placed into the 3p subshell. This leaves 1 electron. According to the order of orbital energies, the 3d does not fill next. Rather, the 4s orbital is lower in energy than the 3d. The last electron in potassium enters the 4s orbital. The electron configuration of potassium is:

　　K $1s^2 2s^2 2p^6 3s^2 3p^6 4s^1$

or in terms of the argon core abbreviation:

　　K $[Ar]4s^1$

EXAMPLE 7.11   Orbital Diagrams

a. Write the electron configuration for arsenic.
b. Draw its orbital diagram.
c. Are As atoms diamagnetic or paramagnetic?

•Method of Solution

a. From the atomic number of As we see there are 33 electrons per arsenic atom. From Figure 7.23 in the text, the order of filling orbitals is 1s, 2s, 2p, 3s, 3p, 4s, etc. Placing electrons in the lowest energy orbitals until they are filled, we find that the first 18 electrons are arranged $1s^2 2s^2 2p^6 3s^2 3p^6$, which corresponds to an Ar core. The next two enter the 4s, and the next ten enter the 3d. This leaves 3 electrons for the 4p subshell. *Answer:* The electron configuration of As is:

　　As $[Ar]4s^2 3d^{10} 4p^3$

b. All of the orbitals are filled except for the 4p orbitals. The electrons must be placed into the 4p orbitals in accordance with Hund's rule. The orbital diagram for the outer orbitals is:

As   4s   3d   4p

c. *Answer:* Arsenic atoms are paramagnetic because they contain 3 unpaired electrons.

EXERCISES

18. What element has atoms with the electron configuration $[Xe]6s^2 4f^{14} 5d^{10} 6p^2$?

19. What third period element has atoms in the ground state with three unpaired electrons?

20. Write the electron configuration for the following atoms: Sb, V, Pb.

21. How many unpaired electrons do oxygen atoms have?

CONCEPTUAL QUESTIONS

1. Briefly describe Bohr's theory of the hydrogen atom and discuss how the theory explains the production of an emission line spectrum.

2. Which of the following are observable.
    a. the position of an electron in an H atom.
    b. the wavelength of light emitted by H atoms.
    c. the nucleus of an atom.
    d. an s orbital

3. Is the electron density distributed evenly inside an s orbital?

4. In many-electron atoms the 2s orbital has a lower energy than the 2p. Explain.

PRACTICE TEST

1. A certain AM radio station broadcasts at a frequency of 600 kHz. What is the wavelength of these radio waves in meters?

2. How long would it take a radio wave of frequency 5.5×10^5 Hz to travel from the planet Venus to Earth (28 million miles)?

3. Calculate the energy required to remove an electron from a hydrogen atom in the ground state.

    $n_i = 1 \qquad E_f = 0 \Rightarrow n_f = \infty$

4. Which of the following has the greatest ionization energy?
    a. H   b. He$^+$ c. Li^{2+}

    $\llcorner n_f = \infty \rightarrow \Delta E = 0$

    $n_i =$ whatever "n" value

5. The average kinetic energy of a neutron at 25°C is 6.2×10^{-21} J. What is its de Broglie wavelength? The mass of a neutron is 1.008 amu. Hint: kinetic energy $= (1/2)mu^2 = (mu)^2/2m$.

6. Deduce the possible sets of four quantum numbers for an electron that arise when $n = 2$.

7. What is the maximum number of electrons that can occupy an $n = 3$ energy level? The $n = 4$ energy level?

8. Which of the following subshells has a capacity of 10 electrons?
   a. 5s   b. 2p   c. 4p   d. 3d   e. 6s

9. How many orbitals are occupied by one or more electrons in a germanium atom?

10. Complete the sentence: An electron with $l = 2$ must
    a. have $m_l = -2$.
    b. be in an $n = 3$ energy level
    c. be in a p orbital
    d. be in a d orbital

11. Write quantum numbers for:
    a. an electron in a 2s orbital
    b. the outermost electrons in Ge

12. Write the electron configuration for the following atoms: Ar, Se, Ag.

13. How many electrons are represented by the abbreviation [Kr]?

14. Which of the following is the correct orbital diagram for chromium?

15. Which of the following atoms has the greatest number of unpaired electrons?
    a. Ti   b. Ag   c. O   d. P   e. K

General Problems

16. If the energy required to ionize 1 mole of H atoms were used to raise the temperature of water, what mass of water could have its temperature increased by 50°C?

17. If 5 percent of the energy supplied to an incandescent light bulb is radiated as visible light, how many "visible" photons per second are emitted by a 100 watt bulb? Assume the wavelength of all visible light to be 560 nm. Given: 1 watt = 1 J/s.

18. Silver bromide (AgBr) is the light-sensitive compound in most photographic films. Assume, when film is exposed, that the light energy absorbed dissociates the molecule into atoms. (The actual process is more complex.) If the energy of dissociation of AgBr is 100 kJ/mol, find the wavelength of light that is just able to dissociate AgBr.

ANSWERS

Exercises

1. 600 nm
2. 4.0×10^{14} Hz
3. 4.11×10^{-19} J
4. 2.97×10^{-19} J/photon, 179 kJ/mol
5. 9.70×10^{-7} m
6. 2.62×10^{-19} J
7. a. 6   b. n = 4 to n = 1   c. n = 4 to n = 3
8. 95.2 nm, ultraviolet
9. n = 5
10. 4.09×10^{-19} J
11. 1.32×10^{-14} m
12. a, b
13. b
14. a, c
15. a. 1   b. 5   c. 3   d. 9
16. a. 3d, m_l = –2, –1, 0, 1, 2.   5 orbitals
    b. 4f, m_l = –3, –2, –1, 0, 1, 2, 3.   7 orbitals
    c. 5p, m_l = –1, 0, 1.   3 orbitals
17. a. n = 2   l = 0   b. n = 3   l = 1   c. n = 4   l = 3
18. Pb
19. P
20. Sb $[Kr]5s^2 4d^{10} 5p^3$;   V $[Ar]4s^2 3d^3$;   Pb $[Xe]6s^2 4f^{14} 5d^{10} 6p^2$
21. 2

Conceptual Questions

1. Essentially each orbit in the H atom has a certain energy associated with it. The emission of radiation by an excited H atom is explained by the electron dropping from a higher to a lower energy orbit. Since only a few orbits are permitted, then only a few transitions are possible. Each transition produces one line in the spectrum. With the number of transitions restricted, one observes a line spectrum rather than a continuous spectrum. See Section 7.3 of the text for details.
2. Only b. The wavelength of light emitted by H atoms can be observed.
3. No, in the 1s orbital the electron density is more dense near the nucleus. See Figure 7.17 of the text.
4. In the hydrogen atom the 2s and 2p orbitals have the same energy. The energy of an electron in a many-electron atom depends not only on its principle quantum number (n) but also on its angular momentum quantum number (*l*). Electrons are assigned to subshells in order of increasing "n + *l*" value. For the 2s subshell n + *l* = 2, while for the 2p subshell n + *l* = 3. If two subshells have the same value of n + *l*, then electrons are assigned to the one with the lower n value first.

Practice Test

1. 500 m

2. 150 s

3. 2.18×10^{-18} J

4. Li^{2+}

5. $\lambda = 95.2$ nm; ultraviolet

6.

n	l	m_l	m_s
2	0	0	$-1/2$
2	0	0	$+1/2$
2	1	-1	$-1/2$
2	1	-1	$+1/2$
2	1	0	$-1/2$
2	1	0	$+1/2$
2	1	$+1$	$-1/2$
2	1	$+1$	$+1/2$

7. 18

8. d. 3d

9. 16

10. d.

11. a. $n = 2, l = 0, m_l = 0, m_s = 1/2$

   b. $n = 4, l = 0, m_l = 0, m_s = 1/2$

   $n = 4, l = 0, m_l = 0, m_s = -1/2$

   $n = 4, l = 1, m_l = -1, m_s = 1/2$

   $n = 4, l = 1, m_l = 0, m_s = 1/2$

12. Ar $1s^2 2s^2 2p^6 3s^2 3p^6$

   Se $[Ar]4s^2 3d^{10} 4p^4$

   Ag $[Kr]5s^1 4d^{10}$

13. 36

14. d.

15. d. P

16. 6280

17. 1.4×10^{19} photons/s

18. 1200 nm

Chapter Eight
PERIODIC RELATIONSHIPS AMONG THE ELEMENTS

- Periodic Classification of the Elements (8.1 – 8.2)
- Periodic Variation in Physical Properties (8.3)
- Ionization Energy and Electron Affinity (8.4 – 8.5)
- Types of Elements (8.6)

PERIODIC CLASSIFICATION OF THE ELEMENTS (8.1 – 8.2)

STUDY OBJECTIVES

1. Discuss the basis for the arrangement of elements in the periodic table.
2. Write the electron configuration of an element and determine to what group of the periodic table it belongs.
3. Give the locations in the periodic table of the representative, transition, and inner transition elements.
4. Write the electron configurations of cations and anions.

The Periodic Table. A modern version of the periodic table is shown inside the cover of the text. The table organizes a vast array of information about the chemical and physical properties of the elements. In the periodic table the elements are arranged by atomic number. The horizontal rows of elements with atomic number increasing by one from left to right are called **periods**. The vertical columns with atomic number and atomic mass increasing from top to bottom are called **groups** or **families.** Elements within a group have similar, but not identical, properties. Table 8.1 shows how properties of the halogens vary gradually within a group.

Table 8.1 Properties of the Halogens

Group 7A Element	Melting Point (°C)	Boiling Point (°C)	Formula of Sodium Salt	Reactivity Toward H_2
F_2	–223	–188	NaF	Explosive
Cl_2	–101	– 34	NaCl	Initiated by light at 25°C
Br_2	–7	+ 59	NaBr	Slow at 25°C
I_2	+114	+187	NaI	Requires high temperatures

Among their similar chemical properties are the following: the free elements are all diatomic molecules, they all react with sodium, for instance, and form salts that have the same formula, that is, NaX, and they all react with hydrogen to form compounds with the formula HX. The ease of reaction with hydrogen, and the physical properties of melting point and boiling point all vary in a regular way as you read down in the group.

Element groups are often referred to collectively by their group number: Group 1A, 2A, and so on. Some groups have acquired special names which are given below:

Group 1A elements are called **alkali metals.**
Group 2A elements are called **alkaline earth metals.**
Group 7A elements are called **halogens.**
Group 8A elements are called **noble gases.**

Element Groups and Electron Configurations. The modern theory of atomic structure suggests an explanation as to why elements in a group have similar chemical properties. First let's examine the electron configurations of elements in any vertical column in the periodic table. Table 8.2 shows the electron configurations of atoms of the Group 7A elements.

Table 8.2 Group 7A
Electron Configurations

F	$1s^2 2s^2 2p^5$
Cl	$[Ne]3s^2 3p^5$
Br	$[Ar]4s^2 3d^{10} 4p^5$
I	$[Kr]5s^2 4d^{10} 5p^5$

The table shows that the electron configurations of the outermost principal energy levels of all Group 7A elements contain seven electrons arranged $ns^2 np^5$, where n is the principal quantum number of the outermost energy level. Furthermore, the halogen group is not unique in this regard. The electron configuration of the outermost electrons of elements in any group is the same. See Figure 8.2 in the text.

The chemical similarities of the Group 7A elements can be explained quite simply by their electron configurations. *All the elements in a group have the same electron configuration in their outermost principal energy level.* The electrons in the outermost energy level are called **valence electrons.** The outer electrons are the ones involved in chemical bonding and so deserve a special name.

Representative Elements. The modern periodic table is arranged according to the type of subshells being filled with electrons (Figure 8.2 textbook). Groups 1A through 7A of the periodic table include elements that have an incompletely filled *set* of s or p orbitals. These are called **representative elements**, or **main group** elements. The groups are given numerals from one to seven, followed by the letter A, which stands for a group of representative elements. The **noble gases** have completely filled outer ns and np subshells, $ns^2 np^6$, except for helium.

Electron Configurations. The chemical properties of an atom are related to the configuration of the atom's valence electrons. For representative elements the group number gives the number of electrons in the outermost principal energy level (valence electrons), as shown in Table 8.3.

Table 8.3 Valence Electron Configurations of Representative Elements

Group	1A	2A	3A	4A	5A	6A	7A	8A
Configuration	ns^1	ns^2	$ns^2 np^1$	$ns^2 np^2$	$ns^2 np^3$	$ns^2 np^4$	$ns^2 np^5$	$ns^2 np^6$

Transition Elements. The elements in the center of the periodic table, whose groups are labeled B, are the **transition metals.** Transition metals have incompletely filled d subshells or readily form cations having incompletely filled d subshells. In the fourth period, either the atoms or the ions of the elements Sc through Cu have incompletely filled 3d subshells. In the fifth period, the atoms or the ions of the elements Y to Cd have incompletely filled 4d subshells. Because their d subshells are completed, the elements Zn, Cd, and Hg (Group 2B) are often not considered true transition elements.

Below the main part of the periodic table, there are two sets of 14 elements each depicted as horizontal rows. These groups are generally known as **inner transition** elements, but are also referred to as **lanthanides** (atomic numbers 58–71) and **actinides** (atomic numbers 90–103). For these elements an inner f subshell is incompletely filled. Figure 8.1 shows the positions in the periodic table of the groups of elements discussed above.

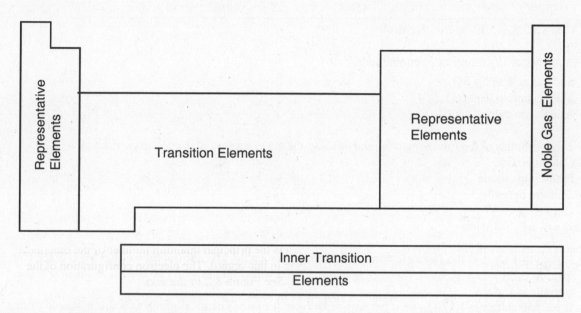

Figure 8.1 General groups of elements.

Electron Configurations of Cations and Anions. The ions of the representative elements have noble gas electron configurations, ns^2np^6. The representative metals form positive ions by losing one or more electrons until a noble gas core is achieved.

$$Na \ [Ne]3s^1 \rightarrow Na^+ \ [Ne] + e^-$$
$$Mg \ [Ne]3s^2 \rightarrow Mg^{2+} \ [Ne] + 2e^-$$
$$Al \ [Ne]3s^23p^1 \rightarrow Al^{3+} \ [Ne] + 3e^-$$

The electron configurations of the above ions (Na^+, Mg^{2+}, and Al^{3+}) are all the same, namely $1s^22s^22p^6$ or [Ne]. All three have 8 electrons in the outer shell ($n = 2$), and 10 electrons all together. Atoms and ions that have the *same number* of electrons and hence the *same electron configurations* are called **isoelectronic** species. All three ions are isoelectronic with the noble gas neon. *Ions of the representative metals have electron configurations that are isoelectronic with the previous noble gas atom.*

The nonmetal atoms form negative ions by acquiring electrons until a noble gas core is achieved.

$$S \ [Ne]3s^23p^4 + 2e^- \rightarrow S^{2-} \ 3s^23p^6 \ \text{ or } \ [Ar]$$
$$Cl \ [Ne]3s^23p^5 + e^- \rightarrow Cl^- \ 3s^23p^6 \ \text{ or } \ [Ar]$$

Sulfide and chloride ions then have net charges of –2 and –1 because the atoms needed $2e^-$ and $1e^-$, respectively, to complete an octet. *Negative ions of the representative nonmetals have electron configurations that are isoelectronic with the next noble gas atom.*

Transition Metal Cations. When forming a cation from an atom of a transition metal, recall that *most* of the transition metal cations do not have noble gas electron configurations. Atoms of transition metal elements generally have 2 electrons in the s orbital of the highest principle energy level n, and a partially filled d subshell of the n − 1 principle energy level (see Table 7.3 of the text). The electrons most easily lost are those in the outermost principle energy level, the ns. Indeed many of the transition metals form +2 ions. Loss of additional electrons from the (n − 1)d subshell yields ions with charges greater than a +2 charge. See Example 8.5, parts c and d.

EXAMPLE 8.1   Element Groups

Given the following formulas of compounds:

Sodium sulfate Na_2SO_4
Magnesium oxide MgO
Aluminum chloride $AlCl_3$

Write the formulas of the following compounds using the the given formulas and the periodic table:
a.   Calcium oxide
b.   Potassium sulfate
c.   Gallium bromide

•Method of Solution

a.   Notice that Ca is in the same element group as Mg. Among the given formulas we find MgO Since one Mg ion combines with one oxide ion, then one Ca ion will combine with one oxide ion. *Answer:* The formula is CaO.
b.   If sodium sulfate is Na_2SO_4, then potassium, being in the same chemical group as sodium, should form a sulfate with the same ratio of potassium atoms to sulfate ions. *Answer:* The formula is K_2SO_4.
c.   Gallium is in Group 3A along with Al, and bromine is a halogen, as is Cl. Therefore, since aluminum chloride has the formula $AlCl_3$, gallium will combine with any three halogen atoms. *Answer:* The formula of gallium bromide is $GaBr_3$.

EXAMPLE 8.2   Electron Configurations and the Periodic Table

Referring only to the periodic table:
a.   Write the electron configuration for the outermost principal energy level (the valence shell) for any Group 5A element.
b.   Identify the group whose outermost electrons have the configuration ns^2np^1.

•Method of Solution

a.   The elements in Group 5A are representative elements and so have all inner subshells filled, and the outermost electrons occupy s and p orbitals. The numeral 5 corresponds to five outermost electrons. The first two occupy the ns subshell, and the next three the np subshell. *Answer:* The valence electron configuration of any Group 5A element is ns^2np^3.
b.   Among the representative elements, those in the first group to the right of the transition metals, Group 3A, have the electron configuration ns^2np^1.

EXAMPLE 8.3   Electron Configurations of Ions

Write the electron configurations of the following ions:
   a. Mg^{2+} b. S^{2-} c. Fe^{2+} d. Fe^{3+}

•**Method of Solution**

a. A Mg atom has Z = 12, so the atom has 12 electrons, but the Mg^{2+} ion contains only 10 electrons. Write the configuration for Mg.

Mg $1s^2 2s^2 2p^6 3s^2$ or $[Ne]3s^2$

Then remove the two electrons in the outermost energy level to form the 2+ ion.

Mg^{2+} $1s^2 2s^2 2p^6$ or [Ne]

Mg^{2+} ions are *isoelectronic* with a Ne atom. That is, they have the same number of electrons.

b. A S atom has 16 electrons and has the electron configuration $1s^2 2s^2 2p^6 3s^2 3p^4$. The sulfur atom becomes the negative S^{2-} ion by gaining two electrons. Adding two electrons to the outermost subshell gives

S^{2-} $1s^2 2s^2 2p^6 3s^2 3p^6$ or [Ar]

c. As before, write the configuration of the atom.

Fe $[Ar]4s^2 3d^6$

When forming a cation from an atom of a transition metal, electrons are always removed first from the ns orbital, and then from the (n − 1)d orbital. The first two electrons lost are removed from the outermost principal energy level (highest n value), which is the 4s orbital. This gives:

Fe^{2+} $[Ar]3d^6$

d. Next, remove one more electron from Fe^{2+}. This gives:

Fe^{3+} $[Ar]3d^5$

•**Comment**

The stable ions of all but a few representative elements are isoelectronic to a noble gas. Keep in mind that most transition metals can form more than one cation, and that for the most part these ions are not isoelectronic with the preceding noble gases.

EXERCISES

1. What is a representative element? Give three examples of representative elements.

2. What is a transition element? Give three examples of transition elements.

3. What is an inner transition element? Give two examples of inner transition elements.

4. How many valence electrons does an atom of phosphorus have?

5. Write the outer electron configurations for:
   a. the Group 3A elements. b. the Group 6A elements.

6. Write the symbol of an element with 4 valence electrons.

7. What group of elements has the outer electron configuration ns^2np^5?

8. Write the electron configuration of scandium (Sc). Classify the element as a representative, transition, or inner transition element. Classify the element as a metal or nonmetal.

9. Name the groups of the periodic table in which each of the following elements are found.
   a. $[Ar] 4s^2$ b. $[Ar]4s^24p^3$ c. $[Kr]5s^24d^7$

10. Which of the following species are isoelectronic with each other?
   a. Kr   b. Cl^- c. Rb^+ d. K^+ e. Cu^{2+} f. Se^{2-} g. Zn^{2+}

11. Write the ground state electron configurations of the following ions:
   a. K^+ b. O^{2-} c. Sr^{2+} d. I^- e. Li^+ f. Mg^{2+}

12. Write the ground state electron configurations of the following transition metal ions.
   a. Fe^{2+} b. Fe^{3+} c. Cu^{2+} d. Cu^+ e. Ti^{2+} f. Ti^{4+}

13. Write the ground state electron configurations of the following ions.
   a. Sc^{3+} b. V^{5+} c. Pb^{2+} d. Pb^{4+}

PERIODIC VARIATION IN PHYSICAL PROPERTIES (8.3)

STUDY OBJECTIVES

1. Account for the observed trends in the atomic radii of the elements in a period, and those within a group of the periodic table.
2. Predict the relative sizes of atomic and ionic radii within an isoelectronic series.
3. Estimate certain physical properties of an element given properties of elements above and below the given element in a group of elements.

Atomic Radius. Figure 8.5 of the textbook shows that the atomic radius decreases, in general, as the atomic number increases across any period of the periodic table. Thus, the alkali metals have the largest atoms and the noble gases the smallest. Furthermore, within a group the atomic radius increases as the atomic number increases.

Another trend in chemical behavior is the **diagonal relationship.** This refers to similarities in some properties between a second row element, and the third row element that is one space to the right in the table. Two such elements are on a diagonal (Figure 8.13 in the text). For example, Li and Mg are on a diagonal, and they have similar atomic radii. If you start at Li and go first to Be, the atomic radius decreases. Next, on moving down to Mg, the radius will increase again. If the decrease and increase are about the same, then Li and Mg will have approximately the same radius (155 pm for Li versus 160 pm for Mg). This relationship does not hold for all properties of two diagonally related elements.

Figure 8.2 Periodic trends in the atomic radius.

Ionic Radii. Because atoms and ions of the same element have different numbers of electrons, we should expect atomic radii and ionic radii to be different. *The radii of cations are smaller than those of the corresponding neutral atoms.* This should be expected because positive ions are formed by removing one or more electrons from the outermost shell. Since these electrons are furthest from the nucleus, their absence will make the cation significantly smaller. In addition, loss of an electron causes a decrease in the amount of electron-electron repulsion which also causes the cation to be smaller than the neutral atom.

In contrast, *the radii of anions are larger than those of the corresponding neutral atom.* When an electron is added to an atom to form an anion, there is an increase in the electron-electron repulsion. This causes electrons to spread out as much as possible, and so anions have a larger radius than the corresponding atoms. The radii of several metal and nonmetal ions are compared with their atomic radii in Table 8.4.

Table 8.4 Atomic and Ionic Radii of Elements in Groups 1A and 7A

	Group 1A Radius (pm)			Group 7A Radius (pm)	
Element	Atomic	Ionic	Element	Atomic	Ionic
Li	123	60	F	64	136
Na	154	95	Cl	99	181
K	203	133	Br	111	195
Rb	216	148	I	128	216
Ca	174	99	O	73	140

We can also compare radii within an isoelectronic series. In the series in Table 8.5 *the radii decrease steadily as atomic number increases.*

Table 8.5 Atomic and Ionic Radii in an Isoelectronic Series

Species	S^{2-}	Cl^-	Ar	K^+	Ca^{2+}
Radius (pm)	219	181	154	133	99

Each species has 18 electrons arranged in an argon configuration, and so the electron-electron repulsion is about the same in each ion. The reason for the decrease in radius is that the nuclear charge increases steadily within this series. This causes the electrons to be attracted more strongly toward the nucleus, and the ionic radius to contract as the nuclear charge increases.

Effective Nuclear Charge. The explanation of the trends in atomic size and ionization energy (to be discussed) is based on the concept of the *effective nuclear charge* (Z_{eff}). The outermost electrons in an atom do not feel the full positive charge of the nucleus. Electrons in inner levels, lying between the nucleus and the outermost electrons, tend to shield the outermost electrons from the nuclear charge. In a sodium atom, for example, the inner 10 electrons ($1s^2 2s^2 2p^6$) shield the outer 3s electron from the positive charge of 11 protons. The Z_{eff} experienced by the 3s electron is only about +1. The Z_{eff} is equal to the nuclear charge Z minus the shielding constant σ. The shielding constant is essentially equal to the number of inner shell electrons. Therefore, $Z_{eff} = Z - \sigma$.

Atomic Radii. The Z_{eff} affects the atomic radius in the following way. In proceeding across a period, one proton at a time is added to the nucleus, and one electron is added to the outermost orbital. *Electrons within the same energy level do not effectively shield each other from the nucleus.* Thus, for magnesium the inner 10 electrons shield the electrons in the 3s orbitals from the 12+ charge of the nucleus, resulting in an effective nuclear charge of +2. One of the 3s electrons does not effectively shield the other 3s electron because they are in the same orbital. Continuing from left to right across the third period, all the outermost electrons are in the n = 3 energy level, and the Z_{eff} increases by about one unit per element.

Increasing Z_{eff}

$\longrightarrow$

Na(+1) Mg(+2) Al(+3) Si(+4) P(+5) S(+6) Cl(+7) Ar(+8)

This means a greater force of attraction is experienced by the outermost electrons of atoms of elements on the right side of the periodic table than for those on the left side. Consequently, the atoms gradually decrease in size as we proceed from left to right across a period.

In proceeding from one element to another down a group, each successive element has its outer electrons in a principal energy level with a larger n value. The effective nuclear charge experienced by the outermost electrons is essentially the same for all elements within a group. The result of this is that the size of the outer orbital is affected most by the value of n. As you will recall, the size of an orbital increases as n increases. Therefore, atomic radius increases from top to bottom in a group.

Transition Elements. Within a row of transition elements, in contrast to the representative elements, there is only a slight decrease in atomic radius when reading from left to right, as shown in Table 8.6.

This is so because electrons are being added to an inner d subshell. In period four, for instance, the outermost electrons occupy a 4s orbital, but each successive electron is added to an inner 3d subshell. As we read across the period, the increasing nuclear charge is effectively shielded by the increase in the number of 3d electrons. Therefore, the outer 4s electrons within a series of the transition elements experience almost a constant Z_{eff}. This causes the atomic radius to change only a little across the period.

Table 8.6 Atomic Radius (pm)

Sc	Ti	V	Cr	Mn	Fe	Co	Ni	Cu	Zn
144	132	122	119	118	117	116	115	118	121

EXAMPLE 8.4 Trends in Atomic Radius

Which one of the following has the smallest atomic radius?
a. Li   b. Na   c. Be   d. Mg

•Method of Solution

Atomic radii increase from top to bottom in a group, therefore Li atoms are smaller than Na atoms, and Be atoms are smaller than atoms of Mg. Next compare Be and Li. The atomic radius decreases from left to right within a period of the periodic table, thus Be atoms are smaller than Li atoms, as well as the other choices given.

EXAMPLE 8.5   Trends in Ionic Radii

In each of the following pairs, choose the ion with the *largest* ionic radius.
a. K^+ or Na^+
b. K^+ or Ca^{2+}
c. K^+ or Cl^-

•Method of Solution

a. K^+ and Na^+ are in the same group in the periodic table. The outer electrons in K^+ occupy the third principal energy level, and those in Na^+ occupy the second. *Answer:* K^+ has the greater ionic radius.
b. K^+ and Ca^{2+} belong to an isoelectronic series; both have 18 electrons. Calcium ions with their greater nuclear charge attract their electrons more strongly than K^+ ions and so are smaller. *Answer:* K^+ is larger.
c.   Again K^+ and Cl^- are isoelectronic species. The ion with the greater nuclear charge (K^+) will be smaller. *Answer:* Cl^- has the greater ionic radius.

EXAMPLE 8.6   Predicting Properties of Elements

Predict the missing value for the following:

a.
Element	Density (g/cm³)
Ca	1.55
Sr	?
Ba	3.5

b.
Element	Density (g/cm³)
Cl	0.099
Br	0.114
I	?

•Method of Solution

Within a group of the periodic table the physical properties vary in a regular fashion as you read down in the group.
a.   From the positions of these elements in Group 2A of the periodic table, the density of Sr could be estimated to be half-way between the values for Ca and Ba. The average of 1.55 and 3.5 g/cm³ is 2.5 g/cm³. The observed value is 2.6 g/cm³.

b.   In this case we must extrapolate rather than interpolate as above because I is farther down in Group 7A than Cl and Br. We could assume the change in the radius from Br to I will be the same as the change from Cl to Br. The difference between Cl and Br is 0.015 nm. Adding 0.015 to 0.114 gives the estimated radius for I as 0.129 nm. The observed radius of I is 0.133 nm.

EXERCISES

14. Which of the following has the largest radius?
    a. Na, Cl, Ne, S, Li
    b. C, Si, Sn, Pb

15. Identify any isoelectronic pairs among the following.
    Na^+, Ar, Cl^-, Ne, Se^{2-}

16. Which of the following has the largest radius?
    S^{2-}, Ar, Se^{2-}, O^{2-}, Al^{3+}

17. Which of the following has the largest radius?
    Na^+, Mg^{2+}, Al^{3+}, Ne

18. Which species is larger: a. Co^{2+} or Co^{3+} b. S or S^{2-}

19. Estimate the melting point of $Br_2(s)$ given that the melting points of $Cl_2(s)$ and $I_2(s)$ are
    $-101°C$ and $114°C$, respectively.

IONIZATION ENERGY AND ELECTRON AFFINITY (8.4 – 8.5)

STUDY OBJECTIVES

1. Describe the terms ionization energy and electron affinity.
2. Describe the trends in magnitudes of the ionization energy and electron affinity while reading across a period and down a group in the periodic table.
3. Know what types of elements have low ionization energies and high electron affinities.
4. Account for the observed trends in ionization energy within a period and within a group of the periodic table.

   Ionization Energy. The minimum energy required to remove an electron from the ground state of an atom in the gas phase is its *ionization energy*. The magnitude of the ionization energy is a measure of how strongly the outermost electron is held by an atom. The greater the ionization energy, the stronger the electron is held.
   For a many-electron atom, the energies required to remove a second and a third electron are called the second ionization energy and the third ionization energy, respectively. The third ionization energy is always greater than the second, which in turn is always greater than the first ionization energy. The explanation of this trend is related to the electron repulsion among the remaining electrons. The first electron removed comes from a neutral atom. With an electron missing the repulsion among the remaining electrons decreases, and they move closer to the nucleus. This means that to remove the second electron requires a higher ionization energy. The third ionization energy is higher yet due to a further reduction in electron repulsion.
   Table 8.3 in the textbook shows that the ionization energy increases when moving from left to right across a period. Thus, the alkali metals have the lowest ionization energy and the noble gases the highest. Within a group the ionization energy decreases as atomic number increases (Figure 8.3).

Figure 8.3 Periodic trends in ionization energy.

Ionization Energy Trends. The importance of the ionization energy is that it correlates closely with electron configuration and chemical properties. When considering ionization energy, it helps to recall that the first electron removed is the one farthest from the nucleus. *Within a group of elements the ionization energy decreases with increasing atomic number.* This trend is explained in the following way. When reading down a group of elements the effective nuclear charge is essentially constant. However, the outermost electron resides in increasingly higher energy levels. Therefore, as atomic number increases within a group, the outermost electron is held more weakly and the ionization energy decreases.

The trend within a period should also be related to the effective nuclear charge. In moving across the periodic table from left to right, the size of the atoms decreases due to the increase in effective nuclear charge. The outer electrons become more tightly held as we move from left to right, and the ionization energy must increase.

One characteristic of metals is the relative ease with which electrons can be removed from their atoms. Thus, the lower the ionization energy, the more metallic the element. The metals are located on the left hand side of the periodic table. Also, elements in all groups exhibit increased metallic behavior with increased atomic number.

Electron Affinity. Another property of atoms is one that relates to their ease in forming negative ions. The **electron affinity** of an element is the energy given off when a gaseous atom gains an additional electron to form a negatively charged ion. A fluorine atom, for example, gives off energy when it gains an electron to form a fluoride ion.

$$F(g) + e^- \rightarrow F^-(g) \qquad \Delta H = -328 \text{ kJ/mol}$$

The electron affinity of fluorine is +328 kJ/mol because that is the amount of energy *given off.*

$$F(g) + e^- \rightarrow F^-(g) + 328 \text{ kJ/mol}$$

The greater the attraction of an atom for an electron, the greater the electron affinity value. The more energy released, the more stable the ion is compared to the atom.

Table 8.4 of the text shows the electron affinity values of elements arranged according to their positions in the periodic table. Figure 8.12 (textbook) is a plot of electron affinity values for the first 20 elements. A clear trend in electron affinity values of elements within a period is not evident. Some trends in electron affinity can be summarized.

1. The electron affinities of the Group 6A and 7A elements are much greater than those of other elements.
2. The electron affinities of the noble gases are only slightly negative. Therefore the anions of these gases would be unstable.
3. The Group 1A metals have small positive electron affinities. The Group 2A metals have very small positive values and some negative electron affinities.
4. The electron affinities of metals are generally lower than those of nonmetals.

Electron Affinity Trend. The trend in electron affinities is also related to effective nuclear charge, and to the energy of the orbital that the electron will enter. The Group 6A and 7A elements have the highest electron affinities. These elements have high Z_{eff} and the added electron(s) enters the valence shell. The noble gases with even higher Z_{eff} have no affinity for an additional electron. With filled s and p subshells, noble gas atoms have no tendency to add an electron because the next available orbital is in the next higher energy level beyond the valence shell.

EXAMPLE 8.7   Trends in Ionization Energy

Which one of the following has the highest ionization energy?
a. K   b. Br   c. Cl   d. S

•Method of Solution

Ionization energy increases from left to right within a period. Thus, the value for Cl is greater than for S, and the value for Br is greater than for K. In comparing Cl and Br, Cl has the higher ionization energy value because ionization energy decreases from top to bottom within a group.

EXAMPLE 8.8   Ionization Energy

Why does sodium have a lower first ionization energy than lithium?

•Method of Solution

When a lithium atom is ionized the electron most easily removed comes from a 2s orbital, whereas in sodium the electron most easily removed comes form a 3s orbital. Because of shielding effects, the effective nuclear charge Z_{eff} experienced by these electrons is about +1 in both atoms. The main reason for a difference in ionization energy in these atoms is the larger distance of separation between the electron and the nucleus in the case of Na atoms. As the principal quantum number n increases, so does the average distance of the electron from the nucleus. Consequently, the electron becomes easier to remove as you read down a group. *Answer:* Na will have a lower ionization energy than lithium.

EXAMPLE 8.9   Ionization Energies

The first, second, and third ionization energies for calcium are:

$$I_1 = 590 \text{ kJ/mol} \qquad I_2 = 1145 \text{ kJ/mol} \qquad I_3 = 4900 \text{ kJ/mol}$$

Explain why so much more energy is required to remove the third electron from Ca, as compared to removal of the first and second electrons.

Answer: The successive ionization energy values follow the usual trend $I_3 > I_2 > I_1$. However, we must explain the very large difference between I_2 and I_3. The first two electrons are valence electrons and are removed from the 3s orbital. But the third electron must be removed from the inner 2p subshell. The n = 2 principal energy level lies much closer to the nucleus than the n = 3 energy level; therefore, its electrons are held much more strongly. Consequently, $I_3 \gg I_2$.

EXERCISES

20. What group of elements in the periodic table have:
    a. the highest ionization energies   b. the lowest ionization energies

21. Based on periodic trends, which one of the following elements has the lowest ionization energy?
    K, S, Se, Li, Br

22. Based on periodic trends, which one of the following elements has the largest ionization energy?
    Cl, K, S, Se, Br

23. For silicon atoms, which ionization energy value (I) will show an exceptionally large increase over the preceding ionization energy value?
    I_2 I_3 I_4 I_5 I_6

24. Define electron affinity. What group of elements have negative electron affinities?

25. Which one of the following elements has both a low ionization energy and a positive electron affinity?
    K, Ne, Br, Fe, N

26. Which of the following atoms has both a high ionization energy and a large negative electron affinity?
    K, Ne, Br, Fe, N

TYPES OF ELEMENTS (8.6)

STUDY OBJECTIVES

1. Describe typical properties of metals, nonmetals, and metalloids.
2. Identify the following chemical groups in the periodic table: alkali metals, alkaline earths, halogens, noble gases, and coinage metals.

Metals. All elements are classified in three broad groups according to their chemical and physical properties. These are the metals, nonmetals, and metalloids. Within this classification, certain groups of elements are known by common names, such as the alkali metals, alkaline earth metals, coinage metals, halogens, and noble gases. In the section below, the three broad groups and their characteristics are described first followed by the groups with common names and their characteristics.

Most of the elements are metals. In general, they appear in the left and center of the periodic table. All metallic elements except mercury are solids at 25°C. They have a lustrous appearance, are good conductors of heat and electricity, can be hammered or rolled into sheets (a property referred to as malleability), and can be drawn into wires (a property referred to as ductility). Metals generally have high densities and high melting points.

Metals have low ionization energies, low electron affinities, and relatively large atomic radii. All metallic elements combine with nonmetals such as oxygen and chlorine to form salts. The most reactive metals are at the left of the periodic table. The transition metals are less reactive than the Group 1A and 2A metals. Elements within a chemical group are more metallic as atomic mass increases.

Nonmetals. Eighteen of the elements are nonmetals. These elements are on the right side of the periodic table. At 25°C, eleven are gases, one (bromine) is a liquid, and the rest are brittle solids. Typically, their densities and melting points are low.

The atoms of nonmetal elements have high ionization energies and high electron affinities. Nonmetals combine with metals to form ionic compounds, and with other nonmetal elements to form molecular compounds. The nonmetals, except hydrogen, are located on the upper right hand side of the periodic table. Recall that hydrogen is a gas and exists as diatomic molecules, as do a number of nonmetals.

Metalloids. Several elements have some properties that are characteristic of metals and some that are like those of nonmetals. These elements are called *metalloids*. Many periodic tables show a zig-zag line separating the metals from nonmetals. The elements that border this line on both sides are metalloids (except for Al, which is a metal). They include boron, silicon, germanium, arsenic, antimony, and tellurium. The metalloids have ionization energies and electron affinity values intermediate between metals and nonmetals.

Alkali Metals. Certain groups of elements are known by common names (Figure 8.4). Some of the characteristics of the alkali metals, alkaline earth metals, coinage metals, halogens, and noble gases are described below.

The elements in Group 1A, with the exception of hydrogen, are called alkali metals. The ns electron in the highest principal energy level is well shielded from the nucleus and is easily lost. Their low ionization energies make the alkali metals the most active family of metals. These elements are found in nature as +1 ions in chemical combination with nonmetal ions and polyatomic ions. The densities of these elements are low, in part, because of their large radii. Li, Na, and K are even less dense than water.

Figure 8.4 Common names of several groups of elements.

Alkaline Earths. Group 2A elements are called *alkaline earth* metals. The outermost electron configuration is ns^2. Although these two are not as easily lost as the ns^1 electron in alkali metal atoms, they are readily given up. Alkaline earth metals exist in nature as +2 ions in chemical combination with negative ions.

Halogens. Group 7A elements are called the *halogens*. None of these elements is ever found free in nature. All have high electron affinities and so tend to acquire one electron to form −1 ions. Ionic compounds containing these anions are salts. Indeed, the name halogen means "salt-former." These nonmetals exist in the elemental form as diatomic molecules. At 25°C, F_2 and Cl_2 are gases, Br_2 is a liquid, and I_2 is a solid.

Noble Gases. Group 8A elements make up the *noble gas* elements. the term noble here means nonreactive. He, Ne, and Ar do not form any chemical compounds. Since 1960, a number of Kr and Xe compounds have been synthesized. The chemical inactivity of the noble gases is the result of the ns^2np^6 electron configuration of the outermost electrons in their atoms. The noble gas elements have negative electron affinities and high ionization energies.

Coinage Metals. The elements of Group 1B, copper, silver, and gold, are generally nonreactive and are called the *noble metals* or *coinage metals*. These metals are excellent conductors of heat and electricity.

Atoms of the coinage metals have one electron in the outer s subshell and 10 electrons in the underlying d subshell. The electron configurations of their outermost electrons are $ns^1 (n-1)d^{10}$. Note that this is an exception to the configuration expected from the Aufbau principle, which predicts $ns^2(n-1)d^9$. It is as if the $(n-1)d$ orbitals "borrowed" an electron from the higher energy ns orbital. For the coinage metal atoms the "borrowed" electron is used to complete an inner subshell. Apparently the completed d subshell has an enhanced stability that corresponds to a lowering of the energy of the atom. Thus $ns^1(n-1)d^{10}$ is a lower energy configuration than $ns^2(n-1)d^9$.

In copper, silver, and gold this one outer electron is held much more tightly than the outermost electron in the alkali metal atoms. This can be most easily seen by comparing ionization energies as shown in Table 8.7.

Table 8.7 Comparison of Ionization Energies of Coinage Metals and Alkali Metals

Period	Element	I (kJ/mol)
4	K	419
	Cu	745
5	Rb	403
	Ag	731
6	Cs	375
	Au	890

EXAMPLE 8.10   Properties of Group 4A Elements

Write the electron configurations for the atoms of all elements in Group 4A. Specify whether the element is a metal, a nonmetal, or a metalloid.

•**Method of Solution**

First see Figure 2.8 (text) for the type of element. The outermost principal energy level of atoms in Group 4A contains four electrons. The general electron configuration for elements in this group is ns^2np^2. Their electron configurations are

C $1s^22s^22p^2$	nonmetal
Si $[Ne]3s^23p^2$	metalloid
Ge $[Ar]3d^{10}4s^24p^2$	metalloid
Sn $[Kr]4d^{10}5s^25p^2$	metal
Pb $[Xe]4f^{14}5d^{10}6s^26p^2$	metal

EXAMPLE 8.11   Properties of Potassium and Chlorine

Compare the magnitudes of electron affinity and ionization energy for the alkali metal potassium with the halogen chlorine. Comment on their relative abilities to form ions.

•Method of Solution

From Table 8.4 (text) the electron affinities of K and Cl are 48 and 349 kJ/mol, respectively. Table 8.3 (text) gives the ionization energies of K and Cl as 419 and 1251 kJ/mol, respectively. These values suggest that the metal atom K can lose an electron and form a K^+ ion much more easily than can the nonmetal Cl. And the nonmetal atom Cl can attract an electron to form a Cl^- much more readily than can the metallic K atom.

EXERCISES

27. Of the following, which is the most metallic element?
    V, Ge, Se, As, Zn

28. Which of the following is the least metallic element?
    V, Ge, Al, As, Ca

29. The first ionization energies of boron and silicon are 801 and 786 kJ/mol , respectively. The similarities of these values is an example of what relationship?

30. What charges do you expect for the ions of the alkali metals and the alkaline earth metals?

31. What charges do you expect for the ions of the Group 6A and the halogen elements?

CONCEPTUAL QUESTIONS

1. Explain why the radii of atoms increases when proceeding down a group of the periodic table.

2. Explain why the ionization energy of atoms increases when proceeding from left to right across a period of the periodic table.

3. Why is an emphasis placed on the valence electrons in an atom when discussing its atomic properties.

PRACTICE TEST

1. The element francium is extremely rare, and very little is known about its chemical and physical properties. Use the following data to estimate its density and ionization energy.

	K	Rb	Cs
Density (g/cm³)	0.86	1.53	1.87
Ionization energy (kJ/mol)	419	403	375

2. The element technetium (Z = 43) does not occur naturally on earth. The densities of Mo and Ru are 10.2 and 12.4 g/cm³, respectively. Estimate the density of Tc.

3. The periodic table has been extended to include the transuranium elements (Z > 92). What elements would element number 104 be similar to in chemical properties?

4. Which is the general electron configuration for the outermost electrons of elements in Group 4A?
   a. ns^1 b. ns^2 c. ns^2np^4 d. ns^2np^2 e. $ns^2np^6nd^7$

5. In what group of the periodic table is each of the following elements found?
   a. $1s^22s^22p^6$ b. $[Ar]4s^1$ c. $[Xe]6s^24f^{14}5d^5$ d. $[Ne]3s^23p^5$

6. Use the periodic table to write the electron configurations of Cr, Sb, and Pb.

7. How many valence electrons does an arsenic atom have?

8. Successive ionization energies—1st, 2nd, 3rd, etc.—always show an increasing trend. $I_1 < I_2 < I_3 < I_n$. For aluminum atoms, which ionization energy value will show an exceptionally large increase over the preceding ionization energy value?
   a. 2nd   b. 3d   c. 4th   d. 5th   e. 6th

9. Write the electron configurations for the following ions.
   a. Ca^{2+} b. Se^{2-} c. Cl^- d. Mn^{2+} e. Co^{3+} f. Sc^{3+}

10. An Ar atom is isoelectronic with which one of the following?
    a. Ne   b. K   c. Sc^{3+} d. Cl^{2-} e. Na^+

11. Which atom should have the largest radius?
    a. Br   b. Cl   c. Se   d. Ge   e. C

12. Which is the larger ion or atom in each pair?
    a. I^- or Cs^+
    b.   Ne or K^+
    c.   Mg or Mg^{2+}

13. Which atom should have the greatest ionization energy?
    a. Se   b. Te   c. Na   d. Si   e. S

14. Why does atomic radius decrease in going from left to right across a row in the periodic table?

15. Which two from the following would be most likely to have similar ionization energies?
    B, C, Si, Al, Ar

ANSWERS

Exercises

1. Elements that have an incompletely filled set of s or p orbitals. Pick any from Groups 1A through 7A
2. Transition metals have incompletely filled d subshells or readily form cations having incompletely filled d subshells. Pick any from Groups 2B through 8B in the periodic table.
3. These elements have an incompletely filled inner f subshell. Pick any between atomic numbers 58 and 71.
4. 5
5. a. ns^2np^1 b. ns^2np^4
6. Pick any element from Group 4A.
7. The halogen elements. Group 7A.
8. $[Ar] 4s^2 3d^1$ Sc is a metal.
9. a. Group 7A b. Group 5A c. Group 8B
10. a, c and f have 36 electrons. b and d have 18 electrons.
11. a. K^+ [Ar] b. O^{2-} [Ne] c. Sr^{2+} [Kr] d. I^- [Xe] e. Li^+ [He] f. Mg^{2+} [Ne]
12. a. $[Ar] 3d^6$ b. $[Ar] 3d^5$ c. $[Ar] 3d^9$ d. $[Ar] 3d^{10}$ e. $[Ar] 3d^2$ f. [Ar]
13. a. [Ar] b. [Ar] c. $[Xe] 6s^2 5d^{10}$ d. $[Xe] 5d^{10}$
14. a. Na b. Pb
15. Ar and Cl^- are one isoelectronic pair, and Na^+ and Ne are another.
16. Se^{2-}
17. Ne
18. a. Co^{2+} b. S^{2-}
19. 6.5°C
20. a. Noble gases b. Alkai metals
21. K
22. Cl
23. I_5
24. Noble gases
25. K
26. Br
27. V
28. As
29. Diagonal relationship
30. +1 and +2, respectively
31. −2 and −1, respectively

Conceptual Questions

1. As the principal quantum number n increases, the size of the orbital increases, and the farther on average electrons will be from the nucleus. When reading down a group, the outermost electrons have an increasingly higher n value, and so the outermost electrons are farther from the nucleus and the atomic radius increases.

2. When going across a period from left to right protons are being added one at a time to the nucleus and the corresponding electrons are added simultaneously to the outermost principle energy level. The first ionization energy increases across the period because the electrons in the same principle energy level do not completely shield the increasing nuclear charge due to the added protons. As the effective nuclear charge increases from left to right so will the ionization energy.

3. The number of outermost electrons and the type of orbital occupied is what determines the chemical behavior of an element.

Practice Test

1. $I \cong 350$ kJ; density $\cong 2.0$ g/cm^3

2. Density $\cong 11.3$ g/cm^3

3. Ti, Zr, and Hf

4. d. ns^2np^2

5. a. Group 8A, noble gases   b. Group 1A, alkali metals

    c. A transition metal   d. Group 7A, halogens

6. Cr $[Ar]4s^23d^4$; the observed configuration is Cr $[Ar]4s^13d^5$, which is an exception to the rules we have developed.

    Sb $[Kr]5s^24d^{10}5p^3$

    Pb $[Xe]6s^24f^{14}5d^{10}6p^2$

7. 5

8. c. 4th

9. a. Ca^{2+} [Ar]   b. Se^{2-} [Kr]   c. Cl^- [Ar]

    d. Mn^{2+} $[Ar]3d^5$ e. Co^{3+} $[Ar]3d^6$ f. Sc^{3+} [Ar]

10. c. Sc^{3+}

11. d. Ge

12. a. I^- b. K^+ c. Mg

13. e. S

14. The effective nuclear charge increases from left to right because the nuclear charge increases by one proton at a time, and as electrons are added to the atom they go into the same principal energy level where they do not shield each other effectively. As Z_{eff} increases, the electrons are pulled closer to the nucleus and the atomic radius decreases correspondingly.

15. B($I = 801$ kJ) and Si($I = 786$ kJ) diagonal relationship.

Chapter Nine
CHEMICAL BONDING I: BASIC CONCEPTS

- Lewis Dot Symbols (9.1 – 9.2)
- Ionic Bonding and the Lattice Energy (9.3)
- Covalent Bonding and Lewis Structures (9.4 & 9.6)
- Electronegativity (9.5)
- Formal Charge (9.7)
- The Concept of Resonance (9.8)
- Exceptions to the Octet Rule (9.9)
- Bond Energies (9.10)

LEWIS DOT SYMBOLS (9.1 – 9.2)

STUDY OBJECTIVES

1. Write the Lewis symbols for atoms and ions.

Lewis Symbols. Ionic and covalent bonds are important types of chemical bonds. In this chapter you will learn to use electron configurations and the periodic table to predict the type and the number of bonds an atom of a particular element can form. Electron configurations can be used to write the Lewis dot symbols of the representative elements. A **Lewis dot symbol** of an element consists of the chemical symbol with one or more dots placed around it. Each dot represents a valence electron. The orbital diagram and the Lewis symbol for the fluorine atom are shown in Figure 9.1. Fluorine has 7 electrons in its outermost principal energy level (n = 2), and therefore has 7 dots in its Lewis symbol.

Figure 9.1 Orbital diagram and Lewis symbol for the fluorine atom.

The Lewis dot symbols and the electron configurations of the outermost electrons of elements in the third period are as follows:

$$Na\cdot \quad \cdot Mg\cdot \quad \cdot \overset{\cdot}{Al}\cdot \quad \cdot \overset{\cdot}{Si}\cdot \quad \cdot \overset{\cdot\cdot}{P}\cdot \quad \cdot \overset{\cdot\cdot}{S} \quad :\overset{\cdot\cdot}{Cl}\cdot \quad :\overset{\cdot\cdot}{Ar}:$$

$$3s^1 \qquad 3s^2 \qquad 3s^2 3p^1 \quad 3s^2 3p^2 \quad 3s^2 3p^3 \quad 3s^2 3p^4 \quad 3s^2 3p^5 \quad 3s^2 3p^6$$

For representative elements, the number of valence electrons is the same as the group number.

The representative metals form ions by *losing all of their valence electrons*. Lewis symbols can be used to represent the formation of positive magnesium and aluminum ions as follows:

$$\cdot Mg\cdot \ \rightarrow \ Mg^{2+} + 2e^-$$

$$\cdot \overset{\cdot}{Al}\cdot \ \rightarrow \ Al^{3+} + 3e^-$$

The nonmetal elements form negative ions by acquiring electrons until they are *isoelectronic with a noble gas atom*. Ions of nonmetal have 8 electrons in their valence shell.

$$:\overset{\cdot}{S}: + 2e^- \ \rightarrow \ :\overset{\cdot\cdot}{S}:^{\ 2-}$$

$$:\overset{\cdot\cdot}{Cl}\cdot + e^- \ \rightarrow \ :\overset{\cdot\cdot}{Cl}:^-$$

Sulfide and chloride ions, for example, have net charges of −2 and −1 because the atoms needed 2e⁻ and 1e⁻, respectively, to become isoelectronic with the noble gas argon.

EXAMPLE 9.1   Lewis Dot Symbols

Write Lewis symbols for the following elements: a. Ca   b. O, S and Se

•Method of Solution

The Lewis symbol of an element consists of the element symbol surrounded by between 1 and 8 dots, representing valence electrons.
a.   Ca is in Group 2A. It has two electrons in the outermost energy level. These are its valence electrons. The Lewis symbol is ·Ca·
b.   O, S, and Se are all in the same group, 6A. *Answer:* Each has 6 valence electrons.

$$\cdot \overset{\cdot\cdot}{O}\cdot \qquad \cdot \overset{\cdot\cdot}{S}\cdot \qquad \cdot \overset{\cdot\cdot}{Se}\cdot$$

EXAMPLE 9.2   Lewis Dot Symbols for Ions

Write Lewis symbols for the following ions: a. Ca²⁺   b. Se²⁻

•Method of Solution

a.   Removing two electrons (dots) from ·Ca· gives simply Ca²⁺ as the Lewis symbol of a calcium ion. No electrons are shown around the Ca²⁺ symbol because this ion has lost its valence shell electrons.

b.   Adding two electrons to $\cdot \overset{\cdot\cdot}{Se} \cdot$ gives $\overset{\cdot\cdot}{\underset{\cdot\cdot}{:}Se:}^{2-}$ as the Lewis symbol for a selenide ion.

EXERCISES

1.   Write Lewis dot symbols for atoms of the following elements:
     a. Mg   b. Se   c. Al   d. Br   e. Xe

2.   Write Lewis dot symbols for the following ions:
     a. K^+ b. S^{2-} c. N^{3-} d. I^- e. Sr^{2+}

IONIC BONDING AND LATTICE ENERGY (9.3)

STUDY OBJECTIVES

1.   Predict the relative strength of ionic bonds using Coulomb's law.
2.   List three energy terms that influence the tendency for two elements to form an ionic compound.
3.   Use the Born-Haber cycle to calculate the magnitude of the lattice energies of ionic solids.

   Coulomb's Law.   The force that gives rise to the ionic bond is the electrical attraction existing between a positive ion and a negative ion. Chemical bonding results when the energy of two interacting atoms (or ions) is lowered. Coulomb's law states that the potential energy (E) of interaction of two ions is directly proportional to the product of their charges and inversely proportional to the distance between them:

$$E = k \, \frac{Q_+ Q_-}{r}$$

where Q_+ and Q_- are the charges of the two ions, r is their distance of separation, and k is a proportionality constant (its value will not be needed). When one ion is positive and the other negative, E will be negative. Bringing two oppositely charged particles closer together lowers their energy. The lower the value of the potential energy, the more stable is the pair of ions. Energy would need to be added to separate the two ions.
   The factors that govern the stability of ion pairs are the magnitude of their charges, and the distance between the ion centers. The distance between ionic centers (r) is the sum of the ionic radii of the individual ions.

$$r = r_+ + r_-$$

> **The energy of attraction between two oppositely charged ions depends:**
>
> •Directly on the magnitude of the ion charges. The greater the ion charges, the stronger the attraction. The interaction of Mg^{2+} and O^{2-} ions is much greater than that of Na^+ and Cl^- ions.
>
> •Inversely on the distance between ion centers. The distance between ions depends on the sizes of the ions involved. As the sum of the ionic radii increases the interaction energy decreases.

Lattice Energy. In an ionic crystal the cations and anions are arranged in an orderly three-dimensional array, as illustrated in Figure 2.11 of the textbook. In a sodium chloride or rock salt type of crystal, the anion has six nearest neighbor cations. The ions are packed in such a way as to maximize attraction and minimize repulsion.

The lattice energy provides a measure of the attraction between ions and the strength of the ionic bond. The **lattice energy** is the energy required to separate the ions in 1 mole of a solid ionic compound into gaseous ions. Gaseous ions are far enough apart from one another that they do not interact. For example, the lattice energy of NaF(s) is equal to 908 kJ/mol. This means 908 kJ are required to vaporize one mole of NaF(s) and form one mole of Na^+ ions and one mole of F^- ions in the gas phase.

$$NaF(s) \rightarrow Na^+(g) + F^-(g) \quad \text{lattice energy} = 908 \text{ kJ}$$

The lattice energy is related to the stability of the ionic solid. The larger the lattice energy, the more stable the solid.

The value of the lattice energy depends on the charges of the ions and the ionic radii, in accordance with Coulomb's law. This can be seen in Table 9.1 by comparing the sodium halides. As the sum of the two ionic radii increases in going from NaF to NaI, the separation of ionic centers ($r = r_+ + r_-$) increases and the lattice energy decreases. The effect of ionic charge on lattice energy can be seen by comparing the lattice energies of NaF and CaO in Table 9.1. The larger value for CaO is due to the stronger attraction of doubly charged cations (Ca^{2+}) for doubly charged anions (O^{2-}).

The lattice energies are reflected in the melting points of ionic crystals. During melting the ions gain enough kinetic energy to overcome the potential energy of attraction, and they move away from each other. The higher the melting point, the more energy the ions need to separate from one another. Therefore as the lattice energy increases so does the melting point.

Table 9.1 Lattice Energies of Several Ionic Compounds

	$r_+ + r_-$* (pm)	Lattice Energy (kJ/mol)	Melting Point (°C)
NaF	231	908	1012
NaCl	276	788	801
NaBr	290	736	747
NaI	314	686	660
CaO	239	3540	2580

* See Figure 8.9 (textbook) for ionic radii values.

Factors Favoring Formation of Ionic Bonds. It is important to identify any properties of atoms that affect their ability to form ionic compounds. In the formation of ions from atoms, the metal atom loses an electron and the nonmetal atom gains an electron.

$$Li(g) \rightarrow Li^+(g) + e^-$$
$$e^- + F(g) \rightarrow F^-(g)$$

The overall change is:

$$Li(g) + F(g) \rightarrow Li^+(g) + F^-(g)$$

One factor favoring the formation of an ionic compound is the ease with which the metal atom loses an electron. A second factor is the tendency of the nonmetal atom to gain an electron. Thus ionic compounds tend to form between elements of low ionization energy and those of high electron affinity. The alkali metals and alkaline earth metals have low ionization energies. They tend to form ionic compounds with the halogens and Group 6A elements both of which have high electron affinities.

A third factor is the lattice energy. Electrostatic attraction of the ions results in large amounts of energy being released when two kinds of gaseous ions are brought together to form a crystal lattice.

$$Li^+(g) + F^-(g) \rightarrow LiF(s)$$

In general, the smaller the ionic radius and the greater the ionic charge, the greater the lattice energy.

Calculation of the Lattice Energy Using the Born-Haber Cycle.

Lattice energies cannot be measured directly and must be calculated using Hess's law. In Chapter 6 you learned that if a reaction can be broken down into a series of steps, the overall enthalpy of reaction is equal to the sum of the enthalpy changes for the individual steps. The series of steps used to calculate the lattice energies of ionic solids is called the Born-Haber cycle. Each step is one you've seen previously in relation to the properties of atoms. Here we will illustrate the calculation of the lattice energy for potassium chloride.

$$KCl(s) \rightarrow K^+(g) + Cl^-(g) \qquad \text{lattice energy} = ?$$

The Born-Haber cycle starts by taking the overall equation to be the reaction in which the ionic compound is formed from the elements in their standard states. In this example the enthalpy change is the same as the enthalpy of formation of potassium chloride from potassium and chlorine.

$$K(s) + \frac{1}{2}Cl_2(g) \rightarrow KCl(s) \qquad \Delta H^\circ_{overall} = \Delta H^\circ_f(KCl)$$

This reaction is then envisioned to occur by a number of steps.

1. First potassium sublimes:

    $$K(s) \rightarrow K(g) \qquad \Delta H^\circ_1 = \Delta H_{subl}$$

2. Next, diatomic chlorine is dissociated into Cl atoms. The required energy is the bond energy for 1/2 mole of Cl—Cl bonds.

    $$\frac{1}{2}Cl_2(g) \rightarrow Cl(g) \qquad \Delta H^\circ_2 = \frac{1}{2}BE(Cl—Cl)$$

Up until now the focus has been on making the atoms of the elements potassium and chlorine. Next these will be made into the appropriate ions.

3. Now ionize 1 mole of potassium atoms:

    $$K(g) \rightarrow K^+(g) + e^- \qquad \Delta H^\circ_3 = I_1$$

4. To form a Cl$^-$ ion, an electron is added to the chlorine atom. This will release an energy equal to the electron affinity of chlorine, EA. Because energy is given off, ΔH has the opposite sign of the electron affinity.

    $$Cl(g) + e^- \rightarrow Cl^-(g) \qquad \Delta H^\circ_4 = -EA$$

5. Finally, 1 mole of gaseous K$^+$ ions, and 1 mole of gaseous Cl$^-$ ions are combined to make 1 mole of KCl(s). An amount of energy equal to the lattice energy will be released. The lattice energy must have the same magnitude as ΔH°_5, but an opposite sign.

    $$K^+(g) + Cl^-(g) \rightarrow KCl(s) \qquad \Delta H^\circ_5 = - \text{lattice energy}$$

The summation of the steps gives the overall reaction above.

$$K(s) + \frac{1}{2}Cl_2(g) \rightarrow KCl(s) \qquad \Delta H^\circ_{overall} = \Delta H^\circ_f(KCl)$$

Therefore, in general:

$$\Delta H_f^o (KCl) = \Delta H_{subl} + \frac{1}{2}BE + I(\text{metal element}) - EA(\text{nonmetal}) - \text{lattice energy}$$

or

$$\Delta H_{overall}^o = \Delta H_1^o + \Delta H_2^o + \Delta H_3^o + \Delta H_4^o + \Delta H_5^o$$

where $\Delta H_5^o = -\text{lattice energy}$

$$-\text{lattice energy} = \Delta H_5^o = \Delta H_{overall}^o - [\Delta H_1^o + \Delta H_2^o + \Delta H_3^o + \Delta H_4^o]$$

EXAMPLE 9.3   Predicting Melting Points

Which member of the pair will have the higher melting point?
a.   NaCl or CaO
b.   NaCl or NaI

•Method of Solution

a.   According to Coulomb's law, the doubly charged ions in CaO will attract each other more strongly than do the singly charged ions in NaCl. *Answer:* CaO will have a higher melting temperature.
b.   According to Coulomb's law, the closer the centers of two ions can approach each other the stronger the attraction between them. The sum of the ionic radii in NaCl is smaller than in NaI. *Answer:* NaCl will have the higher melting point.

EXAMPLE 9.4   The Born-Haber Cycle

Given the following data calculate the lattice energy of potassium chloride:

Enthalpy of sublimation of potassium	= 90.0 kJ
Bond energy (BE) of Cl—Cl	= 242.7 kJ
Ionization energy (I) of K	= 419 kJ
Electron affinity (EA) of Cl	= +349 kJ
ΔH_f^o (KCl)	= –435.9 kJ

•Method of Solution

From the discussion in the above section, we can calculate the lattice energy (or $-\Delta H_5^o$) from the equation:

$$\Delta H_5^o = \Delta H_{overall}^o - [\Delta H_1^o + \Delta H_2^o + \Delta H_3^o + \Delta H_4^o]$$

Next we identify the appropriate energy changes:

$K(s) \rightarrow K(g)$	$\Delta H_1^o = \Delta H_{subl} = 90.0$ kJ
$\frac{1}{2}Cl_2(g) \rightarrow Cl(g)$	$\Delta H_2^o = \frac{1}{2}BE = 121.4$ kJ
$K(g) \rightarrow K^+(g) + e^-$	$\Delta H_3^o = I = 419$ kJ
$Cl(g) + e^- \rightarrow Cl^-(g)$	$\Delta H_4^o = -EA = -349$ kJ
$K(s) + \frac{1}{2}Cl_2(g) \rightarrow KCl(s)$	$\Delta H_{overall}^o = -435.9$ kJ

Next substitute into the equation:

$$\Delta H_5^o = \Delta H_{overall}^o - [\Delta H_1^o + \Delta H_2^o + \Delta H_3^o + \Delta H_4^o]$$

$$\Delta H_5^o = -435.9 \text{ kJ} - [90.0 \text{ kJ} + 121.4 \text{ kJ} + 419 \text{ kJ} + (-349 \text{ kJ})]$$

$$\Delta H_5^o = -435.9 \text{ kJ} - 281.4 \text{ kJ} = -717 \text{ kJ}$$

lattice energy = 717 kJ

EXERCISES

3.   Write a chemical equation for the process that corresponds to the lattice energy of MgO.

4.   Which member of the pair will have the higher melting point?
     a. KCl or $CaCl_2$ b. RbI or NaI

5.   Which member of the pair will have the higher lattice energy?
     a. NaCl or CaO   b. KI or KCl

6.   Calculate the lattice energy of sodium bromide from the following information.

     ΔH_{subl} (Na) = 109 kJ
     Ionization Energy (Na) = 496 kJ
     Bond Energy (Br—Br) = 192 kJ
     Electron Affinity (Br) = 324 kJ
     ΔH_f^o (NaBr) = -359 kJ kJ

COVALENT BONDING AND LEWIS STRUCTURES (9.4 & 9.6)

STUDY OBJECTIVES

1.   State the octet rule and describe its basis.
2.   Describe covalent bonding in molecules by drawing their Lewis structures.

Octet Rule. In our study of the periodic table we saw that the valence electron configuration was related to the chemical and physical properties of an element. The noble gas elements are the least reactive. and therefore the most stable group of elements. The Lewis dot symbols of the noble gas elements show eight valence electrons corresponding to filled s and p subshells. G. N. Lewis reasoned that when atoms enter into chemical combination they become more stable. He proposed that atoms gain or lose electrons until they have the same number of valence electrons as noble gas atoms, that is, eight. The **octet rule** states that when forming bonds atoms of the representative elements tend to gain, lose, or, share electrons until they have eight electrons in the valence shell.

Molecules are held together by bonds resulting from the sharing of electrons between two atoms in a manner that is consistent with the octet rule. A simple **covalent bond** is formed when two atoms in a molecule share a pair of electrons.

The formation of a covalent bond in hydrogen chloride can be represented with Lewis structures:

$$H\cdot + \cdot \ddot{\underset{\cdot\cdot}{Cl}}: \rightarrow H : \ddot{\underset{\cdot\cdot}{Cl}}: \quad \text{or} \quad H—\ddot{\underset{\cdot\cdot}{Cl}}:$$

where the dash represents a covalent bond, or a pair of electrons shared by both the H atom and the Cl atom. By sharing the electron pair, the Cl atom has eight valence shell electrons. The stability of this bond results from both atoms acquiring a noble gas configuration. Notice that hydrogen is an exception to the octet rule. Rather than achieving an octet, it needs only two electrons to achieve a filled outer energy level. The H atom becomes isoelectronic with helium. The electron pairs on the Cl atom that are not involved in bonding are called **lone pairs**, **unshared pairs**, and **nonbonding electrons**. The sharing of valence electrons in methane and carbon tetrachloride is shown in Figure 9.2. The circles represent the valence shells of the atoms. They help to point out that each atom achieves an octet of valence electrons by sharing one or more pairs of electrons.

Figure 9.2 Sharing of electron pairs in CH_4 and CCl_4.

In some cases two or three pairs of electrons are shared by two atoms in order to reach an octet. In these molecules, *multiple bonds* exist. A *double bond* is a covalent bond in which two pairs of electrons are shared between two atoms, as between C and O in formaldehyde.

$$H_2CO \text{ (formaldehyde)} \qquad H : \overset{\cdot\cdot}{C} :: \overset{\cdot\cdot}{\underset{\cdot\cdot}{O}} \qquad \text{or} \qquad \overset{\textstyle H}{\underset{\textstyle |}{H—C}} = \overset{\cdot\cdot}{\underset{\cdot\cdot}{O}}$$

In general, atoms joined by a double bond lie closer together than atoms joined by a single bond. The $C = O$ bond length is shorter than the $C—O$ bond length.

Nitrogen molecules (N_2) contain a triple bond.

$$: N ::: N : \qquad \text{or} \qquad : N \equiv N :$$

Lewis Structures. Lewis structures represent the covalent bonding and location of unshared electron pairs within molecules and polyatomic ions. The steps for writing Lewis structures are as follows.

1. Arrange the atoms in a reasonable skeletal form, placing the unique atom in the center. Determine what atoms are bonded to each other.
2. Count the valence electrons. For polyatomic anions remember to add one electron for each unit of charge.
3. Connect the central atom to the surrounding atoms with single bonds. First add unshared pairs to the atoms bonded to the central atom to complete their octets (except for hydrogen, of course). Then, keeping in mind that the maximum number of electrons is the number counted in step 2, add the remaining unshared pairs to the central atom.
4. If the octet rule is satisfied for each atom, and the total number of electrons is correct, stop here since the structure can be considered correct. If the octet rule is not met for the central atom, go on to step 5.

5. In some cases there is a shortage of valence electrons. To complete the octet of the central atom we must write double or triple bonds between the central atom and the surrounding atoms. To make a double bond, move one of the unshared pairs from the surrounding atom to make the additional bond.
6. Repeat steps 4 and 5.

For application of this procedure see Example 9.5 and 9.6.

EXAMPLE 9.5 Drawing a Lewis Structure

Draw the Lewis structure for hydrazine, N_2H_4. How many unshared electron pairs (lone pairs) are there on each N atom?

•Method of Solution

1. Arrange the atoms in a reasonable skeletal form. H atoms form only one bond and so must be located on the outside of the atom.

   H N N H
     H H

2. Count the valence electrons. Each N atom has 5 valence electrons and each H atom has 1. There are 2(5) + 4(1) = 14 valence electrons.

3. Connect the atoms with single bonds:

   H—N—N—H
      | |
      H H

   Normally we would add unshared pairs to complete all octets of surrounding atoms, but in this case the H atoms only need the two electrons shared in the bond to the N atom. Counting the number of electrons used; 5 pairs = 10 valence electrons. Now add unshared pairs to complete the octets of the N atoms.

      .. ..
   H—N—N—H
      | |
      H H

4. Count the electrons: 7 pairs = 14 valence electrons. This is the same number as given in step 2.

EXERCISES

7. How many lone pairs are on the underlined atoms in the following compounds?
   a. $\underline{P}H_3$ b. $\underline{S}Cl_2$ c. $H_2C\underline{O}$

8. Write the Lewis structures for the following species.
   a. NH_4^+ b. NCl_3 c. CF_2Cl_2

ELECTRONEGATIVITY (9.5)

STUDY OBJECTIVES

1. Describe the trends in electronegativity within the periodic table.
2. Explain what is meant by a polar bond and predict the relative polarity of bonds.
3. Classify bonds in given substances as ionic, polar covalent, or covalent.

Electronegativity. Chemical bonds are rarely purely covalent or completely ionic. Rather, most bonds exhibit some characteristics of both. In the previous chapter we saw that atoms of the elements exhibit varying tendencies in their ability to attract and hold free electrons in the gas phase. In other words, electron affinity values show periodic variations. The term **electronegativity** is used to describe *the ability of an atom within a molecule to attract a shared electron pair toward itself.*

Linus Pauling developed a method for determining the relative electronegativities of the elements. These values are given in Figure 9.5 (text). Pauling assigned the value 1.0 to Li and 4.0 to F. The values for second and third row elements are given in Table 9.2. Electronegativity values exhibit periodic behavior. In general, electronegativities increase from left to right across a period, and decrease within a group from top to bottom as shown in Figure 9.3.

Table 9.2 Electronegativities of Second- and Third-Row Representative Elements.

Second Row	Li	Be	B	C	N	O	F
	1.0	1.5	2.0	2.5	3.0	3.5	4.0
Third Row	Na	Mg	Al	Si	P	S	Cl
	0.9	1.2	1.5	1.8	2.1	2.5	3.0

Figure 9.3 Periodic trends in electronegativity.

As a consequence of the differing abilities of atoms in a bond to attract the shared electron pair, most electron pairs are not shared equally. This imbalance causes an electron pair to shift slightly toward the more electronegative atom, giving rise to a *polar covalent bond*. In the HCl molecule, for instance, the electronegativity χ of Cl is 3.0, and for H it is 2.1. The electronegativity difference $\Delta\chi$ is:

$$\Delta\chi = \chi_{Cl} - \chi_H$$
$$= 3.0 - 2.1 = 0.9$$

The chlorine atom with its higher electronegativity attracts the electron pair more strongly. This makes the Cl atom slightly negative and H slightly positive.

$$\delta+ \quad \delta-$$
$$\text{H---Cl}$$

Here δ denotes a partial charge, that is, a charge less than 1.0, as it would be in an ion.

Pure covalent bonding, which is the equal sharing of electron pairs, occurs only in homonuclear diatomic molecules. Examples are H_2, N_2, and Cl_2. In a diatomic molecule with both atoms the same, $\Delta\chi$ must be zero, and the bonding electron pair is shared equally. Bonds of this type are described as **nonpolar covalent bonds**, or **pure covalent bonds.**

In bonds involving different atoms, the electronegativity difference will depend on the relative electronegativities of the two atoms. When $\Delta\chi \geq 2.0$ a bond is mostly **ionic**, for in this case one atom so outdoes the other at attracting electron pairs that electrons can be considered to be completely transferred to the more electronegative atom. Bonds between atoms such that $\Delta\chi < 2.0$ are classified as **polar covalent bonds**, or simply *polar* bonds. *The "2.0 rule" is an approximation and does not apply to all ionic compounds.*

Nonpolar covalent bonds and completely ionic bonds represent extreme situations in bonding. To refer to a bond as being "ionic" or "covalent" is an oversimplification. Sometimes the term *percent ionic character* is used to describe the polar nature of a bond. A pure covalent bond has zero percent ionic character. Bond with $\Delta\chi \geq 2.0$ will have at least 50% ionic character and will be called ionic. There are no 100% ionic bonds. Compounds of Group 1A or 2A metals with a nonmetal usually have ionic bonds, whereas compounds of two nonmetals usually have polar covalent bonds. See Table 9.3 for examples.

Table 9.3 Bond Character of Some Common Bonds

Ionic Compounds	$\Delta\chi$
NaCl	2.1
NaBr	1.9
NaI	1.6
KCl	2.2
KBr	2.0
$MgCl_2$	1.8

Polar Covalent Bonds	$\Delta\chi$
C—Cl	0.5
N—H	0.9
O—H	1.4
S—H	0.4
S—O	1.0
N—O	0.5

EXAMPLE 9.6   Bond Polarity

Arranging the following bonds in order of increasing ionic character: C—O, C—H, and O—H.

•Method of Solution

As the electronegativity difference increases the bond becomes more polar and its ionic character increases. Using Figure 9.5 (textbook), we can determine the differences.

for C—O $\Delta\chi = 3.5 - 2.5 = 1.0$
for C—H $\Delta\chi = 2.5 - 2.1 = 0.4$
for O—H $\Delta\chi = 3.5 - 2.1 = 1.4$

The ionic character increases in the order
　　　　C—H < C—O < O—H

•Comment

The electronegativity of hydrogen is unlike that of the other elements of Group 1A. In terms of electronegativity, hydrogen is similar to the nonmetal elements boron and carbon. Bonds of H to nonmetal atoms are polar covalent bonds, rather than ionic bonds, such as the bonds in LiCl and NaCl.

EXAMPLE 9.7   Electronegativity Trends

Using the trends within the periodic table, determine which of the following is the most electronegative element: As, Se, or S.

•Method of Solution

Se and As are in the same period, and so the one further to the right has the higher electronegativity. That one is Se. Now compare S and Se. They are in the same group. The one nearer the top of the group has the greater electronegativity which is sulfur.

EXAMPLE 9.8   Types of Bonds

For the following pairs of elements label all bonds between them as ionic, polar covalent, or pure covalent.
a.　Rb and Br
b.　S and S
c.　C and N

•Method of Solution

a.　Bond type depends on electronegativity differences. The value of $\Delta\chi$ for a Rb—Br bond is 2.0. *Answer:* RbBr is ionic.
b.　For S—S, $\Delta\chi = 0.0$, and so the bond is a pure covalent bond.
c.　In the periodic table carbon and nitrogen are adjacent to each other in period 2. The one on the right is N; it is more electronegative and so we expect a polar covalent bond. Electronegativity values give $\Delta\chi = 0.5$.

EXERCISES

9.　Which atom is the most electronegative?
　　　Li　Cs　P　As　Ge

10.　List the following bonds in order of increasing ionic character:
　　　N—O,　Na—O,　O—O,　S—O

11.　Classify the O—H bond in CH_3OH as ionic, polar covalent, or nonpolar covalent.

12. Which of the following is a nonpolar covalent bond (pure covalent)?
    H—Cl   Li—Br   Se—Br   Br—Br

13. Classify the following bonds as ionic, polar covalent, or nonpolar covalent.
    Se—Cl   Al—Cl   K—F   Cl—Cl

FORMAL CHARGE (9.7)

STUDY OBJECTIVES

1. Define the formal charge of an atom in a molecule.
2. Determine the formal charges of atoms in a Lewis structure.

Formal Charges on Lewis Structures. The concept of formal charge provides a rational basis for choosing the more plausible Lewis structure from among several possibilities. The **formal charge** is the charge that an atom seems to have in a Lewis structure. When determining the formal charge all nonbonding electrons count as belonging entirely to the atom in which they are found. All bonding electrons are divided equally between the bonded atoms. Thus the formal charge of an atom in a Lewis structure is the number of valence electrons in an isolated atom minus the number of electrons assigned to that atom in a molecule. The formula for the formal charge of an atom is:

$$\begin{matrix} \text{formal} \\ \text{charge} \end{matrix} = \begin{pmatrix} \text{number of} \\ \text{valence electrons} \end{pmatrix} - \begin{pmatrix} \text{number of} \\ \text{nonbonding electrons} \end{pmatrix} - \tfrac{1}{2} \begin{pmatrix} \text{number of} \\ \text{bonding electrons} \end{pmatrix}$$

It will be good to keep in mind that the formal charge is really more a property of a structural formula than that of the species the formula represents. Formal charges do not indicate actual charges on atoms in the real molecule.

Two possible Lewis structures for BF_3 are

$$
\begin{array}{cc}
\overset{\cdot\cdot}{:}\text{F}\overset{\cdot\cdot}{-}\text{B}\overset{\cdot\cdot}{-}\text{F}\overset{\cdot\cdot}{:} & \qquad\text{and}\qquad & \overset{\cdot\cdot}{:}\text{F}-\text{B}=\overset{\cdot\cdot}{\text{F}} \\
\overset{\cdot\cdot}{\;}\;|\;\overset{\cdot\cdot}{\;} & & \overset{\cdot\cdot}{\;}\;|\;\overset{\cdot\cdot}{\;} \\
:\overset{\;}{\text{F}}: & & :\overset{\;}{\text{F}}: \\
\overset{\cdot\cdot}{\;} & & \overset{\cdot\cdot}{\;} \\
(1) & & (2)
\end{array}
$$

The formal charges in (1) are:

        The boron atom: formal charge = 3 – 0 – 1/2 (6) = 0
        The fluorine atom: formal charge = 7 – 6 – 1/2 (2) = 0

The formal charges in (2) are:

        The boron atom: formal charge = 3 – 0 – 1/2 (8) = –1
        The fluorine atom (double bonded):   formal charge = 7 – 4 – 1/2 (4) = +1
        The other fluorine atoms: formal charge = 7 – 6 – 1/2 (2) = 0

The rule that is used to establish the more plausible structure is: *A Lewis structure in which there are no formal charges is preferred over one where formal charges are present.* Therefore structure (1) is preferred over structure (2).

When formal charges are assigned in a Lewis structure, *the sum of the formal charges must be zero in a neutral molecule.* For a polyatomic ion *the formal charges must add up to the charge of the ion.* For the chlorite ion (ClO_2^-), for instance:

$$[: \overset{..}{\underset{..}{O}} - \overset{..}{Cl} - \overset{..}{\underset{..}{O}} :]^-$$

the sum of the formal charges is –1:

$$[: \overset{-}{\overset{..}{O}} - \overset{+}{\overset{..}{Cl}} - \overset{-}{\overset{..}{\underset{..}{O}}} :]^-$$

The example brings out another feature of the formal charge concept. *The most plausible Lewis structures will be those with negative formal charges on the more electronegative atoms.* Since oxygen is more electronegative than chlorine, the formal negative charges on O atoms in this structure are more reasonable than in some other structure that would place a positive charge on the O atoms, and a negative charge on the Cl atoms.

EXAMPLE 9.9   Assigning Formal Charges

Assign formal charges to the atoms in the following Lewis structures:

a. $: C \equiv O :$

b. $\overset{..}{O} = \overset{..}{S} - \overset{..}{\underset{..}{O}} :$

•Method of Solution

The formula used to calculate the formal charge of an atom is:

$$\begin{pmatrix} \text{formal} \\ \text{charge} \end{pmatrix} = \begin{pmatrix} \text{number of} \\ \text{valence electrons} \end{pmatrix} - \begin{pmatrix} \text{number of} \\ \text{nonbonding electrons} \end{pmatrix} - \frac{1}{2} \begin{pmatrix} \text{number of} \\ \text{bonding electrons} \end{pmatrix}$$

a.   The carbon atom: formal charge = $4 - 2 - 1/2 \, (6) = -1$
     The oxygen atom: formal charge = $6 - 2 - 1/2 \, (6) = +1$

b.   The sulfur atom: formal charge = $6 - 2 - 1/2 \, (6) = +1$
     The oxygen atom on the right: formal charge = $6 - 6 - 1/2 \, (2) = -1$
     The oxygen atom on the left: formal charge = $6 - 4 - 1/2 \, (4) = 0$

•Comment

Some chemists don't approve of the CO structure given in part (a) because it places a positive formal charge on the more electronegative oxygen atom.

EXERCISES

14. Assign formal charges to the atoms in the following Lewis structures:

a. H—N—H
      |
      H

b. Ö=S=Ö

15. Assign formal charges to the atoms in the following possible Lewis structures.

a. : O≡C—Ö :

b. F̈=Be=F̈

THE CONCEPT OF RESONANCE (9.8)

STUDY OBJECTIVES

1. Draw resonance structures for molecules or polyatomic ions.
2. Determine the formal charges on atoms in a resonance structure.

Resonance Structures. It sometimes happens that a satisfactory electron dot structure for a molecule or polyatomic ion cannot be drawn. When such a situation arises a special procedure is invoked to arrive at a Lewis structure. For the nitrite ion NO_2^-, for instance, the following structure shows the correct number of valence electrons and satisfies the octet rule. The brackets are used to indicate that the -1 charge belongs to the entire nitrite ion, and not to just one atom in the structure.

$$[\,:\ddot{O}\!-\!\ddot{N}\!=\!\ddot{O}\,]^{\,-}$$

However, the structure does not accurately represent what is known about the bond lengths of the N—O bonds in NO_2^-. Both bond lengths are known to be the same, whereas according to the structure we expect the double bond to be shorter than the single bond. An N—O single bond length should be about 136 pm, and an N=O double bond length about 122 pm. However, the two bond lengths in NO_2^- are equal and are intermediate between these two values.

It turns out to be impossible to draw a single satisfactory Lewis structure for NO_2^-. Situations like this can be handled by using the concept of resonance. First draw two structures for NO_2^- that reflect different choices of electron arrangements.

$$[\,:\ddot{O}\!-\!\ddot{N}\!=\!\ddot{O}\,]^{\,-} \longleftrightarrow [\,\ddot{O}\!=\!\ddot{N}\!-\!\ddot{O}\,:\,]^{\,-}$$

The nitrite ion is not adequately represented by either structure; but it may be described by a composite of these structures. This composite structure cannot be drawn by the rules for writing Lewis structures. Each of the structures that contributes to the composite structure is called a **contributing** or **resonance structure**. The symbol, $\longleftrightarrow$ indicates that the structures shown are resonance structures.

In applying the concept of resonance we assume that NO_2^- is a composite or an average of the two structures. Thus, the N—O bonds are intermediate between single and double bonds. The term resonance was perhaps a poor choice because the word implies to some that the real molecule flips from one structure to the other. *This is not what is implied here.* The point is that the properties of NO_2^- cannot be accounted for by a single Lewis structure. The term **resonance** means the that there is a need for two or more Lewis stuctures to represent a particular molecule.

These structures have the same placement of atoms but different positions of electron pairs. The electron pairs do not actually move from one position to another within the molecule. Therefore resonance structures are generated by moving electron pairs within the same skeletal structure.

EXAMPLE 9.10   Drawing a Lewis Structure with Resonance

Draw three resonance structures for N_2O. The skeletal structure is N—N—O.

•Method of Solution

1. Arrange the atoms

    N—N—O

2. Count the valence electrons.

    $2(5) + 6 = 16$ electrons

3. Add unshared pairs to the terminal atoms:

    $: \overset{..}{\underset{..}{N}} — N — \overset{..}{\underset{..}{O}} :$

4. Count the electron pairs used. 8 electron pairs = 16 electrons. There are not enough electrons to add any to the central atom.

5. Move one of the unshared pairs from the terminal N atom to make a double bond, and move an unshared pair from the terminal O atom to make a double bond.

    $\overset{..}{N} = N = \overset{..}{O} :$

This is a satisfactory structure because it completes the octets of all three atoms and uses the correct number of electrons. Additional resonance structures can be generated by moving electron pairs in such a way that the octet rule for each atom is always satisfied. The positions of the atoms in N_2O cannot be altered. To possibilites are

$: N \equiv N — \overset{..}{\underset{..}{O}} :$ and $: \underset{..}{N} — N \equiv O :$

•Comment

The three structures for N_2O are called resonance structures. The actual bonding in N_2O is a composite of these three structures.

EXERCISE

16. Draw resonance structures for nitric acid: H—O—N—O

    $\overset{\displaystyle |}{O}$

EXCEPTIONS TO THE OCTET RULE (9.9)

STUDY OBJECTIVES

1. Describe three types of molecules that are exceptions to the octet rule.
2. Draw Lewis structures for molecules that do not follow the octet rule.

Types of Exceptions. Lewis structures can be drawn for many compounds with the aid of the octet rule; however, structures of some compounds do not follow the rule. The text points out three types of molecules that are exceptions to the octet rule:

- Molecules that have an incomplete octet.
- Molecules with an odd number of electrons.
- Molecules in which the central atom has an expanded octet.

The boron halides BX_3 are well-known examples of molecules with an incomplete octet. They are all planar molecules in which the boron atom has only six valence electrons. The boron atom has incomplete octet. This situation is typical in boron chemistry.

$$:\ddot{F}:$$
$$:\ddot{F}-B-\ddot{F}:$$

Normally when there is a shortage of electrons, we can draw a double-bonded structure as shown below for BF_3.

$$:\ddot{F}:$$
$$:\ddot{F}-B=\ddot{F}:$$

Two common oxides of nitrogen, NO and NO_2, have **odd numbers of electrons**. Since an even number of electrons is required for complete pairing, the octet rule cannot be satisfied. Two additional odd-electron molecules that are known to exist in our atmosphere for very short periods of time are OH (hydroxyl radical) and HO_2 (hydroperoxyl radical).

$$\cdot\ddot{O}-H \qquad\qquad H-\ddot{O}-\ddot{O}\cdot$$

However, experiments indicate that each B—F bond is a single bond as shown in the first dot structure. Also, the assignment of formal charges indicates that a structure with a double bond would have adjacent formal charges. As was discussed above, this is unfavorable. In addition, the more electronegative F atom would have a positive charge, rather than the preferred negative charge for fluorine.

Molecules exhibiting an **expanded octet** (having more than eight valence electrons) require the presence of nonmetal atoms from *the third period or beyond in the periodic table*. Second period elements never exceed the octet rule. Third-period elements are just as likely to exceed the octet rule as they are to follow it. Where for example, PCl_3 obeys the octet rule, gaseous PCl_5 has a phosphorus atom that is joined by single bonds to five chlorine atoms. The phosphorus atom has 10 electrons in its valence shell. This is called an expanded octet.

The central atoms in SF_4 and SF_6, and in the interhalogen compounds ClF_5, BrF_5, and IF_7 exhibit expanded octets.

In Chapter 10, you will need to draw a number of dot structures of molecules exhibiting expanded octets. When the central atom is from the third period or beyond, complete the octets of the surrounding atoms first, and then complete the central atom. *If extra electron pairs remain, place them on the central atom.* See Example 9.11 below.

These exceptions seem to be telling us that an atom with a completed octet is not necessary for covalent bonding to occur. Within the Lewis framework, it is really the sharing of electron pairs that leads to the covalent bond. A shared pair of electrons acts to attract both atoms.

EXAMPLE 9.11   Exceptions to the Octet Rule

Draw Lewis structures for:
a. GaI_3
b. NO_2 (all bonds are equivalent)
c. ClF_3

•Method of Solution

a. Gallium is in Group 3A of the periodic table, a group well known for its electron deficient elements. GaI_3 has $3 + 3(7) = 24$ valence electrons.

$$
\begin{array}{c}
\ddot{\mathbf{:}}\,\ddot{\mathrm{I}}\,\ddot{\mathbf{:}} \\
| \\
\ddot{\mathbf{:}}\,\ddot{\mathrm{I}}\!-\!\mathrm{Ga}\!-\!\ddot{\mathrm{I}}\,\ddot{\mathbf{:}}
\end{array}
$$

This structure shows 24 electrons, the correct number. Ga, with three electron pairs, is electron deficient.

b. NO_2 has 17 valence electrons. With an odd number of electrons, it cannot obey the octet rule. The best we can do is start with 18 valence electrons as in NO_2^- and then remove one from the nitrogen atom (because it is the unique atom). Two contributing structures are necessary.

$$
\ddot{\mathrm{O}}\!=\!\overset{\cdot}{\mathrm{N}}\!-\!\ddot{\mathrm{O}}\ddot{\mathbf{:}} \longleftrightarrow \ddot{\mathbf{:}}\ddot{\mathrm{O}}\!-\!\overset{\cdot}{\mathrm{N}}\!=\!\ddot{\mathrm{O}}
$$

c. ClF_3 has 28 valence electrons. 26 electrons are required to complete the octets of the four atoms. The remaining two electrons are placed on the central Cl atom because chlorine is in the third period, and has vacant 3d orbitals that can hold electrons in addition to an octet. Chlorine is said to have an expanded octet.

$$
\begin{array}{c}
\ddot{\mathbf{:}}\,\ddot{\mathrm{F}}\!-\!\ddot{\mathrm{Cl}}\!-\!\ddot{\mathrm{F}}\,\ddot{\mathbf{:}} \\
| \\
\ddot{\mathbf{:}}\,\ddot{\mathrm{F}}\,\ddot{\mathbf{:}}
\end{array}
$$

EXERCISE

17. Write Lewis structures for the following molecules:
    a. $AlBr_3$ b. SF_4 c. NO

BOND ENERGIES (9.10)

STUDY OBJECTIVES

1. Determine the average bond energy of a bond using given enthalpies of reaction.
2. Use a table of bond energies to estimate ΔH for a reaction.

Average Bond Energies. Occasionally the enthalpy change ΔH is needed for a reaction for which enthalpy of formation data does not exist. One approach that allows an estimate of $\Delta H°$ uses the concept of bond energy. Consider the gas phase atomization process:

$$CH_4(g) \rightarrow C(g) + 4H(g) \qquad \Delta H°_{rxn} = 1664 \text{ kJ}$$

In this reaction, four C—H bonds are broken; therefore we can define the average C—H bond energy as one-fourth of $\Delta H°_{rxn}$ for the reaction. Hence the average C—H bond energy in CH_4 is 416 kJ/mol.

The actual bond energies of the individual C—H bonds in CH_4 are not the same as the average value. Even so, the use of average bond energies makes it possible to estimate the enthalpy changes of certain reactions. Some of the bond energies in Table 9.4 of the textbook are listed here in Table 9.4.

Table 9.4 Bond Energies of Some Common Bonds

Bond	Bond Energy (kJ/mol)
H—H	436
C—H	414
N—H	392
O=O	499
C=O	802
O—H	460
I—Cl	210
N≡N	941
N=O	630

We can estimate ΔH for the combustion of methane as follows:

$$CH_4(g) + 2O_2(g) \rightarrow CO_2(g) + 2H_2O(g)$$

$$CH_4 + 2O_2 \xrightarrow{\Sigma BE \text{ (reactants)}} C + 4H + 4O \xrightarrow{-\Sigma BE \text{ (products)}} CO_2 + 2H_2O$$

First we break 4 C—H bonds and 2 O=O bonds, where ΣBE (reactants) is the sum of the bond energies of bonds broken. Then let the atoms recombine to form the 2 C=O and 4 O—H bonds of the products.

(products) is the sum of the bond energies of all bonds formed from the atoms. Keep in mind that breaking bonds is an endothermic process, and making new bonds is an exothermic process. Therefore

$$\Delta H = \Sigma BE \text{ (reactants)} - \Sigma BE \text{ (products)}$$

$$\Delta H = [4 \text{ BE (C—H)} + 2 \text{ BE (O}=\text{O)}] - [2 \text{ BE (C}=\text{O)} + 4 \text{ BE (O—H)}]$$

Inserting values from Table 9.5 gives:

$$\Delta H = [4(414) + 2(499)] - [2(802) + 4(460)]$$
$$= 2654 \text{ kJ} - 3444 \text{ kJ}$$
$$\Delta H = -790 \text{ kJ}$$

If we compare this value with the standard enthalpy of combustion of methane, we get

$$CH_4(g) + 2O_2(g) \rightarrow CO_2(g) + 2H_2O(l) \qquad \Delta H^o_{rxn} = -890 \text{ kJ}$$

At first it seems there is a large discrepancy of 100 kJ. However, in our calculation, H_2O is present as a gas! But ΔH° refers to the standard state of H_2O as a liquid. Therefore, the 100 kJ difference is largely due to the ΔH of vaporization of two moles of H_2O, which is 81.4 kJ. Two important conclusions can be drawn here; (1) Bond energies are used to estimate enthalpies of reaction of *gas phase reactions*, and (2) enthalpies of reaction calculated from average bond energies are only approximate values.

EXAMPLE 9.12 Use of Bond Energies to Estimate ΔH_{rxn}

Estimate ΔH_{rxn} for the reaction:

$$Cl_2(g) + I_2(g) \rightarrow 2ICl(g)$$

using bond energies given in Table 9.4 (text), and given that BE (I—Cl) = 210 kJ/mol.

•**Method of Solution**

Recall that
$$\Delta H = \Sigma BE \text{ (reactants)} - \Sigma BE \text{ (products)}$$
$$\Sigma BE \text{ (reactants)} = BE \text{ (Cl—Cl)} + BE \text{ (I—I)}$$
$$= 243 + 151 = 394 \text{ kJ}$$

and
$$\Sigma BE \text{ (products)} = 2BE \text{ (I—Cl)} = 2(210) = 420 \text{ kJ}$$

Subtraction yields:

$$\Delta H = 394 \text{ kJ} - 420 \text{ kJ}$$
$$= -26 \text{ kJ}$$

EXERCISES

18. Given the $N\equiv N$ and H—H bond energies in Table 9.5 and standard enthalpy of reaction:

$$\tfrac{1}{2}N_2(g) + \tfrac{3}{2}H_2(g) \rightarrow NH_3(g) \qquad \Delta H^o_{rxn} = -46.3 \text{ kJ/mol}$$

calculate the average N—H bond energy in ammonia.

19. Given the bond energies in Table 9.5, calculate the enthalpy of formation for ammonia, NH_3.

$$\tfrac{1}{2}N_2(g) + \tfrac{3}{2}H_2(g) \rightarrow NH_3(g) \qquad \Delta H_f^0(NH_3) = ?$$

CONCEPTUAL QUESTIONS

1. Explain the difference between electron affinity and electronegativity.

2. Do formal charges represent actual charges? What do they mean?

3. Do resonance structures actually exist? Describe why the concept of resonance is needed.

PRACTICE TEST

1. Arrange the following ionic compounds in order of increasing lattice energy: RbI, MgO, $CaBr_2$.

2. Which compound in each pair has the higher melting point?
   a. RbI or KI
   b. $MgCl_2$ or $MgBr_2$

3. Draw Lewis dot structures for
   a. IBr   b. CS_2 c. PCl_5 d. SO_2
   e. CO   f. P_2Cl_4 g. SO_4^{2-} h. PO_4^{3-}

4. Draw resonance structures for SO_3.

5. Which of the following elements is the most electronegative?
   a. Ca   b. P   c. As   d. K   e. Si

6. Which one of the following is the most polar bond?
   a. B—C   b. B—N   c. B—S   d. C—C

7. Which bond has the the greatest percent ionic character?
   a. CO   b. SO   c. NaI   d. NaBr

8. Which molecule has a Lewis structure that does not obey the octet rule?
   a. NO   b. PF_3 c. CS_2 d. HCN   e. BF_4^-

9. Which molecule has covalent bonding between atoms?
   a. NaF   b. K_2O c. ICl   d. SrI_2 e. none of these

10. Indicate the molecule in each pair that does not follow the octet rule.
    a. PCl_5 or $AsCl_3$ b. BCl_3 or CCl_4 c. NH_3 or CH_3 d. $SbCl_5$ or NCl_3

11. Consider the following Lewis structures for sulfate ion. Which is the more reasonable structure in terms of formal changes?

a.

$$
\left[\quad \ddot{\text{O}} \quad \right]^{2-}
$$

b.

$$
\left[\quad \ddot{\text{O}} \quad \right]^{2-}
$$

12. Assign formal changes to the atoms in the Lewis structures shown.

a. $: \text{N} = \text{N} = \text{O} :$ b. $[: \text{O} — \text{H}]^-$ c. $[: \text{O} — \text{Cl} :]^-$

13. Write a mathematical expression that would allow you to calculate the lattice energy of $CaCl_2$ using Hess's law.

14. Calculate the $\text{N} \equiv \text{N}$ bond energy, given that the standard enthalpy of formation of atomic nitrogen is 473 kJ/mol.

15. Estimate the enthalpy change for the following reaction

$N_2(g) + O_2(g) \rightarrow 2NO(g)$

Hint: NO has a double bond. BE $(\text{N} = \text{O}) = 630$ kJ

ANSWERS

Exercises

1. a. $\cdot \text{Mg} \cdot$ b. $: \ddot{\text{Se}} :$ c. $\cdot \text{Al} \cdot$ d. $: \ddot{\text{Br}} \cdot$ e. $: \ddot{\text{Xe}} :$

2. a. K^+ b. $: \ddot{\text{S}} :^{2-}$ c. $: \ddot{\text{N}} :^{3-}$ d. $: \ddot{\text{I}} :^-$ e. Sr^{2+}

3. $MgO(s) \rightarrow Mg^{2+}(g) + O^{2-}(g)$ $\Delta H =$ lattice energy

4. a. $CaCl_2$ b. NaI
5. a. CaO   b. KCl
6. 736 kJ
7. a. 1   b. 2   c. 2

8.

9. P

10. O—O < S—O < N—O < Na—O

11. polar covalent

12. Br—Br

13. polar covalent, ionic, ionic, nonpolar covalent

14. a. The formal charges in NH₃ are: The N atom: 0.
                                           The H atoms: 0.

    b. The formal charges in SO₂ are: The S atom: 0.
                                           The O atoms: 0.

15. a. $\overset{+}{O}\!\equiv\!C\!-\!\overset{-}{O}$ b. $\overset{+}{F}\!=\!\overset{2-}{Be}\!=\!\overset{+}{F}$

16.

17.

18. 391 kJ

19. −50.7 kJ

Conceptual Questions

1. Electron affinity is a property of a gaseous atom, while electronegativity is a property of an atom within a molecule. Both have to do with attracting electrons but they are very different properties. Electron affinity is related to the energy released by an uncombined atom in the gaseous state when it adds an additional electron. Electronegativity is a measure of the ability of an atom within a molecule to attract an electron pair in a bond it has formed.

2. Formal charges do not represent actual charges of the atoms in a molecule. However they do give an indication of the sign (+ or −), if not the magnitude, of the charge and we use formal charges to choose from several possible Lewis structures.

3. When a molecule cannot be accurately represented by a single Lewis structure we resort to resonance structures. These structures have the same placement of atoms but different positions of electron pairs. The electron pairs do not actually move from position to position within the molecule. Therefore none of the resonance structures actually exist. Rather the real molecule is said to be a composite of the resonance structures.

Practice Test

1. $RbI > CaBr_2 > MgO$
2. a. KI   b. $MgCl_2$

3. a. $:\ddot{I}—\ddot{Br}:$ b. $\ddot{S}=C=\ddot{S}$

   c. (structure: P centered with five Cl)

   d. $\ddot{O}=\ddot{S}—\ddot{O}: \leftrightarrow :\ddot{O}—\ddot{S}=\ddot{O}$

   e. $:C\equiv O:$ f. (structure: P_2Cl_4)

   g. $\left[\ddot{O}—S(\ddot{O})(\ddot{O})—\ddot{O}\right]^{2-}$ h. $\left[\ddot{O}—P(\ddot{O})(\ddot{O})—\ddot{O}\right]^{3-}$

4. $O=\overset{..}{S}—\ddot{O}: \leftrightarrow :\ddot{O}—\overset{..}{S}—\ddot{O}: \leftrightarrow :\ddot{O}—S=\ddot{O}$

5. b.
6. b.
7. d.
8. a.
9. c.
10. a. PCl_5 b. BCl_3 c. CH_3 d. $SbCl_5$
11. b. Because fewer formal charges are present. Particularly the formal charge on sulfur is zero compared to +2 in a.
12. a. $:N=N=O:$ b. $[:\ddot{O}—H]^-$ c. $[:\ddot{O}—\ddot{Cl}:]^-$

13. Lattice energy $= -\Delta H_f^o (CaCl_2) + \Delta H_{subl}(Ca) + I_1(Ca) + I_2(Ca) + BE(Cl_2) + 2(-EA(Cl))$
14. 946 kJ
15. 180 kJ

Chapter Ten

CHEMICAL BONDING II: MOLECULAR GEOMETRY AND HYBRIDIZATION OF ATOMIC ORBITALS

- Molecular Geometry and the VSEPR Model (10.1)
- Polar Molecules (10.2)
- Valence Bond Theory and Hybrid Orbitals (10.3 – 10.5)
- Molecular Orbital Theory (10.6)

MOLECULAR GEOMETRY AND THE VSEPR MODEL (10.1)

STUDY OBJECTIVES

1. Predict geometrical shapes of molecules and polyatomic ions.
2. Predict approximate bond angles between atoms within a molecule or polyatomic ion.

Electron-Pair Repulsion. The major features of the geometry of molecules and polyatomic ions can be predicted by applying a simple principle: "The valence shell electron pairs surrounding an atom are arranged such that they are as far apart as possible."

According to the valence-shell electron-pair repulsion (VSEPR) theory, a particular molecular geometry results from the orientation of electron pairs around a central atom. For example, consider the BeH_2 molecule in which Be is the central atom. Drawing its Lewis structure, we count two electron pairs in the valence shell of the Be atom. There are no lone pairs.

$$H : Be : H$$

In order to minimize repulsion the two electron pairs are located as far apart as possible. The angle between two bonds to the same atom is called the **bond angle**. The H—Be—H bond angle is 180°. BeH_2 is an example of a molecule in which the atoms lie in a straight line. BeH_2 has a *linear geometry*.

When the central atom of a molecule has no lone pairs in its valence shell, the possible orientations in space around the central atom are determined by the number of bonding pairs of electrons. Table 10.1 gives the shapes of molecules depending on the number of valence electron pairs about the central atom.

Three Electron Pairs. When there are three valence electron pairs around a central atom they avoid each other best if they are positioned at the corners of a triangle. The central atom sits in the middle of the triangle, and all four atoms lie in the same plane. All bond angles are 120°. Such a molecule is called *trigonal planar*.

Table 10.1 Bond Angles and Molecule Shapes

Number of BondingElectron Pairs about Central Atom	Predicted Bond Angles	Shape of Molecule
2	180°	Linear
3	120°	Trigonal planar
4 (octet)	109.5°	Tetrahedral
5	120°, 90°	Trigonal bipyramid
6	90°	Octahedral

Four Electron Pairs. In Chapter 9 we discussed the use of the octet rule to determine the correct formulas of covalent compounds. Lewis structures describe only the bonding within a molecule in a two-dimensional manner. They are not geometrical structures. For instance, the Lewis structure for methane CH_4 does not tell us whether the molecule is flat (planar) or tetrahedral. When the central atom has four bonding electron pairs (as C does in CH_4) these pairs try to keep as far as possible from each other, while maintaining their distance from the nucleus. The **tetrahedral** arrangement has less repulsion between the four electron pairs than the flat, two-dimensional geometry. This geometry is illustrated for methane in Figure 10.1. Four electron pairs around a central atom will be arranged at the corners of a tetrahedron. In methane the H—C—H bond angle is the *tetrahedral angle*, 109.5°.

Figure 10.1 The tetrahedral geometry of the methane molecule. The C atom is at the center of a four-sided three-dimensional figure called a tetrahedron. The four H atoms are found at the four corners. The H–C–H bond angle is the "tetrahedral" angle, 109.5°.

Five Electron Pairs. Central atoms with expanded octets have 5 or 6 electron pairs occupying their valence shells. The geometrical figure defined by the preferred orientation of five electron pairs around the central atom is called a *trigonal bipyramid* (Table 10.1). The electron pairs that lie around the "equator" are called equatorial electrons. The electrons at the "north and south poles" are called axial electrons. The equatorial electron pairs lie in the same plane. In PF_5 the equatorial F—P—F bond angles are 120°. The axial atoms are at right angles to the equatorial plane, and so any bond angle formed by an axial atom, the central atom, and an equatorial atom is 90°.

Six Electron Pairs. The most stable arrangement of 6 electron pairs around the central atom is called an **octahedron** (Table 10.1). The example of SF_6 is shown in the textbook.

Effect of Lone Electron Pairs. When the central atom contains lone pairs (nonbonding pairs) in addition to bonding pairs, the situation becomes slightly more complicated. The position of the lone electron pairs helps to establish the geometry of a molecule. For instance, in ammonia (NH_3) the nitrogen atom has one lone electron pair.

$$\begin{array}{c} .. \quad \leftarrow \text{lone pair} \\ H : N : H \\ .. \\ H \end{array}$$

There are 4 electron pairs in the valence shell of the central N atom. Therefore, when the electron pairs are oriented as far from each other as possible, they will be positioned at the corners of a tetrahedron. But when describing the shape of a molecule we consider *only the positions of the atoms.* Ammonia is trigonal pyramidal, not tetrahedral. Its geometry is described as a trigonal pyramid because the pyramid has three sides. The N atom sits at the top of the pyramid.

trigonal pyramid

Table 10.2 lists the molecular shapes that result when the central atom has the symbol A, a pair of bonding electrons is B, and a pair of nonbonding electrons is E.

Table 10.2 Molecule Shapes As Determined by VSEPR Theory

Molecular Type	Shape of Molecule
AB_2	Linear
AB_3	Trigonal planar
AB_2E	Bent (nonlinear)
AB_4	Tetrahedral
AB_3E	Trigonal pyramidal
AB_2E_2	Bent (nonlinear)
AB_5	Trigonal bipyramidal
AB_4E	Distorted tetrahedron (see saw)
AB_3E_2	T-shaped
AB_2E_3	Linear
AB_6	Octahedral
AB_5E	Square pyramidal
AB_4E_2	Square planar

Multiple Bonds. The idea of electron pair repulsion can be extended to predict geometries of molecules or ions containing multiple bonds by assuming that *as far as molecular shape is concerned, electron pairs in a multiple bond affect the structure in the same way as those in a single bond.*

Consider SO_3, for example. One of its resonance structures is shown on the left below.

There are four electron pairs in the valence shell of the central atom (S), but remember two of these are used in a double bond. Thus, only three regions of electron density exist around the central atom. The SO_3 molecule has a geometry as if three electon pairs existed around the central S atom; that is, it is a planar triangle as shown on the right.

Bond Angles. VSEPR theory also explains why many observed bond angles are not quite equal to the predicted angle. For instance, a true tetrahedral angle, as in CH_4, is 109.5°, but the measured H—N—H bond angle in NH_3 is 106.7°, and in water the H—O—H bond angle is 104.5°.

The deviation from 109.5° arises because there are three types of electron pair repulsion represented: (1) repulsion between bonding pairs, (2) repulsion between lone pairs, and (3) repulsion between a bonding pair and a lone pair. The force of repulsion is not the same among these three, but decreases as follows:

| lone-pair vs. lone-pair | > | lone-pair vs. bonding-pair | > | bonding-pair vs. bonding-pair |
| repulsion | | repulsion | | repulsion |

In other words, lone pairs require more space than bonding pairs and tend to push the bonding pairs closer together. Bonding electron pairs are attracted by two nuclei at the same time. These pairs require less space than lone electron pairs.

In ammonia, the lone pair on the nitrogen atom repels the bonding pairs more strongly than the bonding pairs repel each other. The result is that the H—N—H bond angle is several degrees less than a tetrahedral angle.

In water the oxygen atom has two bonding pairs and two lone pairs. The greater lone-pair versus bonding-pair repulsion causes the two O—H bonds to be pushed in toward each other. The H—O—H angle in water should be less than the H—N—H angle in ammonia.

In general, when nonbonding electrons are present on the central atom the shape will be close to, but not exactly, as predicted in Table 10.2.

EXAMPLE 10.1   VSEPR Theory

Predict the geometrical shapes and bond angles for the following compounds using the VSEPR theory.
a. $SnCl_4$
b. O_3
c. IF_5
d. XeF_2

•Method of Solution

a. First determine the number of valence shell electrons for the central Sn atom. A tin atom has four valence electrons, and each Cl atom contributes one electron which it shares with the Sn atom, so $4 + 4(1) = 8$. The eight valence shell electrons are arranged as four electron pairs around the Sn atom. The Lewis structure is:

$$: \overset{\cdot\cdot}{\underset{}{Cl}} :$$
$$|$$
$$: \overset{\cdot\cdot}{\underset{\cdot\cdot}{Cl}} - \overset{}{Sn} - \overset{\cdot\cdot}{\underset{\cdot\cdot}{Cl}} :$$
$$|$$
$$: \overset{}{\underset{\cdot\cdot}{Cl}} :$$

This structure shows that there are no lone pairs about the Sn atom. Thus, the four bonding electron pairs and the Cl atoms, as well, are oriented at the corners of a tetrahedron. $SnCl_4$ is an AB_4 type molecule, and has a tetrahedral structure. The Cl—Sn—Cl bond angles are 109.5°.

b. Ozone (O_3) has as one contributing structure

$$\overset{\cdot\cdot}{\underset{\cdot\cdot}{O}} = \overset{\cdot\cdot}{\underset{}{O}} - \overset{\cdot\cdot}{\underset{\cdot\cdot}{O}} : \qquad\qquad \overset{\cdot\cdot}{\underset{\cdot\cdot}{O}} = \overset{}{\underset{}{O}}$$

Lewis structure   nonlinear geometry

When applying VSEPR theory to a molecule requiring resonance structures we can use *any one of the contributing structures*. Here we will take the second O atom as the central atom. In VSEPR theory *the double bond is counted as if it were a single bond with one effective electron pair*. The double bond acts as one center of electron density to repel other electron pairs. Therefore, the central O atom has the equivalent of 3 pairs of valence electrons. These three regions of charge are oriented at the corners of a triangle. However, one of these pairs is a lone pair. The molecule is of the type AB_2E. The positions of the three O atoms are described as *nonlinear*. The O—O—O bond angle should be close to 120°. However, the presence of the lone pair compresses the bond angle because the lone-pair vs. bonding-pair repulsion is greater than the bonding-pair vs. bonding-pair repulsion. The observed angle in ozone is 117°.

c. Now that we have had some practice we will abbreviate the explanation. To determine the shape of IF_5 count the valence shell electrons around the central I atom.

from the I atom	= 7
from the F atoms (5×1)	= 5
total valence shell electrons	= 12

The Lewis structure is

valence shell electron pairs = 6
number of lone pairs   = 1
number of bonding pairs = 5
molecular type AB_5E
molecular shape: square pyramid

The F—I—F bond angles are 90°.

d. To predict the shape of XeF_2. First count the valence shell electrons around Xe.

from the Xe atom   = 8
from the F atoms 2 × 1   = 2
total valence shell electrons = 10

Lewis structure

: F — Xe — F :

valence shell electron pairs   = 5
number of lone pairs   = 3
number of bonding pairs   = 2

molecular type AB_2E_3
molecular shape: linear

F—Xe—F

EXERCISES

1. Predict the shape of the following molecules.

    a. $BeCl_2$ b. NF_3 c. SF_4 d. ClF_3 e. KrF_4

2. Predict the shape of the following ions.
    a. ClO_2^- b. CO_3^{2-} c. $ZnCl_4^{2-}$

3. Which of the following molecules and ions have tetrahedral geometry?
    a. CCl_4 b. SCl_4 c. $AlCl_4^-$

4. Predict the following bond angles.

    a. O—C—O in CO_2 b. F—C—F in CF_4 c. H—O—O in H_2O_2

POLAR MOLECULES (10.2)

STUDY OBJECTIVE

1. Use electronegativity values and molecular geometries to predict whether or not a molecule possesses a dipole moment.

Diatomic Molecules. In the preceding chapter you learned that polar bonds are those in which the centers of positive charge and negative charge do not coincide. This charge separation is called a dipole, and it results from electronegativity differences between two bonded atoms. A molecule with an electric dipole is said to be polar and to possess a dipole moment.

The dipole moment μ is represented by an arrow or vector with the tail at the positive center and the head at the negative center. The length of the arrow represents the magnitude of the dipole moment.

$$\begin{array}{cc} \delta+ \quad \delta- & +\rightarrow \\ \text{H—Cl} & \text{H—Cl} \end{array}$$

Quantitatively the dipole moment is calculated from the equation

$$\mu = Qr$$

where Q is the magnitude of the charge in coulombs from either end of the dipole and r is the charge separation (the distance between the centers of positive and negative charge) in meters. In polar molecules this charge is never as great as one unit of charge. Rather the charge is a partial charge, $\delta+$ or $\delta-$, which signifies a charge of less than one unit. The SI unit for a dipole moment is the coulomb meter. Traditionally, dipole moments have been measured in debye units, where 1 debye (D) = 3.33×10^{-30} C m.

Molecules that possess a dipole moment are called **polar molecules**. Molecules without a dipole moment are called **nonpolar molecules**. All homonuclear diatomic molecules are nonpolar. In general, heteronuclear diatomic molecules are polar.

Polyatomic Molecules. The dipole moment of a molecule containing more than one bond depends both on *bond polarity and molecular geometry.* In a polyatomic molecule the arrows representing polar bonds may add together to yield a polar molecule with a resultant dipole moment μ. Conversely, the arrows may cancel each other when added, producing a zero resultant dipole moment. Table 10.3 lists dipole moments of some small molecules in Debye units.

Table 10.3 Dipole Moments
of Some Molecules

Molecule	μ (D)
H_2O	1.87 D
H_2S	1.10 D
NH_3	1.46 D
CO_2	0
SO_2	1.60 D
BCl_3	0

An important property of the water molecule is its dipole moment of 1.85 D. In the water molecule each bond is polar. The resultant of these bond dipoles yields a dipole moment for the molecule. The center of negative charge lies closer to the O atom than does the center of positive charge.

In CO_2, the two bond dipoles cancel and $\mu = 0$.

resultant $\mu = 0$

When the bond dipoles are equal and point in opposite directions the centers of + and − charge are both on the central atom. Thus, the absence of a molecular dipole in CO_2 suggests that the molecule is linear. Conversely, water has a dipole moment, and so it cannot be linear.

EXAMPLE 10.2   Polar and Nonpolar Molecules

Predict whether the following molecules are polar or nonpolar:
a.   CO
b. H_2CO
c. CCl_4

•Method of Solution

The polarity of a diatomic molecule depends on the electronegativity difference of the two bonded atoms.
a.   Oxygen is a more electronegative element than carbon. As a result the C≡O bond is polar with partial positive and negative charges on C and O, respectively.

The dipole moment of carbon monoxide has been found to be 0.1 D.

b. The dipolar nature of a polyatomic molecule depends on both the bond polarity and molecular geometry. Formaldehyde is a planar molecule with a polar $C \equiv O$ bond ($\Delta = 1.0$) and two C—H bonds of very low polarity ($\Delta = 0.4$). Neglecting the low polarity C—H bond dipoles, we get:

The net effect of these bond dipoles is that the center of positive charge is located approximately on the carbon atom and the center of negative charge lies near the oxygen atom.

Formaldehyde has a measured dipole moment of about 2.5 D.

c. Chlorine is more electronegative than carbon. Therefore CCl_4 has four bond dipoles. However, these polar bonds are arranged in a symmetric tetrahedral fashion about the central carbon atom.

In this situation the centers of positive and negative charge are on the carbon atom. The bond dipoles have canceled each other and the molecule as a whole does not possess a dipole moment.

EXERCISES

5. What two requirements must be met in order for a molecule to be polar?

6. Which of the following molecules are polar?
    a. BCl_3 b. SO_3 c. PF_3 d. SF_6

7. Which of the following molecules are polar?
    a. BeH_2 b. CO_2 c. SO_2 d. NO_2

VALENCE BOND THEORY AND HYBRID ORBITALS
(10.3 – 10.5)

STUDY OBJECTIVES

1. Describe covalent bond formation in terms of overlap of atomic orbitals.
2. Describe the formation of hybrid orbitals from atomic orbitals.
3. Assign a hybridization to the central atom of a given molecule.
4. Describe sigma and pi bond formation.

Covalent Bonds. The **valence bond theory** is one of two quantum mechanical descriptions of chemical bonding currently in use. The other is the molecular orbital theory which will be discussed in the next section. In the valence bond theory, the atomic orbitals of each bonded atom are essentially the same as in separate isolated atoms. A covalent bond forms when the orbitals from two atoms *overlap* and a pair of electrons occupies the region between the atoms. This overlap produces a region between the two nuclei where the probability of finding an electron is greatly enhanced. The presence of greater electron density between the two atoms tends to attract both atoms and bonding occurs. In HCl, for example, bonding results from the overlap of the hydrogen 1s orbital with the half-filled $2p_x$ orbital of the chlorine atom. A chlorine atom has only the $2p_z$ orbital available for bonding because its other orbitals are already filled. The valence electron configuration of Cl is $2p_x^2 2p_y^2 2p_z^1$.

Handwritten notes: Steps: 1) Draw Lewis Structure 2) Determine # of bonds 3) Draw and Tabulate

Hybridization. The concept of hybridization is used to account for bonding and geometry in molecules with trigonal planar, tetrahedral, and more complex shapes. The process of orbital mixing is called **hybridization,** and the new atomic orbitals are called **hybrid orbitals**. We can illustrate the main features of hybridization by examining the linear $BeCl_2$ molecule. To start with an isolated Be atom has the ground state electron configuration $1s^2 2s^2$. The orbital diagram for the *valence electrons* is

Handwritten note: ↳ $2s^2$

In a Be atom the 2p orbitals lie quite close in energy to the 2s orbital. Orbitals on the same atom that lie close together in energy have an ability to combine with one another to form *hybrid orbitals*. The presence of two Be—Cl bonds can be accounted for if, prior to bond formation, a 2s electron in a Be atom is promoted into an empty 2p orbital.

Now there are two unpaired electrons that could participate in two covalent bonds to two chlorine atoms. Since one electron is in an s orbital and one is in a p orbital, we would expect two types of Be—Cl bonds. However, both Be—Cl bonds are observed to be the same.

This equivalence is explained by hybridization or "mixing" of the 2s and the 2p orbitals to create two new equivalent orbitals called sp hybrid orbitals.

These two hybrid orbitals point in opposite directions from the Be nucleus. The Be—Cl covalent bonds result from overlap of beryllium sp hybrid orbitals that contain one electron each with the half-filled $3p_z$ orbitals of the Cl atoms.

We can summarize the rules of hybridization as follows:

1. Hybridization is a process of mixing orbitals on a single atom in a molecule.
2. Only orbitals of similar energies can be mixed to form hybrid orbitals.
3. The number of hybrid orbitals obtained always equals the number of orbitals mixed together.
4. Hybrid orbitals are identical in shape and they project outward from the nucleus with characteristic orientations in space.

Table 10.4 shows several of the more important hybrid orbitals discussed in the textbook and their characteristic geometrical shapes.

Table 10.4 Hybrid Orbitals and Their Geometries

Designation	Arrangement	Example
sp	linear (180° angle)	$BeCl_2$
sp^2	planar triangle (120° angle)	BF_3
sp^3	tetrahedral (109.5° angle)	CH_4, NH_4^+
sp^3d	trigonal bipyramidal (90° & 120°)	PCl_5
sp^3d^2	octahedral (90° angle)	SF_6, $SbCl_6^-$

Multiple Bonds. Double and triple bonds can also be understood in terms of overlap of atomic orbitals. The $C=C$ double bond in the ethylene molecule (see Figure 10.15 and Section 10.5 of the textbook), consist of one sigma (σ) bond and one pi (π) bond.

A **sigma bond** is a covalent bond in which the electron density is concentrated in a cylindrical pattern around the line of centers between the two atoms. In the $C=C$ bond, the σ bond results from overlap of an sp^2 hybrid orbital from one of the carbon atoms with a sp^2 hybrid orbital from the other carbon atom (Figure 10.15 text).

In a **pi bond** the electron density is concentrated above and below the C—C internuclear axis, but not cylindrically as in a sigma bond (Figure 10.2). The pi bond has two regions of electron density, but it should be kept in mind that it is only *one* bond. The π bond in ethylene results from sideways overlap of the unhybridized $2p_z$ atomic orbitals of the two carbon atoms.

(handwritten note in right margin:) Rules for sigma & pi bonds: • double bond: 1 sigma 1 pi • tuble bond: 1 sigma 2 pi

Figure 10.2 The pi bond has electron density above and below the internuclear axis. The electron density is not symetrical apound the internuclear axis as in a σ bond. The two parts represent one π bond.

EXAMPLE 10.3   Valence Bond Theory

Using unhybridized atomic orbitals, describe what atomic orbitals are used in the following molecules to form sigma bonds.
a. F_2
b. H_2S

•Method of Solution

a.   From the orbital diagram of fluorine we can see that a F atom has one half-filled 2p orbital that can be used for bonding. If this orbital overlaps with the half-filled 2p orbital of another F atom, a pair of electrons (one from each atom) spend most of the time between the two atoms and bonding occurs.

b.   Hydrogen atoms have a $1s^1$ electron configuration. Therefore, they can use the half-filled 1s orbital for bonding. The orbital diagram of the valence electrons in a S atom is

Overlap of the two half-filled 3p orbitals of a sulfur atom with the half-filled 1s orbitals of the two H atoms results in the formation of two covalent bonds.

•Comment

The directional properties of two p orbitals (say the p_y and p_z) are such that the orbitals project out from the S atom at right angles to each other. The H—S—H bond angle should be 90°. The measured angle is 92.2°.

EXAMPLE 10.4 Hybrid Orbitals

Using orbital diagrams, outline the steps used to describe the formation of a set of sp^2 hybridized orbitals on a carbon atom.

•Method of Solution

The ground state orbital diagram for the *valence electrons* of a carbon atom is

First promote one 2s electron to the empty 2p orbital. Mixing or hybridization of the 2s orbital with two of the 2p orbitals creates three equivalent sp^2 hybrid orbitals. These hybrid orbitals point out from the carbon atom to the corners of a triangle. One electron remains in an unhybridized 2p^2 atomic orbital.

EXAMPLE 10.5 Hybrid Orbitals

Describe the hybridization of the central atom in:
a. PH$_3$
b. PCl$_5$

•Method of Solution

First use VSEPR theory to predict a geometry, then match the geometry to the corresponding hybridization.
a.   According to VSEPR theory the PH$_3$ molecule has trigonal pyramidal geometry and a lone pair. There are four electron pairs around the P atom that occupy the corners of a teterahedron.

Therefore, four equivalent orbitals project out from the central P atom at 109° from each other. The bonding orbitals of the P atom must be sp^3 hybrid orbitals.
b.   According to VSEPR theory, PCl$_5$ has the geometrical shape of a trigonal bipyramid.

The type of hybridization that gives five equivalent orbitals that point to the corners of a trigonal bipyramid is sp^{3}d.

EXAMPLE 10.6   Sigma And Pi Bonds

Account for the bonding in H_2CO, which is a triangular molecule.
a.   What is the hybridization of the carbon atom?
b.   What are the approximate bond angles about the carbon atom?
c.   How many sigma and pi bonds are there in the molecule?

•Method of Solution

a.   The central carbon atom of H_2CO must be using sp^2 hybrid orbitals, since the three hybrid orbitals of this
     type point to the corners of a triangle. There is also one electron in an unhybridized 2p atomic orbital.

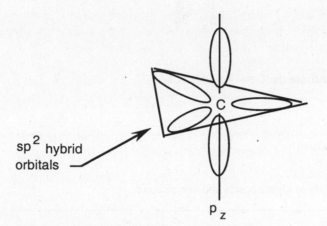

sp^2 hybrid
orbitals

p$_z$

b.   Overlap of the half-filled 1s orbital of a H atom with the half-filled sp^2 hybrid orbitals leads to formation of
     a C—H sigma bond. Two C—H bonds are formed in this way. The H—C—H bond angle should be 120°.
c.   Oxygen atoms have the valence electron configuration

Overlap of one half-filled 2p orbital with the remaining sp^2 hybrid yields a C—O sigma bond. The two
C—H sigma bonds and the C—O sigma bond are shown below.

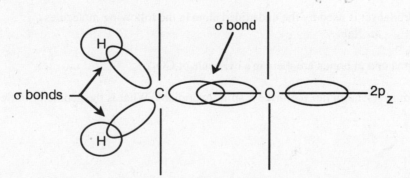

The sideways overlap of the carbon $2p_z$ orbital with the $2p_z$ orbital of the oxygen atom produces a C—O pi
bond.

The C—O sigma and pi bonds constitute the C=O double bond.

The complete molecule consists of three sigma bonds and one pi bond.

EXERCISES

8. Using only atomic orbitals describe the bond formation in F_2 and HBr.

9. What is the angle between the following two hybrid orbitals on the same atom?
   a. two sp orbitals   b. two sp^2 orbitals   c. two sp^3 orbitals

10. What hybrid orbital set is used by the underlined atom in the following molecules?
   a. C$\underline{S}_2$ b. $\underline{N}O_2^-$ c. $\underline{B}F_3$ d. $\underline{Cl}F_3$ e. $H_2\underline{C}O$

11. What hybrid orbital set is used by the underlined atom in the following molecules?
   a. $\underline{S}F_2$ b. $\underline{S}F_4$ c. $\underline{Xe}F_4$

12. How many sigma and pi bonds are there in a molecule of CO_2?

13. The sulfur atom in selenium hexachloride is sp^3d^2 hybridized. What is the geometry of $SeCl_6$?

MOLECULAR ORBITAL THEORY (10.6)

STUDY OBJECTIVES

1. Describe the formation of bonding and antibonding molecular orbitals.
2. Use molecular orbital theory to predict for diatomic molecules the electron configurations, bond orders, magnetic properties, and relative bond lengths.

Molecular Orbitals. Until now we have considered covalent bonds in terms of the *valence bond theory*. In that theory the electrons in a *molecule* occupy atomic orbitals of the individual *atoms*. Covalent bonds are visualized in terms of the overlap of two half-filled atomic orbitals. This overlap increases the electron density between both atomic centers.

The *molecular orbital model* has a different approach. In this model the available atomic orbitals are combined to form orbitals that belong to the *entire molecule*. Like atomic orbitals, these molecular orbitals have different shapes, sizes, and energies. Once the molecular orbitals have been derived, electrons are allotted to them in a manner analogous to allotting electrons to the orbitals of atoms.

According to the molecular orbital (MO) theory, the result of the interaction of two 1s orbitals is the formation of two molecular orbitals. One of these is a **bonding molecular orbital**, which has a lower energy than the original 1s orbitals. The other molecular orbital is **an antibonding orbital**, which is of higher energy than the original 1s orbitals. It is the greater stability of the electrons in the bonding MO that results in covalent bond formation. In the bonding MO, electron density is concentrated between the two atoms. While the antibonding MO actually has a node (that is, a zero value) in the electron density midway between the two nuclei. There is no contribution to bonding when electrons enter antibonding MOs. An energy level diagram for the combination of two atomic orbitals of hydrogen to form the MOs in the H_2 molecule is shown in Figure 10.3.

Figure 10.3 Energy levels of bonding and antibonding molecular orbitals in the H_2 molecule. The two electrons of the molecule enter the bonding molecular orbital (σ_{1s}). Sigma antibonding and bonding molecular orbitals are formed from the combination of 1s atomic orbitals. The bonding MO is lower in energy than the atomic orbitals, and the antibonding MO is higher in energy than the atomic orbitals.

The antibonding and bonding MOs that result from combining 1s orbitals are examples of **sigma molecular orbitals**. The electron density of sigma MOs is distributed symmetrically about the line of centers between the two atoms. The sigma bonding MO and sigma antibonding MO resulting from combination of 1s atomic orbitals are designated σ_{1s} and σ_{1s}^*, respectively. Sigma MOs also result from the combination of certain other atomic orbitals. For instance, the $2p_x$ orbitals on different atoms combine to form σ_{2p} bonding and σ_{2p}^* antibonding orbitals as shown in Figure 10.23 (a) of the text.

Two $2p_y$ orbitals and two $2p_z$ orbitals on different atoms must approach each other sideways as shown in Figure 10.23 (b) of the text. In the resulting molecular orbital, the electron density is concentrated above and below the line joining the two bonded atoms, rather than around the line of centers. Such an MO is called a **pi molecular orbital**. The symbol π_{2p_z} stands for a bonding pi orbital formed by combination of two $2p_z$ atomic orbitals. The pi antibonding orbital is designated, $\pi_{2p_z}^*$.

Also, combination of two $2p_y$ atomic orbitals from different atoms forms π_{2p_y} bonding and $\pi_{2p_y}^*$ antibonding orbitals as well.

Molecular Orbital Configurations.

For homonuclear diatomic molecules any two similar atomic orbitals can merge to form two MOs, one bonding and one antibonding orbital. For H_2 we can illustrate the formation of the σ_{1s} and σ_{1s}^* MO as the separate atoms are brought closer together in Figure 10.3. The H_2 molecule has two electrons that are positioned in the lowest energy orbital, the σ_{1s}. With both electrons in the bonding MO the molecule is stable. We write the ground state MO configuration

$$H_2 \ (\sigma_{1s})^2$$

where the superscript "2" means two electrons occupy the sigma 1s orbital.

The molecular orbital theory predicts He_2 to be unstable. A diatomic helium molecule would have four electrons. Using the same diagram as for H_2 we can see that the third and fourth electrons would enter the σ_{1s}^* orbital, resulting in the configuration:

$$He_2 \ (\sigma_{1s})^2(\sigma_{1s}^*)^2$$

In this case the stability gained by having two electrons in a bonding orbital is cancelled by the presence of two electrons in the antibonding orbitals. A stable He_2 molecule should not exist.

When comparing the stabilities of molecules, we use the concept of **bond order**.

$$\text{bond order} = \frac{1}{2} \left(\begin{array}{c} \text{number of electrons} \\ \text{in bonding MOs} \end{array} - \begin{array}{c} \text{number of electrons} \\ \text{in antibonding MOs} \end{array} \right)$$

The value of the bond order is a measure of the stability of the molecule. For H_2:

$$\text{bond order} = \frac{1}{2} \ (2 - 0) = 1$$

and for He_2:

$$\text{bond order} = \frac{1}{2} \ (2 - 2) = 0$$

A bond order of 1 represents a single covalent bond, and a bond order of 0 means the molecule is not stable. Bond orders of 2 and 3 result for molecules with double bonds and triple bonds, respectively.

For most homonuclear diatomic molecules containing atoms of second period elements, an approximate order of molecular energy levels is:

$$\sigma_{1s} < \sigma_{1s}^* < \sigma_{2s} < \sigma_{2s}^* < \pi_{2p_y} = \pi_{2p_z} < \sigma_{2p_x} < \pi_{2p_y}^* = \pi_{2p_z}^* < \sigma_{2p_x}^*$$

The electron configurations of diatomic molecules of the second period elements and some of their known properties are summarized in Table 10.5 in the textbook. The table shows the bond order of these molecules as predicted by MO theory and the corresponding bond energies and bond lengths. The number of unpaired electrons also correlates with the magnetic properties of these molecules. The molecular orbital theory satisfactorily predicts the magnetic properties of molecules, something that the valence bond theory does not do. Example 10.7 illustrates writing an MO electron configuration and calculation of a bond order.

A number of rules that apply to molecular orbitals and the stability of molecules are summarized below.

1. The number of molecular orbitals formed is always equal to the number of atomic orbitals combined.
2. The more stable the bonding molecular orbital, the less stable is the corresponding antibonding molecular orbital.
3. In a stable molecule, the number of electrons in bonding molecular orbitals is always greater than that in antibonding molecular orbitals.
4. As with atomic orbitals, each molecular orbital can accommodate up to two electrons with opposite spins, in accordance with the Pauli exclusion principle.
5. When electrons are added to molecular orbitals having the same energy, the most stable arrangement is that predicted by Hund's rule. That is, as far as possible, electrons occupy these orbitals singly, and with parallel spins, rather than in pairs.
6. The number of electrons in the molecular orbitals is equal to the sum of all the electrons on the atoms.

EXAMPLE 10.7   Molecular Orbital Diagram

Represent bonding in the O_2^- ion by means of a molecular orbital diagram. What is the bond order?

•Method of Solution

The O_2 molecule has 16 electrons. The O_2^- ion will have 17 electrons. The energy level diagram is shown below. The 4 electrons in the σ_{1s} and $\sigma_{1s}*$ orbitals are not shown because they are not involved in bonding.

$$\underline{\quad\quad}\ \sigma_{2p}^{*}$$

$$\underline{\uparrow\downarrow}\ \pi_{2p}^{*}\quad \underline{\uparrow}\ \pi_{2p}^{*}$$

$$\underline{\uparrow\downarrow}\ \sigma_{2p}$$

$$\underline{\uparrow\downarrow}\ \pi_{2p}\quad \underline{\uparrow\downarrow}\ \pi_{2p}$$

$$\underline{\uparrow\downarrow}\ \sigma_{2s}^{*}$$

$$\underline{\uparrow\downarrow}\ \sigma_{2s}$$

The electron configuration of O_2^- ion is

$$O_2^-\ (\sigma_{1s})^2(\sigma_{1s}*)^2(\sigma_{2s})^2(\sigma_{2s}*)^2(\pi_{2p_y})^2(\pi_{2p_z})^2(\sigma_{2p_x})^2(\pi_{2p_y}*)^2(\pi_{2p_z}*)^1$$

$$\text{bond order} = \frac{1}{2}\left(\begin{array}{l}\text{number of electrons}\\ \text{in bonding MOs}\end{array} - \begin{array}{l}\text{number of electrons}\\ \text{in antibonding MOs}\end{array}\right)$$

$$= \frac{1}{2}(8-5) = 1.5$$

EXAMPLE 10.8 Paramagnetism and Diamagnetism

Which of the following species are paramagnetic and which are diamagnetic:

a. N_2

b. N_2^+

c. B_2

•Method of Solution

a. The electron configuration of N_2 is:

$$N_2 \ (\sigma_{1s})^2(\sigma_{1s}*)^2(\sigma_{2s})^2(\sigma_{2s}*)^2(\pi_{2p_y})^2(\pi_{2p_z})^2(\sigma_{2p_x})^2$$

All orbitals contain pairs of electrons. Because of the lack of unpaired electrons, N_2 is diamagnetic.

b. N_2^+ will have one less electron than N_2. Therefore it has one unpaired electron and is paramagnetic.

c. The electron configuration of B_2 should have 10 electrons:

$$B_2 \ (\sigma_{1s})^2(\sigma_{1s}*)^2(\sigma_{2s})^2(\sigma_{2s}*)^2(\pi_{2p_y})^1(\pi_{2p_z})^1$$

Because of the unpaired electrons, B_2 is paramagnetic.

EXERCISES

14. a. Which has the higher energy, a bonding molecular orbital or its corresponding antibonding molecular orbital? b. Which has the higher energy, a bonding molecular orbital or the atomic orbitals from which it is created?

15. Compare the bond order in Be_2 to that in Be_2^+.

16. Give the electron configurations for Li_2 and Li_2^+. Which has the stronger Li—Li bond?

17. Which of the following diatomic species are paramagnetic?
    a. O_2 b. O_2^- c. Li_2

18. What orbitals overlap to form a delocalized π bond in ozone (O_3)?

CONCEPTUAL QUESTIONS

1. In molecules having the trigonal bipyramidal geometry why does a lone pair occupy an equitorial position rather than an axial position?

2. Experimental evidence shows that carbon dioxide has no dipole moment. What does this suggest about its molecular shape?

3. Both N_2 and O_2 can be ionized to form N_2^+ and O_2^+, respectively. Explain why the bond order of N_2 is greater than that of N_2^+, but the bond order of of O_2 is less than that of O_2^+.

PRACTICE TEST

1. Use VSEPR theory to predict the shapes of the following molecules:
   a. BrF_5 b. HCN   c. BF_3 d. SO_2 e. SCl_2

2. What type of hybrid orbital is used by the central atom of each of the molecules in problem 1?

3. Use VSEPR theory to predict the shapes of the following polyatomic ions:
   a. BeF_3^- b. AsF_4^- c. ClF_4^- d. NO_3^- e. SO_4^{2-}

4. What type of hybridization is used by the central atom of each of the ions in problem 3?

5. Which of the following molecules would be tetrahedral?
   a. SO_2 b. SiH_4 c. SF_4 d. BCl_3 e. XeF_4

6. Predict the approximate bond angles in
   a. $GeCl_2$ b. IF_4^- c. $TeCl_4$

7. CCl_4 is a perfect tetrahedron, but $AsCl_4^-$ is a distorted tetrahedron. Explain.

8. $BeCl_2$ and $TeCl_2$ are both covalent molecules, yet $BeCl_2$ is linear while $TeCl_2$ is nonlinear (bent). Explain.

9. What types of hybrid orbitals can be formed by elements of the third period that cannot be formed by elements of the second period?

10. When we describe the formation of hybrid orbitals on a central atom, must all the available atomic orbitals enter into hybridization?

11. Which molecule should have the larger dipole moment, HBr or HI?

12. Which one of the following molecules has a dipole moment?
    a. CCl_4 b. H_2S c. CO_2 d. BCl_3 e. Cl_2

13. Which of the following molecules have no dipole moment?
    a. ClF   b. BCl_3 c. $BeCl_2$ d. Cl_2O e. H_2CO

14. Choose the best answer. For the water molecule:
    a. The bonds are polar, and the molecule is nonpolar.
    b. The bonds are nonpolar, and the molecule is polar.
    c. The bonds are polar, and the molecule is polar.
    d. The bonds are nonpolar, and the molecule is nonpolar.

15. Hydrogen peroxide (H_2O_2) has a dipole moment of 2.1 D. Which of the bonds in H_2O_2 are polar? Is the molecule linear?

16. Draw a molecular orbital energy level diagram and determine the bond order for each of the following.
    a. H_2^+ b. HHe   c. He_2^+

17. The compound calcium carbide, CaC_2, contains the acetylide ion C_2^{2-}.
    a. Write molecular orbital electron configurations for C_2 and C_2^{2-}.
    b. Compare their bond orders.

18. The bond length in N_2 is 109 pm and in N_2^+ it is 112 pm. Explain why the bond lengths differ in this way.

ANSWERS

Exercises

1. a. linear   b. trigonal pyramidal   c. seesaw   d. T shaped   e. square planar
2. a. bent   b. trigonal planar   c. tetrahedral
3. a. and c.
4. a. 180°   b. 109.5°   c. less than 109.5°
5. The two requirements are polar bonds and a molecular geometry in bonds dipoles cancel.
6. c.
7. c. and d.
8. Each fluorine atom has the same electron configuration: $1s^2 2s^2 2p_x^2 2p_y^2 2p_z^1$. The $2p_z$ orbitals of each F atom are only half full. Therefore, a sigma bond can be formed by overlap of the $2p_z$ orbital from each atom. In HBr the 1s orbital of the H atom overlaps with the $4p_z$ orbital of Br to form a sigma bond.
9. a. 180°   b. 120°   c. 109.5°
10. a. sp   b. sp^2 c. sp^2 d. sp^3d e. sp^2
11. a. sp^2 b. sp^3d c. sp^3d^2
12. 2 σ and 2 π
13. octahedral
14. The antibonding MO.
15. The bond order for Be_2 is zero, while for Be_2^+ it is 0.5.
16. Li_2 $(\sigma_{1s})^2(\sigma_{1s}^*)^2(\sigma_{2s})^2$ Li_2^+ $(\sigma_{1s})^2(\sigma_{1s}^*)^2(\sigma_{2s})^1$ Li_2 has the stronger bond.
17. a. and b.
18. The central oxygen atom is sp^2 hybridized and forms σ bonds to the two terminal oxygen atoms. The central O atom has an unhybridized $2p_z$ orbital that overlaps "sideways" with the $2p_z$ orbitals of both terminal O atoms.

Conceptual Questions

1. Recall that lone-pair electrons take up more space than electrons in a bond. A lone pair in an axial position has three close neighbors in the bonding electron pairs in the equatorial positions. A lone pair in an equatorial position has two close neighbors in the bonding electron pairs in the axial positions. (The neighboring equatorial electron pairs are too far away). This means that a lone pair will have more room and have less repulsion when it is in a equatorial position.

2. We know that the C—O bonds are polar. For CO_2 to be nonpolar, the bond dipoles must be pointing in opposite directions so that they cancel. The molecule must be linear.

3. Compare the electron configurations.

    $$N_2 \ (\sigma_{1s})^2(\sigma_{1s}^*)^2(\sigma_{2s})^2(\sigma_{2s}^*)^2(\pi_{2p_y})^2(\pi_{2p_z})^2(\sigma_{2p_x})^2$$

    $$N_2^+ \ (\sigma_{1s})^2(\sigma_{1s}^*)^2(\sigma_{2s})^2(\sigma_{2s}^*)^2(\pi_{2p_y})^2(\pi_{2p_z})^2(\sigma_{2p_x})^1$$

    $$O_2 \ (\sigma_{1s})^2(\sigma_{1s}^*)^2(\sigma_{2s})^2(\sigma_{2s}^*)^2(\pi_{2p_y})^2(\pi_{2p_z})^2(\sigma_{2p_x})^2(\pi_{2p_y}^*)^1(\pi_{2p_z}^*)^1$$

    $$O_2^+ \ (\sigma_{1s})^2(\sigma_{1s}^*)^2(\sigma_{2s})^2(\sigma_{2s}^*)^2(\pi_{2p_y})^2(\pi_{2p_z})^2(\sigma_{2p_x})^2(\pi_{2p_y}^*)^1$$

    The bond order in N_2 is 3. Loss of 1 electron gives N_2^+. The electron came out of a bonding orbital and so the bond order is less than in N_2, b.o. = 2.5. The bond order in O_2 is 2. Loss of 1 electron gives O_2^+. The electron came out of an antibonding orbital and so the bond order is greater than in O_2, b.o. = 2.5.

Practice Test

1. a. Square pyramid   b. linear   c. triangular   d. bent   e. bent
2. a. sp^3d^2 b. sp   c. sp^2 d. sp^2 e. sp^2
3. a. triangular   b. distorted tetrahedral   c. square planar
    d. triangular   e. tetrahedral
4. a. sp^2 b. sp^3d c. sp^3d^2 d. sp^2 e. sp^3
5. b. SiH_4
6. a. 120°   b. 90°   c. 90° and 120°
7. CCl_4 has four electron pairs about the central carbon atom, while $AsCl_4^-$ has 5 pairs about the As atom with only 4 Cl atoms attached. The extra pair of unshared electrons prevents the formation of a tetrahedron by $AsCl_4^-$.
8. There are two electron pairs in the valence shell of Be in $BeCl_2$, but there are four pairs in the valence shell of Te in $TeCl_2$. These four pairs are at the corners of a tetrahedron. When two chlorine atoms bond to a Te atom via two of the pairs, the Cl—Te—Cl bond angle is 109.5°.
9. sp^3d, sp^3d^2
10. No
11. HBr
12. b. H_2S
13. b. BCl_3 c. $BeCl_2$
14. c.
15. The O—H bonds are polar. Nonlinear.

16.

	H_2^+	HHe	He_2^+
σ_{1s}^*	—	$\uparrow$	$\uparrow$
σ_{1s}	$\uparrow$	$\uparrow\downarrow$	$\uparrow\downarrow$
Bond order	0.5	0.5	0.5

17. a. C_2 $(\sigma_{1s})^2(\sigma_{1s}^*)^2(\sigma_{2s})^2(\sigma_{2s}^*)^2(\pi_{2p_y})^2(\pi_{2p_z})^2$

C_2^{2-} $(\sigma_{1s})^2(\sigma_{1s}^*)^2(\sigma_{2s})^2(\sigma_{2s}^*)^2(\pi_{2p_y})^2(\pi_{2p_z})^2(\sigma_{2p_x})^2$

b. For C_2 the bond order = 2. For C_2^{2-} the bond order = 3.

18. The bond order in N_2 is 3, while in N_2^+ it is only 2.5. The greater the bond order, the shorter the bond distance.

Chapter Eleven
INTERMOLECULAR FORCES AND LIQUIDS AND SOLIDS

- The Kinetic Molecular Theory of Liquids and Solids (11.1)
- Intermolecular Forces (11.2)
- Properties of Liquids (11.3)
- The Solid State (11.4 – 11.7)
- Phase Changes and Phase Diagrams (11.8 – 11.9)

THE KINETIC MOLECULAR THEORY OF LIQUIDS AND SOLIDS (11.1)

STUDY OBJECTIVE

1. Describe the characteristic properties of solids, liquids, and gases.

The Molecular View of Liquids and Solids. In this chapter we will examine the structure of liquids and solids, and discuss some of the fundamental properties of these two states of matter. Matter can exist in three states, or phases: gas, liquid, and solid. Each of these states will be described from a molecular viewpoint in terms of four characteristics: distance between molecules, attractive forces between molecules, the motion of molecules, and the orderliness of the arrangement of molecules. Figure 1.6 (text) gives a molecular representation of a solid, a liquid, and a gas.

In Chapter 5 you learned that a gas can be pictured as a collection of molecules that are far apart, are in constant random motion, and exert almost no attractive forces on each other. The kinetic molecular theory was applied to explain the behavior of gases.

Liquids and solids are quite different from gases. They are referred to as the condensed states because their molecules are very close together. Gases have low density, high compressibility, and completely fill a container. The condensed states have relatively high density, are almost incompressible, and have definite volumes. These properties indicate that the "molecules" of solids and liquids are close together and are held by strong intermolecular forces.

One difference between solids and liquids is that liquids like gases are *fluids*, but solids are not. In liquids, molecules can move past each other even though they cannot get very far from each other. In solids, molecules are fixed in position, and at most can only vibrate about that position. Many solids are characterized by long-range order of their constituent "molecules." This order gives rise to crystal structures. Table 11.1 summarizes some of the properties of the three states of matter.

Table 11.1 Characteristic Properties of the States of Matter

Gas	Liquid	Solid
Assumes the volume and shape of its container	Has a definite volume but assumes the shape of its container	Has a definite volume and shape
Is a fluid (flows readily)	Is a fluid	Is not a fluid
Very compressible	Only slightly compressible	Virtually incompressible
Low density	High density	High density
Molecules far apart	Molecules close together	Molecules close together

INTERMOLECULAR FORCES (11.2)

STUDY OBJECTIVES

1. Describe the different types of intermolecular attractive forces.
2. Relate the strength of dispersion forces to molecular mass.

van der Waals Forces. The forces that hold atoms together such as covalent bonds exist within molecules. In this chapter you will learn about the types of forces that act between individual molecules that hold them close together as in a solid or a liquid. Forces of attraction between molecules are called **intermolecular forces (IMF)**. Intermolecular forces can be grouped for convenience into *van der Waals forces, ion-dipole forces, and hydrogen bonds.* The van der Waals forces cause a gas to deviate from ideal gas behavior and are also responsible for gases condensing to form liquids. They are attractive forces resulting from *dipole-dipole forces, dipole-induced dipole forces, and dispersion forces.* We shall discuss each of these interactions briefly.

Dipole-Dipole Forces. Some molecules have a dipole moment μ, and are referred to as polar molecules. In a polar substance, molecules tend to become aligned with the positive end of one molecule directed toward the negative ends of neighboring dipoles. See Figure 11.1 in the text. The attraction is electrostatic. The larger the dipole moment, the stronger the force of attraction. **Dipole-dipole forces** are attractive forces between polar molecules.

OMIT

Dipole-Induced Dipole Forces. The presence of a polar molecule in the vicinity of another molecule (usually nonpolar) has the effect of polarizing the second molecule. This means that a dipole moment has been induced in the previously nonpolar molecule. The induced dipole can then interact with the dipole moment of the first molecule, and the two molecules are attracted toward each other. The induced dipole is a temporary dipole. Its interaction with a permanent dipole such as that in a polar molecule is called a **dipole-induced dipole force**.

Dispersion Forces. Nonpolar molecules such as O_2, SF_6, and even the noble gases He, Ne, Ar, Kr, and Xe all show deviations from the ideal gas equation, and all condense at low temperatures to form liquids. We can ask: What kind of attractive forces exist between nonpolar molecules? In 1930, Fritz London proposed that although these molecules have no permanent dipole moments, their electron clouds are fluctuating. In a helium atom, for instance, the two electrons occupy a 1s orbital, which has a spherical shape. The electrons are in constant motion, and if for an instant they should both move to the same side of the nucleus, a short-lived dipole will exist. In an instant the electron will continue its motion and the dipole will be gone, but a new one will be formed. These short-lived dipoles are called *instantaneous dipoles or temporary dipoles*. An instantaneous dipole can polarize a neighboring molecule, thereby producing an induced dipole. The two dipoles will tend to stick together. The attractive interactions caused by instantaneous dipoles are known as **dispersion forces**.

The strength of dispersion forces depends on the polarizability of the molecule and can be as large as, or larger than, dipole-dipole forces. **Polarizability** is the tendency of an electron cloud to be distorted by the presence of an electrical charge such as that of an ion or the partial charge of a dipole. In general, the polarizability increases as the total number of electrons in a molecule increases. Since molecular mass and number of electrons are related, the polarizability of molecules and the strength of dispersion forces increases with increasing molecular mass.

Ion-Dipole Forces. The electrostatic attraction between an ion and a polar molecule is called an **ion-dipole force**. Hydration of ions, which was discussed in Section 6.6, is a good example of ion-dipole interactions. When an ionic compound such as NaBr dissolves its cations and anions are attracted to water molecules. Water molecules being polar have a negative end and a positive end. The Na^+ ions attract the negative end of the water molecule, and the Br^- ions attract the positive end of the water molecule. As the charge of an ion increases, it attracts polar molecules more strongly. Thus Mg^{2+} ions attract water molecules more strongly than Na^+ ions, and S^{2-} ions attract water molecules more strongly than Br^- ions.

Hydrogen Bonding. An additional type of intermolecular force is necessary to explain certain properties of ammonia, water, and hydrogen fluoride. A **hydrogen bond** is a special type of dipole-dipole interaction between the hydrogen atom in a polar bond and an O, N, or F atom. In water, for instance, the attractions are more than just the attractions of one dipole for another. Each hydrogen atom with its partial positive charge is attracted to one of the lone electron pairs of an oxygen atom of a *neighboring molecule*. This interaction is called a hydrogen bond, and is usually represented by the three dots in the following:

$$
\begin{array}{c}
\text{H} \\
\quad \quad \quad \overset{..}{} \quad \quad \delta+ \ | \ \delta- \\
\text{H—O} : \cdots \text{ H—O} : \\
\quad \quad | \quad \quad \quad \quad \overset{..}{} \\
\quad \quad \text{H}
\end{array}
$$

Hydrogen bonding is limited to compounds containing nitrogen, oxygen, and fluorine. Hydrogen bonds are the strongest of the intermolecular forces, with energies of the order 10–40 kJ/mol. Van der Waals interactions correspond to between 2 and 20 kJ/mol.

The requirements for hydrogen bond formation are:

1. The element that is covalently linked to hydrogen must be sufficiently electronegative to attract bonding electrons and leave the hydrogen atom with a significant δ+ charge. Only N, O, and F are sufficiently electronegative.
2. The electronegative atom, bound to hydrogen by the hydrogen bond, must have a lone pair of electrons.
3. The small size of the hydrogen atom allows it to approach N, O, and F atoms in neighboring molecules very closely. It is significant that hydrogen bonding is limited to these three elements of the second period. Both sulfur and chlorine are highly electronegative, but do not form hydrogen bonds. These third period atoms are apparently too large and do not present a highly localized nonbonding electron pair for the H atom to be attracted to.

EXAMPLE 11.1   Types of Intermolecular Forces

Indicate all the different types of intermolecular forces that exist in each of the following substances.
a. $CCl_4(l)$
b. $HBr(l)$
c. $CH_3OH(l)$

•Method of Solution

First you must use the concepts of bond polarity and molecular structure to decide whether the molecule is polar or not.
a. CCl_4 is nonpolar. The only type of IMF between nonpolar molecules are dispersion forces.
b. HBr is a polar molecule. The types of IMF are dipole-dipole and dispersion forces. There is no hydrogen bonding in HBr. The Br atom is not electronegative enough.
c. CH_3OH is polar and has a hydrogen atom bound to an oxygen atom. Dipole-dipole attractions, dispersion forces, and H bonds exist in liquid methanol.

EXAMPLE 11.2   Types of Intermolecular Forces

Which of the following substances should have the strongest intermolecular attractive forces:
N_2, Ar, F_2, or Cl_2?

•Method of Solution

Note that none of these molecules is polar and that there is no chance for H bonding to occur. The only intermolecular forces existing in these molecules then are dispersion forces. Dispersion forces increase as the polarizability of the molecule increases, while polarizability increases with molecular mass. *Answer:* The most polarizable molecule will be Cl_2 (70.9 amu).

EXAMPLE 11.3   Intermolecular Forces

The dipole moment μ in HCl is 1.03 D, and in HCN it is 2.99 D. Which one should have the higher boiling point?

•Method of Solution

The larger the dipole moment, the stronger the IMF. The stronger the IMF, the higher the temperature needed to provide molecules of the liquid with enough kinetic energy to overcome these attractive forces. The boiling point of HCN should be greater than that of HCl. Observed values are b.p. (HCl) = –85°C, and b.p. (HCN) = 26°C.

EXERCISES

1. Indicate the types of intermolecular forces that exist between molecules (or basic units) in each of the following.
    a. PCl_3 b. CO_2 c. Cl_2 d. ICl   e. KCl

2. Which of the following should have the strongest intermolecular forces?
    CH_4 Cl_2 CO CS_2

3. Which of the folowing species are capable of hydrogen bonding?
   a. CH_3F b. HF   c. CH_3CH_2OH d. CH_3NH_2 e. CH_4

4. Which member of each pair should have the higher boiling point?
   a. O_2 or CO   b. Br_2 or ICl   c. H_2O or H_2S d. PH_3 or AsH_3

5. What type of intermolecular forces must be overcome in order to:
   a. vaporize water   b. dissociate H_2 into H atoms   c. boil liquid O_2

PROPERTIES OF LIQUIDS (11.3)

STUDY OBJECTIVES

1. Describe surface tension and viscosity.
2. List six unusual properties of water.

Surface Tension. The surface tension of a liquid is a property that has no direct counterpart in solids or gases. Surface tension is what makes water bead up on a freshly waxed surface and what makes soap bubbles round. Surface tension is a force that tends to minimize the surface area of a drop of liquid. Energy is required to expand the surface of a liquid, and the **surface tension** is the amount of energy required to increase the surface area of a liquid by a unit area. Liquids in which strong intermolecular forces exist exhibit high surface tensions.

Water beads up on a freshly waxed surface because its cohesive forces are stronger than its adhesive forces. **Cohesion** is the intermolecular attraction between like molecules in the drop. **Adhesion** is the intermolecular attractions between unlike molecules, such as between water and wax. Since there is very little attraction between water and wax (adhesion), the strong cohesive forces tend to maintain the drop. The drop adopts spherical shape because a sphere has the least surface area.

When water comes in contact with glass, adhesion is stronger than cohesion. Water is pulled against the glass surface, and we say that water "wets" the surface. Wetting results in the spreading of a thin film of water on the glass surface. The strength of the cohesive forces overcomes the ability of the surface tension to make a spherical drop.

Viscosity. One characteristic of liquids that we have all observed is related to how freely they flow. Water pours much more freely than motor oil, and motor oil more readily than glycerol, for instance. The unique pouring characteristics of each liquid are the result of its viscosity. **Viscosity** is a measure of a fluids resistance to flow. Liquids whose molecules have strong intermolecular forces have greater viscosities than liquids that have weaker intermolecular forces. Table 11.3 (in the text) lists viscosities of some common liquids.

The viscosity of water and glycerol is the result of strong hydrogen bonds. The viscosity of motor oil results from strong London dispersion forces.

Properties of Water. Given its abundance and familiarity, we often fail to realize just how unusual water actually is. Some of the unusual properties of water are:

1. Water has a considerably greater surface tension than most other liquids.
2. At 4.184 J/g·°C, the specific heat of water is one of the highest of all substances.
3. The heats of fusion and vaporization are 6.0 and 40.8 kJ/mol, respectively. H_2O is unusual because 40.8 kJ is a huge vaporization energy. Compare values in Table 11.6 (text).

4. Boiling point. The boiling point of water is about 200°C higher than might be reasonably expected. Boiling point tends to be related to the molar mass. Compare the boiling point of water with that of other low molar mass liquids in Table 11.2.

Table 11.2 Boiling Points of Some Small Molecules

Liquid	Molar Mass (g/mol)	Boiling Pt (°C)
H_2O	18	100°
CH_4	16	−159°
NH_3	17	−33°

5. Density versus temperature. As liquids cool they become more and more dense. As water is cooled from 100°C down to 4°C it does indeed get more dense. But from 4°C (actually 3.98°C) down to 0°C just the opposite happens: it gets less dense (see Table 11.3).

Table 11.3 Density of H_2O Near 0°C

Temperature °C	Density (g/cm³)
10.0	0.9997
5.0	0.99999
3.98	1.00000
2.0	0.99997
0	0.9998

6. Density of liquid versus density of solid. There are over eight million known chemical compounds. All, except a dozen or so, have a solid state that is more dense than the liquid. Water, of course, is one of the exceptions. The density of solid water at 0°C is 0.917 g/cm³, which compares with 0.9998 g/cm³ for the liquid. Because of its lower density, ice floats on water.

EXAMPLE 11.4   Intermolecular Forces in Liquids

The viscosities of ether and water are 0.000233 and 0.00101 N s/m². Discuss their relative values in terms of the following molecular structures.

```
            O                              CH3  O  CH3
water      / \              ethyl ether     \ /  \ /
          H   H                             CH2  CH2
```

•Method of Solution

Molecules with strong intermolecular forces have greater viscosities than those that have weak intermolecular forces. Here both molecules are dipolar: water because of the O—H bonds, and ether because of the polar C—O bond. However, water has a higher viscosity than ether because of its ability to form hydrogen bonds between molecules.

EXERCISES

6. What property of a liquid makes it possible to fill a glass of water even above the rim?

7. Which of the two liquids has the higher viscosity at room temperature?
    C_2H_5OH or CH_3OCH_3

THE SOLID STATE (11.4 – 11.7)

STUDY OBJECTIVES

1. Determine the number of atoms per unit cell given the type of unit cell.
2. Calculate the spacing between planes in a crystal using Bragg's equation.
3. Predict whether the type of crystal formed by a substance is ionic, covalent, molecular, or metallic.

Crystal Structure. The atoms, molecules, or ions of a **crystalline solid** occupy specific positions in the solid, and possess long-range order. The **unit cell** is the smallest unit that when repeated over and over again generates the entire crystal. Three types of unit cell are the simple cubic cell (scc), the body-centered cubic cell (bcc), and the face-centered cubic cell (fcc). These types are illustrated in Figure 11.17 of the text.

In a **simple cubic cell**, particles are located only at the corners of each unit cell. In a **body-centered cubic cell**, particles are located at the center of the cell as well as at the corners. In a **face-centered cubic cell**, particles are found at the center of each of the six faces of the cell as well as at the corners. In many calculations involving properties of a crystal, it is important to know how many atoms or ions are contained in each unit cell. Atoms at the corners of unit cells are shared by neighboring unit cells. For a cubic cell, each corner atom is shared by eight unit cells and is counted as 1/8 particle for each unit cell. A face-centered atom is shared by two unit cells and is counted as 1/2 of an atom for each unit cell. An atom located at the center belongs wholly to that unit cell.

The packing efficiency is the percentage of the cell space occupied by spheres. The empty space is the interstices or "holes" between spheres. There are many substances whose atomic arrangement can be pictured as a result of packing together identical spheres so as to achieve maximum density. Many metallic elements and many molecular crystals display these "closest packed" structures. The text describes these two "closest packed" structures in Section 11.4. Alternating layers of spheres in ABABAB… fashion yields a hexagonal close-packed (hcp) structure, while alternating layers of spheres in the ABCABC… fashion generates the cubic close-packed (ccp) structure. In both, structures each sphere has a coordination number of 12 which is the maximum coordination number. For both structures the packing efficiency is 74%.

X-Ray Diffraction of Crystals. Crystal structures are analyzed experimentally by the technique of X-ray diffraction. Diffraction is a wave property. When an X-ray beam encounters a single crystal, the beam is scattered from the crystal at only a few angles, rather than randomly. The scattering angle θ depends on the spacing d between layers or planes of atoms in the crystal, the wavelength of the X rays, and the reflection order ($n = 1, 2, 3, \ldots$). These quantities are related by the Bragg equation:

$$n\lambda = 2d \sin \theta$$

The spacing (d) between atomic planes in crystals can be determined by using X rays of known wavelength (λ) and by experimental measurement of the angle θ.

Types of Crystals. In a solid, atoms, molecules, or ions occupy specific positions called lattice points. Crystalline solids are characterized by a regular three-dimensional arrangement of lattice points. In *ionic solids* the lattice points are occupied by positive and negative ions. In all ionic compounds there are continuous three-dimensional networks of alternating positive and negative ions held together by strong electrostatic forces (ionic bonds). Most ionic crystals possess high lattice energies, high melting points, and high boiling points. In ionic solids there are no discrete molecules as are found in molecular or covalent substances.

In **covalent crystals** the lattice points are occupied by atoms that are held by a network of covalent bonds. Diamond, graphite, and quartz are well known examples. Materials of this type have high melting points and are extremely hard because of the large number of covalent bonds that have to be broken to melt or break up the crystal. The entire crystal can be thought of as one giant molecule.

Covalent compounds form crystals in which the lattice positions are occupied by molecules. Such solids are called **molecular crystals**. Molecular crystals are soft and have low melting points. These properties are the result of the relatively weak intermolecular forces (van der Waals forces and hydrogen bonding) that hold the molecules in the crystal.

Crystals of polar compounds are held together by dipole-dipole forces and dispersion forces. Only dispersion forces occur in the lattices of nonpolar compounds. As a rule polar compounds melt at higher temperatures than nonpolar compounds of comparable molecular mass.

Metallic crystals are quite strong. Most transition metals have high melting points and densities. In metals the array of lattice points is occupied by positive ions. The outer electrons of the metal atoms are loosely held and move freely from ion to ion throughout the metallic crystal. This mobility of electrons in metals accounts for one of the characteristic properties of metals, namely the ability to conduct electricity.

Amorphous solids lack a well-defined arrangement and long range order. Glass is considered to be an amorphous solid because it lacks a regular three dimensional arrangement of atoms.

EXAMPLE 11.5   Number of Atoms Per Unit Cell

If atoms of a solid occupy a face-centered cubic lattice, how many atoms are there per unit cell.

•Method of Solution

In a face-centered cubic cell there are atoms at each of the eight corners, and there is one in each of the six faces.

•Calculation

8 corners (1/8 atom per corner) + 6 faces (1/2 atom per face) = 4 atoms per unit cell

---

EXAMPLE 11.6   Dimensions of a Unit Cell

Potassium crystallizes in a body-centered cubic lattice, and has a density of 0.856 g/cm^3 at 25°C.
a.   How many atoms are there per unit cell?
b.   What is the length of the side of the cell?

•Method of Solution

a.   A body-centered cubic structure has one K atom in the center, and eight other K atoms, one at each corner of the cube. The corner atoms, however, are shared by 8 adjoining cells. Each corner atom contributes 1/8 of an atom to the unit cell. The total number of K atoms per unit cell is

$$1 + \frac{1}{8}(8) = 2 \text{ atoms}$$

b. Ideally the crystal of potassium is made up of a large number of unit cells repeated over and over again. Thus, the density of the unit cell will be the same as the density of metallic K. Recall that density is an intensive property.

$$\text{density} = \frac{\text{mass}}{\text{volume}}$$

$$\text{density (unit cell)} = \frac{\text{mass of 2K atoms}}{a^3}$$

where a is the length of the side of the unit cell.

•Calculation

The mass of a K atom is:

$$\frac{39.1 \text{ g}}{1 \text{ mol}} \times \frac{1 \text{ mol}}{6.02 \times 10^{23} \text{ atoms}} = 6.50 \times 10^{-23} \text{ g/atom}$$

Substituting into the density equation yields:

$$0.856 \text{ g/cm}^3 = \frac{2(6.50 \times 10^{-23} \text{ g/atom})}{a^3}$$

Rearranging yields: $a^3 = 1.52 \times 10^{-22} \text{ cm}^3$

$$a = 5.34 \times 10^{-8} \text{ cm} = 534 \text{ pm}$$

EXAMPLE 11.7   Atomic Radius of a Potassium Atom

Determine the radius of a K atom for the metallic potassium crystal in Example 11.8.

•Method of Solution

From Figure 11.22 in the text, it is clear that we need the diagonal distance of the body-centered cubic cell because only along the diagonal do the atoms actually touch. The diagonal c will equal 4r where r is the atomic radius. From the figure we see that

$$c = 4r \qquad \text{and} \qquad c = \sqrt{3} \; a$$

Therefore

$$4r = \sqrt{3} \; a$$

where a is the length of an edge. Solving for r, we get

$$r = \frac{a\sqrt{3}}{4} = 534 \text{ pm} \times \frac{\sqrt{3}}{4}$$

$$r = 231 \text{ pm}$$

EXAMPLE 11.8   Using the Bragg Equation

When X rays with a wavelength of 0.154 nm are diffracted by a crystal of metallic copper, the angle corresponding to the first order diffraction (n = 1) is found to be 37.06°. What is the spacing between planes of Cu atoms that gives rise to this diffraction angle?

•**Method of Solution**

The spacing between Cu atoms corresponds to d in the Bragg equation, which we rearrange to solve for d.

$$d = \frac{n\lambda}{2 \sin \theta}$$

•**Calculation**

For a first-order diffraction n = 1. The sine of 37.06° can be determined with your calculator.

$$d = \frac{0.154 \text{ nm}}{2 \sin 37.06°} = \frac{0.154 \text{ nm}}{2 (0.603)}$$

$$= 0.128 \text{ nm (128 pm)}$$

EXAMPLE 11.9   Types of Crystals

What type of force must be overcome in order to melt crystals of the following substances?
a. Mg   b. Cl_2 c. $MgCl_2$ d. SO_2 e. Si

•**Method of Solution**

a.   Magnesium is a metal and so it forms metallic crystals. Since the mobile electrons are shared between positive ions, the forces between Mg atoms could be described as covalent bonds, in this case called metallic bonds. The melting point is 1105°C.
b. Cl_2 forms molecular crystals that are held together by intermolecular forces, more specifically dispersion forces. The melting point is –101°C.
c. $MgCl_2$ is an ionic compound. When it melts, Mg^{2+} ions and Cl^- ions break away from their lattice positions. To melt magnesium chloride, ionic bonds must be broken. The melting point is 1412°C.
d. SO_2 is a polar molecule. Thus, dipole-dipole attractions and dispersion forces must be overcome. The melting point is –73°C.
e.   Si crystallizes in a diamond structure with each silicon atom bound to four others by covalent bonds. It is a covalent crystal, and covalent bonds must be broken in order for it to melt. The melting point is 1410°C.

EXERCISES

8. Nickel crystallizes in a face-centered cubic lattice with an edge length of 352 pm. Calculate the density of Ni.

9. Nickel crystallizes in a face-centered cubic lattice with an edge length of 352 pm. Given that the density of Ni is 8.94 g/cm³, calculate Avogadro's number.

10. Copper crystallizes in a cubic system and the edge of the unit cell is 361 pm. The density of copper is 8.96 g/cm³. How many atoms are contained in one unit cell? What type of cubic cell does Cu form?

11. Copper crystallizes in a face-centered cubic lattice. The density of copper is 8.96 g/cm³. What is the length of the edge of a unit cell?

12. In a diffraction experiment, X rays of wavelength 0.154 nm were reflected from a gold crystal. The first-order reflection was at 22.2°. What is the distance between the planes of gold atoms in pm?

13. The distance between the planes of gold atoms in a gold crystal is 204 pm. In a diffraction experiment the first-order reflection was observed at 22.2°. What was the wavelength of the X-rays?

PHASE CHANGES AND PHASE DIAGRAMS (11.8 – 11.9)

STUDY OBJECTIVES

1. Describe the dynamic equilibrium between a liquid and its vapor state and list the factors that affect the vapor pressure of a liquid.
2. Describe the significance of the critical point.
3. Define the molar heats of fusion, vaporization, and sublimation.
4. Using a phase diagram, predict what changes a substance will undergo as it is heated or cooled, or subjected to pressure.
5. Use the Clausius-Clapeyron equation to relate vapor pressure, temperature, and the heat of vaporization of a liquid.

Liquid-Vapor Equilibrium. The different physical states of a substance that are present in a system are referred to as phases. A **phase** is a homogeneous part of the system that is in contact with other parts of the system, but is separated by a noticeable boundary. Ice, liquid water, and water vapor can exist together in a suitable container. Each of these is a phase. When a liquid substance is placed in a closed container, it will not be long before molecules of this substance can be found in the gas phase above the liquid. Vaporization is the process in which liquids become gases. When a portion of a liquid evaporates the gaseous molecules exert a pressure called the **vapor pressure**. The term **vapor** is often applied to the gaseous state of a substance that is normally a liquid or solid at the temperature of interest.

In a closed container the vapor pressure does not just continually increase, rather a state of equilibrium is reached in which the vapor pressure becomes constant and the amount of liquid remains constant. This equilibrium vapor pressure results from two opposing processes. The opposing process to vaporization is condensation. As the concentration of the molecules in the vapor phase increases, some will strike the liquid surface and condense.

$$\text{liquid} \underset{\text{condensation}}{\overset{\text{vaporization}}{\rightleftarrows}} \text{vapor}$$

When the rates of the opposing processes become equal, the vapor pressure remains constant and is called the **equilibrium vapor pressure**, or just vapor pressure. Note that in the state of equilibrium, while the amounts of vapor and liquid do not change, there is considerable activity on the molecular level. Vaporization and condensation are constantly occurring, but *at the same rates*. Such an equilibrium state is referred to as a *dynamic equilibrium*.

Vapor pressure is a function of the temperature. As temperature increases the rate of vaporization increases, but the rate of condensation remains the same. Therefore, more molecules exist in the vapor phase at higher temperatures than at lower temperatures.

Table 11.4 Vapor Pressure of Water

Temp (°C)	Pressure (mm Hg)
20	17.54
30	31.82
40	55.32
50	92.51
80	355.1
86.5	460.0
100	760.0

Boiling of a liquid is marked by bubble formation. In a container open to the atmosphere, boiling of a liquid will occur when the temperature is raised high enough. In order for a bubble to form, the vapor inside the bubble must be able to push back the atmosphere. This will not occur until a temperature is reached at which the vapor pressure is greater than the atmospheric pressure. Therefore, the **boiling point** of a liquid is the temperature at which the vapor pressure is equal to the atmospheric pressure. Since the boiling point of a liquid depends on the atmospheric pressure, and the atmospheric pressure varies daily, the boiling point is not a constant. A normal boiling point is defined for purposes of comparing liquid substances. The **normal boiling point** is the boiling temperature when the external pressure is 1 atm. Figure 11.1 shows the vapor pressures and normal boiling points of three liquids.

Figure 11.1 The effect of temperature on the vapor pressure of chloroform and water. Notice that the normal boiling points for the liquids are the temperatures at which their vapor pressures are equal to 1 atm.

The energy required to vaporize one mole of a liquid is called the **molar heat of vaporization** (ΔH_{vap}). The value of ΔH_{vap} is directly proportional to the strength of intermolecular forces. Liquids with relatively high heats of vaporization have low vapor pressures and high normal boiling points.

The quantity ΔH_{vap} can be determined experimentally. The vapor pressure of a liquid is proportional to the temperature. The quantitative relationship between the vapor pressure (P) and the absolute temperature (T) is given by the *Clausius-Clapeyron equation:*

$$\ln P = -\frac{\Delta H_{vap}}{RT} + C$$

where ln P is the natural logarithm of P, R is the ideal gas constant (8.314 J K^{-1} mol^{-1}), and C is a constant.

A useful form of this equation is

$$\ln\left(\frac{P_1}{P_2}\right) = \frac{\Delta H_{vap}}{R}\left(\frac{T_1 - T_2}{T_1 T_2}\right)$$

The ΔH_{vap} can be calculated if the vapor pressures P_1 and P_2 are measured at two temperatures T_1 and T_2. Alternatively if ΔH_{vap} is already known and one vapor pressure P_1 is known at temperature T_1, you can calculate the vapor pressure P_2 at some new temperature T_2. See Example 11.10.

Boiling Point as a Measure of IMF. The boiling point and the heat of vaporization are convenient measures of the strength of intermolecular forces. In the liquid, molecules are very close together and are strongly influenced by intermolecular forces. In the gas phase, molecules are widely spaced, move rapidly, and have enough energy to overcome intermolecular forces. During vaporization, molecules are completely separated from one another. The heat of vaporization is the energy necessary to overcome intermolecular forces. The boiling point reflects the amount of kinetic energy that liquid molecules must achieve in order to overcome intermolecular forces and escape into the gas phase. When comparing molecules, as the strength of IMFs increase, the ΔH_{vap} and the normal boiling points increase.

Critical Temperature and Pressure. When a liquid is heated in a closed container the vapor pressure increases, but boiling does not occur. The vapor cannot escape, and the vapor pressure continually rises. Eventually a temperature is reached at which the meniscus between liquid and vapor disappears. This temperature is called the critical temperature. The **critical temperature** is the highest temperature at which the substance can exist as a liquid. The **critical pressure** is the lowest pressure that will liquefy a gas at the critical temperature. Critical temperatures and pressures are listed in Table 11.7 (in the text).

As we saw previously, intermolecular forces are dominant in liquids and effectively determine many of their properties. The critical temperature and pressure are no exceptions. *Substances with high critical temperatures have strong intermolecular forces of attraction.*

Liquid-Solid Equilibrium. The temperature at which the solid and liquid are in dynamic equilibrium at 1 atm pressure is called the **normal melting point** or freezing point. Melting is also called fusion. During melting the average distance between molecules is increased slightly as evidenced by the approximately 10% decrease in density.

$$\text{solid} \underset{\text{freezing}}{\overset{\text{fusion}}{\rightleftharpoons}} \text{liquid}$$

The energy required to melt one mole of a solid is called the **molar heat of fusion**, ΔH_{fus}. Upon freezing, the substance will release the same amount of energy. The ΔH_{fus} is always much less than ΔH_{vap} because during vaporization molecules are completely separated from each other, while melting only separates molecules to a small extent.

Heating Curves. A heating curve is a convenient way to summarize the solid-liquid-gas transitions for a compound. The heating curve of water is shown in Figure 11.2. It is the result of an experiment in which a given amount of ice, 1 mole for instance, at some initial temperature below 0°C is slowly heated at *a constant rate*.

The curve shows that the temperature of ice increases on heating (line 1) until the melting point is reached. No rise in temperature occurs while ice is melting (line 2); as long as some ice remains, the temperature stays at 0°C. The length of line 2 is a measure of the heat necessary to melt 1 mole of ice, which is ΔH_{fus}. Along line 3 the temperature of liquid water increases from 0°C to 100°C. The slope of the line Δt per joule of heat depends inversely on the specific heat of liquid water. The greater slope for line 1 than for line 3 means that less heat is required to raise the temperature of ice than of water. This is evidenced by the difference in specific heats of ice and liquid water. The specific heat of ice is 2.09 J/g·°C versus 4.18 J/g·°C for liquid H_2O.

Figure 11.2 The heating curve for water.

At 100°C the liquid begins to boil, and the heat added is used to bring about vaporization. The temperature does not rise until all the liquid has been transformed to gas. The length of line 4 is the heat required to vaporize 1 mole of liquid. Line 4 will always be longer than line 2 because $\Delta H_{vap} > \Delta H_{fus}$. Line 5 corresponds to the heating of steam. Again the slope of line 5 depends on the specific heat of steam, which is about 1.98 J/g·°C.

Solid-Vapor Equilibrium. **Sublimation** is the evaporation of a solid directly into the vapor phase. Dry ice and iodine are substances that sublime readily. Ice also sublimes to some extent. As with liquids, the vapor pressure of a solid increases as the temperature increases. The direct conversion of solid to vapor is equivalent to melting the solid first and then vaporizing the liquid. From Hess's law we obtain:

$$\Delta H_{subl} = \Delta H_{fus} + \Delta H_{vap}$$

Phase Diagrams. From discussions in preceding sections we can see that the phase in which a substance exists depends on its temperature and pressure. In addition, two phases may exist in equilibrium at certain temperatures and pressures. Information about the stable phases for a specific compound is summarized by a **phase diagram** such as that shown in Figure 11.3.

Figure 11.3 A typical phase diagram..

A phase diagram is a graph of pressure versus temperature. The lines represent P and T at which two phases can coexist. The regions between lines represent P and T at which only one phase can exist. The line between the solid and liquid regions is made up of points of P and T at which the solid and liquid phases are in equilibrium. The line between the solid and gas regions is made up of points of P and T at which solid and gas are in equilibrium. And the line between the liquid and gas regions gives temperatures and pressures at which liquid and gas are in equilibrium. All three lines intersect at the **triple point**. This one point describes the conditions under which gas, liquid, and solid are all in equilibrium.

The phase diagram shows at a glance several properties of a substance: melting point, boiling point, and triple point. If a point (P and T) describing a system falls in the solid region, the substance exists as a solid. If the point falls on a line such as that between liquid and gas regions, the substance exists as liquid and vapor in equilibrium.

EXAMPLE 11.10   Clausius-Clapeyron Equation

The vapor pressure of water is 55.32 mm Hg at 40.0°C, and 92.51 mm Hg at 50.0°C. (See Table 5.3 in the text). Calculate the molar heat of vaporization of water.

•Method of Solution

The Clausius-Clapeyron equation relates the vapor pressure of a liquid to its molar heat of vaporization.

$$\ln\left(\frac{P_1}{P_2}\right) = \frac{\Delta H_{vap}}{R}\left(\frac{T_1 - T_2}{T_1 T_2}\right)$$

•Calculation

Substituting the value given above

$$\ln\left(\frac{55.32 \text{ mm Hg}}{92.51 \text{ mm Hg}}\right) = \frac{\Delta H_{vap}}{8.314 \text{ J K}^{-1}\text{mol}^{-1}} \times \left(\frac{313 \text{ K} - 323 \text{ K}}{(313 \text{ K})(323 \text{ K})}\right)$$

$$\ln 0.5980 = \frac{\Delta H_{vap}}{8.314 \text{ J K}^{-1}\text{mol}^{-1}}\left(\frac{-10 \text{ K}}{1.01 \times 10^5 \text{ K}^2}\right)$$

After taking the logarithm and rearranging yields:

$$\Delta H_{vap} = \frac{-0.5142 \, (8.314 \text{ J K}^{-1}\text{mol}^{-1})(1.01 \times 10^5 \text{ K})}{-10 \text{ K}}$$

$$\Delta H_{vap} = 43,000 \text{ J/mol}$$

•Comment

Compare this result to the ΔH_{vap} listed in Table 11.6 of the text, 40,790 J/mol. The observed difference is real. The ΔH_{vap} is less at the boiling point than it is at a lower temperature. Over the entire temperature range of a liquid, ΔH_{vap}, is close to, but not really, a constant.

EXAMPLE 11.11   Vapor Pressure and Boiling Point

Using Figure 11.35 in the text estimate the boiling point of water in a pressure cooker at 2.0 atm pressure.

•Method of Solution

The boiling point of water at 2.0 atm pressure is the temperature at which the vapor pressure of water is 2.0 atm. According to the figure this should be approximately 120°C.

EXAMPLE 11.12   Heat of Vaporization

How much heat is given off when 1.0 g of steam condenses at 100°C?

•Method of Solution

The heat of vaporization of water at 100°C is 40.8 kJ/mol. Condensation is the reverse of vaporization so it releases the same amount of heat.

$$H_2O(g) \rightarrow H_2O(l) \quad \Delta H_{cond} = -40.8 \text{ kJ/mol}$$

•Calculation

Since ΔH_{cond} is in molar units, first convert 1.0 g H_2O into moles. The heat evolved when 1.0 g of water vapor condenses is:

$$q = \text{mol } H_2O \times \Delta H_{cond}$$

$$= 1.0 \text{ g} \times \frac{1 \text{ mol}}{18.0 \text{ g}} \times \frac{40.8 \text{ kJ}}{1 \text{ mol}}$$

$$= -2.27 \text{ kJ}$$

•Comment

This is a lot of heat and shows us why burns from steam can be very serious.

EXAMPLE 11.13   Heat of Fusion

The heat of fusion of aluminum is 10.7 kJ/mol. How much energy is required to melt one ton of Al at the melting point, 660°C.

•Method of Solution

The heat of fusion is the energy required to melt 1 mole of a substance. Therefore, first we need to convert 2000 lb of Al to moles.

•Calculation

$$q = \text{mol Al} \times \Delta H_{fus}$$

$$= 2000 \text{ lb} \times \frac{454 \text{ g}}{1 \text{ lb}} \times \frac{1 \text{ mol Al}}{27.0 \text{ g Al}} \times \frac{10.7 \text{ kJ}}{1 \text{ mol Al}} = 3.60 \times 10^5 \text{ kJ (360 MJ)}$$

EXAMPLE 11.14   The Critical Temperature

Discuss the possibilities of liquefying oxygen and carbon dioxide at 25°C by increasing the pressure.

•Method of Solution

Above its critical temperature a substance cannot be liquefied by increasing the pressure. Substances having critical temperatures above 25°C can be liquefied at 25°C by application of sufficient pressure. With a critical temperature of 31°C, CO_2 can be liquefied at 25°C with application of enough pressure. The pressure required can be read off the phase diagram for CO_2 (Figure 11.41 in the text). Reading straight up from 25°C you will cross the liquid-vapor equilibrium point at a pressure of 67 atm. This is the pressure required to liquefy CO_2 at 25°C. Oxygen has a critical temperature of –119°C; therefore, at any temperature above –119°C oxygen cannot be liquefied, no matter how much pressure is applied. To liquefy O_2, its temperature must first be lowered below –119°C and pressure applied.

EXERCISES

14. The molar heat of vaporization of bromine (Br_2) is 30.04 kJ/mol. What is the heat of vaporization per gram of bromine?

15. The heat of vaporization of iodine is 41.7 kJ/mol at the boiling point of iodine, 456 K. How much heat is required to vaporize 20.0 g of I_2?

16. The molar heats of fusion and vaporization of potassium are 2.4 and 79.1 kJ/mol, respectively. Estimate the molar heat of sublimation of potassium metal.

17. The heat of vaporization of ammonia is 23.2 kJ/mol at its boiling point. How many grams of ammonia can be vaporized with 12.5 kJ of heat?

18. How much heat is needed to convert 1.0 kg of ice at 0°C completely into steam at 100°C?

19. Indicate whether each of the following processes is exothermic or endothermic.
    a. melting of ice
    b. evaporation of ethanol
    c. condensation of steam
    d. freezing of water
    e. sublimation of iodine

20. Which substance in each pair has the higher vapor pressure at a given temperature?
    a. C_2H_5OH or CH_3OH
    b. Cl_2 or Br_2
    c. CH_3Br or CH_3Cl

21. The vapor pressure of liquid potassium is 1.00 mm Hg at 341°C and 10.0 mm Hg at 443°C. Calculate the molar heat of vaporization of potassium.

22. The vapor pressure of ethanol is 100 mm Hg at 34.9°C, and 400 mm Hg at 63.5°C. What is the molar heat of vaporization of ethanol?

23. The vapor pressure of liquid potassium is 10.0 mm Hg at 443°C. The heat of vaporization is 82.5 kJ/mol. Calculate the temperature where the vapor pressure is 760 mm Hg. This is the boiling point of liquid potassium.

24. The vapor pressure of ethanol is 400 mm Hg at 63.5°C. Its molar heat of vaporization is 41.7 kJ/mol. Calculate the vapor pressure of ethanol at 34.9°C.

CONCEPTUAL QUESTIONS

1. Consider a steel cylinder containing 20 kg of CO_2 in the form of liquid and gas. The valve is opened and CO_2 gas flows rapidly from the cylinder. Soon, however, the flow slows, and a coating of ice frost forms on the outside of the cylinder. When the valve is closed the cylinder is weighed and is found to contain 14 kg of CO_2.
   a. Why is the flow faster at first?
   b. Why does the flow almost stop before the cylinder is empty?
   c. Why does the outside become icy?

2. On top of Pikes Peek water boils at 86°C. Why do liquids boil at lower temperatures in the mountains as compared to sea level?

3. The viscosities of liquids generally decrease with increasing temperature. Water has the following viscosities in units of N s/m^2: 0.0018 at 0°C; 0.0010 at 20°C; 0.0005 at 55°C; and 0.0003 at 100°C. Interpret this trend on a molecular basis.

4. Why does your skin feel cool when water evaporates from it? Why will you feel warmer on a more humid summer day than on a less humid day?

PRACTICE TEST

1. Identify the types of intermolecular forces for each of the following substances:
   a. H_2O_2 b. H_2S c. SF_6 d. $NOCl$ e. NH_3

2. The enthalpies of vaporization of H_2S and H_2Se are 18.7 and 23.2 kJ/mol, respectively. Why is the ΔH_{vap} of water (40.6 kJ/mol) so much higher than the values for H_2S and H_2Se?

3. Which member of each pair has the stronger intermolecular forces of attraction?
   a. CO_2 or SO_2 b. F_2 or Br_2 c. H_2O or H_2S d. HCl or C_8H_{18}

4. In the laboratory, glassware is considered clean when water wets the glass in a continuous film. If droplets of water stand on the glass surface, the glass is considered dirty. Discuss the basis of this cleanliness test.

5. The meniscus for mercury in a glass tube is concave downward. Explain.

6. At what angle would a first-order reflection be observed in the diffraction of 0.090 nm X rays by a set of crystals planes for which d = 0.500 nm?

7. Calcium oxide like NaCl crystallizes in a face-centered cubic cell.
   a. How many Ca^{2+} and O^{2-} ions are in each unit cell?
   b. Given the ionic radius of Ca^{2+} (99 pm) and O^{2-} (140 pm), what is the length of an edge of the unit cell?
   c. Calculate the density of CaO. See Figure 11.22 in the textbook.

8. From the following data calculate Avogadro's number. Potassium crystallizes in a body-centered cubic lattice with an edge of 530 pm. The density of K is 0.86 g/cm^3.

9. Aluminum crystallizes in a face-centered cubic lattice with the length of an edge equal to 405 pm. Assume that a face-centered atom touches each of the surrounding corner atoms. Calculate the length of a diagonal of a face.

10. Pick the substance in each pair that has the higher melting point.
    a. ICl or KCl   b. K or Fe   c. Fe or Cl_2
    d. SiO_2 (quartz) or SiF_4 e. CO_2 or SiO_2

11. Distinguish between the boiling point of a liquid and the normal boiling point of a liquid.

12. Use the following data to draw a rough phase diagram for methane: triple point −183.0°C; normal melting point −182.5°C; normal boiling point −161.5°C; critical point −83.0°C; critical pressure 45.6 atm.
    a. What is the stable state of methane at −160°C and 2.0 atm pressure?
    b. At a constant temperature of −182°C the pressure on a sample of methane is gradually increased from 1.0 mm Hg to 1.0 atm. What phases would be observed, and in what order would they occur?

13. Given the following data for N_2:

Normal melting point: −210°C	Specific heat of liquid: 2.0 J/g·°C
Normal boiling point: −196°C	Specific heat of gas: 1.0 J/g·°C
Heat of fusion: 25 J/g	Specific heat of solid: 1.6 J/g·°C
Heat of vaporization: 200 J/g	

    How much energy in kilojoules is required to convert 1000 g of N_2 at −206°C to N_2 gas at 20°C?

14. In the following pairs indicate which member would have the higher heat of vaporization?
    a. He or Kr   b. CH_3OH or CH_3OCH_3

15. Which one of the following compounds will exhibit hydrogen bonding?
    a. CH_4 b. HBr   c. CH_3OH d. CCl_4

16. The vapor pressure of mercury is 17.3 mm Hg at 200°C. What is its vapor pressure at 300°C if its molar heat of vaporization is 59.0 kJ/mol?

ANSWERS

Exercises

1. a. dipole-dipole forces and dispersion forces   b. dispersion forces   c. dispersion forces   d. dipole-dipole forces and dispersion forces   e. ionic bonds
2. CS_2 because of its much larger number of electrons and therefore polarizability.
3. b, c, and d.
4. a. CO   b. ICl   c. H_2O d. AsH_3
5. a. Hydrogen bonds and dispersion forces   b. covalent bonds   c. dispersion forces
6. Surface tension
7. C_2H_5OH
8. 8.94 g/cm³
9. 6.02×10^{23}
10. 4 atoms, face-centered cubic
11. 361 pm
12. 204 pm
13. 0.154 nm

14. 0.188 kJ/g
15. 3.29 kJ
16. 81.5 kJ/mol
17. 9.16 g
18. 3020 kJ
19. a. endothermic   b. endothermic   c. exothermic   d. exothermic   e. endothermic
20. a. CH_3OH b. Cl_2 c. CH_3Cl
21. 82.5 kJ/mol
22. $\Delta H_{vap} = 41.7$ kJ
23. 1052 K
24. 100 mm Hg

Conceptual Questions

1. a. The flow is faster at first because the vapor pressure of CO_2 is higher at first. The temperature of the CO_2 is higher at first. Therefore more of the CO_2 is in the gas phase and the vapor pressure is higher at first. The liquid CO_2 cools down because it supplies some of the heat of vaporization.
    b. The flow almost stops before the cylinder is empty because as the liqiud cools off the vapor pressure of CO_2 drops so low that very little of the liquid CO_2 can vaporize.
    c. As the liquid CO_2 evaporates the heat of vaporization must be continually supplied. Heat is taken form the steel cylinder and then from the surrounding air. As the temperature of the steel drops below the freezing point of water the water vapor in the air condenses (deposition) as ice on the cold cylinder.
2. The atmospheric pressure decreases as altitude increases. Therefore the higher one goes the lower the vapor pressure needed to form the bubbles observed during boiling. The lower the vapor pressure needed for boiling, the lower the temperature.
3. Viscosity is the resistance to flow and is caused by intermolecular forces. As the temperature of water is raised, more and more H bonds are broken, and the viscosity decreases.
4. Heat is required for water to evaporate. When water evaporates from your skin, the heat of vaporization is absorbed from the skin, and the skin cools. You feel warmer on a humid day because the condensation of water vapor on your skin slows the net rate of evaporation (cooling).

Practice Test

1. a. Dipole-dipole, dispersion forces, and hydrogen bonding b. Dipole-dipole and dispersion forces
    c. Dispersion forces d. Dipole-dipole and dispersion forces e. Dipole-dipole, dispersion forces and hydrogen bonding.
2. H_2O molecules are attracted to each other by hydrogen bonding, whereas H_2S and H_2Se are not.
3. a. SO_2 b. Br_2 c. H_2O d. C_8H_{18}
4. Grease and water do not mix, thus if grease remains on the glass surface, water will bead up. When the surface is clean, the cohesive forces between water and glass cause water to spread out in a thin film and "wet" the glass.
5. The cohesive forces between Hg atoms are greater than the adhesives forces between glass and mercury.
6. $\theta = 5.2°$
7. a. 4   b. 478 pm   c. 3.41 g/cm^3
8. 6.10×10^{23}
9. 573 pm

10. a. KCl   b. Fe   c. Fe   d. SiO_2 e. SiO_2
11. The bubble formation that signifies boiling cannot occur until the vapor pressure is great enough to push back the atmosphere. Thus the boiling point varies with atmospheric pressure. The normal boiling point of a liquid is the temperature at which its vapor pressure is 1.0 atm.
12. a. Liquid   b. Gas, then liquid
13. 436 kJ
14. a. Kr   b. CH_3OH
15. c. CH_3OH
16. 237 mm Hg

Chapter Twelve
PHYSICAL PROPERTIES OF SOLUTIONS

- The Solution Process (12.1 – 12.2)
- Concentration Units (12.3)
- Temperature and Pressure Effects on Solubility (12.4 – 12.5)
- Colligative Properties (12.6 – 12.7)

THE SOLUTION PROCESS (12.1 – 12.2)

STUDY OBJECTIVES

1. Describe the role of intermolecular forces in the solution process.
2. Given a particular substance, predict what solvent will be suitable.
3. Describe the exothermic and endothermic steps that occur during the solution process.

Some Definitions. The main subject of this chapter concerns the formation and properties of liquid solutions. Recall from Chapter 4 that a **solution** is a homogeneous mixture of two or more substances. The component in greater quantity is called the **solvent,** and the component in lesser amount is called the **solute.** In aqueous solutions water is the solvent. Solutes can be liquids, solids, or gases. Several terms are used to describe the degree to which a solute will dissolve in a solvent:

1. When two liquids are completely soluble in each other in all proportions, they are said to be **miscible.** For example, ethanol and water are miscible. If the liquids do not mix, they are said to be **immiscible.** Oil and water, for example, are immiscible.
2. A solution that contains the maximum concentration of a solute at a given temperature is called a **saturated solution**. For example, the solubility of NaCl is 6.1 moles per liter of water. An **unsaturated solution** has a concentration of solute that is less than the maximum concentration. **Supersaturated solutions** have a concentration of solute that is greater than that of a saturated solution.

The Solution Process. Dissolving is a process that takes place at the molecular level and can be discussed in molecular terms. When one substance dissolves in another, the particles of the solute disperse uniformly throughout the solvent. The solute particles occupy positions that are normally taken by solvent molecules. The ease with which a solute particle may replace a solvent molecule depends on the relative strengths of three types of interactions:

- Solvent-solvent interaction
- Solute-solute interaction
- Solvent-solute interaction

Imagine the solution process as taking place in three steps as shown in Figure 12.1. Step 1 is the separation of solvent molecules. Step 2 is the separation of solute molecules. These steps require inputs of energy to overcome attractive intermolecular forces. Step 3 is the mixing of solvent and solute molecules; it may be exothermic or endothermic. According to Hess's law (see Section 6.6 of the text) the heat of solution is given by the sum of the enthalpies of the three steps:

$$\Delta H_{soln} = \Delta H_1 + \Delta H_2 + \Delta H_3$$

The solute will be soluble in the solvent if the solute-solvent attraction is stronger than the solvent-solvent attraction and solute-solute attraction. Such a solution process is exothermic. Only a relatively small amount of the solute will be dissolved if the solute-solvent interaction is weaker than the solvent-solvent and solute-solute interaction; then the solution process will be endothermic. **Solvation** is the process in which a solute particle (an ion or molecule) is surrounded by solvent molecules due to strong solute-solvent attractive forces. When water is the solvent the process is called *hydration*.

Figure 12.1. A molecular view of the solution process. Think of the solute molecules and solvent molecules first being spread apart, and then being mixed together. The relative strength of forces holding solvent molecules together ΔH_1, solute particles together ΔH_2, and the forces between solvent and solute molecules ΔH_3 in the solution are important in determining the solubility.

Here we see that the solution process is assisted by an exothermic heat of solution. Yet there are a number of soluble compounds with endothermic heats of solution. In addition every substance is somewhat soluble no matter what the value of ΔH_{soln}. It turns out that all chemical processes are governed by two factors. The first of these is the energy factor. In other words, does the solution process absorb energy or release energy?

Disorder or randomness is the other factor that must be considered. Processes that increase randomness or disorder are also favored. In the pure state, the solvent and solute possess a fair degree of order. Here we mean the ordered arrangement of atoms, molecules, or ions in a three-dimensional crystal. Where order is high, randomness is low.

The order in a crystal is lost when the solute dissolves and its molecules are dispersed in the solvent. The solution process is accompanied by an increase in disorder or randomness. It is the increase in disorder of the system that favors the solubility of any substance.

A General Solubility Rule. The most general solubility rule is that "like dissolves like." In this rule, the term "like" refers to molecular polarity. "Like dissolves like" means that substances of like polarity will mix to form solutions. And substances of different polarity will be immiscible, or will tend only slightly to form solutions. The rule predicts that two polar substances will form a solution, and two nonpolar substances will form a solution, but a polar substance and a nonpolar substance will tend not to mix. For example, water and oil, being polar and nonpolar substances, respectively, are immiscible. Water and ethanol, both being polar, are miscible. Oil will dissolve in carbon tetrachloride because both substances are nonpolar. Ionic compounds tend to be quite soluble in water (a polar solvent).

EXAMPLE 12.1   Solubility

Which would be a better solvent for molecular $I_2(s)$: CCl_4 or H_2O?

•**Method of Solution**

Using the like-dissolves-like rule, first we identify I_2 as a nonpolar molecule. Therefore it will be more soluble in the nonpolar solvent CCl_4 than in the polar solvent H_2O.

EXAMPLE 12.2   Solubility

In which solvent will NaBr be more soluble, benzene or water?

•**Method of Solution**

Benzene is a nonpolar solvent and water is a polar solvent. Since NaBr is an ionic compound, it cannot dissolve in nonpolar benzene, but is very soluble in water.

EXERCISES

1. Isopropyl alcohol and water dissolve in each other regardless of the proportions of each. What term describes the solubilities of these liquids in each other?

2. Explain why hexane (C_6H_{14}) even though a liquid, is not miscible with water.

3. Indicate whether each compound listed is soluble or insoluble in water.
    a. CH_3OH b. LiBr   c. C_8H_{18} d. CCl_4 e. $BaCl_2$ f. $HOCH_2CH_2OH$

4. Explain why ammonia gas, NH_3, is very soluble in water, but not in hexane, C_6H_{14}.

5. What is the difference between solvation and hydration?

6. For each of the following pairs, predict which substance will be more soluble in water.
    a. NaCl(s) or $I_2(s)$
    b. $CH_4(g)$ or $NH_3(g)$
    c. $CH_3OH(l)$ or C_6H_6(benzene)

7. For each of the following pairs, predict which substance will be more soluble in $CCl_4(l)$.
    a. $I_2(s)$ or KBr(s)
    b. $CS_2(l)$ or $CH_3OH(l)$
    c. $CO_2(g)$ or HCl(g)

CONCENTRATION UNITS (12.3)

STUDY OBJECTIVES

1. Calculate concentrations in units of molarity, mass percent, mole fraction, and molality.
2. Convert concentrations expressed in one unit into any of the other units.

Concentration. The term concentration refers to how much of one component of a solution is present in a given amount of solution. In Chapter 4 you were introduced to the concentration unit *molarity*:

$$\text{molarity} = \frac{\text{moles of solute}}{\text{liters of solution}}$$

Three new concentration units are introduced in this chapter.

The percent by mass of solute

$$\text{Percent by mass of solute} = \frac{\text{mass of solute}}{\text{mass solute} + \text{mass solvent}} \times 100\%$$

$$= \frac{\text{mass of solute}}{\text{mass of soln}} \times 100\%$$

The mole fraction of a component of the solution

$$\text{mole fraction of component A} = X_A = \frac{\text{moles of A}}{\text{sum of moles of all components}}$$

The molality

$$\text{molality} = \frac{\text{moles of solute}}{\text{mass of solvent(kg)}}$$

EXAMPLE 12.3 Concentration Units

The dehydrated form of Epsom salt is magnesium sulfate.
a. What is the percent $MgSO_4$ by mass in a solution made from 16.0 g $MgSO_4$ and 100 mL of H_2O at 25°C? The density of water at 25°C is 0.997 g/mL.
b. What is the mole fraction of each component?
c. Calculate the molality of the solution.

•**Method of Solution for (a)**

Write the equation for percent by mass.

$$\text{percent } MgSO_4 = \frac{\text{mass } MgSO_4}{\text{mass } MgSO_4 + \text{mass water}} \times 100\%$$

•**Calculation for (a)**

$$\text{mass } H_2O = 100 \text{ mL} \times \frac{0.997 \text{ g}}{1 \text{ mL}} = 99.7 \text{ g } H_2O$$

$$\text{percent } MgSO_4 = \frac{16.0 \text{ g}}{16.0 \text{ g} + 99.7 \text{ g}} \times 100\% = \frac{16.0 \text{ g}}{115.7 \text{ g}} \times 100\%$$

$$\text{percent } MgSO_4 = 13.8\%$$

•Method of Solution for (b)

Write the formula for the mole fraction of $MgSO_4$.

$$X_{MgSO_4} = \frac{mol\ MgSO_4}{mol\ MgSO_4 + mol\ H_2O}$$

•Calculation for (b)

Convert the masses in part (a) into moles to substitute into the equation.

$$16.0\ g\ MgSO_4 \times \frac{1\ mol\ MgSO_4}{120.4\ g\ MgSO_4} = 0.133\ mol\ MgSO_4$$

$$99.7\ g\ H_2O \times \frac{1\ mol\ H_2O}{18.0\ g\ H_2O} = 5.54\ mol\ H_2O$$

The mole fraction of $MgSO_4$ is

$$X_{MgSO_4} = \frac{0.133\ mol}{0.133\ mol + 5.54\ mol} = 0.0235$$

The mole fraction of H_2O is

$$X_{H_2O} = \frac{5.54\ mol}{5.67\ mol} = 0.977$$

Notice that the sum of the two mole fractions is 1.000. The sum of the mole fractions of all solution components is always 1.00.

•Method of Solution for (c)

Write the formula for molality:

$$molality = \frac{moles\ of\ MgSO_4}{kilograms\ of\ H_2O}$$

Substitute into this equation the quantities previously calculated in parts (a) and (b).

•Calculation for (c)

$$molality = \frac{0.133\ mol\ MgSO_4}{99.7\ g} \times \frac{10^3\ g}{1\ kg} = 1.33\ m$$

EXAMPLE 12.4   Molality and Molarity of the Same Solution

Concentrated hydrochloric acid is 36.5 percent HCl by mass. Its density is 1.18 g/mL. Calculate:
a.   The molality.
b.   The molarity of HCl.

•Method of Solution for (a)

Write the formula for molality.

$$molality = \frac{mol\ HCl}{kg\ H_2O}$$

Next find the number of moles of HCl per kilogram of H_2O. We take 100 g of solution, and determine how many moles of HCl and how many kilograms of the solvent it contains. A solution that is 36.5 percent HCl by mass corresponds to 36.5 g HCl/100 g solution. Since 100 g of solution contains 36.5 g HCl, the difference 100 g – 36.5 g must equal the mass of water which is 63.5 g. We have two ratios:

$$\frac{36.5 \text{ g HCl}}{100 \text{ g soln}} \quad \text{and} \quad \frac{36.5 \text{ g HCl}}{63.5 \text{ g H}_2\text{O}}$$

•**Calculation for (a)**

The moles of HCl is given by:

$$\text{moles HCl} = 36.5 \text{ g HCl} \times \frac{1 \text{ mol HCl}}{36.5 \text{ g HCl}} = 1.00 \text{ mol HCl}$$

The kilograms of water is given by:

$$\text{kg H}_2\text{O} = 63.5 \text{ g H}_2\text{O} \times \frac{1 \text{ kg}}{1 \times 10^3 \text{ g}} = 0.0635 \text{ kg H}_2\text{O}$$

Then we calculate molality:

$$\text{molality} = \frac{1.00 \text{ mol HCl}}{0.0635 \text{ kg H}_2\text{O}} = 15.7 \text{ m}$$

•**Method of Solution for (b)**

Write the formula for molarity:

$$\text{molarity} = \frac{\text{moles HCl}}{\text{liters soln}}$$

Take 100 g of solution and determine how many moles of HCl and how many liters of solution are present.

$$\frac{36.5 \text{ g HCl}}{100 \text{ g soln}}$$

•**Calculation for (b)**

We know that 100 g of solution contains 36.5 g HCl, which is 1.00 mol HCl. Convert 100 g of solution to volume of solution by using the density.

$$\text{Volume of soln} = (100 \text{ g soln}) \times \frac{1 \text{ mL}}{1.18 \text{ g soln}} \times \frac{10^{-3} \text{ L}}{1 \text{ mL}} = 0.0847 \text{ L}$$

$$\text{molarity} = \frac{1.00 \text{ mol HCl}}{0.0847 \text{ L soln}} = 11.8 \text{ M}$$

EXERCISES

8. Calculate the percent by mass of the solute in each of the following aqueous solutions.
   a. 6.50 g NaCl in 75.2 g of water.
   b. 27.2 g ethanol in 250 g of solution.
   c. 2.0 g I_2 in 125 g methanol.

9. Calculate the molality of each of the following solutions.
   a. 6.50 g NaCl in 75.2 g of water.
   b. 27.5 g glucose ($C_6H_{12}O_6$) in 425 g of water.

10. Calculate the molarity of a 2.44 m NaCl solution given that its density is 1.089 g/mL.

11. Calculate the percent $AgNO_3$ by mass in a 0.650 m $AgNO_3$ solution.

12. Calculate the molality of a 5.5% $AgNO_3$ solution.

TEMPERATURE AND PRESSURE EFFECTS ON SOLUBILITY (12.4 – 12.5)

STUDY OBJECTIVES

1. Discuss the effects of temperature on the solubility of salts in water.
2. Discuss the effects of temperature and pressure on the solubilities of gases.

Temperature Effect on the Solubility of Solids. Temperature changes strongly affect the solublilty of most solid solutes. The effects of temperature on the solubilities of some common salts is shown in Figure 12.3 (text). In most cases the solubility of a solid in water increases with increasing temperature. However, this is not always true. Table 12.1 list several solids whose solubility *decreases* with increasing temperature. In general, it is not practical to predict just how the temperature will affect the solubility of a given substance. Temperature effects must be determined experimentally.

TABLE 12.1 Compounds With Solubilities That Decrease with Increasing Temperature

$CaSO_4$	$Ca(OH)_2$
$Na_2SO_4 \cdot 10H_2O$	$Ca(C_2H_3O_2)_2$
$Ce_2(SO_4)_3$	Li_2SO_4

Pressure and the Solubility of Gases. The solubility of gases in liquids is directly proportional to the gas pressure. **Henry's law** relates gas concentration c, in moles per liter, to the gas pressure p in atmospheres.

$$c = kP$$

where k is a constant for each gas and has units of mol/(L·atm). The greater the value of k, the greater the solubility of the gas. A few of Henry's law constants for gases dissolved in water at 25°C are listed in Table 12.2

TABLE 12.2 Some Henry's Law Constants at 25°C

Gas	k(mol/L·atm)
O_2	1.3×10^{-3}
N_2	6.8×10^{-4}
CO_2	3.4×10^{-2}

Gas Solubility and Temperature. A common observation is that a glass of cold water when warmed to room temperature shows the formation of many small air bubbles. This phenomenon results in part from the decreased solubility of gases in water with increasing temperature.

The solubility of an unreactive gaseous solute is due to intermolecular attractive forces between solute molecules and solvent molecules. As the temperature of such a solution is increased, more solute molecules attain sufficient kinetic energy to break away from these attractive forces and enter the gas phase, and so the solubility decreases.

EXAMPLE 12.5   Henry's Law

What is the concentration of O_2 at 25°C in water that is saturated with *air* at an atmospheric pressure of 645 mm Hg? Assume the mole fraction of oxygen in air is 0.209.

•Method of Solution

Henry's law states that the concentration of dissolved O_2 (C_{O_2}) is proportional to its partial pressure (P_{O_2}) in atm.

$$C_{O_2} = k\, P_{O_2}$$

The partial pressure of O_2 in air is found by using Dalton's law of partial pressures.

$$P_{O_2} = X_{O_2}P_T = 0.209\ (645\ \text{mm Hg}) \times \frac{1\ \text{atm}}{760\ \text{mm Hg}}$$

$$= 0.177\ \text{atm}$$

Taking k from Table 12.2, the concentration of dissolved oxygen is:

$$C_{O_2} = kP_{O_2} = (1.28 \times 10^{-3}\ \text{mol/L·atm})\ (0.177\ \text{atm})$$

$$= 2.27 \times 10^{-4}\ \text{mol/L}$$

EXERCISES

13. The solubility of KNO_3 at 70°C is 135 g per 100 g of water. At 10°C the solubility is 21 g per 100 g of water. What mass of KNO_3 will crystallize out of solution if exactly 100 g of its saturated solution at 70°C is cooled to 25°C?

14. As temperature increases the solubility of all gases in water increases / decreases.

15. The Henry's law constant for argon is 1.5×10^{-3} mol/L·atm at 20°C. Calculate its solubility in water at 20°C and 7.6 mm Hg.

COLLIGATIVE PROPERTIES (12.6 – 12.7)

STUDY OBJECTIVES

1. Calculate for a given solution the magnitude of the colligative properties: vapor-pressure lowering, boiling-point elevation, freezing-point depression, and osmotic pressure.
2. Determine the molar mass of a solute from freezing-point depression and osmotic pressure measurements.
3. Calculate the van't Hoff factor (i) for electrolyte solutions

Colligative Properties. The properties of solutions that depend on the number of solute particles in solution are called **colligative properties**. The four colligative properties of interest here are:

1. The **vapor-pressure lowering** ΔP is given by:

$$\Delta P = X_2 P_1^{\circ}$$

where X_2 is the mole fraction of the solute, and P_1° is the vapor pressure of the pure solvent. Recall that the vapor pressure of a liquid is the pressure exerted by a vapor in equilibrium with its liquid phase. For a solution containing a nonvolatile solute the vapor pressure due to the solvent is less than it is for the pure solvent. The amount of lowering of the vapor pressure can be seen to depend on X_2, the mole fraction of the solute.

2. The **boiling-point elevation**. The boiling point of a solution is higher than that of pure solvent because the vapor pressure of a solution is always less than the vapor pressure of pure solvent. Thus, a solution must be hotter than a pure solvent, if both vapor pressures are to be 1 atm. Figure 12.11 in the text shows the effect that lowering the vapor pressure has on the boiling point. The boiling-point elevation ΔT_b of a solution of molality m is given by:

$$\Delta T_b = K_b m$$

where K_b is a constant called the molal boiling-point elevation constant and has the units °C/m. There is a K_b for each solvent, and Table 12.1 (text) lists K_b values for six solvents. The magnitude of ΔT_b, the increase in boiling point of the solution over the boiling point of the pure solvent, is proportional to the solute concentration m. The preceding equation is independent of the solute, but requires that the solute be nonvolatile.

3. The **freezing-point depression** ΔT_f is given by:

$$\Delta T_f = K_f m$$

where K_f is the molal freezing-point depression constant and has the units °C/m. As is the case of K_b, K_f is different for each solvent, and its value must be found in a table such as Table 12.2 (textbook). The equation applies to all solutes, even volatile ones, because freezing points are rather insensitive to vapor pressure.

4. The **osmotic pressure** π is given by:

$$\pi = MRT$$

where M is the molar concentration, R is the ideal gas constant (0.0821 L·atm/K·mol), and T is the absolute temperature. Osmosis is the net flow of solvent molecules through a semipermeable membrane from a more dilute solution to a more concentrated solution. Note that in this process the solvent flows from a region of greater solvent concentration to a region of lesser solvent concentration. The osmotic pressure of a solution is the pressure required to stop osmosis.

Molar Mass Determination. Colligative properties provide a means to determine the molar mass of the solute. For instance, the freezing-point depression equation is

$$\Delta T_f = K_f m = K_f \times \frac{\text{mol solute}}{\text{kg solvent}}$$

Since ΔT_f can be measured and K_f is known, then the molality, m, of the solution can be calculated. Recall that molality is

$$\text{molality} = \frac{\text{mol solute}}{\text{kg solvent}} = \frac{\text{mass solute/molar mass of solute}}{\text{kg solvent}}$$

If you already know the molality of a solution and a known mass of solute is dissolved in a known mass of solvent, this leaves the *molar mass* of the solute as the only unknown quantity.

Freezing-point depression is more sensitive to the number of moles of solute than the boiling-point elevation because for the same solvent K_f is larger then K_b. The most sensitive method for the determination of the molar mass of a solute is to measure the osmotic pressure. The equation for osmotic pressure then becomes useful in the following form:

$$\pi = MRT = \frac{(\text{g solute/molar mass of solute})RT}{\text{L soln}}$$

Colligative Properties of Electrolyte Solutions. For the same solute concentrations, solutions of electrolytes always have greater colligative properties than those of nonelectrolyte solutions. For a 0.1 m methanol solution, ΔT_f is the same as the calculated value. For 0.1 m NaCl, ΔT_f is 1.9 times greater than calculated. And for Na_2SO_4, ΔT_f is 2.7 times greater than the calculated value. Table 12.3 compares values of ΔT_f (obs)/ΔT_f (calc)for several solutes. This ratio is always greater than one for solutions of electrolytes. Since colligative properties depend on the number of solute particles, the most likely explanation of this behavior is that NaCl and Na_2SO_4 dissociate into ions and this produces a greater concentration of solute particles. Therefore, the effective concentration of solute particles in 0.10 m NaCl is essentially 2×0.10 m, and in Na_2SO_4 it is almost 3×0.10 m. The freezing-point depression depends on the total concentration of all solute particles. Each ion acts as an independent particle.

TABLE 12.3 van't Hoff Factor i for Several Solutes

Solution	ΔT_f	ΔT_f (calc)	$\Delta T_f/\Delta T_f$ (calc)
0.10 m CH_3OH	0.186°C	0.186°C	1.0
0.10 m NaCl	0.353°C	0.186°C	1.9
0.10 m Na_2SO_4	0.502°C	0.186°C	2.7

The ratio $\Delta T_f/\Delta T_f$(calc) is called the **van't Hoff factor *i*.** For 1:1 electrolytes, such as NaCl, KCl, and $MgSO_4$, i ~ 2; for 2:1 electrolytes, such as $MgCl_2$ and K_2SO_4, *i* is a little less than 3. The deviation from a whole number results from the temporary formation of some ion pairs such as "NaCl" in the solution.

EXAMPLE 12.6   Boiling Point Elevation

What is the boiling point of an "antifreeze/coolant" solution made from a 50-50 mixture (by volume) of ethylene glycol, $C_2H_6O_2$ (density 1.11 g/mL), and water?

•**Method of Solution**

The boiling point depends on the molality of the 50-50 mixture. $\Delta T = K_b m$. For simplicity assume 100 mL of the solution. Then using the density of water, the mass of 50 mL (50 vol %) of H_2O is 50 g. The mass of 50 mL (50 vol %) of ethylene glycol using the density given above is 55.5 g.

•**Calculation**

$$m = \frac{mol\ C_2H_6O_2}{kg\ H_2O}$$

$$mol\ C_2H_6O_2 = 55.5\ g \times \frac{1\ mol}{62.0\ g} = 0.895\ mol$$

$$m = \frac{0.895\ mol\ C_2H_6O_2}{0.050\ kg} = 17.9\ mol/kg = 17.9\ m$$

$$\Delta T_b = K_b m = (0.52°C/m) \times 17.9\ m\ = 9.3°C$$

Therefore the boiling point of the solution is 9.3°C above the normal boiling point of water. Boiling point = 109.3°C.

EXAMPLE 12.7   Vapor Pressure Lowering

Calculate the vapor pressure of an aqueous solution at 30°C made from 100 g of sucrose ($C_{12}H_{22}O_{11}$), and 100 g of water. The vapor pressure of water at 30°C is 31.8 mm Hg.

•**Method of Solution**

Sucrose is a nonvolatile solute, so the vapor pressure over the solution will be due to H_2O molecules. The problem can be worked in two ways.

a. The vapor-pressure lowering is proportional to the mole fraction of sucrose X_2, and P_1^o, the vapor pressure of pure water at 30°C.

$$\Delta P = X_2 P_1^o$$

First let's calculate the mole fraction of sucrose, X_2.

$$X_2 = \frac{n_2}{n_1 + n_2}$$

$$n_2 = 100\ g \times \frac{1\ mol}{342\ g} = 0.292\ mol$$

$$n_1 = 100\ g \times \frac{1\ mol}{18.0\ g} = 5.55\ mol$$

$$X_2 = \frac{0.292}{0.292 + 5.55} = \frac{0.292}{5.84}$$

$$X_2 = 0.050$$

The vapor pressure lowering is:

$$\Delta P = (0.050) \ (31.8 \text{ mm Hg}) = 1.6 \text{ mm Hg}$$

The vapor pressure, P_1, is $P_1^o - \Delta P$.

$$P_1 = 31.8 - 1.6$$
$$= 30.2 \text{ mm Hg}$$

b.   Alternatively Raoult's law can be used to calculate the vapor pressure of the solvent:

$$P_1 = X_1 \ P_1^o$$

where P_1^o and P_1 are the vapor pressures of pure solvent and of the solvent in solution, respectively. From part a, $X_2 = 0.050$, therefore $X_1 = 1.00 - X_2 = 0.95$, and

$$P_1 = (0.95)(31.8 \text{ mm Hg})$$
$$= 30.2 \text{ mm Hg}$$

EXAMPLE 12.8   Finding the Molar Mass of a Solute

Benzene has a normal freezing point of 5.51°C. The addition of 1.25 g of an unknown compound to 85.0 g of benzene produces a solution with a freezing point of 4.52°C. What is the molecular mass of the unknown compound?

•Method of Solution

We need to find the number of moles in 1.25 g of unknown compound X. The freezing-point depression is proportional to the number of moles of X per kilogram of solvent.

•Calculation

The freezing-point depression is:

$$\Delta T_f = 5.51°C - 4.52°C$$
$$= 0.99°C$$

We can calculate the molality of the solution because ΔT_f is given above.

$$m = \frac{\Delta T_f}{K_f}$$

where K_f is the freezing-point depression constant for benzene (Table 12.2 textbook).

$$m = \frac{0.99°C}{5.12°C/m} = 0.19 \ m$$

This means that there are 0.19 mol of X per kg of benzene. The number of moles of X in 0.085 kg of the solvent benzene (the given amount) is:

$$0.085 \text{ kg} \times \frac{0.19 \text{ mol X}}{1 \text{ kg benzene}} = 1.6 \times 10^{-2} \text{ mol X}$$

Therefore 1.25 g X = 1.6×10^{-2} mol X, and

$$\text{molar mass of X} = \frac{1.25 \text{ g}}{1.6 \times 10^{-2} \text{ mol}} = 77.4 \text{ g/mol}$$

EXAMPLE 12.9   Colligative Propereties of Electrolytes and Nonelectrolytes

List the following aqueous solutions in the order of increasing boiling points: 0.10 m glucose ($C_6H_{12}O_6$); 0.10 m $Ca(NO_3)_2$; 0.10 m NaCl.

•Method of Solution

Ethanol is a nonelectrolyte, whereas NaCl and $Ca(NO_3)_2$ are electrolytes. NaCl dissociates into 2 ions per formula unit, and $Ca(NO_3)_2$ dissociates into three ions per formula unit. The effective molalities are approximately:

> glucose 0.10 m
> NaCl ≅ 0.20 m
> $Ca(NO_3)_2$ ≅ 0.30 m

Therefore, the boiling-point elevation and the boiling point are greatest for a solution of $Ca(NO_3)_2$, second highest for NaCl, and lowest for glucose.

•Comment

If you included the van't Hoff factors, the effective molalities would be affected slightly, but not enough to change the predicted results.

EXERCISES

16. What is the freezing point of a solution made from 1.00 g of $C_6H_{12}O_6$ (glucose) and 100.0 g of H_2O?

17. Calculate the approximate osmotic pressure at 25°C of a solution that has a freezing point of –18°C.

18. Calculate the molar mass of naphthalene given that a solution of 2.11 g of naphthalene in 100 g of benzene has a freezing-point depression of 0.85°C.

19. A solution contains 1.00 g of a compound (a nonelectrolyte) dissolved in 100.0 g of water. The freezing point of the solution is –0.103°C. What is the molar mass of the compound?

20. Calculate the freezing point of an aqueous solution that boils at 100.5°C.

21. How many solute particles does each basic unit of the following compounds give in aqueous solution?
    a. $C_2H_6O_2$ (ethylene glycol)   b. $(H_2N)_2CO$ (urea)   c. HBr   d. $(NH_4)_3PO_4$ e. NaOH   f. $Ca(OH)_2$

22. Arrange the following aqueous solutions in order of increasing freezing point.
    0.10 m $MgCl_2$, 0.20 m $C_6H_{12}O_6$, 0.15 m NaCl, 0.10 m KI.

23. Which of following aqueous solutions has the lower freezing point?
    0.10 m HCl or 0.10 m acetic acid

CONCEPTUAL QUESTIONS

1. List three things that happen when a salt dissolves in water. Give the sign of ΔH for each step and identify the forces involved in each.
2. Many salts with ΔH_{soln} near zero still dissolve in appreciable amounts. Explain.
3. When a "seed" crystal is placed in a supersaturated solution at constant temperature, precipitation is induced. When precipitation stops is the solution saturated, unsaturated, or superaturated?
4. What are ion pairs? What effect does ion-pair formation have on the osmotic pressure?

PRACTICE TEST

1. For each of the following pairs predict which substance will be more soluble in water:
   a. $HCl(g)$ or $H_2(g)$
   b. CCl_4 or $AlCl_3$
   c. CH_3Cl or CCl_4
   d. CH_3OH or $CH_3—O—CH_3$

2. For each of the following pairs predict which substance will be more soluble in $CCl_4(l)$.
   a. H_2O or oil
   b. C_6H_6 (benzene) or CH_3OH
   c. I_2 or $NaCl$

3. Calculate the molality of a solution prepared by dissolving 28.0 g of urea, $CO(NH_2)_2$, in 450 g of H_2O.

4. How many grams of KI must be dissolved in 300 g of water to make a 0.500 m solution?

5. What is the molarity of a 3.0 percent hydrogen peroxide (H_2O_2) aqueous solution? The density is essentially 1.0 g/cm³.

6. Concentrated sulfuric acid is 96 percent H_2SO_4 by weight. Its density is 1.83 g/mL. Calculate the molarity of concentrated H_2SO_4.

7. What are the mole fractions of methanol and water in a solution that contains 40.0 g CH_3OH and 40.0 g H_2O?

8. An aqueous solution contains 167 g $CuSO_4$ in 820 mL of solution. Calculate the following:
   a. Molarity
   b. Percent by mass
   c. Mole fraction
   d. Molality of the solution
   The density of the solution is 1.195 g/mL.

9. The Henry's law constant for CO is 9.73×10^{-4} mol/L·atm at 25°C. What is the concentration of dissolved CO in water if the partial pressure of CO in the air is 0.015 mm Hg?

10. How many grams of isopropyl alcohol, C_3H_7OH, should be added to 1.0 L of water to give a solution that will not freeze above −16°C.

11. What is the boiling point of an aqueous solution of a nonvolatile solute that freezes at −3.0°C?

12. When 48 g of glucose (a nonelectrolyte) is dissolved in 500 g of H_2O, the solution has a freezing point of −0.94°C. What is the molar mass of glucose?

13. 7.85 g of a compound with an unknown formula is dissolved in 300 g of benzene. The freezing point of the solution is 2.10°C below that of pure benzene. What is the molar mass of the compound?

14. The dart poison in root extracts used by the Peruvian Indians is called curare. It is a nonelectrolyte. The osmotic pressure at 20°C of an aqueous solution containing 0.200 g of curare in 100 mL is 56.2 mm Hg. Calculate the molar mass of curare.

15. What is the concentration of a NaCl solution with an osmotic pressure of 10 atm at 25°C?

16. The walls of red blood cells are semipermeable membranes, and the solution of NaCl within those walls exerts an osmotic pressure of 7.82 atm at 37°C. What concentration of NaCl must a *surrounding* solution have so that this pressure is balanced and cell rupture (hemolysis) is prevented?

17. Arrange the following aqueous solutions in order of increasing freezing points, lowest to highest.
    0.100 m ethanol
    0.050 m $Ca(NO_3)_2$
    0.100 m NaBr
    0.050 m HCl

18. Calculate the value of i for a 1:1 electrolyte, if a 1.0 m aqueous solution of the electrolyte freezes at −3.28°C.

General Problem

19. In the course of research a chemist isolated a new compound. An elemental analysis gave the following: C 50.7%, H 4.25%, O 45.1%. When 5.01 g of the compound was dissolved in exactly 100 g of water, it produced a solution with a freezing point of −0.65°C. What is the molecular formula of the compound?

ANSWERS

Exercises

1. miscible
2. Hexane is a nonpolar liquid and water is a polar liquid.
3. a. soluble   b. soluble   c. insoluble   d. insoluble   e. soluble   f. soluble
4. Ammonia NH_3 can form hydrogen bonds to water, but hexane cannot form hydrogen bonds..
5. Solvation involves surrounding a solute molecule or ion with solvent molecules. The process is called hydration when the solvent is water. Solvation is the more general term.
6. a. NaCl(s)   b. NH_3(g)   c. CH_3OH(l)
7. a. I_2(s)   b. CS_2(l)   c. CO_2(g)
8. a. 7.96%   b. 9.81%   c. 1.6%
9. a. 1.47 m   b. 0.359 m
10. 2.32 M NaCl
11. 9.94% $AgNO_3$
12. 0.34 M $AgNO_3$
13. 115 g KNO_3
14. decreases
15. 1.5×10^{-5} mol/L
16. −0.103°C

17. 237 atm
18. 127 g/mol
19. 181 g/mol
20. −1.8°C
21. a. 1   b. 1   c. 2   d. 4   e. 2   f. 3
22. 0.20 m $C_6H_{12}O_6$ = 0.10 m KI < 0.15 m NaCl = 0.10 m $MgCl_2$
23. The T_f of 0.10 m HCl (a strong acid) is lower than that of 0.10 m acetic acid (a weak acid).

Conceptual Questions

1. First, the water molecules in the solvent must spread apart a bit to make room for the solute particle. This requires a positive ΔH. Energy must be added in order to break hydrogen bonds between water molecules. Second, the ions in the lattice of the solid salt must be separated from all their neighbors. The energy for this step is related to the lattice energy. The forces holding the crystal together are ionic bonds. ΔH is positive. Next, water molecules and the anions and cations from the salt are mixed. Hydration occurs and heat is released (ΔH is negative) as ion-dipole forces bring water molecules and ions together.

2. Energy is only one factor affecting the solution process. A second factor which affects all natural events is a tendency to produce more disordered systems. In the pure state the solute and solvent molecules are arranged in a fairly ordered manner. Much of this order disappears when the solute dissolves in the solvent. The molecules of solvent and solute become mixed together. It is the increase in disorder of the system that favors the solubility of any substance.

3. Saturated

4. Ion pairs (+/−) can form in solutions of electrolytes. The formation of an ion pair will reduce the number of solute particles is the solution. Therefore the osmotic pressure would decrease compared to a solution in which ions remained separated and there were no ion pairs.

Practice Test

1. a. HCl   b. $AlCl_3$ c. CH_3Cl d. CH_3OH
2. a. oil   b. C_6H_6 c. I_2
3. 1.03 m urea
4. 24.9 g KI
5. 0.88 M H_2O_2
6. 18 M
7. X_{CH_3OH} = 0.360, X_{H_2O} = 0.640
8. a. 1.28 M
   b. 17.0% $CuSO_4$, 83.0% H_2O
   c. X_{CuSO_4} = 0.023, X_{H_2O} = 0.977
   d. 1.29 m
9. 1.9×10^{-8} M CO
10. 520 g
11. 100.84°C
12. 190 g/mol
13. 63.8 g/mol
14. 650 g/mol
15. 0.20 M NaCl
16. 0.15 M NaCl
17. NaBr; $Ca(NO_3)_2$; HCl = ethanol
18. i = 1.76
19. $C_6H_6O_4$

Chapter Thirteen
CHEMICAL KINETICS

- The Rate of Reaction (13.1)
- Rate Laws (13.2)
- The Relation Between Reactant Concentration and Time (13.3)
- Activation Energy and Temperature Dependence of Reaction Rates (13.4)
- Reaction Mechanisms (13.5)
- Catalysis (13.6)

THE RATE OF REACTION (13.1)

STUDY OBJECTIVES

1. Define the rate of reaction.
2. Express the rate of a given reaction in terms of the change in concentration with time of a reactant or a product.
3. Calculate the average rate of reaction from given concentration vs. time data.

Expressing the Rate of Reaction. **Chemical kinetics** is the area of chemistry concerned with the study of the rates of chemical reactions. The purposes of kinetic studies are to find the factors that affect reaction rates and determine the reaction mechanism. Knowledge of the factors that affect reaction rates enables chemists to control rates. Finding the reaction mechanism means we can identify the intermediate steps by which reactants are converted into products.

A rate is a change in some quantity with time. The rate of population growth is the change in population per change in time. The change in a quantity such as population is always equal to the the difference, population (after) minus population (before). The symbol for "the change in" is Δ.

$$\Delta(\text{population}) = \text{population}_{\text{final}} - \text{population}_{\text{initial}}$$

The change in time is some appropriate time interval, Δt, where

$$\Delta t = t_{\text{final}} - t_{\text{initial}}$$

The rate of population change is

$$\text{rate} = \frac{\Delta(\text{population})}{\Delta t}$$

As a chemical reaction proceeds the concentrations of reactants and products change with time. For instance, as the reaction $A + B \rightarrow C$ progresses, the concentration of C increases. The rate is expressed as the change in the molar concentration of C, $\Delta[C]$, during the time interval Δt.

$$\text{rate} = \frac{\Delta[C]}{\Delta t}$$

For a specific reaction we need to take into account the stoichiometry; that is, we need the balanced equation. For example, let's express the rate of the following reaction in terms of the concentrations of the individual reactants and products.

$$2NO(g) + O_2(g) \rightarrow 2NO_2(g)$$

These concentrations can be monitored experimentally as a function of time. Notice from the balanced equation, that 2 mol NO reacts with 1 mol O_2; therefore, the concentration of NO will decrease twice as fast as that of O_2.

$$\frac{\Delta[NO]}{\Delta t} = \frac{2\,\Delta[O_2]}{\Delta t}$$

Since the rates of change of individual reactants and products may differ the convention is to make the reaction rate come out to be the same no matter which reactant or product is used to calculate it. First we divide each concentration change by the coefficient from the balanced equation

$$\text{rate} = -\frac{\Delta[NO]}{2\Delta t} = -\frac{\Delta[O_2]}{\Delta t} = \frac{\Delta[NO_2]}{2\Delta t}$$

Second, a negative sign is inserted before terms involving reactants. The change in NO concentration, $\Delta[NO]$, is negative because the concentration of NO *decreases* with time. Inserting a negative sign in the expression makes the rate of reaction a positive quantity.

For a general equation:

$$aA + bB \rightarrow cC$$

the rate can be expressed in terms of any individual reactant or product.

$$\text{rate} = -\frac{\Delta[A]}{a\,\Delta t} = -\frac{\Delta[B]}{b\,\Delta t} = \frac{\Delta[C]}{c\,\Delta t}$$

No matter which reactant or product we use, the reaction rate will be positive and have the same value.

Calculating an Average Rate. The average rate of reaction over any time interval is equal to the change in the concentration of a reactant $\Delta[A]$, or of a product $\Delta[C]$ divided by the time interval, Δt, during which the change occurred.

$$\text{average rate} = -\frac{\text{change in the concentration of A}}{\text{length of time interval}}$$

$$\text{average rate} = -\frac{\Delta[A]}{\Delta t} = -\frac{[A]_2 - [A]_1}{t_2 - t_1}$$

The concentration term $[A]_2$ is the concentration of A at time t_2, and $[A]_1$ is the concentration of A at time t_1.

EXAMPLE 13.1   Expressing the Rate of Reaction

Write expressions for the rate of the following reaction in terms of each of the reactants and products.

$$2N_2O_5(g) \rightarrow 4NO_2(g) + O_2(g)$$

•**Method of Solution**

Recall that the rate is defined as the change in concentration of a reactant or product with time. Each "change-in-concentration" term is divided by the corresponding stoichiometric coefficient. Terms involving reactants are preceded by a minus sign. Therefore, the rate is expressed as

$$\text{rate} = -\frac{\Delta[N_2O_5]}{2\Delta t} = \frac{\Delta[NO_2]}{4\Delta t} = \frac{\Delta[O_2]}{\Delta t}$$

EXAMPLE 13.2   Rate of Reaction

Oxygen gas is formed by the decomposition of nitric oxide:

$$2NO(g) \rightarrow O_2(g) + N_2(g)$$

If the rate of formation of O_2 is 0.054 M/s, what is the rate of change of NO concentration?

•Method of Solution

From the stoichiometry of the reaction, 2 mol NO react for each mole of O_2 that forms. Therefore:

$$rate = -\frac{\Delta[NO]}{2\Delta t} = \frac{\Delta[O_2]}{\Delta t}$$

•Calculation

$$\frac{\Delta[NO]}{\Delta t} = -\frac{2\Delta[O_2]}{\Delta t} = -2(0.054 \text{ M/s})$$

$$= -0.108 \text{ M/s}$$

EXAMPLE 13.3   Calculation of the Average Rate

Experimental data for the hypothetical reaction:

$$A \rightarrow 2B$$

are listed in the following table.

Time (s)	[A] (mol/L)
0.00	1.000
10.0	0.891
20.0	0.794
30.0	0.707
40.0	0.630

a.   Calculate the average rates of change of [A], and the average reaction rates for the two time intervals from 0 to 10 s and from 30 to 40 s.

b.   Why does the rate decrease from one time interval to the next?

•Method of Solution

a.   The average rate of change of [A] is given by:

$$\frac{\Delta[A]}{\Delta t} = \frac{[A]_2 - [A]_1}{t_2 - t_1}$$

For the time interval 0 – 10 s, we get:

$$\frac{\Delta[A]}{\Delta t} = \frac{0.891 \text{ mol/L} - 1.000 \text{ mol/L}}{10.0 \text{ s} - 0 \text{ s}} = \frac{-0.109 \text{ mol/L}}{10.0 \text{ s}}$$

$$\frac{\Delta[A]}{\Delta t} = -0.0109 \text{ mol/L·s}$$

Since the average reaction rate $= -\dfrac{\Delta[A]}{\Delta t}$, then:

average reaction rate $= -(-0.0109 \text{ mol/L·s}) = 0.0109 \text{ mol/L·s}$

For the time interval 30 – 40 s, the rate of change of [A] is:

$$\dfrac{\Delta[A]}{\Delta t} = -\dfrac{(0.630 \text{ mol/L} - 0.707 \text{ mol/L})}{40.0 \text{ s} - 30.0 \text{ s}}$$

$$\dfrac{\Delta[A]}{\Delta t} = \dfrac{-0.077 \text{ mol/L}}{10.0 \text{ s}} = -0.0077 \text{ mol/L·s}$$

And the average reaction rate is:

$$\text{average rate} = -\dfrac{\Delta[A]}{\Delta t} = 0.0077 \text{ mol/L·s}$$

b. The reaction rate decreases as the total reaction time increases because the rate is proportional to the concentration of the reactant A, and the concentration of A decreases as the time of reaction increases.

EXERCISES

1. Write expressions for the rate of reaction in terms of each of the reactants and products.

    $N_2(g) + 3H_2(g) \rightarrow 2NH_3(g)$

2. Thiosulfate ion is oxidized by iodine in aqueous solution according to the equation

    $2 S_2O_3^{2-}(aq) + I_2(aq) \rightarrow S_4O_6^{2-}(aq) + 2I^-(aq)$

    If 0.025 mol of $S_2O_3^{2-}$ is consumed in 0.50 L solution per minute:

    a. Calculate the rate of removal of $S_2O_3^{2-}$ in M/s.
    b. What is the rate of removal of I_2?

3. N_2O_5 is an unstable compound that decomposes according to the following equation.

    $2N_2O_5 \rightarrow 4NO_2 + O_2$

    The following data was obtained at 50°C.

[N$_2$O$_5$] (M)	Time (s)
1.00	0
0.88	200
0.78	400
0.69	600
0.61	800
0.54	1000
0.48	1200
0.43	1400

    a. What is the average rate of N_2O_5 disappearance in the time interval 200-400 s?
    b. What is the average rate of N_2O_5 disappearance in the time interval 800-1000 s?
    c. What is the rate of O_2 production in the time interval 800-1000 s?

RATE LAWS (13.2)

STUDY OBJECTIVES

1. Describe all the terms in "the rate law," and define the order of reaction.
2. Apply the isolation method to derive the rate law for a reaction.

    Effect of Concentration. The rate of a reaction is proportional to the reactant concentrations. For the reaction

$$NO + \frac{1}{2}O_2 \rightarrow NO_2$$

the rate is proportional to the concentrations of NO and O_2.

    The *rate law* (or *rate equation*) for the reaction is:

$$rate = k[NO]^x[O_2]^y$$

The proportionality constant k is called the **rate constant**. The value of k depends on the reaction and the temperature. x and y are exponents which are often, but not always 1 or 2.

    The exponents x and y determine how strongly the concentration affects the rate. The **exponent x** is called the **order with respect to NO**, and **y** is the **order with respect to O_2**. The sum x + y is the overall order. *The values of x and y must be determined from experiment, and cannot be derived by any other means.* We will discuss how to determine the order of reaction in the next section. For now we will just use the results. For the NO reaction with O_2 experiments show that x = 2 and y = 1. Therefore, the rate law for this reaction is:

$$rate = k[NO]^2[O_2]$$

This reaction is second order in nitric oxide, and first order in oxygen. It is third order overall.

    The fact that the reaction is first order in O_2 means that the rate is directly proportional to the O_2 concentration. If $[O_2]$ doubles or triples, the rate will double or triple also. We can show this mathematically. Consider two experiments. In expt 1 the concentration of O_2 is c. In expt 2 the concentration of O_2 is doubled from c to 2c. If the concentration of NO is the same in both experiments, it will have no effect on the rate. Use of the rate law allows us to write the ratio of the two rates:

$$\frac{rate \ (expt \ 2)}{rate \ (expt \ 1)} = \frac{k[NO]^2(2c)}{k[NO]^2(c)} = 2$$

As discussed, we see that doubling the concentration of a reactant that is first order will cause the rate to double.

    If the concentration of O_2 is held constant in two experiments and the concentration of NO doubles (from c to 2c), the rate law predicts that the rate will quadruple.

$$\frac{rate \ (expt \ 2)}{rate \ (expt \ 1)} = \frac{k[O_2](2c)^2}{k[O_2](c)^2} = 2^2 = 4$$

The fact that the reaction is second order in NO means that the rate is proportional to the square of the concentration of NO. Doubling or tripling of [NO] causes the rate to increase four- or nine-fold, respectively.

    In general, if the concentration of one reactant is doubled while the other reactant concentration is unchanged, *and* the rate is:

1. *unchanged,* the order of the reaction is *zero order* with respect to the changing reactant.
2. *doubled,* the order of the reaction is *first order* with respect to the changing reactant.
3. *quadrupled,* the order of the reaction is *second order* with respect to the changing reactant.

The Isolation Method. One procedure used to determine the rate law for a reaction involves the isolation method. In this method the concentration of all but one reactant is fixed, and the rate of reaction is measured as a function of the concentration of the one reactant whose concentration is varied. Any variation in the rate is due to the variation of this reactant's concentration. In practice the experimenter observes the dependence of the initial rate on the concentration of the reactant.

To determine the order with respect to A in the following chemical reaction

$$2A + B \rightarrow C$$

the initial rate would be measured in several experiments in which the concentration of A is varied and the concentration of B is held constant. To determine the order with respect to B, the concentration of A must be held constant and the concentration of B is varied in several experiments. The application of this method is illustrated in Example 13.5.

EXAMPLE 13.4   Concentration Effect on the Rate

The reaction $A + 2B \rightarrow$ products was found to have the rate law: rate $= k[A][B]^3$. By what factor will the rate of reaction increase if the concentration of B is increased from c to 3c, while the concentration of A is held constant?

•Method of Solution

Write a ratio of the rate law expressions for the two different concentrations of B.

$$\frac{\text{rate (expt 2)}}{\text{rate (expt 1)}} = \frac{k[A][3c]^3}{k[A][c]^3} = \frac{3^3 c^3}{c^3}$$

$$= 27$$

The rate will increase 27-fold when [B] is increased three-fold.

EXAMPLE 13.5   Finding the Rate Law

The following rate data were collected for the reaction:

$$2NO + 2H_2 \rightarrow N_2 + 2H_2O$$

Experiment	$[NO]_0$ (M)	$[H_2]_0$ (M)	$\Delta[N_2]/\Delta t$ (M/h)
1	0.60	0.15	0.076
2	0.60	0.30	0.15
3	0.60	0.60	0.30
4	1.20	0.60	1.21

a.   Determine the rate law.
b.   Calculate the rate constant.

•Method of Solution

a.   We want to determine the exponents in the equation

rate $= k[NO]^x[H_2]^y$

To determine the order with respect to H_2, first find two experiments in which [NO] is held constant. This can be done by comparing the data of experiments 1 and 2. When the concentration of H_2 is doubled, the reaction rate doubles. Thus, the reaction is first order in H_2. When the NO concentration is doubled (experiments 3 and 4) the reaction rate quadruples. Therefore, the reaction is second order in NO.

The rate law is

$$\text{rate} = k[NO]^2[H_2]$$

b. Rearrange the rate law from part (a).

$$k = \frac{\text{rate}}{[NO]^2[H_2]}$$

Then substitute data from any one of the experiments. Using experiment 1:

$$k = \frac{0.076 \text{ M/h}}{(0.60 \text{ M})^2(0.15 \text{ M})} = 1.4/M^2 \text{ h}$$

•**Comment**

You should get the same value of k from all four experiments. Note that the units of k are those of a third-order rate constant. Taking a closer look at proving that $x = 2$. Write the ratio of rate laws for experiments 4 and 3.

$$\frac{\text{rate}_4}{\text{rate}_3} = \frac{k[NO]^x[H_2]}{k[NO]^x[H_2]}$$

$$\frac{1.21}{0.30} = \frac{k(1.20)^x(0.60)}{k(0.60)^x(0.60)} = \left(\frac{1.20}{0.60}\right)^x$$

$$4.0 = 2.0^x$$

Therefore $x = 2$

EXERCISES

4. The rate law for the reaction $2A + B \rightarrow C$ was found to be rate $= k[A][B]^2$. If the concentration of B is tripled and the concentration of A is unchanged, by how many times will the reaction rate increase?

5. Use the following data to determine (a) the rate law and (b) the rate constant for the reaction

$$2A + B \rightarrow C$$

Experiment	$[A]_0$	$[B]_0$	Rate (M/s)
1	0.25	0.10	0.012
2	0.25	0.20	0.048
3	0.50	0.10	0.024

THE RELATION BETWEEN REACTANT CONCENTRATION AND TIME (13.3)

STUDY OBJECTIVES

1. Determine for first- and second-order reactions the concentration of a reactant at any time after the reaction has started, or the time required for a given fraction of the sample to react.
2. Determine the half-life of a first-order reaction.
3. Apply rate law equations and graphs to determine whether a reaction is first order or second order.

First-Order Reactions. One of the most widely encountered kinetic forms is the first-order rate equation. In this case the exponent of [A] in the rate law is 1.

$$A \rightarrow \text{products}$$

$$\text{rate} = -\frac{\Delta[A]}{\Delta t} = k[A]$$

For a first-order reaction the unit of the rate constant is reciprocal time, $1/t$. Convenient units are s^{-1}, h^{-1}, etc.

The equation that relates the concentration of A remaining to the time since the reaction started is

$$\ln \frac{[A]}{[A]_0} = -kt$$

This is a very useful equation called the **integrated first-order equation**. Here $[A]_0$ is the concentration of A at time = 0, and [A] is the concentration of A at time = t. The rate constant k is the first-order rate constant. The concentration [A] decreases as the time increases. This equation allows the calculation of the rate constant k when $[A]_0$ is known, and [A] is measured at time t. Also, once k is known, [A] can be calculated for any future time.

To determine whether a reaction is first order, we rearranged the first-order equation into the form:

$$\ln [A] = -kt + \ln [A]_0$$

corresponding to the linear equation

$$y = mx + b$$

Here m is the slope of the line and b is the intercept on the y axis. Comparing the last two equations, we can equate y and x to experimental quantities.

$$y = \ln [A] \quad \text{and} \quad x = t$$

Therefore, the intercept $b = \ln [A]_0$, and the slope of the line $m = -k$. Thus a plot of $\ln [A]$ versus t for a first-order reaction gives a straight line with a slope of $-k$ as shown in Figure 13.1 below. If a plot of $\ln [A]$ versus t yields a curved line, rather than a straight line, the reaction is not a first-order reaction. This graphical procedure is the method used by most chemists to determine whether or not a given reaction is first order.

Figure 13.1 A plot of [A] versus time for a first-order reaction gives a curved line. A plot of ln [A] versus time gives a straight line for a first-order reaction.

Half-life. The half-life of a reaction, $t_{1/2}$, is a useful concept. For a first-order reaction, the half-life is given by:

$$t_{1/2} = \frac{0.693}{k}$$

where 0.693 is a constant and k is the rate constant. Knowledge of the half-life allows the calculation of the rate constant k. The **half-life** of a reaction is the time required for the concentration of a reactant to decrease to half of the initial value. After one half-life, the ratio $[A]/[A]_0$ is equal to 0.5.

If the reaction continues, then [A] will drop by 1/2 again during the second half-life period as shown in Figure 13.2. After two half-life periods the fraction of the original concentration of A remaining, $[A]/[A]_0$, will be 1/2 of the concentration remaining after the first half-life, so $[A]/[A]_0 = 0.5 \times 1/2 = 0.25$.

Figure 13.2 A plot of [A] versus time for a first-order reaction gives a curved (exponential) line. Over each half-life period, [A] drops in half.

Second-Order Reactions. In a second-order reaction, the rate is proportional either (1) to the square of the concentration of one reactant,

$A \rightarrow$ product

$$\text{rate} = - \frac{\Delta[A]}{\Delta t} = k[A]^2$$

or (2) to the product of the concentrations of two reactants, each raised to the first power:

$A + B \rightarrow$ product

$$\text{rate} = - \frac{\Delta[A]}{\Delta t} = k[A][B]$$

This reaction is first order in A and first order in B, and so it is second order overall. For a second-order reaction, the rate constant has units 1/(molarity × time) which is 1/M·s or $M^{-1}\ s^{-1}$.

An important equation called **the integrated second-order equation** is

$$\frac{1}{[A]} = \frac{1}{[A]_0} + kt$$

where $[A]_0$, $[A]$, and t have their usual meaning, and k is the second-order rate constant. This equation allows the calculation of the concentration of A at any time t after the reaction has begun. Alternatively, if $[A]_0$, $[A]$, and t are known, the rate constant can be calculated. Example 13.6 in the text shows an example calculation using this equation.

The half-life for a second-order reaction is given by

$$t_{1/2} = \frac{1}{k[A]_0}$$

Here we see that the half-life is inversely proportional to the initial concentration, $[A]_0$. This situation is different from that for a first-order reaction, where $t_{1/2}$ is independent of $[A]_0$.

Zero Order Reactions. A zero-order reaction is one in which the rate does not depend on the concentrations of reactants.

$A \rightarrow$ products

$$\text{rate} = \frac{-[A]}{\Delta t} = k[A]^0 = k$$

A plot of [A] versus time is a straight line for a zero order reaction. The rate does not slow down as the reactant is used up..

Summary. The integrated rate equations and graphical methods allow us to distinguish between the various overall orders of reaction.

1. A reaction is first order when a plot of ln [A] versus t is a straight line. A plot of [A] versus time will be curved.
2. A reaction is zero order when a plot of [A] versus t is a straight line.
3. A reaction is second order when a plot of 1/[A] versus t is a straight line, and ln[A] versus t is curved.
4. The half-life equations also provide a way to distinguish between first- and second-order reactions. The half-life of a first-order reaction is independent of starting concentration, whereas the half-life of a second-order reaction is inversely proportional to the initial concentration.

EXAMPLE 13.6   First-Order Reaction

Methyl isocyanide undergoes a rearrangement to form methyl cyanide that follows first-order kinetics.

$$CH_3NC(g) \rightarrow CH_3CN(g)$$

The reaction was studied at 199°C. The initial concentration of CH_3NC was 0.0258 mol/L, and after 11.4 min analysis showed the concentration of product was 1.30×10^{-3} mol/L.
a.   What is the first-order rate constant?
b.   What is the half-life of methyl isocyanide?
c.   How long will it take for 90 percent of the CH_3NC to react?

•Method of Solution for (a)

This problem illustrates the use of the integrated first-order rate equation

$$\ln \frac{[CH_3NC]}{[CH_3NC]_0} = -kt$$

Where k is the rate constant and $[CH_3NC]$ is the reactant concentration at time t. The initial concentration is $[CH_3NC]_0 = 0.0258$ M. After 11.4 min the product concentration is 1.30×10^{-3} M. This implies that the concentration of CH_3NC remaining unreacted is $0.0258 - 0.0013 = 0.0245$ M.

•Calculation for (a)

Substitution gives

$$\ln \frac{0.0245 \text{ M}}{0.0258 \text{ M}} = -k \text{ (11.4 min)}$$

$$\ln 0.950 = -k \text{ (11.4 min)}$$

$$k = 4.54 \times 10^{-3} \text{ min}^{-1}$$

•Method of Solution for (b)

For a first-order reaction the half-life equation is

$$t_{1/2} = \frac{0.693}{k}$$

•Calculation for (b)

$$t_{1/2} = \frac{0.693}{4.54 \times 10^{-3} \text{ min}^{-1}} = 153 \text{ min}$$

•Method of Solution for (c)

If 90 percent of the initial CH_3NC is consumed, then 10 percent remain. Therefore:

$$[CH_3NC] = 0.10 \, [CH_3NC]_0$$

•Calculation for (c)

Substitution into the rate equation gives

$$\ln \frac{0.10 \, [CH_3NC]_0}{[CH_3NC]_0} = -(4.54 \times 10^{-3} \text{ min}^{-1})t$$

Solving for t, we get:

$$t = -\frac{\ln 0.10}{4.54 \times 10^{-3} \text{ min}^{-1}} = -\frac{-2.303}{4.54 \times 10^{-3} \text{ min}^{-1}}$$

$$t = 507 \text{ min}$$

EXAMPLE 13.7   First-Order   Reaction

At 230°C the rate constant for methyl isocyanide isomerization is $9.25 \times 10^{-4} \text{ s}^{-1}$.

$$CH_3NC \rightarrow CH_3CN$$

a.   What fraction of the original isocyanide will remain after 60.0 min?
b.   What is the half-life of methyl isocyanide at this temperature?

•**Method of Solution for (a)**

a.   Again applying the first-order equation:

$$\ln \frac{[CH_3NC]}{[CH_3NC]_0} = -kt$$

The fraction of methyl isocyanide remaining is given by the fraction $\dfrac{[CH_3NC]}{[CH_3NC]_0}$.

•**Calculation for (a)**

Plug in the rate constant and the time, and solve for the fraction remaiming:

$$\ln \frac{[CH_3NC]}{[CH_3NC]_0} = -9.25 \times 10^{-4} \text{ s}^{-1} (60 \text{ min}) \times \frac{60 \text{ s}}{1 \text{ min}} = -3.33$$

Taking the antilog of both sides:gives:

$$\frac{[CH_3NC]}{[CH_3NC]_0} = e^{-3.33}$$

$$\frac{[CH_3NC]}{[CH_3NC]_0} = 0.0358$$

•**Calculation for (b)**

The half-life is

$$t_{1/2} = \frac{0.693}{9.25 \times 10^{-4} \text{ s}^{-1}}$$

$$= 749 \text{ s or } 12.4 \text{ min}$$

•**Comment**

Note the much shorter half-life at 230°C than at 199°C (Example 13.6).

EXERCISES

6.   A certain first-order reaction A $\rightarrow$ B is 40% complete (40% of the reactant is used up) in 75 s. What is (a) the rate constant and (b) the half-life of this reaction?

7. a. Write the integrated rate equation for a first-order reaction? Define the variables.

    b. How can the equation be plotted to give a straight line? Define x, y, and the slope.

    c. What information do you need in order to calculate the first-order rate constant?

8. The half-life of a certain first order reaction is 112 min. What percent of the initial concentration of reactant will remain after 89 min?

9. Consider the reaction in exercise 3 above for the reaction:

    $$2N_2O_5 \rightarrow 4NO_2 + O_2$$

    a. Use the data to calculate the first-order rate constant.
    b. How long will it take for the concentration of N_2O_5 to fall to 0.25 M?
    c. What is the half-life?

10. What are the units of a first-order and a second-order rate constant?

11. A decompositon reaction has a rate constant of 0.12 y^{-1} at a certaian temperature.
    a. What is the half-life of the reaction at the same temperature?
    b. How long will it take for the concentration of the reactant to reach 20% of its initial value?

ACTIVATION ENERGY AND TEMPERATURE DEPENDENCE OF REACTION RATES (13.4)

STUDY OBJECTIVES

1. Describe three fundamental tenents to the collision theory.
2. Calculate the activation energy for a reaction when given rate constants at several different temperatures.
3. Describe the concept of activation energy and show how it explains the variation of reaction rate with temperature.
4. Describe a reaction energy profile, including the activation energy and energy change of reaction.

Collision Theory. The collision theory of chemical reactions provides a general explanation of how reaction rates are affected by reactant concentrations, and temperature. The basic ideas of the theory are:

1. In order for atoms, molecules, or ions to react they must first collide with each other. The rate of reaction is proportional to the rate of collisions, called the **collision frequency**. The more concentrated the reactants, the greater the collision frequency, and the reaction rate.
2. For molecules to react they must come together in the proper *orientation*. Molecules can be complex and often it is just one atom in a molecule that reacts upon collision with another molecule.
3. When reactant molecules collide, they must possess a minimum amount of kinetic energy in order for an effective collision —a reaction— to occur. Without this necessary energy two molecules will just bump each other and bounce back without reacting. The minimum amount of energy required to initiate a chemical reaction is called the **activation energy**.

Effect of Temperature. The temperature of a reaction system is an important variable because of its strong effect on reaction rates. As a rough rule, reaction rates approximately double with a 10°C rise in temperature. In general, the rate equation is rate = $k[A]^x[B]^y$. Since the concentrations [A] and [B] are unaffected by temperature, it is the rate constant that changes with temperature. In 1889, S. Arrhenius found that a plot of the natural logarithm of the rate constant (ln k) versus the reciprocal of the absolute temperature 1/T gave a straight line. See Figure 13.16 (text). Arrhenius identified the slope of the line as being related to an energy term

$$m = -\frac{E_a}{R}$$

where R is the ideal gas constant (in units of joules) and E_a is the activation energy. The logarithmic form of the Arrhenius equation is

$$\ln k = -\frac{E_a}{RT} + \ln A$$

where ln A is the intercept. Both the Arrhenius A and E_a are constants for a particular reaction. Taking the antilog of both sides gives the Arrhenius equation which relates the rate constant to the temperature.

$$k = A\, e^{-E_a/RT}$$

The Meaning of A and E_a. The parameter A is called the **frequency factor**. It is related to the frequency of molecular collisions, and the fraction of collisions that have the correct orientation. As discussed above, the collision frequency is important because molecules must collide in order to react.

The activation energy E_a is related to the formation of the activated complex. The **activated complex** is the high-energy intermediate species that dissociates into the products. Figure 13.3 shows that the *activation energy* is the difference in energy between the reactants and the activated complex. The activation energy is provided by the kinetic energy of rapidly moving molecules during collisions. Reactants must "get over the barrier" before they become products. The factor $e^{-E_a/RT}$ that appears in the Arrhenius equation is the fraction of molecules with energies equal to or greater than the activation energy. This factor changes significantly with temperature. As temperature increases a greater fraction of molecules has an energy equal to, or greater than, the activation energy.

Figure 13.3 Molecules of A and B with average energies must first acquire an energy E_a before they can react.

From the Arrhenius equation we can point out *that reactions for which* E_a *is large will be much slower than those for which* E_a *is small.* As E_a increases, the negative exponent increases, and so k decreases.

A convenient equation that can be used to calculate the activation energy is

$$\ln\left(\frac{k_1}{k_2}\right) = \frac{E_a}{R}\left(\frac{T_1 - T_2}{T_1 T_2}\right)$$

where k_1 is the rate constant at temperature T_1, and k_2 is the rate constant at temperature T_2. Use of this equation requires that rate constants k_1 and k_2 be measured at two temperatures T_1 and T_2, respectively.

EXAMPLE 13.8 Calculating Activation Energy

For the reaction

$$NO + O_3 \rightarrow NO_2 + O_2$$

the following rate constants have been obtained:

Temperature, °C	$k(M^{-1} s^{-1})$
10.0°	9.30×10^6
30.0°	1.25×10^7

Calculate the activation energy for this reaction.

•Method of Solution

Here we are given two rate constants corresponding to two temperatures. The activation energy (E_a) can be calculated by substituting into the equation given earlier that shows the temperature dependence of the rate constant.

$$\ln\left(\frac{k_1}{k_2}\right) = \frac{E_a}{R}\left(\frac{T_1 - T_2}{T_1 T_2}\right)$$

Let $k_1 = 9.3 \times 10^6 \ M^{-1} s^{-1}$ at $T_1 = 273 + 10.0 = 283$ K
$k_2 = 1.25 \times 10^7 \ M^{-1} s^{-1}$ at $T_2 = 273 + 30.0 = 303$ K

Recall R = 8.31 J/mol·K. Substitute into the preceding equation to solve for E_a:

$$\ln\left(\frac{9.3 \times 10^6 \ M^{-1} s^{-1}}{1.25 \times 10^7 \ M^{-1} s^{-1}}\right) = \frac{E_a}{8.314 \ J/mol \cdot K}\left(\frac{283 \ K - 303 \ K}{(283 \ K)(303 \ K)}\right)$$

Solving for E_a yields:

$$(\ln 0.744)(8.314 \ J/mol \cdot K) = E_a\left(\frac{-20 \ K}{85749 \ K^2}\right)$$

$$(-0.296)(8.314 \ J/mol \ K) = E_a(-2.33 \times 10^{-4} \ K^{-1})$$

$$E_a = 1.06 \times 10^4 \ J/mol \quad \text{(or 10.6 kJ/mol)}$$

EXAMPLE 13.9 Reaction Energy Profile

Draw a reaction energy profile for the following endothermic reaction:

$$2HI(g) \rightarrow H_2(g) + I_2(g) \qquad \Delta H^o_{rxn} = 12.5 \ kJ$$

Given the activation energy $E_a = 185$ kJ/mol. Label the activation energy and the activated complex. What is the activation energy for the reverse reaction?

•**Method of Solution**

The reaction is endothermic so the products are at a higher level than the reactants.

Using the diagram we can see that the activation energy for the reverse reaction is 185 kJ – 12.5 kJ = 173 kJ.

EXERCISES

12. Calculate the activation energy for a reaction given that the rate constant is 4.60×10^{-4} s^{-1} at 350°C and 1.87×10^{-4} s^{-1} at 320°C.

13. The rate constant for a first-order reaction is 4.60×10^{-4} s^{-1} at 250°C. If the activation energy is 100.0 kJ/mol, calculate the rate constant at 300°C.

14. A substance decomposes according to first-order kinetics, the rate constants at various temperatures being as follows:

Temp (°C)	k (s^{-1})
15.0	4.41×10^{-6}
25.0	1.80×10^{-5}
30.0	2.44×10^{-5}

a. If one were to make an Arrhenius plot what would its slope be equal to according to the Arrhenius theory?
b. Estimate the numerical value of the slope by using the first and third points.
c. Calculate the activation energy from your value for the slope.

15. The reaction 2NOCl → 2NO + Cl$_2$ has an E$_a$ of 102 kJ/mol and a ΔH_{rxn} of 75.5 kJ. Sketch the reaction energy diagram and determine the activation energy for the reverse reaction.

REACTION MECHANISMS (13.5)

STUDY OBJECTIVES

1. Describe what is meant by a reaction mechanism and distinguish between elementary steps and balanced chemical reactions.
2. Derive the rate law for a reaction from its proposed mechanism.

Elementary Steps. The purpose of a reaction mechanism is to indicate how the reactants are converted into products. A reaction mechanism consists of a set of so-called *elementary steps*. A **mechanism** indicates just what molecules must collide with each other and in what sequence. Each elementary step marks the progress of the conversion of molecules of reactants into products. A mechanism points out intermediates that are formed and consumed along the way, as well as, which steps are fast and which are slow. This information is not supplied by the balanced chemical equation, because it is intended only to indicate the number of moles of one reactant that are consumed per mole of the other reactant.

Elementary steps have a property called **molecularity**, which pertains to the number of molecules that must collide in a single step. An elementary step that involves three molecules is called *termolecular*. A step that involves two molecules is *bimolecular*, and a step in which one molecule decomposes or rearranges is called *unimolecular*. Thus the equation

$$A \rightarrow B$$

represents a unimolecular reaction in which a single molecule of A reacts to form a single molecule of B. The following equation represents a bimolecular reaction.

$$A + B \rightarrow C$$

That is one in which a molecule of A collides with a molecule of B, and as a result a molecule of C is formed.

Reaction Order for Elementary Steps. The rate of a unimolecular elementary step

$$A \rightarrow B$$

will depend on the number of molecules of A per unit volume of the container. Therefore, unimolecular reactions follow first-order rate laws.

$$rate = k[A]$$

The rate of a bimolecular step ($A + B \rightarrow C$) will depend on the number of molecules of A per unit volume times the number of molecules of B per unit volume.

$$rate = k[A][B]$$

Therefore, bimolecular steps follow second-order rate laws.

This can be understood also by realizing that A and B must collide with each other if they are to react, and so the rate of a bimolecular process depends on the rate of collisions of A and B. Doubling the concentration of A will double the probability of collision between A and B. Similarly, doubling the concentration of B will also double the rate of collisions of A and B. See Figure 13.14 in the textbook. Reasoning further along these lines, we conclude that termolecular reactions are third order.

Rate-Determining Step. It often turns out that one of the elementary steps in a mechanism is much slower than all the rest. This step determines the overall rate of reaction much as the slowest person ahead of you a cafeteria line determines how fast you and others move through the line. The slowest step in a sequence of elementary steps is called the **rate-determining step**. This allows us to say that the overall law predicted by a mechanism will be the one corresponding to the rate-determining step.

Testing a Mechanism. The sequence of events in a kinetic study of a reaction that leads to a proposed mechanism is as follows:

1. Determine the rate law experimentally.
2. Propose a mechanism for the reaction.
3. Test the mechanism.

To test a proposed mechanism, assume one step to be the rate-determining step. Then establish the rate law for that step. This gives the rate law predicted by the mechanism. If the mechanism is adequate, then when the predicted rate law is compared with the experimental rate law, the two will match. On the other hand, if the predicted and experimental rate laws do not match, the mechanism has been proved inadequate. Only those mechanisms consistent with all the data can be considered adequate.

EXAMPLE 13.10   Reaction Mechanism

The rate law for the following reaction has been experimentally determined to be third order:

$$2NO(g) + O_2(g) \rightarrow 2NO_2(g) \qquad \text{rate} = k[NO]^2[O_2]$$

Which of the two proposed mechanisms that follow is more satisfactory?

a. $NO + NO \xrightarrow{k_1} N_2O_2$ slow   b. $NO + NO \underset{k_{-1}}{\overset{k_1}{\rightleftharpoons}} N_2O_2$ fast

$N_2O_2 + O_2 \xrightarrow{k_2} 2NO_2$ fast $N_2O_2 + O_2 \xrightarrow{k_2} 2NO_2$ slow

•**Method of Solution**

Testing mechanism (a) first, we note that the first step is the rate-determining step. The rate law for this bimolecular step is:

$$\text{rate} = k_1[NO]^2 \quad \text{I}$$

This mechanism predicts a rate law that is second order in NO concentration and zero order in O_2. By comparison, the experimental rate law is second order in NO and first order in O_2. The predicted rate law and the experimental rate law do not match, and so mechanism (a) is not satisfactory.

Testing mechanism (b), we find that the rate-determining step is the second step. Since it is bimolecular, its rate law should be:

$$\text{rate} = k_2[N_2O_2][O_2] \quad \text{II}$$

It is not possible to compare this predicted rate law directly with the experimental rate law because of the unique term, which is the concentration of N_2O_2. In this mechanism, N_2O_2 is an *intermediate*. Being formed in step 1 and consumed in step 2, its concentration is always small and usually not measurable.

The way around this situation is to find a mathematical substitution for $[N_2O_2]$. Note that step 1 is reversible and equilibrium can be established. This means that the rates of the forward and reverse reactions are equal:

$$k_1[NO]^2 = k_{-1}[N_2O_2]$$

where k_1 is the rate constant for the forward reaction, and k_{-1} for the reverse.

Rearranging terms gives us:

$$[N_2O_2] = \frac{k_1}{k_{-1}}[NO]^2$$

We now have an expression for $[N_2O_2]$, which we can substitute into equation II. This yields:

$$rate = \frac{k_2k_1}{k_{-1}}[NO]^2[O_2]$$

Whenever a reaction step is reversible, we can use the equality of the forward and reverse rates to form an expression to substitute for the concentration of an intermediate.

Now we compare this rate law with the experimental rate law, which is

$$rate = k[NO]^2[O_2]$$

We see that this mechanism predicts that the reaction will be second order in NO and first order in O_2, just as observed. Also, the collection of constants k_2k_1/k_{-1} will equal the rate constant.

$$k = \frac{k_2k_1}{k_1}$$

Therefore, the second mechanism predicts a rate law that matches the experimental order of reaction.

EXERCISES

16. Which of the following species can be isolated from a chemical reaction?
    product, activated complex, intermediate

17. Carbon monoxide can be converted to carbon dioxide by the following overall reaction.

    $$NO_2(g) + CO(g) \rightarrow NO(g) + CO_2(g)$$

    The experimentally determined rate law is rate $= k[NO_2]^2$. The suggested mechanism involves two bimolecular elementary steps.

    $$NO_2(g) + NO_2(g) \rightleftharpoons NO_3(g) + NO(g) \qquad \text{step 1}$$
    $$NO_3(g) + CO(g) \rightarrow NO_2(g) + CO_2(g) \qquad \text{step 2}$$

    a. What is the rate law for each step?
    b. Derive the predicted rate laws when step 1 is rate determining, and when step 2 is rate determining.

18. The reaction of nitric oxide and chlorine has the following rate law as determined by experiment:

    $$2NO(g) + Cl_2(g) \rightarrow 2NOCl(g) \qquad rate = k[NO][Cl_2]$$

    Is the following proposed mechanism consistent with the experimental rate law?
    $$NO(g) + Cl_2(g) \rightarrow NOCl_2(g) \qquad \text{slow}$$
    $$NO(g) + NOCl_2(g) \rightarrow 2NOCl(g) \qquad \text{fast}$$

CATALYSIS (13.6)

STUDY OBJECTIVES

1. Describe in general how a catalyst increases the rate of a reaction without being consumed.
2. Describe the essentials of heterogeneous, homogeneous, and enzyme catalysis.

Mechanism of a Catalyzed Reaction. A **catalyst** is a substance that increases the rate of a chemical reaction without being consumed in the reaction. For this reason a catalyst does not appear in the stoichiometric equation. Chromium(III) oxide, Cr_2O_3, is the catalyst present in catalytic converters in American automobiles. This compound accelerates the reaction of carbon monoxide with oxygen. By converting CO to CO_2 in the exhaust stream, CO emissions are reduced.

$$CO(g) + \frac{1}{2}O_2(g) \xrightarrow{Cr_2O_3} CO_2(g)$$

The lifetime of catalytic converters is quite long; the Cr_2O_3 and other catalysts do not need to be replaced.

Catalysts accelerate reaction rates by providing a new reaction pathway (mechanism) which has a lower activation energy. See Figure 13.17 (textbook). In the catalytic converter both CO and O_2 are chemically adsorbed on the catalyst's surface. Adsorption of O_2 by Cr_2O_3 weakens the O—O bond enough so that oxygen atoms can react with adsorbed CO. The reaction path involving a weakened O—O bond has a significantly lower activation energy than that of the reaction occurring purely in the gas phase. Recall that as the activation energy is lowered the rate constant for a reaction increases. After CO_2 is formed, it desorbs from the surface, leaving the Cr_2O_3 catalyst chemically unaltered. The catalyst is not consumed in the reaction. Catalysts are regenerated in one of the last steps of the mechanism. The textbook describes three types of catalysts, which we will review here.

Heterogeneous Catalysis. In heterogeneous catalysis the reactants are in one phase and the catalyst is in another phase. The catalyst is usually a solid, and the reactants are gases or liquids. Ordinarily the site of the reaction is the surface of the solid catalyst. Many industrially important reactions involve gases and are catalyzed by solid surfaces. For example, the Haber process for the synthesis of ammonia is catalyzed by iron plus a few percent of the oxides of potassium and aluminum. The famous Ziegler-Natta catalyst for polymerizing ethylene gas (C_2H_4) into polyethylene polymer contains triethylaluminum and titanium or vanadium salts.

Homogeneous Catalysis. In homogeneous catalysis the reactants, products, and catalysts are all in the same phase, which is usually the gas or the liquid phase. Many reactions are catalyzed by acids. The decomposition of formic acid (HCO_2H) is an example.

$$HCO_2H \rightarrow H_2O + CO$$

Formic acid is normally stable and lasts a long time on the shelf. However, when sulfuric acid is added, bubbles of carbon monoxide gas can be observed immediately. The hydrogen ions from the sulfuric acid initiate a new reaction path. The mechanism is

$$HCO_2H + H^+ \rightarrow HCO_2H_2^+$$
$$HCO_2H_2^+ \rightarrow H_2O + HCO^+$$
$$HCO^+ \rightarrow CO + H^+$$

$$HCO_2H \rightarrow CO + H_2O$$

Note that the H^+ ion is consumed in the first step, but, as for all catalysts, H^+ is regenerated. The net reaction is the sum of the three steps. The intermediate $HCO_2H_2^+$ cancels out as does the catalyst.

Enzyme Catalysis. Nearly all chemical reactions that occur in living organisms require catalysts. **Enzymes** are biological catalysts. All enzymes are protein molecules with molecular masses well over 10,000 amu. All enzyme molecules are highly specific. That is, they can only affect the reaction rates of a few specific reactant molecules. Typically, an enzymes catalyzes a single reaction, or a group of closely related reactions. The molecule on which an enzyme acts is called a **substrate**.

The simplest mechanism which explains enzyme activity, and is consistent with these trends is one in which the substrate S and the enzyme E form an enzyme-substrate complex, ES. This complex has a lower energy than the activated complex without the enzyme (Figure 13.25 of the text). The enzyme-substrate complex can either dissociate back into E and S, or break apart into the products P and the regenerated enzyme E. The simplest mechanism has the two steps shown below.

$$E + S \rightleftharpoons ES$$
$$ES \rightarrow E + P$$

EXAMPLE 13.11   Intermediates and Catalysts

Given the following mechanism for the decomposition of ozone in the stratosphere, identify the intermediate and the catalyst:

$$O_3 + Cl \rightarrow O_2 + ClO$$
$$ClO + O \rightarrow Cl + O_2$$

•Method of Solution

Adding the two steps gives the overall reaction

$$O_3 + O \rightarrow 2O_2$$

An intermediate is formed in an early step and removed in a later step and does not appear in the overall equation; therefore ClO is the intermediate. A catalyst is consumed in an early step and is later regenerated. Atomic chlorine, Cl, is the catalyst.

EXERCISES

19. Define a catalyst. Why is a catalyst not consumed in the reaction?

20. Identify the catalyst, any intermediates, and the overall equation for the following mechanism.

$$O_3 + NO \rightarrow O_2 + NO_2$$
$$NO_2 + O \rightarrow NO + O_2$$

CONCEPTUAL QUESTIONS

1. Explain why the rate of reaction increases rapidly with increasing temperature.

2. Cooking an egg involves the denaturation of a protein called albumin from a soluble globular protein to an insoluble fibrous protein. This reaction has an activation energy of roughly 75 kJ/mol. At sea level it takes three minutes to prepare a "soft-boiled egg." However on top of Pikes Peak at 4452 m elevation it takes 7.5 minutes to cook a "soft-boiled egg." Explain why its takes longer to cook an egg on the tops of mountains.

3. Describe the difference between an overall reaction (balanced chemical equation) and an elementary reaction.

PRACTICE TEST

1. The following experimental data were obtained for the reaction $2A + B \rightarrow$ products. What is the rate law for this reaction?

Expt.	Initial Concentration (M)		Rate
	$[A]_0$	$[B]_0$	M/s
1	0.40	0.20	5.6×10^{-3}
2	0.80	0.20	5.5×10^{-3}
3	0.40	0.40	22.3×10^{-3}

2. For the reaction

$$30CH_3OH + B_{10}H_{14} \rightarrow 10B(OCH_3)_3 + 22H_2$$

express the rate in terms of the change in concentration with time for each of the reactants and each of the products.

3. The hydrolysis of sucrose ($C_{12}H_{22}O_{11}$) yields the simple sugars, glucose ($C_6H_{12}O_6$) and fructose ($C_6H_{12}O_6$), which just happen to be isomers.

$$C_{12}H_{22}O_{11} + H_2O \rightarrow C_6H_{12}O_6 + C_6H_{12}O_6$$

The rate follows the rate equation, rate = $k[C_{12}H_{22}O_{11}]$. At 27°C the rate constant is 2.1×10^{-6}/s.
a. Starting with the sucrose solution with a concentration of 0.10 M at 27°C, what would the concentration of sucrose be 24 hours later? (The solution is kept at 27°C).
b. What is the half-life of sucrose at 27°C?

4. N_2O_5 decomposes according to first-order kinetics:

$$2N_2O_5(g) \rightarrow 4NO_2(g) + O_2(g)$$

a. At a certain temperature 20.0% of the initial N_2O_5 decomposes in 2.10 h. Determine the rate constant.
b. What fraction of the initial N_2O_5 will remain after 13.0 h?
c. What is the half-life?
d. If the initial concentration of N_2O_5 was 0.222 M, then what concentration remains after 24 h?

5. The following data were obtained on the rate of disintegration of a pesticide in soil at 30°C.

Time (days)	11	60	93
% Pesticide remaining	96	80	71

   a. What is the order of reaction with respect to pesticide concentration?
   b. What is the value of the rate constant?

6. a. It takes 30.0 min for the concentration of a reactant in a second-order reaction to drop from 0.40 M to 0.30 M. What is the value of the rate constant for this reaction?
   b. How long will it take for the concentration to drop from 0.40 M to 0.20 M?

7. The reaction A + 2B → C + D has for its experimental rate law $r = k[A]^2[B]$. By what factor will the rate increase if the concentration of A is doubled and the concentration of B is tripled?

8. The rate of the reaction of hydrogen with iodine

   $$H_2 + I_2 \rightarrow 2HI$$

   has rate constants of 1.41×10^{-5} /M·s at 393°C and 1.40×10^{-4} /M·s at 443°C. Calculate the activation energy for this reaction.

9. At 300 K the rate constant is 1.5×10^{-5} /M·s for the reaction:

   $$2NOCl \rightarrow 2NO + Cl_2$$

   The activation energy is 90.2 kJ/mol. Calculate the value of the rate constant at 310 K. By what factor did the rate constant increase?

10. The activation energy for the reaction

    $$CO + NO_2 \rightarrow CO_2 + NO$$

    is 116 kJ/mol. How many times greater is the rate constant for this reaction at 250°C than it is at 200°C?

11. If the reaction $2HI \rightarrow H_2 + I_2$ has an activation energy of 190 kJ/mol and a $\Delta H = 10$ kJ, what is the activation energy for the reverse reaction?

    $$H_2 + I_2 \rightarrow 2HI$$

12. The reaction between nitrite ion and oxygen gas is:

    $$2\,NO_2^- + O_2 \rightarrow 2\,NO_3^-$$

    and proceeds at a rate that is first order in $[NO_2^-]$ and zero order in $[O_2]$. A mechanism has been proposed:

$2\,NO_2^- + O_2 \rightarrow 2\,NO_4^-$	slow
$NO_4^- + NO_2^- \rightarrow 2\,NO_3^-$	fast

    Assume that NO_4^- is a very reactive intermediate and that the first step is the rate determining step. Show that this mechanism is consistent with the experimental rate law.

13. The rate law for the replacement of H_2O by NH_3 in the reaction of a complex ion:

$$Ni(H_2O)_6^{2+}(aq) + NH_3(aq) \rightarrow Ni(H_2O)_5(NH_3)^{2+}(aq) + H_2O(l)$$

is first order in $Ni(H_2O)_6^{2+}$, and zero order in NH_3.

rate = $k[Ni(H_2O)_6^{2+}]$

Show that the following mechanism is consistent with the experimental rate law.

$$Ni(H_2O)_6^{2+} \rightarrow Ni(H_2O)_5^{2+} + H_2O \qquad \text{slow}$$

$$Ni(H_2O)_5^{2+} + NH_3 \rightarrow Ni(H_2O)_5(NH_3)^{2+} \quad \text{fast}$$

14. The rate law for the net reaction

$$2NO + O_2 \rightarrow 2NO_2$$

is rate = $k[NO_2]^2[O_2]$. Could the mechanism of this reaction be a single termolecular process?

15. The rate law for the net reaction

$$H_2(g) + I_2(g) \rightarrow 2HI(g)$$

is rate = $k[H_2][I_2]$. A possible mechanism involves a bimolecular elementary step:

$$H_2 + I_2 \rightarrow 2HI$$

A second possibility has also been proposed:

$$I_2 \rightarrow 2I \qquad \text{fast}$$
$$2I + H_2 \rightarrow 2HI \qquad \text{slow}$$

Show that both mechanisms are consistent with the experimental rate law.

16. The following statements are sometimes made with reference to catalysts. Explain the fallacy in each one of them.
a. A catalyst is a substance that accelerates the rate of a chemical reaction but does not take part in the reaction.
b. A catalyst may increase the rate of a reaction going in one direction without increasing the rate of the reaction going in the reverse direction.

17. a. Identify the catalyst in the following mechanism.
b. What is the overall reaction?
c. Is this reaction an example of homogeneous or heterogeneous catalysis?

$$H_2O_2(aq) + Br_2(aq) \rightarrow 2H^+(aq) + O_2(aq) + 2Br^-(aq)$$
$$H_2O_2(aq) + 2H^+(aq) + 2Br^-(aq) \rightarrow Br_2(aq) + H_2O(l)$$

ANSWERS

Exercises

1. rate $= \dfrac{\Delta[NH_3]}{2\Delta t} = \dfrac{-\Delta[N_2]}{\Delta t} = \dfrac{-\Delta[H_2]}{3\Delta t}$

2. a. $\dfrac{-\Delta[S_2O_3^{2-}]}{\Delta t} = 8.3 \times 10^{-4}$ M/s   b. $\dfrac{-\Delta[I_2]}{\Delta t} = 4.2 \times 10^{-4}$ M/s

3. a. 5.0×10^{-4} M/s   b. 3.5×10^{-4} M/s   c. 1.8×10^{-4} M/s

4. 9 times

5. a. rate $= k\,[A]\,[B]^2$ $k = 4.8\ /M^2\,s$

6. a. $k = 6.8 \times 10^{-3}\ /s$ b. 102 s

7. a. $\ln \dfrac{[A]}{[A]_0} = -kt$ b. Plot $\ln[A]$ vs. t   where $x = t$, $y = [A]$, and $m = -k$

   c. The initial concentration $[A]_0$ and $[A]$ again at a known time t.

8. 58%

9. a.   b.   c.

10. first-order: s^{-1};   second order: $L/mol\cdot s$ or $M^{-1}s^{-1}$

11. a. 5.8 y   b. 13 y

12. 92 kJ/mol

13. $k_2 = 1.4\ s^{-1}$

14. a. According to Arrhenius's interpretation the slope is: $m = -\dfrac{E_a}{R}$

    b. Calculate the slope from $m = \dfrac{\ln k_2 - \ln k_1}{1/T_2 - 1/T_1}$, $m = -1.0 \times 10^4$ K

    c. 83 kJ/mol

15. $E_a(\text{rev}) = 26.5$ kJ/mol

16. products and intermediates

17. a. $\text{rate}_1 = k_1\,[NO_2]^2$; $\text{rate}_2 = k_2\,[NO_3]\,[CO]$

    b. when step 1 is rate determining: rate $= k_1\,[NO_2]^2$

    when step 2 is rate determining: rate $= \dfrac{k_2 k_1}{k_{-1}} [NO_2]^2[CO][NO]^{-1}$

18. yes

19. NO is the catalyst. NO_2 is the intermediate. The overall equation is:
    $O_3 + O \rightarrow 2O_2$

Conceptual Questions

1. Normally only a small fraction of the colliding molecules have enough energy to exceed the activation energy. As temperature increases the average kinetic energy of molecules increases. Therefore, the fraction of the colliding molecules having energies greater than the activation energy increases. More molecules are capable of reacting per second and the rate of reaction increases as temperature increases.

2. First of all cooking an egg is a chemical reaction and as such its rate will increase with temperature in accordance with the Arrhenius equation. At sea level water (used to cook the egg) boils at 100°C, but at higher elevations such as 4452 m water boils at a temperature lower than 100°C (86°C on top of Pikes Peak). Therefore the rate of cooking an egg is much slower on mountain tops than at sea level.

3. An elementary reaction represents a reaction between individual molecules. The reaction occurs when the molecules collide. An overall reaction is the net change due to the sum of all the elementary reactions. It does not represent the collsions and reactions of any particular molecules. Indeed any intermediates have canceled out of the net reaction. The balanced equation shows only the reactants and products in the proper molar amounts.

Practice Test

1. $Rate = k[B]^2$

2. $Rate = -\dfrac{1}{30}\dfrac{\Delta[CH_3OH]}{\Delta t} = -\dfrac{\Delta[B_{10}H_{14}]}{\Delta t} = \dfrac{1}{22}\dfrac{\Delta[H_2]}{\Delta t} = \dfrac{1}{10}\dfrac{\Delta[B(OCH_3)_3]}{\Delta t}$

3. a. [sucrose] = 0.083 M; b. $t_{1/2} = 3.3 \times 10^5$ s
4. a. k = 0.106 h^{-1} b. 0.252   c. 6.54 h   d. 0.0174 M
5. a. first order   b. 3.7×10^{-3} per day
6. a. 2.76×10^{-2} /M $\cdot$ min   b. 90 min
7. Twelve-fold
8. 182 kJ/mol
9. $k = 4.8 \times 10^{-5}$/M s; $k_2/k_1 = 3.21$
10. 16.8
11. 180 kJ/mol
12. The rate law for the rate determining step is
    rate = $k[NO_2^-][O_2]$
    The concentration of dissolved O_2 is a constant according to Henry's law (Chapter 12).
    Therefore $k[O_2] = K$ (a constant), and rate = $K[NO_2^-]$.
13. The rate law for the rate determining step is
    rate = $k[Ni(H_2O)_6^{2+}]$
    This rate law matches the experimental rate law.
14. Yes
15. The rate law for the rate determining step is
    rate = $k_2[I]^2[H_2]$
    Obtaining a mathematical substitution for [I] from
    $k_1[I_2] = k_{-1}[I]^2$

    yields   rate = $\dfrac{k_2 k_1}{k_{-1}}[H_2][I_2]$

    This predicted rate law matches the experimental rate law given in the problem.
16. a. The catalyst reacts in one of the first few steps, but then is regenerated in the last step.
    b. If E_a is lowered for the forward reaction it must also be lowered for the reverse.
17. a. Br_2 b. $2H_2O \rightarrow 2H_2O + O_2$ c. homogeneous

Chapter Fourteen
CHEMICAL EQUILIBRIUM

CHEMICAL EQUILIBRIUM AND THE EQUILIBRIUM CONSTANT (14.1)

STUDY OBJECTIVES

1. Describe the affect of a reverse reaction on the amount of product formed.
2. Describe how equilibrium is established in a chemical reaction system.

Reversible Chemical Reactions. In this chapter we will discuss how equilibrium is established in several types of chemical reaction systems. We will define the equilibrium constant and show how to calculate equilibrium concentrations of reactants and products. A state of **chemical equilibrium** exists when the concentrations of reactants and products are observed to remain constant with time. When a mixture of SO_2 and O_2, for instance, is introduced into a reaction vessel at a temperature of 700 K, a reaction that produces SO_3 occurs:

$$2SO_2 + O_2 \rightarrow 2SO_3$$

When a specific concentration of SO_3 is reached, no additional SO_3 is formed even though some SO_2 and O_2 remain. From that time on, the concentrations of SO_2, O_2, and SO_3 stay constant, and we say the system has reached chemical equilibrium.

The constant concentrations are the result of a **reversible chemical reaction.** In the reverse reaction some SO_3 decomposes back into SO_2 and O_2.

$$2SO_3 \rightarrow 2SO_2 + O_2$$

When the rates of the forward reaction and reverse reaction are the same, no *net* chemical change occurs and a state of chemical equilibrium exists. The equilibrium state is referred to as **dynamic** because of the continual conversions of reactants into products, and products into reactants at the molecular level. A reversible reaction is represented by the opposing arrows in the chemical equation:

$$2SO_2 + O_2 \rightleftharpoons 2SO_3$$

At 700 K all three species exist together in the reaction vessel. The three compounds are said to be in equilibrium. Equilibrium can be reached by the reverse reaction as well. That is, if only SO_3 is added to the reaction vessel at 700 K, it will form an equilibrium amount of SO_2 and O_2.

    The Equilibrium Constant. In quantitative terms, the equilibrium state is described by the *equilibrium constant expression* which is also called the mass action expression.

$$K_c = \frac{[SO_3]^2}{[SO_2]^2[O_2]}$$

Here the brackets [ ] indicate concentration in moles per liter. In this expression, note that the concentration of the product is in the numerator, and the concentrations of the reactants appear in the denominator. Since there are two reactants their concentrations are *multiplied*. Also, the concentration of each component is raised to a power equal to its coefficient in the balanced equation. The value of this expression is called the *equilibrium constant*.

    The equilibrium constant for a reaction can be determined by measuring the concentrations of all components at equilibrium and substituting these values into the K_c expression. The equilibrium concentrations of SO_2, O_2, and SO_3 are related to each other by the equilibrium constant expression, which at 700 K has the following value.

$$K_c = \frac{[SO_3]^2}{[SO_2]^2[O_2]} = 4.3 \times 10^6$$

WRITING EQUILIBRIUM CONSTANT EXPRESSIONS (14.2)

STUDY OBJECTIVES

1. Write the equilibrium constant expressions for homogeneous and heterogeneous chemical reactions, given balanced chemical equations.
2. Given K_c for a reaction, calculate the value of K_c for the same reaction when it is balanced by using some multiple of the coefficients in the given equation. Also, find K_c for the reverse reaction.
3. Convert K_c to K_p for reactions involving gases.

    The Form of K_c and the Equilibrium Equation. In general the equilibrium constant expression has the form of a ratio of product concentrations over reactant concentrations at equilibrium. For the general equation

$$aA + bB \rightleftharpoons cC + dD$$

where a, b, c, and d are the stoichiometric coefficients for the balanced equation. The equilibrium constant expression is

$$K_c = \frac{[C]^c[D]^d}{[A]^a[B]^b}$$

In an equilibrium constant expression each concentration is raised to a power equal to its stoichiometric coefficient. The numerator is found by multiplying together the equilibrium concentrations of the products, each raised to an exponent equal to its stoichiometric coefficient. The denominator is found by multiplying together the equilibrium concentrations of the reactants, each raised to an exponent equal to its stoichiometric coefficient.

    The equilibrium constant expression and its value depend on how the equation is balanced. Often an equation can be balanced with more than one set of coefficients, as shown below:

$$2SO_2 + O_2 \rightleftharpoons 2SO_3$$
$$SO_2 + \tfrac{1}{2}O_2 \rightleftharpoons SO_3$$

How does this affect the equilibrium constant? For the latter equation, the equilibrium constant expression is written:

$$K_c' = \frac{[SO_3]}{[SO_2][O_2]^{1/2}}$$

where the prime (') is just to help us keep track of the constant to which we are referring. Note that K_c' is the square root of K_c from the previous section.

$$K_c = \frac{[SO_3]^2}{[SO_2]^2[O_2]}$$

$$K_c' = \frac{[SO_3]}{[SO_2][O_2]^{1/2}}$$

Therefore, by inspection:

$$K_c' = \sqrt{K_c}$$

The value of K_c' is:

$$K_c' = \sqrt{4.3 \times 10^6} = 2.1 \times 10^3$$

The equilibrium constant expression for the reaction written in the reverse direction

$$2SO_3 \rightleftharpoons 2SO_2 + O_2$$

is

$$K_c'' = \frac{[SO_2]^2[O_2]}{[SO_3]^2}$$

Compare this to K_c for the original forward reaction:

$$K_c = \frac{[SO_3]^2}{[SO_2]^2[O_2]}$$

By inspection you can tell that K_c'' is the reciprocal of K_c for the forward reaction.

$$K_c'' = \frac{[SO_2]^2[O_2]}{[SO_3]^2} = \frac{1}{K_c}$$

Therefore the value of K_c'' is

$$K_c'' = \frac{1}{4.3 \times 10^6} = 2.3 \times 10^{-7}$$

Always use the K_c expression and value that are consistent with the way in which the equation is balanced.

K_p and K_c. For reactions involving gases, the equilibrium constant expression can be written in terms of the partial pressures, P_i, of each gaseous component. For the equation:

$$2SO_2 + O_2 \rightleftharpoons 2SO_3$$

$$K_p = \frac{P_{SO_3}^2}{P_{SO_2}^2 \, P_{O_2}}$$

The pressure, in atmospheres, of each gas at equilibrium is raised to a power corresponding to its coefficient in the balanced equation. The subscript p in K_p indicates that the equilibrium constant has a value that was calculated by using equilibrium partial pressures in atmospheres, rather than in units of moles per liter.

The values of K_p and K_c are related. Recall that for an ideal gas $PV = nRT$, and so the pressure of an ideal gas is proportional to its concentration:

$$P = \frac{n}{V}RT$$

where n/V = moles per liter. Substitution of $(n/V)RT$ for the pressure of each gas gives the equation relating K_p and K_c that was derived in the text.

$$K_p = K_c(RT)^{\Delta n}$$

where R is the ideal gas constant (0.0821 L·atm/K·mol) and Δn is the change in the number of moles of gas when going from reactants to products. For the preceding reaction $\Delta n = -1$, and at 700 K, K_p is given by

$$K_p = K_c(RT)^{-1} = \frac{K_c}{RT} = \frac{4.3 \times 10^6}{(0.0821)(700)} = 7.5 \times 10^4$$

In general, $K_p \neq K_c$ except in the case when $\Delta n = 0$.

Heterogeneous Equilibria. Whenever a reaction involves reactants and products that exist in different phases, it is called a *heterogeneous reaction,* and in a closed container a *heterogeneous equilibrium* will result. For example, when steam is brought into contact with charcoal the following equilibrium is established in a reaction vessel.

$$C(s) + H_2O(g) \rightleftharpoons H_2(g) + CO(g)$$

The equilibrium constant expression for this reaction will *not* be the same as for a homogeneous reaction.

The usual expression for the constant would be

$$K = \frac{[H_2][CO]}{[C][H_2O]}$$

However, the concentration of a pure solid is itself a constant, and is not changed by the addition to, or removal of, some of that solid. Remember that solids have their own volume, and do not fill their containers as gases do. The concentration (mol/L) of a solid such as charcoal depends only on its density.

This means that the equilibrium constant expression can be written

$$K_c = K[C] = \frac{[H_2][CO]}{[H_2O]}$$

where the constant concentration of the pure solid [C] has been combined with the equilibrium constant. Thus the equilibrium constant expression does not contain the solid. In general, concentrations of solids and pure liquids do not appear in equilibrium constant expressions. In this reaction, equilibrium will be maintained as long as some C(s) is present. The *amount* of solid carbon present does not affect the position of equilibrium.

Multiple Equilibria. When the product molecules of one equilibrium reaction become reactants in a second equilibrium process, we have an example of multiple equilibria. In such a case the overall reaction is the sum of the two individual reactions.

For instance, consider the following reactions at 700°C:

$$NO_2 \rightleftharpoons NO + \tfrac{1}{2}O_2 \quad K_1 = 0.012$$

followed by

$$SO_2 + \tfrac{1}{2}O_2 \rightleftharpoons SO_3 \quad K_2 = 20$$

The overall reaction is the sum of the two steps:

$$NO_2 + SO_2 \rightleftharpoons NO + SO_3 \quad K_c = ?$$

The equilibrium constant value for the overall reaction is given by *the product of the equilibrium constants of the individual reactions.*

$$K_c = K_1 K_2 = (0.012)(20) = 0.24$$

Guidelines for Writing Equilibrium Constant Expressions

•The concentrations of the reacting species in the solution phase are expressed in mol/L. In the gaseous phase, the concentrations can be expressed in mol/L or in atm. The constant K_c is related to K_p by a simple equation.
• The concentrations of pure solids, pure liquids, and solvents are constants, and do not appear in equilibrium constant expressions.
• The equilibrium constant (K_c or K_p) is a dimensionless quantity.
• In stating a value for the equilibrium constant, we must specify the balanced equation and the temperature.
• If a reaction can be expressed as the sum of two or more reactions, the equilibrium constant for the overall reaction is given by the product of the equilibrium constants of the individual reactions.

EXAMPLE 14.1   The Equilibrium Constant Expression

Write the equilibrium constant expressions for the following reversible reactions:

a. $4NH_3(g) + 5O_2(g) \rightleftharpoons 4NO(g) + 6H_2O(g)$

b. $BaO(s) + CO_2(g) \rightleftharpoons BaCO_3(s)$

•**Method of Solution**

Remember that the equilibrium constant expression has the concentrations of the products in the numerator and those of the reactants in the denominator. Raise each concentration to a power equal to the coefficient of that substance in the balanced equation. Include concentration terms for gaseous components only; leave out the concentrations of pure liquids and solids as they are not included in the equilibrium constant expression. The expressions are:

a. $K_c = \dfrac{[NO]^4[H_2O]^6}{[NH_3]^4[O_2]^5}$ b. $K_c = \dfrac{1}{[CO_2]}$

EXAMPLE 14.2   The Equilibrium Constant for a Reverse Reaction

The equilibrium constant for the reaction

$$H_2(g) + I_2(g) \rightleftharpoons 2HI(g)$$

at 400°C is

$$K_c = \frac{[HI]^2}{[H_2][I_2]} = 64$$

What is the equilibrium constant value for the reverse reaction?

$$2HI(g) \rightleftharpoons H_2(g) + I_2(g)$$

•**Method of Solution**

Look to see if the reactions are related in some way. Note that the equilibrium constant for the second reaction is the reciprocal of the equilibrium expression for the forward reaction.

$$K_c' = \frac{[H_2][I_2]}{[HI]^2} = \frac{1}{K_c}$$

•**Calculation**

$$K_c' = \frac{1}{K_c} = \frac{1}{64} = 0.016$$

EXAMPLE 14.3 K_p and K_c Values

What are the values of K_p and K_c at 1000°C for the reaction

$$CaCO_3(s) \rightleftharpoons CaO(s) + CO_2(g)$$

if the pressure of CO_2 in equilibrium with $CaCO_3$ and CaO is 3.87 atm?

•**Method of Solution**

Enough information is given to find K_p first. Writing the K_p expression for this heterogeneous reaction:

$$K_p = P_{CO_2} = 3.87$$

Then to get K_c, rearrange the equation,

$$K_p = K_c(RT)^{\Delta n}$$

where Δn, the change in the number of moles of gas in the reaction is +1.

•**Calculation**

$$K_c\,[CO_2] = \frac{K_p}{(RT)^{\Delta n}} = \frac{3.87}{[(0.0821)(1273)]^1}$$

$$K_c = 0.0370$$

EXAMPLE 14.4 K_p and Partial Pressures at Equilibrium

At 400°C, $K_c = 64$ for the reaction

$$H_2(g) + I_2(g) \rightleftharpoons 2HI(g)$$

a. What is the value of K_p for this reaction?
b. If, at equilibrium, the partial pressures of H_2 and I_2 in a container are 0.20 atm and 0.50 atm, respectively, what is the partial pressure of HI in the mixture?

•**Method of Solution**

a. The equation relating K_p to K_c is

$$K_p = K_c(RT)^{\Delta n}$$

Here the change in the number of moles of gas Δn is:

$$\Delta n = 2 \text{ mol HI} - (1 \text{ mol } H_2 + 1 \text{ mol } I_2) = 0$$

Since, $\Delta n = 0$, K_p and K_c are the same.

$$K_p = K_c(RT)^0 = K_c$$

$$K_p = K_c = 64$$

b. Writing the equilibrium constant expression

$$K_p = \frac{P_{HI}^2}{P_{H_2}P_{I_2}} = 64$$

and substituting the given pressures

$$\frac{P_{HI}^2}{(0.20)(0.50)} = 64$$

the partial pressure of HI is

$$P_{HI} = \sqrt{(0.20)(0.50)(64)}$$

$$P_{HI} = 2.5 \text{ atm}$$

EXERCISES

1. Write the equilibrium constant expressions (K_c) for the following equations.

   a. $2HgO(s) \rightleftharpoons 2Hg(l) + O_2(g)$

   b. $Ni(s) + 4CO(g) \rightleftharpoons Ni(CO)_4(g)$

   c. $2NO(g) + Br_2(g) \rightleftharpoons 2NOBr(g)$

2. K_p for the decomposition of ammonium chloride at 427°C is

   $NH_4Cl(s) \rightleftharpoons NH_3(g) + HCl(g)$ $K_p = 4.8$
   Calculate K_c for this reaction.

3. For the reaction

   $H_2(g) + Br_2(g) \rightleftharpoons 2HBr(g)$

   $K_p = 7.1 \times 10^4$ at 700 K. What is the value of K_p for the following reactions at the same temperature?

   a. $2HBr(g) \rightleftharpoons H_2(g) + Br_2(g)$

   b. $\frac{1}{2}H_2(g) + \frac{1}{2}Br_2(g) \rightleftharpoons HBr(g)$

4. The following equilibrium constants were determined at 1123 K:

   $C(s) + CO_2(g) \rightleftharpoons 2CO(g)$ $K_c = 1.4 \times 10^{12}$

   $CO(g) + Cl_2(g) \rightleftharpoons COCl_2(g)$ $K_c = 5.5 \times 10^{-1}$

   Write the equilibrium constant expression K_c and calculate the equilibrium constant at 1123 K for the following reaction:

   $C(s) + CO_2(g) + 2Cl_2(g) \rightleftharpoons 2COCl_2(g)$

CALCULATING EQUILIBRIUM CONSTANTS (14.2)

STUDY OBJECTIVES

1. Determine an equilibrium constant value, given equilibrium concentrations.
2. Given initial concentrations and the equilibrium concentration of at least one component determine the equilibrium constant.

Equilibrium Concentrations. The equilibrium constant for a reaction can be determined by measuring the concentrations of all components at equilibrium and substituting these values into the K_c expression. Two situations arise. In one all of the equilibrium concentrations are given, and a straight forward calculation yields the equilibrium constant. In the second, the initial concentrations of all reactants are given along with the equilibrium concentration of only one product or one reactant. You need to determine the other equilibrium concentrations first, and then K_c. Examples 14.5 and 14.6 show these types of calculations.

EXAMPLE 14.5   Calculating an Equilibrium Constant

Ammonia is synthesized from hydrogen and nitrogen according to the equation:

$$3H_2(g) + N_2(g) \rightleftharpoons 2NH_3(g)$$

An equilibrium mixture at a given temperature was analyzed and the following concentrations were found: 0.31 mol N_2/L; 0.90 mol H_2/L; and 1.4 mol NH_3/L. What is the equilibrium constant value?

•**Method of Solution**

When the concentrations of each component of a chemical system at equilibrium are known, the value of K_c can be determined readily by substituting these concentrations into the equilibrium constant expression.

$$K_c = \frac{[NH_3]^2}{[H_2]^3[N_2]}$$

•**Calculation**

$$K_c = \frac{(1.4)^2}{(0.90)^3(0.31)}$$

$$K_c = 8.7$$

EXAMPLE 14.6   Calculating an Equilibrium Constant

When 3.0 mol of I_2 and 4.0 mol of Br_2 are placed in a 2.0 L reaction chamber at 150°C, the following reaction occurs until equilibrium is reached:

$$I_2(g) + Br_2(g) \rightleftharpoons 2IBr(g)$$

Chemical analysis then shows that the reactor contains 3.2 mol of IBr. What is the equilibrium constant K_c for the reaction?

•Method of Solution

We need to know the equilibrium concentrations of I_2, Br_2, and IBr. These, when substituted into the K_c expression, will give the value of K_c. First find the number of moles of I_2 and Br_2 at equilibrium. Initially, the *amounts* of each component in the reactor are 3.0 moles of I_2, 4.0 moles of Br_2, and zero moles of IBr. The number of moles of I_2 remaining at equilibrium is given by *the initial number of moles of I_2 minus the moles of I_2 reacted*. This is also true for Br_2. The information that 3.2 mol IBr is formed tells us that 1.6 mol I_2 and 1.6 mol Br_2 must have reacted. We know this because the balanced equation states that 2 mol IBr are formed whenever 1 mol I_2 and 1 mol of Br_2 react.

•Calculation

moles of I_2 at equilibrium = 3.0 mol − 1.6 mol = 1.4 mol I_2

moles of Br_2 at equilibrium = 4.0 mol − 1.6 mol = 2.4 mol Br_2

Substituting the *equilibrium concentrations* into the equilibrium constant expression gives the K_c value:

$$K_c = \frac{[IBr]^2}{[I_2][Br_2]} = \frac{\left(\dfrac{3.2 \text{ mol}}{2.0 \text{ L}}\right)^2}{\left(\dfrac{1.4 \text{ mol}}{2.0 \text{ L}}\right)\left(\dfrac{2.4 \text{ mol}}{2.0 \text{ L}}\right)}$$

$$K_c = 3.0$$

•Comment

The key to finding all the equilibrium concentrations was that the changes in I_2 and Br_2 could be related to the change in IBr through the balanced chemical equation. Keep in mind that the equilibrium concentration equals the initial concentration plus the change in concentration due to reaction to reach equilibrium. The change in concentration is positive for a product, and negative for a reactant.

EXERCISES

5. A sample of nitrosyl bromide was heated to 100°C in a 10.0 L container in order to partially decompose it.

   $2NOBr(g) \rightleftharpoons 2NO(g) + Br_2(g)$

   At equilibrium the container was found to contain 0.0585 mol of NOBr, 0.105 mol of NO, and 0.0524 mol of Br_2. Calculate the value of K_c.

6. 1.25 mol NOCl was placed in a 2.50 L reaction chamber at 427°C. After equilibrium was reached, 1.10 mol NOCl remained. Calculate the equilibrium constant K_c for the reaction.

   $2NOCl(g) \rightleftharpoons 2NO(g) + Cl_2(g)$

7. The brown gas NO_2 and the colorless gas N_2O_4 exist in equilibrium.

   $N_2O_4(g) \rightleftharpoons 2NO_2(g)$

   0.625 mol of N_2O_4 was introduced into a 5.00 L vessel and was allowed to decompose until it reached equilibrium with NO_2. The concentration of N_2O_4 at equilibrium was 0.0750 M. Calculate K_c for the reaction.

8. 5.0 mol of ammonia were introduced into a 5.0 L reaction chamber in which it partially dissociated at high temperatures.

$$2NH_3(g) \rightleftharpoons 3H_2(g) + N_2(g)$$

At equilibrium at a particular temperature, 80.0% of the ammonia had reacted. Calculate K_c for the reaction.

WHAT THE EQUILIBRIUM CONSTANT TELLS US (14.4)

STUDY OBJECTIVES

1. Determine whether a reaction is at equilibrium or not, and if it is not, predict the direction in which a net reaction will occur.
2. Predict the relative extents of several reactions given the values of their equilibrium constants.
3. Given K_c and the initial concentrations, calculate the equilibrium concentrations of all components.

Predicting the Direction of Reaction. Equilibrium constants provide useful information about chemical reaction systems. In this section, we will use them to predict the direction a reaction will proceed to establish equilibrium. We will also use them to determine the extent of reaction.

The *reaction quotient*, Q_c, is a useful aid in predicting whether or not a reaction system is at equilibrium. Take, for instance, the reaction

$$2SO_2 + O_2 \rightleftharpoons 2SO_3$$

The reaction quotient is:

$$Q_c = \frac{[SO_3]_0^2}{[SO_2]_0^2[O_2]_0}$$

You will notice that Q has the same algebraic form of the concentrations terms as does K_c. The difference is that these concentrations are not necessarily equilibrium concentrations. We will call them initial concentrations, $[\ ]_0$. When a set of initial concentrations is substituted into the reaction quotient, Q_c takes on a certain value. In order to predict whether the system is at equilibrium or not, the magnitude of Q_c must be compared with that of K_c.

- When $Q_c = K_c$, the reaction is at equilibrium, and no net reaction will occur.
- When $Q_c > K_c$, the system is not at equilibrium, and a net reaction will occur in the reverse direction until $Q_c = K_c$.
- When $Q_c < K_c$, the system is not at equilibrium, and a net reaction will occur in the forward direction until $Q_c = K_c$.

See Example 14.6 for an application of these criteria.

The Magnitude of the Equilibrium Constant. The magnitude of the equilibrium constant is related to the degree of conversion of reactants to products before chemical equilibrium is reached. Since K_c is proportional to the concentrations of products divided by the concentrations of reactants present at equilibrium, the value of K_c tells us the relative quantities of reactants and products present at equilibrium. When $K_c \gg 1$ more products are present than reactants at equilibrium. Conversely, when $K_c \ll 1$ more reactants are present at equilibrium than products. In this case equilibrium is reached before appreciable concentrations of products build up. *The larger the value of K_c, the greater the extent of reaction before equilibrium is reached.* See Example 14.7. In general, we can say:

• When $K_c \gg 1$, the forward reaction will go nearly to completion. The equilibrium will lie to the right and favor the products.

• When $K_c \ll 1$, the reaction will not go forward to an appreciable extent. The equilibrium will lie to the left and favor the reactants.

Calculation of Equilibrium Concentrations. Not only can we estimate the extent of reaction from the K_c value, but also the expected concentrations at equilibrium can be calculated from a knowledge of the initial concentrations and the K_c value. See Example 14.9 for this important type of calculation. In these types of problems it will be very helpful to use the following approach.

1. Express the equilibrium concentrations of all species in terms of the initial concentrations and an .unknown x, which represents *the change in concentration*.
2. Substitute the equilibrium concentrations derived in part 1 into the equilibrium constant expression, and solve for x. The equilibrium concentration is given by:

equilibrium concentration = initial concentration ± the change due to reaction
equilibrium concentration = initial concentration ± x

where the + sign is used for a product, and the – sign for a reactant.
3. Use x to calculate the equilibrium concentrations of all species.

EXAMPLE 14.7   Predicting Direction of Reaction

At a certain temperature the reaction

$$CO(g) + Cl_2(g) \rightleftharpoons COCl_2(g)$$

has an equilibrium constant $K_c = 13.8$. Is the following mixture an equilibrium mixture? If not, in which direction (forward or reverse) will reaction occur to reach equilibrium?

$$[CO]_0 = 2.5 \text{ M}; \; [Cl_2]_0 = 1.2 \text{ M}; \text{ and } [COCl_2]_0 = 5.0 \text{ M}$$

•**Method of Solution**

Recall that for the system to be at equilibrium $Q_c = K_c$. Substitute the given concentrations into the reaction quotient for the reaction, and determine Q_c.

•**Calculation**

$$Q_c = \frac{[COCl_2]_0}{[CO]_0[Cl_2]_0} = \frac{(5.0)}{(2.5)(1.2)} = 1.7$$

Compare Q_c to K_c. Since $Q_c < K_c$ the reaction mixture is not an equilibrium mixture. The product concentrations are too low and a *net forward reaction* will occur until $Q_c = K_c$, and the reaction reaches equilibrium.

EXAMPLE 14.8   Extent of Reaction

Arrange the following reactions in order of their increasing tendency to proceed toward completion (least extent → greatest extent).

a. $CO + Cl_2 \rightleftharpoons COCl_2$ $K_c = 13.8$
b. $N_2O_4 \rightleftharpoons 2NO_2$ $K_c = 2.1 \times 10^{-4}$
c. $2NOCl \rightleftharpoons 2NO + Cl_2$ $K_c = 4.7 \times 10^{-4}$

•Method of Solution

The larger the value of K_c, the more products there are at equilibrium compared to reactants, and the farther a reaction will proceed toward completion (the greater the extent of reaction).
Therefore: b < c < a

EXAMPLE 14.9   Calculating the Equilibrium Concentrations

At 400°C, $K_c = 64$ for the equilibrium

$$H_2(g) + I_2(g) \rightleftharpoons 2HI(g)$$

If 1.00 mol H_2 and 2.00 mol I_2 are introduced into an empty 0.50 L reaction vessel, find the equilibrium concentrations of all components at 400°C.

•Method of Solution

The equilibrium constant expression relates the concentrations H_2, I_2, and HI at equilibrium.

$$K_c = \frac{[HI]^2}{[H_2][I_2]} = 64$$

First we need expressions for the equilibrium concentrations of H_2, I_2, and HI. Begin by tabulating the initial concentrations.

Concentration	H_2	+	I_2	$\rightleftharpoons$	2HI
Initial	1.00 mol/0.50 L		2.00 mol/0.50 L		0
Change	—		—		—
Equilibrium	—		—		—

Since the answer involves three unknowns, we will relate the concentrations to each other by introducing the variable x. Recall, that the equilibrium concentration = initial concentration ± change in concentration. Let x = the change in concentration of H_2. That is, let x = the number of moles of H_2 reacted per liter. From the coefficients of the balanced equation we can tell that if the change in H_2 is –x, then the change in I_2 must also be –x, and the change in HI must be +2x.

The next step is to complete the table in units of molarity:

Concentration	H_2	+	I_2	$\rightleftharpoons$	2HI
Initial (M)	2.0		4.0		0
Change (M)	$-x$		$-x$		$2x$
Equilibrium (M)	$(2.0 - x)$		$(4.0 - x)$		$2x$

Now substitute the equilibrium concentrations from the table into the K_c expression,

$$K_c = \frac{(2x)^2}{(2.0 - x)(4.0 - x)} = 64$$

and solve for x.

$$\frac{(2x)^2}{x^2 - 6.0x + 8.0} = 64$$

Rearranging, we get:

$$4x^2 = 64x^2 - 384x + 512$$

and grouping yields a quadratic equation:

$$60x^2 - 384x + 512 = 0$$

We will use the general method of solving a quadratic equation of the form:

$$ax^2 + bx + c = 0$$

The root x is given by

$$x = \frac{-b \pm \sqrt{b^2 - 4ac}}{2a}$$

In this case, a = 60, b = –384, and c = 512. Therefore,

$$x = \frac{-(-384) \pm \sqrt{(-384)^2 - 4(60)(512)}}{2(60)} = \frac{384 \pm \sqrt{2.5 \times 10^4}}{120}$$

$$x = \frac{384 \pm 158}{120} = 1.9 \text{ and } 4.5 \text{ mol/ L}$$

Recall that x = the number of moles of H_2 (or I_2) reacted per liter. Of the two answers (roots), only 1.9 is reasonable, because the value 4.5 M would mean that more H_2 (or I_2) reacted than was present at the start. This would result in a negative equilibrium concentration, which is physically meaningless. We therefore use the root x = 1.9 M to calculate the equilibrium concentrations:

$$[H_2] = 2.0 - x = 2.0 \text{ M} - 1.9 \text{ M} = 0.1 \text{ M}$$

$$[I_2] = 4.0 - x = 4.0 \text{ M} - 1.9 \text{M} = 2.1 \text{ M}$$

$$[HI] = 2x = 2(1.9 \text{ M}) = 3.8 \text{ M}$$

The results can be checked by plugging these concentrations back into the K_c expression to see if $K_c = 64$:

$$K_c = \frac{[HI]^2}{[H_2][I_2]} = \frac{(3.8)^2}{(0.1)(2.1)} = 68$$

Thus the concentrations we have calculated are correct. The difference between 64 and 68 results from rounding off to maintain the correct number of significant figures. Therefore, our result is correct only to the number of signficant figures given in the problem.

EXERCISES

9. The reaction

   $$PCl_5(g) \rightleftharpoons PCl_3(g) + Cl_2(g)$$

   has the equilibrium constant value $K_c = 0.24$ at 300°C.
   a. Is the following reaction mixture at equilibrium?

   $[PCl_5] = 5.0$ mol/L, $[PCl_3] = 2.5$ mol/L, $[Cl_2] = 1.9$ mol/L

   b. Predict the direction in which the system will react to reach equilibrium

10. At 700 K, the reaction:

    $$2SO_2(g) + O_2(g) \rightleftharpoons 2SO_3(g)$$

    has an equilibrium constant $K_c = 4.3 \times 10^6$.
    a. Is a mixture with the following concentrations at equilibrium?

    $[SO_2] = 0.10$ M; $[SO_3] = 10$ M; $[O_2] = 0.10$ M.

    b. If not at equilibrium, predict the direction in which a net reaction will occur to reach a new equilibrium.

11. The decomposition of NOBr is represented by the equation

    $$2NOBr(g) \rightleftharpoons 2NO(g) + Br_2(g) \qquad K_c = 0.0169$$

    At equilibrium the concentrations of NO and Br_2 are 1.05×10^{-2} M and 5.24×10^{-3} M, respectively. What is the concentration of NOBr?

12. At 400°C, the equilibrium constant for the reaction

    $$H_2(g) + I_2(g) \rightleftharpoons 2HI(g)$$

    is 64. A mixture of 0.250 mol H_2 and 0.250 mol I_2 was introduced into an empty 0.75 L reaction vessel at 400°C, find the equilibrium concentrations of all components.

13. For the equilibrium

    $$N_2O_4(g) \rightleftharpoons 2NO_2(g) \qquad K_c = 0.36 \text{ at } 100°C$$

    a sample of 0.25 mol N_2O_4 is allowed to dissociate and come to equilibrium in a 1.5 L flask at 100°C. What are the equilibrium concentrations of NO_2 and N_2O_4?

FACTORS AFFECTING CHEMICAL EQUILIBRIUM (14.5)

STUDY OBJECTIVE

1. Predict how the equilibrium concentrations of reactants and products are shifted by changes in concentrations of reactants and products, by pressure changes, and by temperature changes.

Le Chatelier's Principle. When a reaction reaches a state of chemical equilibrium under a particular set of conditions, no further changes in the concentrations of reactants and products occur. If a change is made in the conditions under which the system is at equilibrium, chemical change will occur in such a way as to establish a new equilibrium. The factors that can influence equilibrium are change in concentration, change in pressure (or volume), and change in temperature.

What effect does a change in one of these factors have on the extent of reaction? This question can be answered qualitatively by using *Le Chatelier's principle:* When a stress is applied to a system in a state of dynamic equilibrium, the system will, if possible, shift to a new position of equilibrium in which the stress is partially offset. To interpret this statement take, for example, the reaction:

$$2NO_2(g) \rightleftharpoons N_2O_4(g)$$

First, the reaction must be at equilibrium. Let's add more N_2O_4 as an illustration of a change in concentration. The concentration of N_2O_4 increases, and the equilibrium is disturbed. The system will respond by using up part of the additional N_2O_4. In this case, a net reverse reaction will partially offset the increased N_2O_4 concentration. The net reverse reaction brings the system to a new state of equilibrium. When equilibrium is reestablished, more NO_2 will be present than there was before the N_2O_4 was added. Thus, the position of equilibrium has shifted to the left.

Le Chatelier's principle will predict the direction of the net reaction or "shift in equilibrium" that brings the system to a new equilibrium. In this case the stress of adding more N_2O_4 was partially offset by a net reverse reaction that consumed some of the additional N_2O_4. The key to the use of Le Chatelier's principle is to recognize which net reaction, forward or reverse, will partially offset the change in conditions.

Changes in Volume and Pressure. The pressure of a system of gases in chemical equilibrium can be increased by decreasing the available volume. This change causes the concentration of *all components* to increase. The stress will be partially offset by a net reaction that will lower the *total* concentration of gas molecules. Consider our previous reaction:

$$2NO_2(g) \rightleftharpoons N_2O_4(g)$$

When the molecules of both gases are compressed into a smaller volume their total concentration increases (this is the stress). A net forward reaction (shift to the right) will bring the system to a new state of equilibrium, in which the *total concentration* of molecules will be lowered somewhat. This partially offsets the initial stress on the system. Notice that when 2 moles of NO_2 react only 1 mole of N_2O_4 is formed. When equilibrium is reestablished, more moles of N_2O_4 and fewer moles of NO_2 will be present than before the pressure increase occurred. In this case the equilibrium has shifted to the right. In general, *an increase in pressure by decreasing the volume will result in a net reaction that decreases the total concentration of gas molecules.* A special case arises when the total number of moles of gaseous products and of gaseous reactants are equal in the balanced equation. In this case no shift in equilibrium will occur.

Changes in Temperature. If the temperature of a system is changed, a change in the value of K_c occurs. *An increase in temperature always shifts the equilibrium in the direction of the endothermic reaction, while a temperature decrease shifts the equilibrium in the direction of the exothermic reaction.* Therefore, for endothermic reactions the value of K_c *increases* with increasing temperature, and for exothermic reactions the value of K_c *decreases* with increasing temperature. In the case of our example reaction, K_c will decrease as the temperature is increased because the equilibrium will shift in the direction of the endothermic reaction, that is, in the reverse direction.

$$2NO_2(g) \rightleftharpoons N_2O_4(g) \qquad \Delta H^o_{rxn} = -58.0 \text{ kJ}$$

We can explain this in terms of Le Chatelier's principle. As heat is added to the system, it represents a stress on the equilibrium. The equilibrium will shift in the direction that will consume some of the added heat. This partially offsets the stress. In this reaction, the equilibrium shifts to the left, and K_c decreases.

Remember, of these three types of changes; concentration, pressure, and temperature, *only changes in temperature will actually alter the K_c value.*

There are two other factors related to chemical reactions that do not affect the position of equilibrium. The first of these is a catalyst. Catalysts speed up the rate at which equilibrium is reached by lowering the activation energy barrier. Of course, this speeds up the reverse reaction as well. The net result is that changes in the concentration of a catalyst will not affect the equilibrium concentrations of reactants and products, nor will they change the value of K_c.

The second of the two factors is the addition of an inert gas to a system at equilibrium. This will cause an increase in the total pressure within the reactor. However, none of the partial pressures of reactants or products are changed, and so the equilibrium is not upset, and no shifting is needed to bring the system back to equilibrium.

EXAMPLE 14.10 Changing Conditions Affecting Equilibrium

For the reaction at equilibrium

$$2NaHCO_3(s) \rightleftharpoons Na_2CO_3(s) + H_2O(g) + CO_2(g) \qquad \Delta H^o_{rxn} = 128 \text{ kJ}$$

state the effects (increase, decrease, no change) of the following stresses on the number of moles of sodium carbonate, Na_2CO_3, at equilibrium in a closed container. Note that Na_2CO_3 is a solid (this is a heterogeneous equation); its concentration will remain constant, but its amount can change.

a. Removing $CO_2(g)$.
b. Adding $H_2O(g)$.
c. Raising the temperature.
d. Adding $NaHCO_3(s)$.

•Method of Solution

Apply Le Chatelier's principle.

a. If CO_2 concentration is lowered, the system will react in such a way as to offset the change. That is, a shift to the right will replace some of the missing CO_2. The number of moles of Na_2CO_3 increases.
b. Addition of $H_2O(g)$ exerts a stress on the equilibrium that is partially offset by a shift in the equilibrium to the left (net reverse reaction). This consumes Na_2CO_3 as well as some of the extra H_2O. The number of moles of Na_2CO_3 decreases.
c. An increase in temperature will increase the K_c value of an endothermic reaction. There is a shift to the right, and more Na_2CO_3 is formed.
d. The position of a heterogeneous equilibrium does not depend on the amounts of pure solids or liquids present. The same equilibrium is reached whether the system contains 1 g of $NaHCO_3(s)$ or 10 g of $NaHCO_3$. No shift in the equilibrium occurs. No change in the amount of Na_2CO_3 occurs.

EXERCISES

14. Copper can be extracted from its ores by heating Cu_2S in air.

$$Cu_2S(s) + O_2(g) \rightleftharpoons 2Cu(s) + SO_2(g) \qquad \Delta H^o_{rxn} = -250 \text{ kJ}$$

Predict the direction of the shift of the equilibrium position in response to each of the following changes in conditions.
a. Adding more $O_2(g)$.
b. Compressing the vessel volume in half.
c. Raising the temperature.
d. Adding more $SO_2(g)$.
e. Adding more $Cu(s)$.
f. Adding a catalyst.

15. Consider the following reaction at equilibrium:

$$2SO_2(g) + O_2(g) \rightleftharpoons 2SO_3(g)$$

If the volume of the container is decreased at constant temperature, state which way the equilibrium will shift to reach a new equilibium: Left, right, or no change.

16. The following reaction is at equilibrium. Describe what change in the concentration of $NO(g)$, if any, will occur if bromine is added to the reaction chamber.

$$2NOBr(g) \rightleftharpoons 2NO(g) + Br_2(g)$$

17. The following reaction is exothermic:

$$2SO_2(g) + O_2(g) \rightleftharpoons 2SO_3(g)$$

Describe what will happen to the concentration of SO_2 when the temperature is increased.

CONCEPTUAL QUESTIONS

1. Consider the following reaction at 400°C:

$$H_2(g) + I_2(g) \rightleftharpoons 2HI(g)$$

H_2 and I_2 were placed into a flask and allowed to react until equilibrium was reached. Then a very small amount of $H^{131}I$ was added. ^{131}I is an isotope of iodine that is radioactive. Will radioactive ^{131}I atoms stay in the HI molecule, or will some or all of them find their way into the I_2 molecule forming $I^{131}I$? Explain your answer.

2. Consider the following reaction at 400°C:

    $$H_2O(g) + CO(g) \rightleftharpoons H_2(g) + CO_2(g)$$

    Amounts of H_2O, $CO(g)$, $H_2(g)$, and $CO_2(g)$ were put into a flask so that the composition corresponded to an equilibrium mixture. A lab technician added an iron catalyst to the mixture, but was surprised when no additional $H_2(g)$ and $CO_2(g)$ were formed even after waiting for many days. Explain why the technician should not have been surprised.

PRACTICE TEST

1. Write the equilibrium constant expressions for the following reactions:

    a. $4NH_3(g) + 3O_2(g) \rightleftharpoons 2N_2(g) + 6H_2O(g)$
    b. $2N_2O(g) + 3O_2(g) \rightleftharpoons 2N_2O_4(g)$
    c. $2ClO_2(g) + F_2(g) \rightleftharpoons 2FClO_2(g)$
    d. $H_2(g) + Br_2(l) \rightleftharpoons 2HBr(g)$
    e. $C(s) + CO_2(g) \rightleftharpoons 2CO(g)$
    f. $CuO(s) + H_2(g) \rightleftharpoons Cu(s) + H_2O(g)$

2. The equilibrium constant K_c for the reaction:

    $$Ni(s) + 4CO(g) \rightleftharpoons Ni(CO)_4(g)$$

    is 5.0×10^4 at 25°C. What is value of the equilibrium constant for the reaction?

    $$Ni(CO)_4(g) \rightleftharpoons Ni(s) + 4CO(g)$$

3. The decomposition of $HI(g)$ is represented by the equation

    $$2HI(g) \rightleftharpoons H_2(g) + I_2(g) \quad K_c = 64$$

    If the equilibrium concentrations of H_2 and I_2 at 400°C are found to be $[H_2] = 4.2 \times 10^{-4}$ M and $[I_2] = 1.9 \times 10^{-3}$ M, what is the equilibrium concentration of HI?

4. For the reaction

    $$CH_4(g) + 2H_2S(g) \rightleftharpoons CS_2(g) + 4H_2(g) \qquad K_p = 2.05 \times 10^9 \text{ at } 25°C.$$

    Calculate K_p and K_c, at this temperature, for

    $$2H_2(g) + \frac{1}{2}CS_2(g) \rightleftharpoons H_2S(g) + \frac{1}{2}CH_4(g)$$

5. A 1.00 L vessel initially contains 0.777 moles of $SO_3(g)$ at 1100 K. What is the value of K_c for the following reaction, if 0.520 moles of SO_3 remain at equilibrium?

    $$2SO_3(g) \rightleftharpoons 2SO_2(g) + O_2(g)$$

6. Initially a 1.0 L vessel contains 10.0 moles of NO and 6.0 moles of O_2 at a certain temperature. They react until equilibrium is established.

$$2NO(g) + O_2(g) \rightleftharpoons 2NO_2(g)$$

At equilibrium the vessel contains 8.8 moles of NO_2. Determine the value of K_c at this temperature.

7. Given the reaction:

$$N_2 + O_2 \rightleftharpoons 2NO \quad K_c = 2.5 \times 10^{-3} \text{ at } 2130°C$$

Decide whether the following mixture is at equilibrium, or if a net forward or reverse reaction will occur. [NO] = 0.005; [O_2] = 0.25; [N_2] = 0.020 mol/L.

8. Hydrogen iodide decomposes according to the equation:

$$2HI(g) \rightleftharpoons H_2(g) + I_2(g) \quad K_c = 0.0156 \text{ at } 400°C$$

A 0.55 mol sample of HI was injected into a 2.0 L reaction vessel held at 400°C. Calculate the concentration of HI at equilibrium.

9. 2.00 mol of NOCl was placed in a 2.00 L reaction vessel at 400°C. After equilibrium was established, it was found that 24 percent of the NOCl had dissociated according to the equation:

$$2NOCl(g) \rightleftharpoons 2NO(g) + Cl_2(g)$$

Calculate the equilibrium constant K_c for the reaction.

10. For the reaction

$$N_2(g) + O_2(g) \rightleftharpoons 2NO(g) \quad K_p = 3.80 \times 10^{-4} \text{ at } 2000°C$$

what equilibrium pressures of N_2, O_2, and NO will result when a 10 L reactor vessel is filled with 2.00 atm of N_2 and 0.400 atm of O_2 and the reaction is allowed to come to equilibrium?

11. For the equilibrium

$$N_2O_4(g) \rightleftharpoons 2NO_2(g) \quad K_c = 0.36 \text{ at } 100°C$$

Suppose 30.0 g N_2O_4 is placed in a 2.5 L flask at 100°C. (a) Calculate the number of moles of NO_2 present at equilibrium and (b) the percentage of the original N_2O_4 that is dissociated.

12. What changes in the equilibrium composition of the reaction

$$2SO_2(g) + O_2(g) \rightleftharpoons 2SO_3(g) \quad \Delta H^\circ_{rxn} = -197 \text{ kJ}$$

will occur if it experiences the following stresses?
a. The partial pressure of $SO_3(g)$ is increased.
b. Inert Ar gas is added.
c. The temperature of the system is decreased.
d. The total pressure of the system is increased by reducing the available volume.
e. The partial pressure of $O_2(g)$ is decreased.

13. For the chemical equilibrium

$$PCl_5(g) \rightleftharpoons PCl_3(g) + Cl_2(g) \qquad \Delta H^{\circ}_{rxn} = 92.9 \text{ kJ}$$

a. What is the effect on K_c of lowering the temperature?
b. What is the effect on the equilibrium concentration of PCl_3 of adding Cl_2?
c. What is the effect on the equilibrium concentrations of compressing the mixture to a smaller volume?
d. What is the effect on the equilibrium pressure of Cl_2 of removing PCl_3?

14. Consider the equilibrium

$$SO_2(g) + NO_2(g) \rightleftharpoons NO(g) + SO_3(g)$$

where $K_c = 85.0$ at 460°C. The following mixture was prepared in a reactor at 460°C.
$[SO_2] = 0.0200$ M; $[NO_2] = 0.0200$ M; $[NO] = 0.100$ M; $[SO_3] = 0.100$ M
What will the concentrations of the four gases be when equilibrium is reached?

15. Arrange the following reactions in their increasing tendency to proceed toward completion:

a. $2HF \rightleftharpoons H_2F_2 \qquad K_c = 1 \times 10^{-13}$

b. $2H_2 + O_2 \rightleftharpoons 2H_2O \qquad K_c = 3 \times 10^{81}$

c. $2NOCl \rightleftharpoons 2NO + Cl_2 \qquad K_c = 4.7 \times 10^{-4}$

16. For the decomposition of calcium carbonate

$$CaCO_3(s) \rightleftharpoons CaO(s) + CO_2(g) \qquad \Delta H^{\circ}_{rxn} = 175 \text{ kJ}$$

how will the amount (not concentration) of $CaCO_3(s)$ change with the following stresses?
a. $CO_2(g)$ is removed.
b. $CaO(s)$ is added.
c. The temperature is raised.
d. The volume of the container is decreased.

17. At 2000°C, 5.0×10^{-3} mol CO_2 is introduced into a 1.0 L container and the following reaction comes to equilibrium:

$$2CO_2(g) \rightleftharpoons 2CO(g) + O_2(g) \quad K_c = 6.4 \times 10^{-7}$$

a. Calculate the equilibrium concentrations of CO and O_2.
b. What fraction of the CO_2 is decomposed at equilibrium?

18. In a closed container at 1900 K nitrogen reacts with oxygen to yield NO:

$$N_2(g) + O_2(g) \rightleftharpoons 2NO(g) \qquad K_p = 2.3 \times 10^{-4}$$

However, NO(g) quickly reacts with oxygen to produce $NO_2(g)$:

$$2NO(g) + O_2(g) \rightleftharpoons 2NO_2(g) \qquad K_p = 1.3 \times 10^{-4}$$

Calculate K_p and K_c at 1900 K for the net reaction.

$$N_2(g) + 2O_2(g) \rightleftharpoons 2NO_2(g)$$

ANSWERS

Exercises

1. a. $K_c = [O_2]$
    b. $K_c = \dfrac{[Ni(CO)_4]}{[CO]^4}$
    c. $K_c = \dfrac{[NOBr]^2}{[NO]^2[Br_2]}$

2. $K_c = 1.5 \times 10^{-3}$

3. a. $K_p = 1.4 \times 10^{-5}$ b. $K_p = 266$

4. $K_c = \dfrac{[COCl_2]^2}{[CO_2]\,[Cl_2]^2} = 7.7 \times 10^{11}$

5. $K_c = 0.0169$

6. $K_c = 5.6 \times 10^{-4}$

7. $K_c = 0.13$

8. $K_c = 17.3$

9. a. no   b. reverse

10. a. no   b. forward

11. $[NOBr] = 5.85 \times 10^{-3}$ M

12. $[HI] = 0.532$ M, $[H_2] = [I_2] = 6.7 \times 10^{-2}$ M

13. $[NO_2] = 0.17$ M, $[N_2O_4] = 0.080$ M

14. a. right   b. no shift   c. left   d. left   e. no shift   e. no shift

15. right

16. decrease

17. increase

Conceptual Questions

1. Yes some radioactive ^{131}I atoms will become part of I_2 molecules, and form of $I^{131}I$. Some will remain as $H^{131}I$. First of all ^{131}I atoms will react just like any other iodine atom. The reaction system is at equilibrium which is a dynamic equilibrium. This is the key. The forward and reverse reactions are occuring simultaneously and at the same rate. Any $H^{131}I$ will eventually dissociate by the reverse reaction and produce $I^{131}I$ molecules and H_2 molecules. All of the ^{131}I will not end up in the $I^{131}I$ molecules because the forward reaction converts $I^{131}I$ back into $H^{131}I$. Therefore the ^{131}I will be distributed between the two I-containing molecules.

2. The technician should have known that a catalyst cannot change the position of equilibrium. A catalyst lowers the activation energies of *both* the forward and reverse reactions, and so speeds up the rates of both the forward and reverse reactions. This does not change the equilibrium constant.

Practice Test

1. a. $K_c = \dfrac{[N_2]^2[H_2O]^6}{[NH_3]^4[O_2]^3}$ b. $K_c = \dfrac{[N_2O_4]^2}{[N_2O]^2[O_2]^3}$ c. $K_c = \dfrac{[FClO_2]^2}{[ClO_2]^2[F_2]}$

   d. $K_c = \dfrac{[HBr]^2}{[H_2]}$ e. $K_c = \dfrac{[CO]^2}{[CO_2]}$ f. $K_c = \dfrac{[H_2O]}{[H_2]}$

2. 2.0×10^{-5}
3. $[HI] = 1.1 \times 10^{-4}\,M$
4. $K_p = 2.2 \times 10^{-5}$, $K_c = 5.4 \times 10^{-4}$
5. $K_c = 0.031$
6. $K_c = 34$
7. Net reverse reaction.
8. $[HI] = 0.22\,M$
9. $K_c = 0.012$
10. $P_{NO} = 0.017$ atm; $P_{N_2} = 1.99$ atm; $P_{O_2} = 0.39$ atm
11. a. 0.36 mol NO_2 b. 66 % N_2O_4 dissociated
12. a. The partial pressures of SO_2 and O_2 will increase.
    b. No changes in equilibrium partial pressures.
    c. The partial pressures of SO_2 and O_2 decrease while that of SO_3 increases.
    d. The partial pressures of SO_2 and O_2 decrease while that of SO_3 increases.
    e. The partial pressure of SO_2 increases and that of SO_3 decreases.
13. a. K_c will decrease   b. $[PCl_3]$ decreases   c. $[PCl_5]$ increases; $[PCl_5]$ and $[Cl_2]$ decrease   d. Cl_2 pressure increases
14. $[SO_2] = [NO_2] = 0.0118\,M$; $[NO] = [SO_3] = 0.108\,M$
15. a. Least extent of reaction   b. Greatest extent of reaction
    c. Intermediate of the three examples
16. a. Decreases   b. No change   c. Decreases   d. Increases
17. a. $[O_2] = 1.5 \times 10^{-4}\,M$, $[CO] = 3.0 \times 10^{-4}\,M$ b. 6.1 percent
18. $K_p = 3.0 \times 10^{-8}$ $K_c = 4.7 \times 10^{-6}$

Chapter Fifteen
ACIDS AND BASES

- Brønsted Acids and Bases (15.1)
- Acid-Base Properties of Water and the pH Scale (15.2 – 15.3)
- Strengths of Acids and Bases (15.4)
- Weak Acids, Weak Bases, and Ionization Constants (15.5 – 15.7)
- Diprotic and Polyprotic Acids (15.8)
- Molecular Structure and Strengths of Acids (15.9)
- Hydrolysis and Acid-Base Properties of Salts (15.10)
- Lewis Acids and Bases (15.12)

BRØNSTED ACIDS AND BASES (15.1)

STUDY OBJECTIVES

1. Describe a conjugate acid-base pair.
2. Distinguish between a proton, a hydrated hydrogen ion, and a hydronium ion.

In this chapter we will develop further the Brønsted definition of acids and bases previously introduced in Chapter 4. The concept of equilibrium will be applied to acid-base reactions and the pH scale will be introduced. Later in the chapter the structural features of molecules that make them acids or bases will be investigated.

Brønsted Acids and Bases. According to the Brønsted acid-base theory, an **acid** is defined as a substance that can donate a proton to another substance. A **base** is a substance that can accept a proton from another substance. For example, when HBr gas dissolves in water, it donates a proton to the solvent. Therefore, HBr is a Brønsted acid.

$$HBr(aq) + H_2O(l) \rightarrow H_3O^+(aq) + Br^-(aq)$$
acid   base

The water molecule is the proton acceptor. Therefore, water is a Brønsted base in this reaction. All acid-base reactions, according to Brønsted, involve the *transfer of a proton*. When a water molecule accepts a proton, a hydronium ion (H_3O^+) is formed.

There are three forms of the hydrogen ion to keep straight. First is the proton (H^+) that is transferred in any Brønsted acid-base reaction. Second is the hydrated hydrogen ion $H^+(aq)$ which is a general purpose representation of the proton in aqueous solution. Third, when a proton bonds to a water molecule a hydronium ion (H_3O^+) is formed. This is the best representation of the actual structure of the hydrated hydrogen ion.

In the above example, HBr and Br^- are related chemical species. A **conjugate acid-base pair** consists of two species that differ from each other by the presence of one H^+ unit. HBr and Br^- ion are a conjugate acid-base pair. Removing a proton (H^+) from the acid (HBr) gives its conjugate base (Br^-). H_3O^+ and H_2O are also a conjugate acid-base pair.

Substances that can behave as an acid in one proton transfer reaction and as a base in another reaction are called *amphoteric*. Water is an amphoteric substance. When ammonia dissolves in water for example, a proton transfer reaction takes place. Only this time water is the acid and ammonia is the base.

$$NH_3(aq) + H_2O(l) \rightleftharpoons NH_4^+(aq) + OH^-(aq)$$

The conjugate acid-base pairs are:

$$NH_4^+ - NH_3, \text{ and } H_2O - OH^-$$

Such pairs are usually labeled as shown below the following equation:

$$\underset{\text{base}_1}{NH_3(aq)} + \underset{\text{acid}_2}{H_2O(l)} \rightleftharpoons \underset{\text{acid}_1}{NH_4^+(aq)} + \underset{\text{base}_2}{OH^-(aq)}$$

The subscript 1 designates one conjugate acid-base pair, and the subscript 2 the other pair. The double arrow indicates that the reaction is reversible. In the reverse reaction OH^- acts as a base and NH_4^+ acts as an acid.

EXAMPLE 15.1   Conjugate Acid-Base Pair

Write the formula of the conjugate base of H_2SO_4.

•**Method of Solution**

A conjugate base differs from its conjugate acid by the lack of a proton.

$$\underset{\text{acid}}{H_2SO_4} \qquad \underset{\text{conjugate base}}{HSO_4^-}$$

EXAMPLE 15.2   Conjugate Acid-Base Pairs

For the reaction:

$$HSO_4^-(aq) + HCO_3^-(aq) \rightleftharpoons SO_4^{2-}(aq) + H_2CO_3(aq)$$

a.   Identify the acids and bases for the forward and reverse reactions.
b.   Identify the conjugate acid-base pairs.

•**Method of Solution**

a.   In the forward reaction, HSO_4^- ion is the proton donor which makes it an acid. The HCO_3^- ion is the proton acceptor, and is a base. In the reverse reaction, H_2CO_3 is the proton donor (acid), and SO_4^{2-} is the proton acceptor (base).

b. $\underset{\text{acid}_1}{HSO_4^-} + \underset{\text{base}_2}{HCO_3^-} \rightleftharpoons \underset{\text{base}_1}{SO_4^{2-}} + \underset{\text{acid}_2}{H_2CO_3}$
     the subscripts 1 and 2 designate the two conjugate acid-base pairs.

EXERCISES

1. Identify the conjugate base of HCO_3^- in the following acid-base reaction:

$$CO_3^{2-} + CH_3COOH \rightleftharpoons HCO_3^- + CH_3COO^-$$

2. Identify the Brønsted acid and base on the left side of the following equations, and identify the the conjugate partner of each on the right side.

   a. $NH_4^+(aq) + H_2O(l) \rightleftharpoons NH_3(aq) + H_3O^+(aq)$

   b. $HNO_2(aq) + CN^-(aq) \rightleftharpoons NO_2^-(aq) + HCN(aq)$

THE ACID-BASE PROPERTIES OF WATER AND THE pH SCALE (15.2 – 15.3)

STUDY OBJECTIVES

1. Write the ion product constant for the autoionization of water, and use it to relate $[H^+]$ and $[OH^-]$ in aqueous solutions.
2. Describe the pH scale, and calculate pH from a knowledge of $[H^+]$ or $[OH^-]$.
3. Carry out numerical calculations involving the relationships among $[H^+]$, $[OH^-]$, pH and pOH.

Autoionization and the Ion-Product of Water. Pure water is itself a very weak electrolyte and ionizes according to the equation:

$$H_2O \rightleftharpoons H^+(aq) + OH^-(aq)$$

According to the Brønsted theory, the reaction is viewed as a proton transfer from one water molecule to another.

$$H_2O(l) + H_2O(l) \rightleftharpoons H_3O^+(aq) + OH^-(aq)$$
$$\text{acid}_1 \quad\quad \text{base}_2 \quad\quad\quad \text{acid}_2 \quad\quad\quad \text{base}_1$$

Since water can act as both an acid and a base, it is an *amphoteric* substance.

This reaction is reversible and H_2O, H_3O^+ and OH^- are in equilibrium. In pure water at 25°C, $[H^+] = [OH^-]$ $= 1.0 \times 10^{-7}$ M. These are low concentrations and tell us that very few H_2O molecules are ionized, and that the equilibrium lies to the left.

At equilibrium, the product of the hydrogen ion concentration and hydroxide ion concentration equals a constant called **the ion-product constant for water, K_w.**

$$K_w = [H^+][OH^-] = (1.0 \times 10^{-7})(1.0 \times 10^{-7}) = 1.0 \times 10^{-14}$$

Like other equilibrium constants we treat K_w as unitless. The value of K_w applies to all aqueous solutions at 25°C. When an acid is added to water, the $[H^+]$ increases. Therefore, the $[OH^-]$ must decrease in order for K_w to remain constant. In *acidic* solutions $[H^+] > [OH^-]$. Similarly, when a base is added to water, and $[OH^-]$ increases, then $[H^+]$ must decrease. In *basic* solutions $[OH^-] > [H^+]$. Acidic, basic, and neutral solutions are characterized by the following conditions:

neutral $\quad [H^+] = [OH^-]$
acidic $\quad\quad [H^+] > [OH^-]$
basic $\quad\quad\; [H^+] < [OH^-]$

The ion product provides a useful relationship for aqueous solutions. If the value of $[H^+]$ is known, then the concentration of OH^- can be calculated. Similarly, the H^+ ion concentration can be calculated, if the value of $[OH^-]$ is known. Example 15.3 illustrates this type of calculation. In Table 15.1 below each row corresponds to a solution with the given H^+ and OH^- concentrations. The table covers the entire practical range of concentrations found in aqueous solutions. Note that the product of the two concentrations in all aqueous solutions is 1.0×10^{-14}.

The pH Scale. The concentration of $H^+(aq)$ in a solution can be expressed in terms of the pH scale. The pH of a solution is defined as the negative logarithm of the hydrogen ion concentration.

$$pH = -\log [H^+]$$

Recall that the logarithm of a number is the power to which 10 must be raised in order to equal the number. For example, the logarithm of 100 is 2.0 because raising 10 to the 2nd power gives 100.

$$100 = 10^2$$
$$\log 100 = 2$$

The log of a fraction or number less than 1 is a negative number.

$$\frac{1}{100} = 0.01 = 10^{-2}$$
$$\log 0.01 = -2$$

First, let's find the pH of a neutral solution. In pure water at 25°C; $[H^+] = 1 \times 10^{-7}$ M. Using the definition of pH given above, take the log of the H^+ ion concentration first:

$$pH = -\log (1.0 \times 10^{-7})$$
$$pH = -(-7.0) = 7.0$$

The pH of a neutral solution is 7.0.

Likewise, for an acidic solution where, for example, the H^+ ion concentration is 1×10^{-5} M, the pH is 5.0.

$$pH = -\log (1 \times 10^{-5}) = -(-5.0) = 5.0$$

All acidic solutions have a pH < 7.0.

When $[H^+]$ is not an exact power of 10, the pH is not a round number. Take the following basic solution, for example, if $[H^+] = 2.5 \times 10^{-9}$ M, the pH is

$$pH = -\log (2.5 \times 10^{-9}) = -(-8.60)$$
$$pH = 8.60$$

Note that all basic solutions have a pH > 7.0. The pH values corresponding to selected sets of $H^+(aq)$ and $OH^-(aq)$ concentrations are given in Table 15.1. In terms of pH, solutions that are acidic, basic, and neutral are defined as follows:

neutral	pH = 7.0
acidic	pH < 7.0
basic	pH > 7.0

The pOH Scale. A scale just like the pH scale has been devised for the hydroxide ion concentration, where

$$pOH = -\log [OH^-]$$

Just as the H^+ ion and OH^- ion concentrations are related by the ion-product constant of water, K_w, the pH and pOH are also related.

$$K_w = [H^+][OH^-] = 1.0 \times 10^{-14}$$
$$pH + pOH = 14$$

The *sum of the pH and pOH values of any solution is always 14* at 25°C. You can see this in Table 15.1. Sum the pH and pOH for each set of H^+ and OH^- concentrations, and see what you get.

It is also important to notice that *a change in pH of one unit corresponds to a 10-fold change in [H⁺].* As H^+ drops from 10^{-8} to 10^{-9} M, for instance, the pH changes from 8 to 9. A change of 2.0 pH units corresponds to a 100-fold change in H^+ ion concentration. Never say "A pH of 2 is twice as acidic as a pH of 4." It is really 100 times more acidic!

Table 15.1 Relationship of pH and pOH in Aqueous Solutions

$[H^+]$	$[OH^-]$	pH	pOH	Nature of Solution
10^0	10^{-14}	0	14	acidic
10^{-1}	10^{-13}	1	13	acidic
10^{-2}	10^{-12}	2	12	acidic
10^{-3}	10^{-11}	3	11	acidic
10^{-6}	10^{-8}	6	8	acidic
10^{-7}	10^{-7}	7	7	neutral
10^{-8}	10^{-6}	8	6	basic
10^{-11}	10^{-3}	11	3	basic
10^{-12}	10^{-2}	12	2	basic
10^{-13}	10^{-1}	13	1	basic
10^{-14}	10^0	14	0	basic

Changing pH Values to [H⁺]. Given the pH, how do we calculate the $[H^+]$? Rearrange the equation for pH:

$$pH = -\log [H^+]$$

$$\log [H^+] = -pH$$

taking the antilog of both sides:

$$\text{antilog} (\log [H^+]) = \text{antilog} (-pH)$$

gives:

$$[H^+] = 10^{-pH}$$

Any electronic calculator with a 10^x key will easily make the calculation of H^+ ion concentrations from pH values. Just enter –pH for x and push 10^x.

EXAMPLE 15.3 Using the Ion-Product Constant, K_W

The H^+ ion concentration in a certain solution is 5.0×10^{-5} M. What is the OH^- ion concentration?

•Method of Solution

The ion-product constant of water is applicable to all aqueous solutions.

$$K_w = [H^+][OH^-] = 1.0 \times 10^{-14}$$

When $[H^+]$ is known, we can solve for $[OH^-]$.

•Calculation

$$[OH^-] = \frac{1.0 \times 10^{-14}}{[H^+]} = \frac{1.0 \times 10^{-14}}{5.0 \times 10^{-5}}$$

$$[OH^-] = 2.0 \times 10^{-10} \text{ M}$$

EXAMPLE 15.4 pH and pOH

The OH^- ion concentration in a certain ammonia solution is 7.2×10^{-4} M. What is the pOH and pH?

•Method of Solution

The pOH is the negative logarithm of the OH^- ion concentration
$$pOH = - \log [OH^-]$$

•Calculation

$$pOH = - \log (7.2 \times 10^{-4}) = - (-3.14)$$
$$= 3.14$$

The pH and pOH are related by

$$pH + pOH = 14.00$$
$$pH = 14.00 - pOH = 14.00 - 3.14$$
$$= 10.86$$

EXAMPLE 15.5 H^+ Ion Concentration from pH

What is the H^+ ion concentration in a solution with a pOH of 3.9?

•Method of Solution

Recall that $pH + pOH = 14$. With the pOH given, the pH can be calculated, and then the H^+ ion concentration can be determined.

•Calculation

$$pH = 14.0 - pOH$$
$$pH = 14.0 - 3.9 = 10.1$$

Since pH is known, the hydrogen ion concentration is

$$[H^+] = 10^{-pH}$$
$$[H^+] = 10^{-10.1} = 7.9 \times 10^{-11} \text{ M}$$

EXAMPLE 15.6   Comparing pH Values

The pH of many cola-type soft drinks is about 3.0. How many times greater is the H^+ concentration in these drinks than in neutral water?

•**Method of Solution**

First write out the H^+ ion concentrations in the cola drink and in neutral water.

$$[H^+]_{cola} = 1.0 \times 10^{-3} \text{ M, and } [H^+]_{neut} = 1.0 \times 10^{-7} \text{ M.}$$

Then the ratio is

$$\frac{[H^+]_{cola}}{[H^+]_{neut}} = \frac{1.0 \times 10^{-3} \text{M}}{1.0 \times 10^{-7} \text{M}} = 10^4 = 10,000$$

•**Comment**

Here is another approach. Since a change of 1.0 pH unit corresponds to a 10-fold change in H^+ concentration, then a change of 4.0 pH units corresponds to $10 \times 10 \times 10 \times 10 = 10^4$ or a 10,000-fold increase in H^+ concentration.

EXERCISES

3. Calculate the concentration of OH^- ions in an HNO_3 solution where $[H^+] = 0.0010$ M.

4. The OH^- ion concentration in an ammonia solution is 7.5×10^{-3} M. What is the H^+ ion concentration?

5. Calculate the concentration of H^+ ions in an acid solution with a pH of 2.29.

6. What is the concentration of OH^- ions in a NaOH solution which has a pOH of 4.90?

7. The pH of solution A is 2.0 and the pH of solution B is 4.0. How many times greater is the $H^+(aq)$ concentration in solution A than in solution B?

8. The pH of a certain solution is 3.0. How many moles of $H_3O^+(aq)$ ions are there in 0.10 L of this solution?

STRENGTHS OF ACIDS AND BASES (15.4)

STUDY OBJECTIVES

1. Define the terms strong and weak as they are applied to acids and bases.
2. Explain the relationship between the relative strength of an acid and the strength of its conjugate base.
3. Calculate the pH of given solutions of strong acids and strong bases.
4. Predict the direction of a given acid-base reaction.

Strong Acid and Bases. The stronger the acid, the more completely it ionizes in water, producing $H_3O^+(aq)$ and an anion. **Strong acids** are strong electrolytes and ionize completely in water. HBr, for example, is a strong acid.

$$HBr(aq) + H_2O(l) \rightarrow H_3O^+(aq) + Br^-(aq)$$

A solution of hydrobromic acid consists of $H_3O^+(aq)$ and $Br^-(aq)$ ions. The concentration of HBr molecules is zero because all have ionized. The concentration of H_3O^+ is equal to the initial concentration of the acid.

Like a strong acid, a **strong base** is one that ionizes completely in aqueous solution. KOH, a commonly used base, is an example. It is purchased as a white solid. It dissolves readily in water to give a solution of K^+ and OH^- ions:

$$KOH(aq) \rightarrow K^+(aq) + OH^-(aq)$$

A solution of 0.10 M KOH is made by dissolving 0.10 mol of KOH in 1 L of solution. In this solution, $[K^+]$ is 0.10 M, $[OH^-]$ is 0.10 M, and $[KOH]$ is essentially zero. Of the strong bases, only NaOH and KOH are commonly used in the laboratory.

Weak Acids and Bases. Not all acids are strong proton donors in aqueous solutions. A **weak acid** is one that is only partially ionized in water. Hydrofluoric acid is a typical weak acid in water:

$$HF(aq) + H_2O(l) \rightleftharpoons H_3O^+(aq) + F^-(aq)$$

Since HF is only partially ionized the concentration of HF molecules is much greater than the concentrations of $H^+(aq)$ and $F^-(aq)$ ions. For example, a 1.0 M HF solution has the following concentrations: $[H^+] = [F^-] =$ 0.026 M, and $[HF] = 0.974$ M. About 97% of the initial HF molecules are *not* ionized. For weak acids, the ionization equilibrium lies to the left. The extent of reaction is small. A number of weak acids are listed in Table 15.2 of the text. Figure 15.1 compares the relative concentrations of HA, H^+, and A^- in solutions of strong and weak acids.

Figure 15.1 Illustration of the extent of ionization of (a) a strong acid, HA, that ionizes 100%, and (b) a weak acid, HA, that ionizes a few percent. A represents the anion. The height of the bar represents the concentration of the species present.

Some bases are not ionized completely in aqueous solution. A **weak base** ionizes to a limited extent in water. Ammonia is an example of a weak base. In a 0.10 M solution of ammonia, only about 1% of the ammonia molecules are ionized. An important reverse reaction means that the equilibrium lies to the left.

$$NH_3(aq) + H_2O(l) \rightleftharpoons NH_4^+(aq) + OH^-(aq)$$
base   acid

Strength of Conjugate Bases. The strength of an acid is related to the strength of its conjugate base. Let's compare the ionization of HCl, a strong acid, and HF, a weak acid.

$$HCl(aq) + H_2O(l) \rightarrow H_3O^+(aq) + Cl^-(aq)$$

$$HF(aq) + H_2O \rightleftharpoons H_3O^+(aq) + F^-(aq)$$

Chloride ion is the conjugate base of HCl. Here we see that it must be a very weak base because it has no tendency to accept a proton from H_3O^+ in the reverse reaction. In the ionization of the weak acid HF, the reverse reaction occurs to a much greater extent than for HCl. Fluoride ion accepts a proton from H_3O^+. The F^- ion is a much stronger base than Cl^-. *The conjugate bases of strong acids have no measurable base strength.*

When comparing the strength of different acids we see that: *as acid strength increases, the strength of its conjugate base decreases.* Table 15.2 in the text lists several important conjugate acid-base pairs, according to their relative strengths. A short version is reproduced here in Table 15.2.

Table 15.2 Relative Strengths of Conjugate Acid-Base Pairs

	Acid	Conjugate Base	
Increasing Acid Strength ↑	$HClO_4$	ClO_4^-	Increasing Base Strength ↓
	HCl	Cl^-	
	HNO_3	NO_3^-	
	H_3O^+	H_2O	
	HF	F^-	
	HNO_2	NO_2^-	
	HCN	CN^-	
	H_2O	OH^-	

Predicting the Direction of Acid-Base Reactions. The information in Table 15.2 can be used to predict the direction of an acid-base reaction. When an acid reacts with a base will the reaction tend to go from left to right and favor the products, or will it tend to go from right to left and favor the reactants. For instance, when HF reacts with water, on which side will the equilibrium lie?

$$HF(aq) + H_2O \rightleftharpoons H_3O^+(aq) + F^-(aq)$$

To decide compare the strength of HF to that of the acid in the reverse reaction, H_3O^+. According to Table 15.2, H_3O^+ is the stronger acid. Compare the base strength of H_2O to that of the base in the reverse reaction, F^- ion. F^- ion is the stronger base. The reaction can be summarized as:

$$HF(aq) + H_2O \rightleftharpoons H_3O^+(aq) + F^-(aq)$$
weaker   weaker   stronger   stronger
acid   base   acid   base

All proton transfers proceed from the stronger acid and stronger base toward the weaker acid and weaker base. Therefore in this case the reaction proceeds from right to left and equilibrium lies to the left. There will be more HF at equilibrium than H_3O^+ and F^- ion. A simplification is to just compare the strengths of the two bases. *The reaction can be viewed as a competition of two bases for the proton.* Who wins? The stronger base, of course. F^- ions being stronger accept protons more readily than do H_2O molecules, and so HF predominates over H_3O^+ in solution. *The weaker acid and weaker base predominate at equilibrium.* Example 15.8 applies this principle to another reaction.

The pH of Strong Acid and Strong Base Solutions. The pH of a strong acid solution depends on the hydrogen ion concentration. What is the pH of a 0.0052 M HBr solution? When hydrobromic acid is added to water, it ionizes completely.

$$HBr(aq) + H_2O(l) \rightarrow H_3O^+(aq) + Br^-(aq)$$

For simplicity we can write the reaction without the H_2O molecules.

$$HBr(aq) \rightarrow H^+(aq) + Br^-(aq)$$

Before any ionization occurs the HBr concentration is 0.0052 M, but after ionization its concentration is zero. Since each mole of HBr that reacts yields 1 mol of H^+ and 1 mol of Br^-, the concentrations of H^+ and Br^- ions are both 0.0052 M. 0.0052 mol HBr reacts to give 0.0052 mol H^+ ion and 0.0052 mol Br^- ion per liter of solution.

The pH is given by:

$$pH = -\log (5.2 \times 10^{-3})$$
$$= 2.28$$

The calculation of the pH of a strong base is given in Example 15.9.

EXAMPLE 15.7   Increasing Conjugate Base Strength

The strengths of the following acids increase in the order $HCN < HF < HNO_3$. Arrange the conjugate bases of these acids in order of increasing base strength.

•Method of Solution

Recall that as the strengths of acids increase, the strengths of their conjugate bases decrease. The acid strength increases from left to right.

$HCN < HF < HNO_3$
increasing acid strength $\rightarrow$

Therefore, the strengths of their conjugate bases *decrease* from left to right.

$CN^- > F^- > NO_3^-$
decreasing base strength $\rightarrow$
$\leftarrow$ increasing base strength

EXAMPLE 15.8   Direction of an Acid-Base Reaction

To which side does the equilibrium lie in the following acid–base reaction?

$$HF(aq) + NO_2^-(aq) \rightleftharpoons HNO_2(aq) + F^-(aq)$$

•**Method of Solution**

First we need to know which is the stronger acid, the one on the left (HF) or the one on the right (HNO_2) of the reaction. Second we also need to which is the stronger base and which is the weaker base. Look this information up in Table 15.2.

$$HF(aq) + NO_2^-(aq) \rightleftharpoons HNO_2(aq) + F^-(aq)$$

| stronger | stronger | weaker | weaker |
| acid | base | acid | base |

Since proton transfer reactions proceed from the stronger acid and stronger base toward the weaker acid and weaker base, the reaction favors HNO_2 and F^- ion. The equilibrium lies to the right.

EXAMPLE 15.9   The Hydrogen Ion Concentration in a Solution of a Strong Base

What is the H^+ ion concentration in 0.0025 M NaOH?

•**Method of Solution**

Sodium hydroxide is a base and supplies practically all the OH^- ions in the solution. The H^+ ions come from the autoionization of water.

$$H_2O(l) \rightarrow H^+(aq) + OH^-(aq)$$

Sodium hydroxide is a strong base and therefore ionizes 100% in aqueous solution.

$$NaOH(aq) \rightarrow Na^+(aq) + OH^-(aq)$$

This means that we can calculate the hydroxide ion concentration first. Then from the ion-product constant of water (K_w) we can calculate the concentration of H^+ ions in equilibrium with the known concentration of OH^- ion.

•**Calculation**

When 0.0025 mol of NaOH per liter dissociates then 0.0025 mol Na^+ ion and 0.0025 mol OH^- ion are formed per liter. Only the OH^- affects the pH, and $[OH^-] = 0.0025$ M.
Since the concentrations of H^+ and OH^- ions are related to eachother:

$$K_w = [H^+][OH^-] = 1.0 \times 10^{-14}$$

Then

$$[H^+] = \frac{K_w}{[OH^-]} = \frac{1.0 \times 10^{-14}}{2.5 \times 10^{-3}}$$

$$[H^+] = 4.0 \times 10^{-12} \text{ M}$$

EXERCISES

9. What are the H^+ ion and OH^- ion concentrations in a 0.0033 M NaOH solution?

10. Calculate the pH of a 0.00051 M HCl solution.

11. Calculate the pH of a 0.125 M NaOH solution.

12. What is the pH of a 0.0025 M solution of $Ca(OH)_2$?

13. What is the pH of a solution made from 3.0 g HCl dissolved in 3.0 L of solution?

14. Which of the following acids has the strongest conjugate base?
    HNO_2, HCN or H_3O^+

15. Predict the direction in which the equilibrium will lie for the following reaction.
    $$HNO_2(aq) + ClO_4^-(aq) \rightleftharpoons NO_2^-(aq) + HClO_4(aq)$$

WEAK ACIDS, WEAK BASES, AND IONIZATION CONSTANTS (15.5 – 15.7)

STUDY OBJECTIVES

1. Write the *acid ionization constant* for any weak acid and *base ionization constant* for any weak base.
2. Calculate the concentrations of H^+, A^-, and undissociated weak acid HA, given K_a.
3. For a weak base B in water, calculate the concentrations of OH^-, H^+, BH^+, and B, given K_b.
4. Determine the percent ionization of a weak acids and weak bases.
5. Given K_a of a weak acid determine K_b of its conjugate base, and vice versa.

    Ionization Constants. Weak acids and weak bases are ionized only to a small extent in aqueous solution. In both cases the ionization reaction is reversible and equilibrium is established. For instance, hydrocyanic acid ionizes in water as follows:

$$HCN(aq) \rightleftharpoons H^+(aq) + CN^-(aq)$$

The equilibrium constant for the ionization of a weak acid is called the **acid ionization constant, K_a**. The expression for this constant is:

$$K_a = \frac{[H^+][CN^-]}{[HCN]} = 4.9 \times 10^{-10}$$

The value of K_a is experimentally determined, and Table 15.3 (text) and Table 15.3 below list K_a values for a number of weak acids.

    Table 15.3 Values of K_a for Some Monoprotic Acids

Formula	Name	K_a Value
HSO_4^-	Hydrogen sulfate ion	1.1×10^{-2}
HF	Hydrofluoric acid	7.1×10^{-4}
HNO_2	Nitrous acid	4.5×10^{-4}
HCOOH	Formic acid	1.7×10^{-4}
CH_3COOH	Acetic acid	1.8×10^{-5}
HSO_3^-	Hydrogen sulfite ion	6.3×10^{-8}
HCN	Hydrocyanic acid	4.9×10^{-10}

Methylamine is an example of a weak base. It ionizes in water as shown in the following equation:

$$CH_3NH_2(aq) + H_2O(l) \rightleftharpoons CH_3NH_3^+(aq) + OH^-(aq)$$

The equilibrium constant for the ionization of a weak base is called the **base ionization constant, K_b.**

$$K_b = \frac{[CH_3NH_3^+][OH^-]}{[CH_3NH_2]} = 4.4 \times 10^{-4}$$

Table 15.4 in the text and Table 15.4 below list ionization constant values (K_b) for a number of weak bases. Note that the concentration of water is not shown in the K_b expression. The concentration of water is not measurably affected by the reaction and is essentially a constant.

Table 15.4 Values of K_b for Several Weak Bases

Formula	Name	K_b Value
CH_3NH_2	Methylamine	4.4×10^{-4}
CN^-	Cyanide ion	2.0×10^{-5}
NH_3	Ammonia	1.8×10^{-5}
C_5H_5N	Pyridine	1.7×10^{-9}

Acid and Base Strength. The K_a value is the quantitative measure of acid strength. The larger the K_a, the greater the extent of the ionization reaction and the stronger the acid. Therefore, acetic acid with $K_a = 1.8 \times 10^{-5}$ is a stronger acid than hydrocyanic acid with $K_a = 4.9 \times 10^{-10}$. The acids listed in Table 15.3 above are arranged in order of decreasing acid strength as you read down in the table. The K_a values for strong acids are never tabulated. Strong acids are essentially 100 percent ionized. It will help you to be able to recognize 5 or 6 strong acids by their names and formulas, so that you won't confuse them with weak acids.

Percent Ionization. The strength of an acid is also reflected in its percent ionization. The percent ionization is defined as follows:

$$\text{percent ionization} = \frac{\text{concentration of acid ionized}}{\text{initial concentration of acid}} \times 100\%$$

For a weak acid HA, the concentration of the acid that is ionized is equal to the concentration of H^+ ions (or that of A^- ions) at equilibrium. $HA(aq) \rightleftharpoons H^+(aq) + A^-(aq)$

$$\text{percent ionization} = \frac{[H^+]}{[HA]_0} \times 100\%$$

Be sure to work through Examples 15.10 and 15.11 because they illustrate the problem-solving techniques used to calculate $[H^+]$, $[OH^-]$, pH, and percent ionization of weak acids and bases.

The K_b of a Base is Related to the K_a of Its Conjugate Acid. The K_a and K_b for *any conjugate acid-base pair* are always related by the equation:

$$K_a K_b = K_w$$

The K_b for a conjugate base of a weak acid is found by rearranging:

$$K_b = \frac{K_w}{K_a}$$

From this equation you can see that K_a decreases, K_b increases, as *the weaker the acid, the stronger its conjugate base* – a relationship we have noted before.

Fluoride ion is a base. What is its K_b value?

$$F^-(aq) + H_2O(l) \rightleftharpoons HF(aq) + OH^-(aq) \qquad K_b = ?$$

The base ionization constant K_b for F^- is related to the K_a for the conjugate acid HF by:

$$K_b = \frac{K_w}{K_a}$$

Substituting K_a for HF gives

$$K_b = \frac{K_w}{7.1 \times 10^{-4}} = 1.4 \times 10^{-11}$$

EXAMPLE 15.10 Percent Ionization of a Weak Acid

a. Calculate the concentrations of H^+, F^-, and HF in a 0.31 M HF (hydrofluoric acid) solution.
b. What is the percent ionization?

•Method of Solution

a. First write the ionization reaction and the acid ionization constant expression for hydrofluoric acid. The value for K_a can be found in Table 15.3.

$$HF(aq) \rightleftharpoons H^+(aq) + F^-(aq)$$

$$K_a = \frac{[H^+][F^-]}{[HF]} = 7.1 \times 10^{-4}$$

The equilibrium concentrations of all species must be written in terms of a single unknown. Let x equal the moles of HF ionized per liter of solution. This means that x is the molarity of H^+ ion at equilibrium because one H^+ ion is formed for each HF molecule that ionizes. Also $[H^+] = [F^-] = x$ because one F^- ion is formed for each hydrogen ion formed. It is helpful to tabulate the initial and equilibrium concentrations of all species. The initial concentrations are those that existed before any ionization of HF occurred. The *change in concentration* is x. When x moles of HF have ionized per liter, x moles of H^+ and x moles of F^- are formed per liter. The equilibrium concentrations are found by adding algebraically the change in concentration to the initial concentration.

Concentration	HF $\rightleftharpoons$	H^+ +	F^-
Initial (M)	0.31	0	0
Change (M)	– x	+x	+x
Equilibrium (M)	(0.31 – x)	x	x

•Calculation

The value of x can be determined by substituting the equilibrium concentrations into the K_a expression.

$$K_a = \frac{(x)(x)}{(0.31 - x)} = 7.1 \times 10^{-4}$$

The small magnitude of K_a indicates that very little hydrofluoric acid actually ionizes. After all, it is a weak acid. Therefore, the value of x is much less than 0.31, and 0.31 − x _ 0.31 M. This approximation simplifies the equation, and allows us to avoid solving a quadratic equation.

$$K_a = \frac{x^2}{0.31} = 7.1 \times 10^{-4}$$

Solving for x:

$$x = \sqrt{(0.31)(7.1 \times 10^{-4})} = 1.5 \times 10^{-2} \text{ M}$$

Was our approximation valid? As **a rule**, *if x is equal to or less than 5% of the initial concentration [HF]$_0$,* then neglecting x in the calculation does not introduce an unacceptable amount of error into the answer.

$$[HF] = [HF]_0 - x \qquad \text{and} \qquad [HF] \cong [HF]_0$$

To test the validity of the approximation, divide x by the initial concentration of hydrofluoric acid $[HF]_0$;

$$\frac{x}{[HF]_0} \times 100\% = \frac{1.5 \times 10^{-2} \text{ M}}{0.31 \text{ M}} \times 100\% = 4.8\%$$

Since x *is less than 5 percent* of the initial amount of HF, our approximate solution is close enough to the accurate result to be valid. Knowing x we can utilize the bottom row in the table to answer part (a) of the problem.

$$[H^+] = [F^-] = 1.5 \times 10^{-2} \text{ M}$$
$$[HF] = 0.31 \text{ M} - 0.015 \text{ M} = 0.30 \text{ M}$$

b.   The percent ionization was already calculated above.

$$\% \text{ ionization} = \frac{x}{[HF]_0} \times 100\% = 4.8\%$$

•Comment

In this example we have followed **three basic steps** common to all problems of this type.

1.   Write the acid dissociation equilibrium and express the equilibrium concentration of all species in terms of an unknown (x), where x equals the change in the concentration of HA that is required to reach equilibrium.
2.   Substitute these concentrations into the K_a expression and look up the value of K_a in a table.
3.   Solve for x, and calculate all equilibrium concentrations, the pH, and percent ionization.

EXAMPLE 15.11   pH of a Weak Base Solution

What is the pH of a 0.010 M C_5H_5N (pyridine) solution?

•Method of Solution

Pyridine is a weak base. Write the equation for the reaction of pyridine in water and look up its K_b value in Table 15.4.

$$C_5H_5N + H_2O \rightleftharpoons C_5H_5NH^+ + OH^- \qquad K_b = 1.7 \times 10^{-9}$$

Let x equal the moles of C_5H_5N ionized per L to reach equilibrium. Note, when x mol/L of C_5H_5N are ionized, x mol/L of OH^- and x mol/L of $C_5H_5N^+$ are formed. These changes are listed in the following table:

Concentration	H_2O	+	C_5H_5N	$\rightleftharpoons$	$C_5H_5NH^+$	+	OH^-
Initial (M)			0.010		0		0
Change (M)			$-x$		$+x$		$+x$
Equilibrium (M)			$(0.010 - x)$		x		x

•Calculation

Substitute the equilibrium concentrations into the base constant expression.

$$K_b = \frac{[C_5H_5NH^+][OH^-]}{[C_5H_5N]} = \frac{(x)(x)}{(0.010 - x)} = 1.7 \times 10^{-9}$$

Make the approximation $[C_5H_5N] = 0.010 - x \cong 0.010$ M. Substituting

$$\frac{x^2}{0.010} = 1.7 \times 10^{-9}$$

$$x = \sqrt{(0.010)(1.7 \times 10^{-9})}$$

$$x = 4.1 \times 10^{-6} \text{ M}$$

Testing the assumption that the approximation is reasonable

$$\frac{x}{[C_5H_5N]_0} \times 100\% = \frac{4.1 \times 10^{-6} \text{ M}}{0.010 \text{ M}} \times 100\% = 0.041\%$$

Since x is less than 5% of the initial concentration of pyridine, then the assumption is valid. At equilibrium:

$$[OH^-] = x = 4.1 \times 10^{-6} \text{ M}$$

To find the pH, first calculate the pOH, and then use the relationship pH + pOH = 14.

pOH = $-\log(4.1 \times 10^{-6})$ = 5.38
Since, pH = 14.00 - pOH
Then, pH = 14.00 − 5.38
pH = 8.62

EXERCISES

16. What is the percent ionization of a weak acid in a 0.020 M HA solution given that its K_a is 2.3×10^{-5}?

17. What is the pH of a 0.050 M C_6H_5COOH (benzoic acid) solution?

18. HA is a weak acid. If a 0.020 M HA solution is 2.5% dissociated, what is the K_a value of the weak acid?

19. A 0.100 M solution of chloroacetic acid, $ClCH_2COOH$, has a pH of 1.95. Calculate the K_a for chloroacetic acid.

20. A 0.012 M solution of an unknown base has a pH of 10.10.
    a. What are the hydroxide ion and hydronium ion concentrations in the solution?
    b. Is the base a weak base or a strong base?

21. Which of the following is the strongest acid?
    HF, CH_3COOH, or C_6H_5COOH

22. Which of the following solutions has the lowest pH?
    0.10 M HNO_2, 0.10 M CH_3COOH, or 0.10 M HCN

23. Determine accurately the percent ionization of formic acid, HCOOH, in a 0.0050 M solution.

DIPROTIC AND POLYPROTIC ACIDS (15.8)

STUDY OBJECTIVES

1. Write ionic equations for the various stages of ionization of diprotic and polyprotic acids, and predict the relative extent of each stage.
2. Calculate the concentration of the related species present in a solution of a given polyprotic acid.

Stages of Ionization. The acids listed in Table 15.3 of the text are monoprotic acids; that is, they produce one proton per acid molecule. Acetic acid (CH_3COOH) contains four hydrogen atoms per molecule, but it is still monoprotic because only one of them is ionized in solution. The H atoms bonded to the C atom do not ionize because the electronegativity difference between C and H is not great enough to produce sufficient positive charge on the hydrogen atoms.

$$
\begin{array}{c}
\quad \text{H} \quad \text{O} \\
\quad | \quad \| \quad \delta- \;\; \delta+ \\
\text{H—C—C—O—H} \quad \longleftarrow \text{ionizable H atom} \\
\quad | \\
\text{nonionizable} \longrightarrow \quad \text{H} \\
\quad \text{H atoms}
\end{array}
$$

Acids that contain two ionizable H atoms are called **diprotic acids**. Those with three or more ionizable H atoms are called **polyprotic acids**. For example, sulfuric acid (H_2SO_4) is a diprotic acid, and phosphoric acid (H_3PO_4) with its three ionizable H atoms is a polyprotic acid. The ionization reactions occur stepwise; the conjugate base in the first step becomes the acid in the second step of ionization. For oxalic acid, $H_2C_2O_4$:

$$H_2C_2O_4 \rightleftharpoons H^+ + HC_2O_4^- \qquad K_{a1} = 6.5 \times 10^{-2}$$

$$HC_2O_4^- \rightleftharpoons H^+ + C_2O_4^{2-} \qquad K_{a2} = 6.1 \times 10^{-5}$$

This example is typical in that $K_{a1} \gg K_{a2}$. It is always more difficult to remove an H^+ ion from a negative ion such as $HC_2O_4^-$ than from a neutral molecule such as $H_2C_2O_4$. Table 15.5 of the text lists acid ionization constants for several polyprotic acids.

Here are some important observations concerning polyprotic acids:
1. The major product species in solution come from the first ionization step.
2. Since all reactants and products (ions and nonionized acid molecules) exist in the same solution, they are in equilibrium with each other. At equilibrium, each species can have only one concentration.
3. The concentration of all species in solution must be consistent with both K_{a1} and K_{a2}.

EXAMPLE 15.12 Solution Containing a Diprotic Acid

Calculate the concentrations of H_2A, HA^-, A^{2-}, and H^+ in a 1.0 M H_2A solution.

•Method of Solution

First write the equations for the stepwise ionization of H_2A, given the K_a values shown.

$$H_2A \rightleftharpoons H^+ + HA^- \qquad\qquad K_{a1} = 1.3 \times 10^{-2}$$

$$HA^- \rightleftharpoons H^+ + A^{2-} \qquad\qquad K_{a2} = 6.3 \times 10^{-8}$$

•Calculation

Since $K_{a1} \gg K_{a2}$, we assume that essentially all the H^+ comes from step 1. Let x be the concentration of H_2A ionized to reach equilibrium. We solve for x in the usual way.

Concentration	H_2A	$\rightleftharpoons$	H^+	+	HA^-
Initial (M)	1.00		0		0
Change (M)	−x		x		x
Equilibrium (M)	(1.00 − x)		x		x

$$K_{a1} = \frac{[H^+][HA^-]}{[H_2A]} = \frac{(x)(x)}{1.0 - x} = 1.3 \times 10^{-2}$$

$$K_{a1} = \frac{x^2}{1.0 - x} = 1.3 \times 10^{-2}$$

For this acid K_{a1} is quite large, and the percent ionization will be greater than 5%. Rearranging the preceding equation into quadratic form:

$$ax^2 + bx + c = 0$$
$$x^2 + (1.3 \times 10^{-2})x - 1.3 \times 10^{-2} = 0$$

where

$$a = 1, b = 1.3 \times 10^{-2}, \text{ and } c = -1.3 \times 10^{-2}.$$

Substituting into the quadratic formula, and solving for x.

$$x = \frac{-b \pm \sqrt{(b^2 - 4ac)}}{2a}$$

$$x = \frac{-1.3 \times 10^{-2} \pm \sqrt{(1.3 \times 10^{-2})^2 - (4)(1)(-1.3 \times 10^{-2})}}{(2)(1)}$$

$$x = 0.11 \text{ M}$$

Therefore, at equilibrium

$$[H^+] = [HA^-] = 0.11 \text{ M}$$

and

$$[H_2A] = 1.00 - 0.11 = 0.89 \text{ M}$$

which corresponds to 11% ionization.

To determine the concentration of A^{2-}, we must consider the second stage of ionization:

$$HA^- \rightleftharpoons H^+ + A^{2-} \qquad K_{a2} = 6.3 \times 10^{-8}$$

Let y be the equilibrium concentration of A^{2-}, and note that x >> y.

Concentration	HA^- $\rightleftharpoons$	H^+	+	A^{2-}
Initial (M)	0.11	0.11		0
Change (M)	$-y$	$+y$		$+y$
Equilibrium (M)	$(0.11 - y)$	$(0.11 + y)$		$+y$

$$K_{a2} = \frac{[H^+][A^{2-}]}{[HA^-]} = \frac{(0.11+y)(y)}{(0.11-y)} = 6.3 \times 10^{-8}$$

Since K_{a2} is very small, as for a typical weak acid, we can assume that less than 5% of the HA^- dissociates. Therefore,

$$0.11 \text{ M} \pm y \cong 0.11 \text{ M}$$

With this assumption

$$K_{a2} = \frac{(0.11)(y)}{(0.11)}$$

and

$$y = K_{a2} = 6.3 \times 10^{-8} \text{ M}$$

The concentration of A^{2-} is equal to K_{a2}.

$$[A^{2-}] = y = 6.3 \times 10^{-8} \text{ M}$$

Checking the validity of the assumption that

$$0.11 \text{ M} - (6.3 \times 10^{-8} \text{ M}) \cong 0.11 \text{ M}$$

The percent ionization of step 2 is

$$\text{percent ionization} = \frac{y}{[HA^-]} \times 100\%$$

$$\frac{6.3 \times 10^{-8} \text{ M}}{0.11 \text{ M}} \times 100\% = 5.7 \times 10^{-5}\%$$

•Comment

Because $K_{a1} >> K_{a2}$ the species produced in step 1 will be present in much greater concentration than those from step 2. This can be seen by comparing the percent ionization of each step (11% vs. 0.000057%). Also, since H_2A is a weak acid, the concentration of undissociated H_2A should be greater than that of all ionic species. Therefore, we expect the following order of concentrations:

$$[H_2A] > [H^+] \cong [HA^-] > [A^{2-}]$$

EXERCISES

24. Write chemical equations for the three stages of ionization of phosphoric acid, H_3PO_4.

25. Calculate the pH of a 0.50 M solution of phosphoric acid, H_3PO_4.

MOLECULAR STRUCTURE AND STRENGTHS OF ACIDS (15.9)

STUDY OBJECTIVES

1. Predict the relative strengths of binary acids.
2. Predict the relative strengths of oxyacids.

Binary Acids. The acid strengths of nonmetal hydrides are related to two features of these molecules. These are the *polarity* of the X—H bond (X stands for a nonmetal atom), and *the strength of the H—X bond.* The polarity of a X—H bond increases with the electronegativity of the nonmetal atom. The greater the electronegativity of the nonmetal atom, the more it withdraws electrons away from the hydrogen atom. This facilitates the transfer of H^+ as a proton. Thus *acid strength increases as the electronegativity of the nonmetal atom increases.*

The strength of the X—H bond decreases as the atomic radius of X increases. A hydrogen ion is more easily broken away from a larger atom than a smaller one. This happens because the electron cloud in a large atom is more diffuse. The smaller the nonmetal atom, the more dense the electron cloud. This greater electron density results in a greater attraction for the proton. Therefore, *acid strength increases as the atomic radius of the nonmetal atom increases.*

When comparing acid strengths of hydrides in a *group* of the periodic table versus those within a *period*, one must consider both the electronegativity of the nonmetal atom and the bond energy. Within a series of hydrides of elements in a *group* the acid strength increases with increasing atomic radius of the nonmetal. For example, note the following hydrides of the Group 6A elements.

$H_2S < H_2Se < H_2Te$
increasing acid strength →
increasing nonmetal radius →
← increasing electronegativity of X

In this series, the electronegativity of X varies only slightly, and so is outweighed by the more significant increases in atomic radius, as shown below.

	H_2S	H_2Se	H_2Te
Radius of X (pm)	104	117	137
Electronegativity of X	2.5	2.4	2.1

When reading across a row of the periodic table the situation is reversed. The importance of bond polarity in determining acid strength outweighs that of atomic radius. For instance, the hydrides of the elements phosphorus, sulfur, and chlorine show increasing acid strength as shown below. Incxreasing acid strength parallels the increasing electronegativity on the nonmetal atom.

$PH_3 < H_2S < HCl$
increasing acid strength $\rightarrow$
increasing electronegativity of X $\rightarrow$
$\leftarrow$ increasing nonmetal radius

	PH_3	H_2S	HCl
Radius of X (pm)	110	104	99
Electronegativity of X	2.1	2.5	3.0

The smaller decrease in atomic radius is insignificant in comparison to the larger change in electronegativity. The trend in acid strength results from significant increases in bond polarity when going across the row.

Oxoacids. For oxoacids with the same structure, but with different central atoms whose elements are in the same group, acid strength increases with increasing electronegativity of the central atom. Thus, acid strength increases in the following series where the central atom is a halogen element:

$HOI < HOBr < HOCl$

In each molecule, the O—H bond strength is approximately the same. In this series the ability of the halogen atom to withdraw electron density from the O—H bond increases with increasing electronegativity. As the O—H bond in a series of acids becomes more polar, the acid strength increases.

For oxoacids that have the same central atom, but differing numbers of attached oxygen atoms, the acid strength increases with increasing oxidation number of the central atom. Therefore, HNO_3 is stronger than HNO_2, for example.

```
       ..
     : O :
       |
  ..       ..                 ..      ..  ..
  O = N — O — H           O = N — O — H
  ..       ..                 ..      ..
```

The oxygen atoms draw electrons away from the nitrogen atom, making it more positive. The more positive the N atom, the more effective it is in withdrawing electrons from the O—H bond. This increases the polarity of the O—H bond.

EXAMPLE 15.13   Relative Acid Strength

Which member of each of the following pairs is the stronger acid?
a.   HCl or HBr
b.   HCl or H_2S

•Method of Solution

a.   These acids are nonmetal hydrides of elements within Group 6A of the periodic table. Variations in atomic radius are more important within a group than electronegativity variations. The stronger acid is the one with the greater radius of its nonmetal atom. HBr is the stronger acid.

b.   These acids are nonmetal hydrides of elements within the third row of the period table. The electronegativity changes more significantly going across a row than does the atomic radius. HCl with its more polar bond is the stronger acid.

EXAMPLE 15.14   Relative Acid Strength

Which member of each of the following pairs is the stronger acid?
a. $HClO_3$ or $HBrO_3$
b. H_3PO_3 or H_3PO_4

•Method of Solution

a.   These two oxoacids have the same structure:

$$O \qquad\qquad O$$
$$| \qquad\qquad\qquad |$$
$$O=Cl-O-H \qquad O=Br-O-H$$

Within this group the strengths of the acids increase with increasing electronegativity of the central atom. The electronegativity of Cl is greater than that of Br. $HClO_3$ is stronger.
b. H_3PO_3 and H_3PO_4 are oxoacids of the same element. The acid strength increases as the number of O atoms increases. H_3PO_4 is the stronger acid because of the greater number of oxygen atoms bonded to the phosphorus atom. In H_3PO_4, phosphorus has a higher oxidation number than in H_3PO_3.

EXERCISE

26.   Which member of each pair is the stronger acid?
   a. HNO_2 or HNO_3
   b. CH_4 or SiH_4
   c. HOBr or HOI
   d. CH_4 or NH_3

HYDROLYSIS AND ACID-BASE PROPERTIES OF SALTS (15.10)

STUDY OBJECTIVES

1.   Predict which ions of a given salt will hydrolyze, and predict whether a solution will be acidic, basic, or neutral.
2.   Calculate the pH of a given salt solution.

Hydrolysis. A **salt** is an ionic compound formed by the reaction of an acid and a base. In dealing with salts, it will be useful to recall that they are *strong* electrolytes and are *completely dissociated* in water. The term **salt hydrolysis** refers to the reaction of a cation or an anion, or both with water which splits the water molecule into two parts. Salt hydrolysis usually affects the pH of a solution.

   A cation causes hydrolysis by bonding to the hydroxide part of the water molecule, causing a proton to be released. Cations that undergo hydrolysis produce acidic solutions.

$$M^+ + HOH \rightleftharpoons H^+ + MOH$$

Certain metal ions such as Al^{3+}, Cr^{3+}, Fe^{3+}, Be^{2+} and Bi^{3+} are good examples of acidic cations. NH_4^+ is also an acidic cation.

Anions that undergo hydrolysis produce basic solutions. An anion can cause hydrolysis by acting as a Brønsted base. A basic anion can accept a proton from water and releases a OH^- ion.

$$A^- + HOH \rightleftharpoons HA + OH^-$$

Anion Hydrolysis. Anions that cause hydrolysis are Brønsted bases because they can accept a proton from water.

$$A^- + H_2O \rightleftharpoons HA + OH^-$$

Not all anions cause hydrolysis. *You must be able to recognize those that do and those that don't.* Quite simply, those anions with the ability to undergo hydrolysis are conjugate bases of weak acids. The *conjugate base of any weak acid* is a weak base and will raise the pH of an aqueous solution. The extent of hydrolysis is proportional to the base ionization constant, K_b.

Keep in mind that the conjugate base of a *strong acid* has no tendency to accept protons from H_2O. Anions such as Cl^-, NO_3^-, and ClO_4^- do not cause hydrolysis. They do not have acid-base properties.

The Conjugate Bases of Weak Acids. Fluoride ion is the conjugate base of hydrofluoric acid (HF) a weak acid. As a weak base it can accept a proton from water. The resulting OH^- ions make the solution basic.

$$F^- + H_2O \rightleftharpoons HF + OH^- \qquad K_b = \frac{[HF][OH^-]}{[F^-]}$$

The force that makes this reaction possible is the tendency of the weak acid HF to remain undissociated.

Three realms of base strength can be distinguished:

1. The conjugate bases of strong acids ($K_a > 1.0$) are too weak to cause measurable hydrolysis. Thus Cl^-, Br^-, and NO_3^- ions do not hydrolyze in aqueous solutions.

    $$Cl^- + H_2O \rightarrow HCl + OH^- \quad \text{(no reaction)}$$

2. The conjugate bases of acids with a K_a between 10^{-1} and 10^{-5} are weak bases. Thus F^-, NO_2^-, $HCOO^-$(formate), and CH_3COO^- (acetate) ions hydrolyze to a small extent and produce low concentrations of OH^- ions in solution.

    $$NO_2^- + H_2O \rightleftharpoons HNO_2 + OH^-$$

3. The conjugate bases of very weak acids ($K_a < 10^{-5}$) are moderately strong bases. Thus solutions containing CN^-, HS^-, HPO_4^{2-}, and CO_3^{2-} ions can be quite basic.

Cation Hydrolysis. It is important to know which metal ions cause hydrolysis and which do not. Those that cause hydrolysis are *small, highly charged* metal ions such as Al^{3+}, Cr^{3+}, Fe^{3+}, Bi^{3+}, and Be^{2+}. These can split H_2O molecules, yielding acidic solutions. In the case of the hydrated Al^{3+} ion the reaction is

$$Al(H_2O)_6^{3+}(aq) + H_2O(l) \rightleftharpoons Al(OH)(H_2O)_5^{2+}(aq) + H_3O^+(aq)$$

The singly charged metal ions associated with strong bases do not cause hydrolysis. The commonly encountered ions Na^+, K^+, Ca^{2+}, and Mg^{2+} do not have a great enough positive charge to attract an OH^- from a H_2O molecule and bond to it. Recall that NaOH and KOH are strong bases. That is, they ionize completely in solution. These ions, K^+ and OH^- for instance, have no tendency to stay together in solution.

$$KOH(aq) \rightarrow K^+(aq) + OH^-(aq)$$

The hydrolysis reaction that would bond an OH^- to a K^+ ion does not occur!

$$K^+(aq) + H_2O(l) \rightarrow KOH(aq) + H^+(aq) \quad \text{(no reaction)}$$

Another acidic cation that you will need to be familiar with is the ammonium ion (NH_4^+). Solutions of ammonium ion are acidic because NH_4^+ is a Brønsted acid. Recall that NH_4^+ is the conjugate acid of the weak base NH_3.

$$NH_4^+ + H_2O \rightleftharpoons NH_3 + H_3O^+$$
or simply

$$NH_4^+ \rightleftharpoons NH_3 + H^+ \quad K_a = 5.6 \times 10^{-10}$$

Table 15.5 lists the various anions and cations we have discussed according to their acid-base properties. You should review the material above, and in the textbook to see why each ion has the property described.

Table 15.5 Acid-Base Properties of Ions

	Acidic	Basic	Neutral
Cations	Al^{3+}, Cr^{3+}, Fe^{3+} Bi^{3+}, Be^{2+}, NH_4^+		Na^+, K^+ Mg^{2+}, Ca^{2+}
Anions	HSO_4^-	CH_3COO^-, F^-, NO_2^- all conjugate bases of weak acids	Cl^-, Br^-, NO_3^- all conjugate bases of strong acids

Salts That Produce Neutral Solutions. When a salt is dissolved in water the solution will acquire a pH consistent with the acid-base properties of the cation and anion making up the salt. In general, *salts containing alkali metal ions or alkaline earth metal ions (except Be^{2+}) and the conjugate bases of strong acids do not undergo hydrolysis;* their solutions are neutral. Consequently, a solution of KCl, for example, is neutral. A K^+ ion does not accept an OH^- ion from H_2O, and a Cl^- ion does not accept a H^+ ion from H_2O.

Salts That Produce Basic Solutions. Salts producing a basic pH are salts of weak acids and strong bases. These salts have a neutral cation and a basic anion. For example, NaF dissociates into Na^+ and F^- ions. The Na^+ ion has no acidic or basic properties, but the fluoride ion is a conjugate base of a weak acid and therefore is a basic anion.

NaF a salt of the NaOH, a strong base, and HF, a weak acid.

$$F^- + H_2O \rightleftharpoons HF + OH^-$$

Salts That Produce Acidic Solutions. Salts producing acidic solutions are salts of strong acids and weak bases. These salts contain cations that undergo hydrolysis, and anions that do not (the neutral anions). Such a salt has a cation that is a conjugate acid of a weak base. Thus solutions of these salts are acidic. The salt NH_4Cl is a strong electrolyte and dissociates into ammonium ions and chloride ions when dissolved in water. The NH_4^+ ion is a conjugate acid of the weak base NH_3 and releases H^+ ion into solution.

$$NH_4Cl(aq) \rightarrow NH_4^+(aq) + Cl^-(aq) \quad \text{(strong electrolyte)}$$

$$NH_4^+ \rightleftharpoons NH_3 + H^+ \qquad K_a = 5.6 \times 10^{-10}$$

The Cl^- ion is neutral having no acidic or basic properties. NH_4Cl is a salt of a strong acid (HCl) and a weak base (NH_3).

Solutions of certain salts such as $AlCl_3$ and $FeCl_3$ are also acidic. These contain small, highly charged cations, and anions that are conjugate bases of strong acids.

Salts in Which Both the Anion and Cation Hydrolyze. Salts derived from a weak acid and a weak base contain cations and anions that will hydrolyze. Whether a solution containing such a salt is acidic, basic, or neutral depends on the relative strengths of the acidic and basic anions. For example, a solution of ammonium fluoride is *acidic* because NH_4^+ is a stronger acid than F^- is a base.

$$NH_4F(aq) \rightarrow NH_4^+(aq) + F^-(aq) \quad \text{(strong electrolyte)}$$

$$NH_4^+ \rightleftharpoons NH_3 + H^+ \qquad\qquad K_a = 5.6 \times 10^{-10}$$

$$F^- + H_2O \rightleftharpoons HF + OH^- \qquad\qquad K_b = 1.4 \times 10^{-11}$$

NH_4F is a salt of a weak acid and a weak base. On the other hand, another such salt (NH_4CN) is a *basic* salt because CN^- ion is a stronger base than NH_4^+ is an acid.

The pH of Salt Solutions. The pH of a salt solution is calculated just as it is for any weak acid or base. See Example 15.17 for an illustration of this important calculation.

EXAMPLE 15.15   Hydrolysis of Ions

Write net ionic equations to show which of the following ions hydrolyze in aqueous solution:

a. NO_3^- b. NO_2^- c. NH_4^+

•Method of Solution

Strong acids have extremely weak conjugate bases which have no affinity for H^+ ions. They have neutral properties. Weak acids have stronger conjugate bases with significant affinity for an H^+ ion. The conjugate bases of weak acids have measurable basic properties.
a.   Nitrate ion is the conjugate base of a strong acid, HNO_3. Therefore NO_3^- has no acidic or basic properties and has no tendency to react with water to form HNO_3.

$$NO_3^- + H_2O \rightarrow HNO_3 + OH^- \quad \text{(no reaction)}$$

b.   Nitrite ion is the conjugate base of a weak acid, HNO_2. Therefore NO_2^- is a stronger base than NO_3^- and will accept a proton from water to form HNO_2.

$$NO_2^- + H_2O \rightleftharpoons HNO_2 + OH^-$$

c.   Ammonium ion is the conjugate acid of a weak base. Therefore, it is a weak acid:

$$NH_4^+ + H_2O \rightleftharpoons NH_3 + H_3O^+$$

EXAMPLE 15.16   Acid-Base Properties of Ions

Predict whether the following aqueous solutions will be acidic, basic, or neutral.
a.   KI
b. NH_4I
c. CH_3COOK

•Method of Solution

a.   What we must decide is whether either of the ions of the salt undergoes hydrolysis. KI is a salt of a strong acid (HI) and a strong base (KOH). Neither K^+ nor I^- have acidic or basic properties. Thus a solution containing KI remains neutral.

b. NH_4I is the salt of a strong acid (HI) and a weak base (NH_3). As explained in part (a) iodine does not hydrolyze. Since NH_4^+ ion is the conjugate acid of a weak base it will donate protons to water.

$$NH_4^+ + H_2O \rightleftharpoons NH_3 + H_3O^+$$

Therefore, a solution containing NH_4I will be acidic.

c. CH_3COOK is a salt of a weak acid and a strong base. K^+ ion will not hydrolyze, but CH_3COO^- (acetate ion) is the conjugate base of a weak acid, acetic acid. Thus acetate ions will accept protons from water and form acetic acid and OH^- ions.

$$CH_3COO^- + H_2O \rightleftharpoons CH_3COOH + OH^-$$

Solutions containing CH_3COOK will be basic.

EXAMPLE 15.17   pH of a Solution Containing a Salt

Calculate the pH of 0.25 M $C_6H_7O_6Na$ (sodium ascorbate) solution.

•Method of Solution

First decide what ions undergo hydrolysis. Then use the appropriate ionization constant to calculate the equilibrium concentrations. Sodium ascorbate is a strong electrolyte.

$$C_6H_7O_6Na(aq) \rightarrow C_6H_7O_6^-(aq) + Na^+(aq)$$

Ascorbate ion is a conjugate base of a weak acid (ascorbic acid, Table 15.3 text), and so it causes hydrolysis. Na^+ does not and is neutral.

$$C_6H_7O_6^-(aq) + H_2O \rightleftharpoons C_6H_8O_6 + OH^-$$

It is important to use the correct equilibrium constant value. Here we need the K_b for ascorbate ion. From Table 15.3 (text), we need the K_a for ascorbic acid, the conjugate acid of ascorbate ion.

$$K_b = \frac{K_w}{K_a} = \frac{1.0 \times 10^{-14}}{8.0 \times 10^{-5}} = 1.3 \times 10^{-10}$$

•Calculation

The OH^- ion concentration in a solution that is 0.25 M $C_6H_7O_6^-$ is obtained from the equilibrium constant expression

$$K_b = \frac{[C_6H_8O_6][OH^-]}{[C_6H_7O_6^-]}$$

Let x = moles of ascorbate ion that undergo hydrolysis per L of solution.

Concentration	$H_2O + C_6H_7O_6^- \rightleftharpoons C_6H_8O_6 + OH^-$		
Initial (M)	0.25	0	0
Change (M)	−x	+x	+x
Equilibrium (M)	(0.25 − x)	x	x

Substitute into the equilibrium constant expression.

$$K_b = \frac{(x)(x)}{0.25 - x} = 1.3 \times 10^{-10}$$

Apply the approximation that $[C_6H_7O_6^-] = 0.25 - x \cong 0.25$ M

$$\frac{x^2}{0.25} = 1.3 \times 10^{-10}$$

$$x = \sqrt{(0.25)(1.3 \times 10^{-10})} = 5.7 \times 10^{-6} \text{ M} = [OH^-]$$

$$pOH = -\log [OH^-] = 5.24$$

The pH is $14.0 - 5.24 = 8.76$.

EXERCISES

27. Given that K_a for formic acid (HCOOH) is 1.7×10^{-4}, find the K_b value for formate ion, HCOO$^-$.

28. Calculate K_a for ammonium ion NH_4^+ given the K_b value for ammonia in Table 15.4.

29. Predict whether the pH is > 7, < 7, or $= 7$ for aqueous solutions containing the following salts.
    a. K_2CO_3 b. $BaCl_2$ c. $Fe(NO_3)_3$

30. Calculate the pH of 0.021 M NaCN solution.

LEWIS ACIDS AND BASES (15.12)

STUDY OBJECTIVES

1. Define acids and bases according to the Lewis system.
2. Identify Lewis acids and bases in given chemical equations.

Electron-Pair Donors and Acceptors. A more general approach to acids and bases was proposed in 1923 by G. N. Lewis who also developed the use of electron dot structures. Lewis noticed that for a H^+ ion to be accepted by a base, the base must possess at least one unshared electron-pair. The electron pair forms the bond to the proton.

A **Lewis base** is a molecule or ion that can donate a pair of electrons to form a bond. A **Lewis acid** is a molecule or ion that can accept a pair of electrons and form a bond.

$$H^+ + :NH_3 \rightarrow H—NH_3^+ \quad (NH_4^+)$$

electron-pair H^+ accepts
donor   electron-pair

The formation of ammonium ion from H^+ and ammonia is a simple example. The ammonia molecule is the Lewis base because it donates a pair of electrons to H^+ to form a bond. By accepting the pair of electrons the proton is a Lewis acid. Note that a new covalent bond is formed by the donation of the electron pair. Recall when both electrons of the shared pair are donated by the same atom the bond is called a *coordinate covalent bond*. An acid-base reaction in the Lewis system is donation of an electron-pair to form a new covalent bond between the acid and the base.

Metal cations can be Lewis acids because they have lost their valence electrons and so have at least one vacant orbital in which to accept an electron-pair from a base. Consider the beryllium ion which can form four coordinate covalent bonds.

$$Be^{2+}(aq) + 4H_2O(l) \rightarrow [Be(H_2O)_4]^{2+}$$

Water molecules with unshared electron-pairs on their oxygen atoms are Lewis bases and beryllium is a Lewis acid because it accepts electron-pairs to form $Be-OH_2$ bonds. Beryllium can form four $Be-OH_2$ bonds.

EXAMPLE 15.18   Lewis Acids and Bases

Identify the Lewis acids and bases in each of the following reactions.

a. $Ag^+(aq) + Cl^-(aq) \rightarrow AgCl(s)$

b. $Hg^{2+}(aq) + 4I^-(aq) \rightarrow HgI_4^{2-}(aq)$

c. $BF_3(g) + NF_3(g) \rightarrow F_3N-BF_3(s)$

d. $SO_2(g) + H_2O(l) \rightarrow H_2SO_3(aq)$

•**Method of Solution**

Lewis acids are species that accept electron-pairs. In each of the above reactions the first reactant species is the Lewis acid, and the second reactant is the Lewis base (electron-pair donor). The Lewis acids in (a) and (b) are capable of accepting electron-pairs because they are positive ions and have previously lost electrons. In (c) the boron atom has an incomplete octet. And in (d) the S atom in SO_2 is "oxidized," that is, its valence electrons are shifted toward the more electronegative oxygen atoms and away from the S atom. In all four reactions new coordinate covalent bonds are formed.

EXERCISES

31. Why are nitrogen compounds and ions such as NH_3, CH_3NH_2, and NH_2^- good examples of Lewis bases?

32. Why can H^+ ion, Mg^{2+} ion, and Al^{3+} ion act as Lewis acids?

CONCEPTUAL QUESTIONS

1.   What is the strongest acid and the strongest base that can exist in water?

2.   Both oxide ion (O^{2-}) and amide ion (NH_2^-) are stronger bases than hydroxide ion and so do not exist in water. Write equations for the reactions of these ions with water.

3.   Normally we do not need to know the K_a value for a strong acid. Why?

4. Which of the following statements, if any, are true with regard to a 0.10 M solution of a weak acid HA?

   a. $[HA] > [H^+]$
   b. the pH = 1.0
   c. $[H^+] < [A^-]$
   d. the pH > 1.0
   e. $[OH^-] = [H^+]$

5. Why is a solution of NaF basic, whereas a solution of NaCl is neutral?

PRACTICE TEST

1. Identify the conjugate bases of the following acids:
   a. CH_3COOH b. H_2S c. HSO_3^- d. $HClO$

2. Identify the conjugate acid-base pairs in each of the following reactions:
   a. $NH_3 + H_2CO_3 \rightleftharpoons NH_4^+ + HCO_3^-$
   b. $CO_3^{2-} + HSO_3^- \rightleftharpoons HCO_3^- + SO_3^{2-}$
   c. $HCl + HSO_3^- \rightleftharpoons Cl^- + H_2SO_3$
   d. $HCO_3^- + HPO_4^{2-} \rightleftharpoons H_2CO_3 + PO_4^{3-}$

3. What is the $H^+(aq)$ concentration in each of the following?
   a. Neutral water
   b. 1 M HCl
   c. 0.01 M HNO_3

4. What is the $OH^-(aq)$ concentration in each of the following?
   a. Neutral water
   b. 0.01 M NaOH
   c. 0.01 M $Sr(OH)_2$

5. What is the $H^+(aq)$ concentration in 1.0×10^{-3} M NaOH?

6. What is the $OH^-(aq)$ concentration in 2.0 M HCl?

7. What is the pH of the following solutions?
   a. 0.015 M HCl
   b. 0.015 M NaOH

8. What is the pOH of the following solutions?
   a. 0.30 M $Ca(OH)_2$
   b. 2.0×10^{-3} M $HClO_4$

9. The pH of solution A is 2.0, and the pOH of solution B is 10.0. How many times greater is the $H^+(aq)$ concentration in solution A than in solution B?

10. The pH of a solution is 4.45. What is the H^+ ion concentration?

11. Classify each of the following as a weak or strong acid:
    a. H_2SO_4 b. CH_3COOH c. HClO
    d. HCl   e. HNO_2 f. $HClO_4$

12. Use Table 15.2 of the textbook to arrange the following anions in order of increasing base strength.
    a. HSO_4^- b. CH_3COO^- c. Cl^- d. NO_2^-

13. Which member of each of the following pairs is the stronger acid?
    a. H_2CO_3 or H_2SiO_3
    b. H_3AsO_4 or H_3AsO_3
    c. H_3PO_2 or H_3PO_3
    d. $HClO_2$ or $HClO_3$
    e. H_3PO_4 or H_3AsO_4

14. Which member of each of the following pairs is the stronger acid?
    a. HI or HBr
    b. H_2O or H_2S
    c. H_2Te or HI
    d. H_2Se or AsH_3

15. Choose the Lewis acids and bases that could be used to form the following compounds:
    a. HCl
    b. H_3O^+
    c. BCl_3NH_3
    d. $Al(OH)_4^-$
    e. MgF_2

16. What is the pH of a 0.0055 M HA (weak acid) solution that is 8.2% ionized?

17. Predict the direction in which the equilibrium will lie for the following reaction.
    $C_6H_5COO^- + HF \rightleftharpoons C_6H_5COOH + F^-$

18. Acid strength increases in the series: weakest — HCN, HF, to HSO_4^- —strongest. Which of the following is the strongest base?
    A. SO_4^{2-} B. F^- C. CN^- D. HSO_4^-

19. What is the H^+ ion concentration in a 0.82 M HOCl solution? Given $K_a = 3.2 \times 10^{-8}$

20. Calculate the percent ionization in a 1.0 M solution of nitrous acid, HNO_2.

21. What is the pH of a 0.090 M CH_3COOH (acetic acid) solution?

22. A 1.0 M HF solution is only 2.6 percent ionized. What is the K_a value for HF?

23. Calculate the concentration of H^+, OH^-, F^- ions and undissociated HF molecules in 0.010 M HF.

24. What is the pH of a 2.0 M NH_3 solution?

25. The first and second ionization constants of H_2CO_3 are 4.2×10^{-7} and 4.8×10^{-11}. Calculate the concentrations of H^+ ion, HCO_3^-, CO_3^{2-}, and undissociated H_2CO_3 in a 0.080 M H_2CO_3 solution.

26. A 0.1 M solution of Na_2CO_3 is:
    a. acidic   b. basic   c. neutral

27. Classify the solutions of the following salts as acidic, basic or neutral:
    a. $AlCl_3$ b. Na_2SO_4 c. $NaClO_4$
    d. KCN   e. NH_4Cl f. Na_2SO_3

28. Calculate the pH of a 0.30 M CH_3COONa solution. Hint. Use the K_a for CH_3COOH.

29. What is the H^+ ion concentration of a 0.010 M NaCN solution?

30. Which one of the following aqueous solutions (all 0.010 M) will have the highest pH value?
    a. CH_3COOH b. HCl c. $NaCl$ d. KF

ANSWERS

Exercises

1. CO_3^{2-}
2. a. NH_4^+ is the acid on the left and NH_3 is its conjugate base, H_2O is the base on the left and H_3O^+ is its conjugate acid.

   b. HNO_2 is the acid on the left and NO_2^- is its conjugate base, CN^- is the base on the left and HCN is its conjugate acid.
3. 1.0×10^{-11} M
4. 1.3×10^{-12} M
5. 5.1×10^{-3} M
6. 1.2×10^{-5} M
7. 100
8. 1×10^{-4} mol
9. $[OH^-] = 0.0033$ M and $[H^+] = 3.0 \times 10^{-12}$ M
10. 3.29
11. 13.10
12. 11.70
13. pH = 1.56
14. HCN
15. Left. $HNO_2(aq) + ClO_4^-(aq)$ predominate at equilibrium.
16. 3.4%
17. 2.74
18. 1.3×10^{-5}
19. 1.4×10^{-3}
20. a. $[H^+] = 7.9 \times 10^{-11}$ M $[OH^-] = 1.3 \times 10^{-4}$ M   b. weak
21. HF
22. HNO_2
23. 17%
24. 1. $H_3PO_4 \rightarrow H^+ + H_2PO_4^-$ 2. $H_2PO_4^- \rightarrow H^+ + HPO_4^{2-}$ 3. $HPO_4^{2-} \rightarrow H^+ + PO_4^{3-}$
25. 1.21
26. a. HNO_3 b. SiH_4 c. $HOBr$ d. NH_3
27. 5.9×10^{-11}
28. $K_a = 5.6 \times 10^{-10}$
29. a. pH > 7   b. pH = 7   c. pH < 7
30. 10.82
31. Because N atoms in most compounds have an unshared electron pair.
32. Because they have empty orbitals available to accept electron pairs.

Conceputal Questions

1. Any acid stronger than H_3O^+ ion reacts with water to produce H_3O^+. Hydronium ion is the strongest acid existing in water at equilibrium.

    $HA + H_2O \rightleftharpoons H_3O^+ + A^-$

    Any base stronger than OH^- ion accepts a proton from water yielding a OH^- ion. Hydroxide ion is the strongest base that exists in water at equilibrium.

    $B + H_2O \rightleftharpoons HB + OH^-$

2. $O^{2-} + H_2O \rightleftharpoons OH^- + OH^-$

    $NH_2^- + H_2O \rightleftharpoons NH_3 + OH^-$

3. We usually want to know the pH of an acid solution. For a weak acid we use K_a because each weak acid dissociates to a different extent. The K_a relates all the concentrations at equilibrium and allows us to find $[H^+]$. We know that all strong acids are 100% ionized. Therefore, we can calculate $[H^+]$ based on complete reaction.

4. a. and d.

5. NaF(aq) must contain a stronger base than NaCl(aq). Fluoride ion is a stronger base than chloride ion because F^- ion is the conjugate base of a weak acid while Cl^- ion is the conjugate base of a strong acid.

Practice Test

1. a. CH_3COO^- b. HS^- c. SO_3^{2-} d. ClO^-

2. a. $NH_3 + H_2CO_3 \rightleftharpoons NH_4^+ + HCO_3^-$
    base$_1$ acid$_2$ acid$_1$ base$_2$

    b. $CO_3^{2-} + HSO_3^- \rightleftharpoons HCO_3^- + SO_3^{2-}$
    base$_1$ acid$_2$ acid$_1$ base$_2$

    c. $HCl + HSO_3^- \rightleftharpoons Cl^- + H_2SO_3$
    acid$_1$ base$_2$ base$_1$ acid$_2$

    d. $HCO_3^- + HPO_4^{2-} \rightleftharpoons H_2CO_3 + PO_4^{3-}$
    base$_2$ acid$_1$ acid$_2$ base$_1$

3. a. 1×10^{-7} M   b. 1 M   c. 0.01 M

4. a. 1×10^{-7} M   b. 0.01 M   c. 0.02 M

5. $[H^+] = 1.0 \times 10^{-11}$ M

6. $[OH^-] = 5 \times 10^{-15}$ M

7. a. pH = 1.82   b. pH = 12.18

8. a. pOH = 0.22   b. pOH = 11.3

9. 100

10. $[H^+] = 3.55 \times 10^{-5}$ M

11. a. Strong   b. Weak   c. Weak   d. Strong   e. Weak   f. Strong

12. $Cl^- < HSO_4^- < NO_2^- < CH_3COO^-$

13. a. H_2CO_3 b. H_3AsO_4 c. H_3PO_3 d. $HClO_3$ e. H_3PO_4

14. a. HI   b. H_2S c. HI   d. H_2Se

15. a. $H^+ + Cl^- \rightarrow HCl$ b. $H^+ + H_2O \rightarrow H_3O^+$ c. $BCl_3 + NH_3 \rightarrow BCl_3NH_3$

    d. $Al^{3+} + 4OH^- \rightarrow Al(OH)_4^-$ e. $Mg^{2+} + 2F^- \rightarrow MgF_2$

16. 3.34

17. To the right

18. C.

19. $[H^+] = 1.6 \times 10^{-4}$ M

20. 2.1%
21. $pH = 2.9$
22. $K_a = 6.9 \times 10^{-4}$
23. $[H^+] = [F^-] = 2.3 \times 10^{-3}$ M, $[HF] = 7.7 \times 10^{-3}$ M, $[OH^-] = 4.3 \times 10^{-12}$ M
24. $pH = 11.8$
25. $[H^+] = [HCO_3^-] = 1.8 \times 10^{-3}$ M

    $[CO_3^{2-}] = 4.8 \times 10^{-11}$ M

    $[H_2CO_3] = 0.080$ M
26. b. Basic
27. a. Acidic   b. Neutral   c. Neutral   d. Basic   e. Acidic   f. Basic
28. $pH = 9.1$
29. $[H^+] = 2.2 \times 10^{-11}$ M
30. d. KF

Chapter Sixteen

ACID-BASE EQUILIBRIA AND SOLUBILITY EQUILIBRIA

- Buffer Solutions (16.2 – 16.3)
- Titration Curves and Indicators (16.4 – 16.5)
- Solubility and Solubility Product (16.6)
- Predicting Precipitation Reactions and Separation of Ions by Precipitation (16.6 – 16.7)
- Factors Affecting Solubility: Common Ion Effect and pH (16.8 – 16.9)
- Complex Ions and Solubility (16.10)

BUFFER SOLUTIONS (16.2 – 16.3)

STUDY OBJECTIVES

1. Describe the effect of common ions on the percent ionization of weak acids and bases.
2. Calculate the pH of a buffer solution, given the concentrations of weak acid or base and their salts.
3. Determine the pH of a buffer solution after the addition of a small amounts of strong acid base.
4. Describe how to prepare a buffer solution with a specific pH.

The Common Ion Effect. In this chapter we will examine the acid-base equilibria in buffer solutions and in titrations. Then we will discuss two additional types of aqueous equilibria: those involving low solubility salts, and those involving the formation of metal complexes. We will start with the common ion effect.

First consider the ionization of a weak acid such as HF for example. The species in equilibrium are HF molecules, H^+ ions, and F^- ions:

$$HF(aq) \rightleftharpoons H^+(aq) + F^-(aq)$$

The concentration of F^- ion can be increased by adding sodium fluoride, a strong electrolyte:

$$NaF(aq) \rightarrow Na^+(aq) + F^-(aq)$$

According to Le Chatelier's principle, the addition of F^- ions will shift the weak acid equilibrium to the left; this consumes some of the F^- ions and some $H^+(aq)$, and forms more HF. This shift lowers the percent ionization of HF. In effect, the percent ionization of a weak acid (HA) is repressed by the addition to the solution of its conjugate base, A^- ion. The shift in equilibrium caused by the addition of an ion common to one of the products of a dissociation reaction is called the **common ion effect**. The percent ionization of a weak acid in the presence of added common ion is calculated in Example 16.2

Buffer Solutions. Any solution that resists significant changes in its pH, even when a strong acid or strong base is added, is called a **buffered solution** or just a **buffer**. A buffer must contain an acid to react with any OH^- ions that may be added, and a base to react with any added H^+ ions. *A buffer solution may contain a weak acid and its salt* (for example HF and NaF), *or a weak base and its salt* (for example NH_3 and NH_4Cl).

A solution containing HF and NaF is a buffer solution. In a typical buffer the weak acid and its salt are in *approximately equal concentrations*. The weak acid HF and its conjugate base F^- are called the buffer components. If you had a solution of just HF, the concentration of F^- ion would not be high enough to make it a buffer solution.

$$HF(aq) \rightleftharpoons H^+(aq) + F^-(aq)$$

The additional F^- ions needed in order to achieve approximately equal concentrations of the two components are supplied by the addition of NaF which is 100% ionized in solution.

Buffer Action. When a small amount of a strong acid is added to the buffer, it is neutralized by the weak base component. Also, when a small amount of a strong base is added to a buffer, it is neutralized by the weak acid component. In the HF / F^- ion buffer, for example, the following neutralization reactions take place:

• On addition of a strong base, the neutralization reaction is

$$OH^- + HF \rightarrow H_2O + F^-$$

• On addition of a strong acid, the neutralization reaction is

$$H^+ + F^- \rightarrow HF$$

In order for the buffer to be most effective, the amounts of weak acid and weak base used to prepare the buffer must be considerably greater than the amounts of strong acid or strong base that may be added later. **Buffer capacity** refers to the amount of acid or base that can be neutralized by a buffer solution before its pH is appreciably affected.

The pH of a Buffer. To calculate the pH of a buffer requires that the concentrations of the weak acid (such as HF) and a soluble salt of the weak acid (NaF) be substituted into the ionization constant expression for the weak acid.

$$HF(aq) \rightleftharpoons H^+(aq) + F^-(aq)$$

$$K_a = \frac{[H^+][F^-]}{[HF]}$$

$$[H^+] = \frac{[HF]}{[F^-]} K_a$$

The pH is found from $pH = -\log [H^+]$.

Alternatively, the Henderson-Hasselbalch equation, which is introduced in Section 16.3 of the textbook, can be used to calculate buffer pH.

$$pH = pK_a + \log \frac{[\text{conjugate base}]}{[\text{acid}]}$$

where the $pK_a = -\log K_a$.

This equation holds only if the acid is less than 5 percent ionized. This requirement is almost always met because the ionization of the weak acid is lowered by the addition of the conjugate base in accord with the common ion effect. Use of the Henderson-Hasselbalch equation is shown in Example 16.1.

Preparing a Buffer. Each weak acid–conjugate base buffer has a characteristic pH range over which it is most effective. In terms of the Henderson-Hasselbalch equation, *the pH of a buffer depends on the pK_a of the weak acid, and the ratio of conjugate base concentration to weak acid concentration*. A buffer has equal ability to neutralize either added acid or added base when the concentrations of its weak acid and conjugate base components are equal. When [conj base] / [weak acid] = 1, then

$$\log \frac{[\text{conjugate base}]}{[\text{acid}]} = 0$$

and the Henderson-Hasselbalch equation becomes

$$pH = pK_a$$

In order to prepare a buffer of a certain pH, *the weak acid must have a pK$_a$ as close as possible to the desired pH.*

Next we substitute the desired pH and the pK$_a$ of the chosen acid into the Henderson-Hasselbalch equation in order to determine the ratio of [conj base] to [weak acid]. The ratio can then be converted into molar concentrations of the weak acid and its salt. See Example 16.4.

EXAMPLE 16.1   pH of a Buffer Solution

Calculate the pH of a buffer solution that is 0.25 M HF and 0.50 M NaF.

•Method of Solution

The pH of a HF/F$^-$ buffer is given in terms of the Henderson-Hasselbalch equation.

$$HF(aq) \rightleftharpoons H^+(aq) + F^-(aq)$$

$$pH = pK_a + \log \frac{[F^-]}{[HF]}$$

where $pK_a = - \log K_a$

$$pK_a = - \log (7.1 \times 10^{-4}) = 3.15$$

The equilibrium concentrations of F$^-$ ion and HF are determined as follows. The initial concentration of HF is 0.25 M. But at equilibrium [HF] = 0.25 − x. We can assume that x is less than 5 percent of [HF]$_0$ because HF is a weak acid, and because the addition of F$^-$ ion, a common ion, represses the dissociation of HF. We assume that [HF] = [HF]$_0$ = 0.25 M.

The F$^-$ ion is contributed by two sources, sodium fluoride which is a strong electrolyte, and HF, a weak acid.

$$NaF(aq) \rightarrow Na^+(aq) + F^-(aq)$$

Therefore, [F$^-$] = 0.50 M + x. Since 0.50 M >> x, then [F$^-$] = 0.50 M.

Substituting into the Henderson-Hasselbalch equation gives:

$$pH = 3.15 + \log \frac{0.50}{0.25} = 3.15 + \log 2.0 = 3.15 + 0.30$$

$$= 3.45$$

EXAMPLE 16.2   Percent Ionization in a Buffer Solution

Calculate the percent ionization of HF in the buffer solution of the preceding example.

•Method of Solution

$$\text{percent ionization} = \frac{[H^+]}{[HF]_0} \times 100\%$$

where

$$[H^+] = 10^{-pH} = 10^{-3.45}$$

$$[H^+] = 3.55 \times 10^{-4} \text{ M}$$

$$\text{percent ionization} = \frac{3.55 \times 10^{-4} \text{ M}}{0.25} \times 100\% = 0.14\%$$

•Comment

By comparison, the percent ionization of HF in a 0.25 M HF solution is 5.3%. The common ion effect lowers the percent ionization to the value found in the buffer solution. Neglecting x was a valid assumption.

EXAMPLE 16.3   Adding Strong Acid to a Buffer

Suppose 3.0 mL of 2.0 M HCl is added to exactly 100 mL of the buffer described in Example 16.1. What is the new pH of the buffer after the HCl is neutralized?

•Method of Solution

The HCl is neutralized in this buffer by the following reaction which goes to completion.

$$H^+ + F^- \rightarrow HF$$

This consumes some F^- ions and forms more HF. The number of moles of H^+ added as HCl is

$$M \cdot V = 2.0 \text{ mol/L} \times 0.0030 \text{ L} = 0.0060 \text{ mol } H^+$$

The number of moles of HF originally present in 100 mL of buffer was

$$M \cdot V = 0.25 \text{ mol/L} \times 0.100 \text{ L} = 0.025 \text{ mol HF}$$

The number of moles of F^- originally present in 100 mL of buffer was

$$M \cdot V = 0.50 \text{ mol/L} \times 0.100 \text{ L} = 0.050 \text{ mol } F^-$$

•Calculation

After the added H^+ is neutralized by $H^+ + F^- \rightarrow HF$, 0.0060 H^+ reacts with 0.0060 mol of F^- ion to form 0.0060 mol HF.

The number of moles of HF is 0.025 + 0.006 = 0.031 mol.
The number of moles of F^- is 0.050 − 0.006 = 0.044 mol.

The new pH can be found as usual from

$$pH = pK_a + \log \frac{[F^-]}{[HF]}$$

$$= 3.15 + \log \frac{(0.044 \text{ mol/0.103 L})}{(0.031 \text{ mol/0.103 L})} = 3.15 + \log 1.42 = 3.15 + 0.15$$

$$= 3.30$$

•Comment

The pH has dropped only 0.15 unit from 3.45 (Example 16.1) due to the addition of 3.0 mL of 2.0 M HCl; thus the use of the term buffer solution.

EXAMPLE 16.4   Conjugate Base to Weak Acid Ratio in a Buffer

What ratio of $[F^-]$ to $[HF]$ would you use to make a buffer of pH = 2.85?

•Method of Solution

Start with the Henderson-Hasselbalch equation.

$$pH = pK_a + \log \frac{[\text{conjugate base}]}{[\text{acid}]}$$

•Calculation

Substitute in the desired pH and the pK_a value of HF.

$$2.85 = 3.15 + \log \frac{[F^-]}{[HF]}$$

$$\log \frac{[F^-]}{[HF]} = -0.30$$

$$\frac{[F^-]}{[HF]} = 10^{-0.30} = 0.50$$

The ratio of $[F^-]$ to $[HF]$ should be 0.50 to 1. Any amounts of F^- and HF that give a ratio of 0.5 will produce the desired pH.

EXERCISES

1. Compare the percent ionization of acetic acid in 0.10 M CH_3COOH to that in a solution of 0.020 M CH_3COONa and 0.10 M CH_3COOH.

2. a. What is the pH of a buffer solution prepared by dissolving 0.15 mol of benzoic acid (C_6H_5COOH) and 0.45 mol of sodium benzoate (C_6H_5COONa) in enough water to make 400 mL of solution?
    b. Does the volume of solution affect your answer?

3. What mole ratio of benzoate ion to benzoic acid would you need to prepare a buffer solution with a pH of 5.00?

4. a. What is the change in pH when 40.0 mg of NaOH is added to 100 mL of a buffer solution consisting of 0.165 M NH_3 and 0.120 M NH_4Cl?
    b. Write the equation for the buffer equilibrium.
    c. Write the equation for the neutralization of NaOH by the buffer.

5. What is the optimum pH of a H_3PO_4 / $H_2PO_4^-$ buffer?

TITRATION CURVES AND INDICATORS (16.4 – 16.5)

STUDY OBJECTIVES

1. Describe the shape of titration curves for titrations involving a strong acid and a strong base, a weak acid and a strong base, and a strong acid and a weak base.
2. Choose the best indicator for a particular acid-base titration.

Titration Curves. Acid-base titrations were discussed in Chapter 4. With the introduction to pH in the previous chapter, it is informative to follow the pH as a function of the progress of the titration. A graph of pH versus volume of titrant added is called **a titration curve**. Initially, the pH is that of the unknown solution. As titrant is added, the pH becomes that of a partially neutralized solution of unknown plus titrant. The pH at the *equivalence point* refers to the H$^+$ ion concentration when just enough titrant has been added to completely neutralize the unknown. If more titrant is added after the equivalence point has been reached, the pH assumes a value consistent with the pH of excess titrant. We will review three types of titration curves.

Strong acid-strong base titrations. Figure 16.3 of the text shows a titration curve for the addition of a strong base to a strong acid. The main features of this curve can be stated briefly. 1. The pH starts out quite low because this is the pH of the pure acid solution. 2. As base is slowly added, it is neutralized, and the pH is determined by the unreacted excess acid. Near the equivalence point, the pH begins to rise more rapidly. 3. At the equivalence point the pH changes sharply, increasing about 5.0 units upon the addition of only two drops of base. 4. Beyond the equivalence point the pH is determined by the amount of excess base that is added and is higher than 7.

The pH at the equivalence point of an acid-base titration is the pH of the salt solution that is formed by neutralization. NaCl, NaNO$_3$, NaBr, KCl, and KI are examples of salts that can be formed in titrations of strong acids and strong bases. These salts yield ions that do not cause hydrolysis (Chapter 15). Therefore, the pH at the equivalence point in a titration of a strong acid with a strong base is 7.

Weak acid-strong base titrations. Figure 16.4 of the text shows a titration curve for the addition of a strong base to a weak acid. Because the acid is a weak acid, the initial pH is greater than in the titration of a strong acid. At the equivalence point, the pH is above 7.0 because of salt hydrolysis. The anion of the salt is the conjugate base of the weak acid used in the titration. In the titration of acetic acid with sodium hydroxide the salt produced is CH$_3$COONa. The Na$^+$ ion does not hydrolyze, but CH$_3$COO$^-$ does:

$$CH_3COO^- + H_2O \rightleftharpoons CH_3COOH + OH^-$$

Hydrolysis of acetate ions makes the solution basic at the equivalence point.

Strong acid-weak base titrations. Figure 16.6 of the text shows a titration curve for the addition of a weak base to a strong acid. The first part of the curve is the same as in the strong acid versus strong base titration. However, the pH at the equivalence point is below 7.0 because the cation of the salt is the conjugate acid of the weak base used in the titration. Hydrolysis of this salt yields an acidic solution. In the titration of hydrochloric acid with ammonia, the salt produced is NH$_4$Cl. The Cl$^-$ ion does not hydrolyze, but NH$_4^+$ does:

$$NH_4^+ + H_2O \rightleftharpoons NH_3 + H_3O^+$$

Indicators. Indicators are used in the laboratory to reveal the equivalence point of a titration. The abrupt change in color of the indicator signals the *endpoint* of a titration which usually coincides very nearly with the equivalence point.

Indicators are usually weak organic acids that have distinctly different colors in the nonionized (molecular) and ionized forms.

$$HIn(aq) \rightleftharpoons H^+(aq) + In^-(aq) \qquad K_a = \frac{[H^+][In^-]}{[HIn]}$$

color 1   color 2

To determine the pH range in which an indicator will change color we can write the K_a expression in logarithmic form.

$$pH = pK_a + \log \frac{[In^-]}{[HIn]}$$

The color of an indicator depends on which form predominates. Typically, when $[In^-]/[HIn] \geq 10$ the solution will be color 2, and when $[In^-]/[HIn] \leq 0.1$, the solution will be color 1. Thus the color change will occur between

$$pH = pK_a + \log (10/1) = pK_a + 1.0$$

and

$$pH = pK_a + \log (1/10) = pK_a - 1.0$$

At the midpoint of the pH range over which the color changes, $[HIn] = [In^-]$, and $pH = pK_a$.

Like any weak acid each HIn has a characteristic pK_a, and so each indicator changes color at a characteristic pH. Table 16.1 of the text lists a number of indicators used in acid-base titration and the pH ranges over which they change color.

The choice of indicator for a particular titration depends on the expected pH at the equivalence point. For instance, in the titration of acetic acid by sodium hydroxide, which is discussed in Example 16.7 of the text, the pH at the equivalence point is 8.72. According to Table 16.1 both cresol red and phenolphthalein change color over ranges that include pH 8.72. Therefore either of these indicators would show the equivalence point of this titration.

EXAMPLE 16.5   Net Ionic Equations

Write the net ionic equations for the neutralization reactions that occur during the following titrations. Predict whether the pH at the equivalence point will be above, below, or equal to 7.0.
a.   Titration of HI with NH_3.
b.   Titration of HI with NaOH.
c.   Titration of NaOH with HF.

•**Method of Solution**

a.   This is a titration of a strong acid with a weak base. HI is 100 percent dissociated, and NH_3 is a weak base. The net ionic equation for neutralization is

$$H^+ + NH_3 \rightarrow NH_4^+$$

The result of the neutralization reaction is formation of the conjugate acid NH_4^+ of the weak base NH_3. NH_4^+ is an acid. *Answer:* The pH at the equivalence point will be below 7.0.
b.   This is a titration of a strong acid and a strong base. The net ionic equation is

$$H^+ + OH^- \rightarrow H_2O$$

The resulting solution is neutral at the equivalence point. *Answer:* $pH = 7.0$
c.   This titration involves a weak acid and a strong base.

$$HF + OH^- \rightarrow H_2O + F^-$$

The reaction produces the conjugate base (F^- ion) of the weak acid. *Answer:* The solution will be basic at the equivalence point, $pH > 7.0$.

EXAMPLE 16.6   Choosing an Indicator

Choose an indicator for the titration of 50 mL of a 0.10 M HI solution with 0.10 M NH_3.

$$HI(aq) + NH_3(aq) \rightarrow NH_4I(aq)$$

•Method of Solution

First we need to know the pH at the equivalence point. This is the titration of a strong acid with a weak base. The net ionic equation is

$$H^+(aq) \quad + \quad NH_3(aq) \quad \rightarrow \quad NH_4^+(aq)$$
$$5.0 \times 10^{-3} \text{ mol} + 5.0 \times 10^{-3} \text{ mol} \text{ yields } 5.0 \times 10^{-3} \text{ mol}$$

The product NH_4^+ ion is a weak acid and so we expect a slightly acidic solution at the equivalence point.

•Calculation

The concentration of NH_4^+ ion formed at the equivalence point is 5.0×10^{-3} mol/0.100 L = 5.0×10^{-2} M. The $[H^+]$ at the equivalence point is just the pH of a 0.050 M NH_4^+ solution. Consider the ionization of the weak acid, NH_4^+:

$$NH_4^+(aq) \rightleftharpoons NH_3(aq) + H^+(aq) \qquad K_a = 5.6 \times 10^{-10}$$

Concentration	NH_4^+	$\rightleftharpoons$	H^+	+	NH_3
Initial (M)	0.050		0		0
Change (M)	−x		+x		+x
Equilibrium (M)	(0.050 − x)		x		x

$$K_a = \frac{[NH_3][H^+]}{[NH_4^+]} = \frac{x^2}{(0.050 - x)} = \frac{x^2}{0.050} = 5.6 \times 10^{-10}$$

Solving for x:

$$x = [NH_3] = [H^+] = 5.3 \times 10^{-6} \text{ M}$$

and the pH is 5.28 at the equivalence point. According to Table 16.1 of the text, chlorophenol blue and methyl red are indicators that will change color in the vicinity of pH 5.28.

EXERCISES

6.   Approximately what range of pH should be expected at the equivalence points in the titration of weak acids with strong bases?

7.   The pH at the equivalence point in a titration was found to be approximately 5. Give the category of titration in terms of the acid and base strengths.

8. 25.0 mL of 0.222 M HBr was titrated with 0.111 M NaOH.
   a. Write the overall balanced equation for reaction.
   b. After adding 30.0 mL of base solution, what is the pH?

9. Calculate the pH in the following titration after the addition of 12.0 mL of 0.100 M KOH to 20.0 mL of 0.200 M CH_3COOH.

10. Determine the pH at the equivalence point in the titration of 0.100 M NaOH with 0.100 M HNO_3.

SOLUBILITY AND SOLUBILITY PRODUCT (16.6)

STUDY OBJECTIVES

1. Write the solubility product expression for an insoluble salt.
2. Calculate the solubility product constant of a salt given its solubility.
3. Calculate the solubility of a salt given its K_{sp}.

The Solubility Product Constant. The solubility rules were discussed in Chapter 4. Compounds with low solubilities in water are described. as *slightly soluble or insoluble*. In this chapter, solubility is treated quantitatively in terms of equilibrium. In a saturated solution of a slightly soluble compound such as silver bromide, for example, the solubility equilibrium is:

$$AgBr(s) \rightleftharpoons Ag^+(aq) + Br^-(aq)$$

The equilibrium constant for the reaction in which a solid salt dissolves is called the *solubility product constant*, K_{sp}. For the above reaction:

$$K_{sp} = [Ag^+][Br^-]$$

The solubility product expression is always written in the form of the equilibrium constant expression for the solubility reaction. Two additional examples are $Ca(OH)_2$ and $AgCrO_4$:

$$Ca(OH)_2(s) \rightleftharpoons Ca^{2+}(aq) + 2OH^-(aq)$$

$$K_{sp} = [Ca^{2+}][OH^-]^2$$

and

$$Ag_2CrO_4(s) \rightleftharpoons 2Ag^+(aq) + CrO_4^{2-}(aq)$$

$$K_{sp} = [Ag^+]^2[CrO_4^{2-}]$$

The solubility product constant (also called a K_{sp} value) can be calculated by substituting the concentrations of the ions in a *saturated solution* into the solubility product expression. Take AgBr again as an example. Since AgBr is a strong electrolyte, all of the solid AgBr that dissolves is dissociated into Ag^+ and Br^- ions,

$$AgBr(aq) \rightarrow Ag^+(aq) + Br^-(aq)$$

The solubility of AgBr at 25°C is 8.8×10^{-7} moles per liter of solution. Therefore, the ion concentrations in a saturated AgBr solution are

$$[Ag^+] = [Br^-] = 8.8 \times 10^{-7} \text{ M}$$

The solubility product constant is found by substituting these concentrations into the solubility product:

$$K_{sp} = [Ag^+][Br^-] = (8.8 \times 10^{-7})(8.8 \times 10^{-7})$$

$$K_{sp} = 7.7 \times 10^{-13}$$

This calculation points out the difference between the solubility and the K_{sp} value. These two quantities are *not* the same. Here we see that the solubility is used to determine the ion concentrations, and that substitution of these into the solubility product expression gives the solubility product constant (K_{sp} value). Table 16.2 in the text and Table 16.1 in this chapter list solubility product constants for a number of slightly soluble salts, including silver bromide.

The K_{sp} value can be used to calculate the solubility of a compound in moles per liter (the molar solubility), or in grams per liter. For example, the K_{sp} value for $CaSO_4$ is 2.4×10^{-5}. What is the solubility of $CaSO_4$ in mol/L?

Start with the solubility equilibrium:

$$CaSO_4(s) \rightleftharpoons Ca^{2+}(aq) + SO_4^{2-}(aq) \qquad K_{sp} = [Ca^{2+}][SO_4^{2-}]$$

Let s equal the solubility in mol/L. At equilibrium, both $[Ca^{2+}]$ and $[SO_4^{2-}]$ are equal to s because when s moles of $CaSO_4$ dissolve in 1 L, s moles of Ca^{2+} ions and s moles of SO_4^{2-} ions are produced. Substitute into the K_{sp} expression:

$$K_{sp} = [Ca^{2+}][SO_4^{2-}] = (s)(s) = s^2$$

The solubilty product is given above: $K_{sp} = 2.4 \times 10^{-5}$

$$s^2 = 2.4 \times 10^{-5}$$

$$s = \sqrt{2.4 \times 10^{-5}} = 4.9 \times 10^{-3} \text{ M}$$

It is important to notice that the molar solubility and K_{sp} are not the same. *Molar solubility and K_{sp} are related but are not the same quantity.*

Solubility and K_{sp}. Table 16.1 below lists a number of solubility product constants for various salts that will be needed in this chapter. Notice that within a series of salts of the same type of formula (MX, MX, etc.), *the solubility increases as the solubility product constant increases.* Comparing the K_{sp}'s of the three MX salts, we can see that $CaSO_4$ has the highest K_{sp} value. Therefore, its solubility equilibrium lies farthest to the right of the three salts, and it is the most soluble.

Table 16.1 Solubility Products of Some Slightly Soluble Salts at 25°C

Name	Formula	K_{sp}
	MX type	
Calcium sulfate	$CaSO_4$	2.4×10^{-5}
Silver chloride	$AgCl$	1.6×10^{-10}
Silver bromide	$AgBr$	7.7×10^{-13}
	MX_2 and M_2X type	
Lead chloride	$PbCl_2$	2.4×10^{-4}
Calcium hydroxide	$Ca(OH)_2$	8.0×10^{-6}
Lead iodide	PbI_2	1.4×10^{-8}
Magnesium fluoride	MgF_2	6.4×10^{-9}
Magnesium hydroxide	$Mg(OH)_2$	1.2×10^{-11}
Silver chromate	Ag_2CrO_4	1.1×10^{-12}

EXAMPLE 16.7   Calculating a K_{sp} Value

The solubility of magnesium fluoride in water is 7.3×10^{-3} g per 100 mL of solution. What is the solubility product constant for MgF_2?

•**Method of Solution**

Write the solubility equilibrium and the K_{sp} expression:

$$MgF_2(s) \rightleftharpoons Mg^{2+}(aq) + 2F^-(aq)$$

$$K_{sp} = [Mg^{2+}] [F^-]^2$$

Calculate s, the molar solubility of MgF_2, from the given solubility. The molar mass of MgF_2 is 62.31 g/mol.

$$s = \frac{7.3 \times 10^{-3} \text{ g}}{100 \text{ mL}} \times \frac{1 \text{ mL}}{10^{-3} \text{ L}} \times \frac{1 \text{ mol}}{62.31 \text{ g}} = 1.17 \times 10^{-3} \text{ mol/L}$$

The K_{sp} value can be calculated by substituting the ion concentrations in a saturated solution into the solubility product expression. When s mol of MgF_2 dissolves in 1 L of solution, the Mg^{2+} concentration is equal to s, and the F^- ion concentration is 2s.

$$[Mg^{2+}] = s = 1.17 \times 10^{-3} \text{ M}$$

$$[F^-] = 2s = 2(1.17 \times 10^{-3} \text{ M}) = 2.34 \times 10^{-3} \text{ M}$$

• **Calculation**

Now substitute the ion concentrations into the K_{sp} expression.

$$K_{sp} = (1.17 \times 10^{-3}) (2.34 \times 10^{-3})^2$$
$$K_{sp} = 6.4 \times 10^{-9}$$

EXAMPLE 16.8   Calculating a Molar Solubility

The K_{sp} value for Ag_2CrO_4 is 1.1×10^{-12}. Calculate the molar solubility of silver chromate.

•**Method of Solution**

Write the solubility equilibrium

$$Ag_2CrO_4(s) \rightleftharpoons 2Ag^+(aq) + CrO_4^{2-}(aq)$$

and the K_{sp} expression.

$$K_{sp} = [Ag^+]^2[CrO_4^{2-}]$$

Let s = molar solubility of Ag_2CrO_4. Whenever s moles of Ag_2CrO_4 dissolve, 2s moles of Ag^+ and s moles of CrO_4^{2-} are produced.

$$[Ag^+] = 2s \quad [CrO_4^{2-}] = s$$

Summarize the changes in concentrations as follows.

Concentration	$Ag_2CrO_4(s) \rightleftharpoons$	$2Ag^+(aq) +$	$CrO_4^{2-}(aq)$
Initial (M)		0	0
Change (M)		2s	s
Equilibrium (M)		2s	s

•**Calculation**

Substitute these values into the K_{sp} expression.

$$K_{sp} = [2s]^2[s] = 1.1 \times 10^{-12}$$

and solve for s.

$$4s^2(s) = 1.1 \times 10^{-12}$$
$$4s^3 = 1.1 \times 10^{-12}$$

$$s = \sqrt[3]{\frac{1.1 \times 10^{-12}}{4}} = \sqrt[3]{2.75 \times 10^{-13}}$$

$$s = 6.5 \times 10^{-5} M$$

EXERCISES

11. Write balanced equations and solubility product expressions for the solubility equilibria of the following compounds.
    a. $BaCO_3$ b. CaF_2 c. $Al(OH)_3$ d. Ag_3PO_4

12. The solubility of $PbBr_2$ is 0.392 g per 100 mL at 20°C. What is the K_{sp} value for $PbBr_2$?

13. a. What is the solubility of CaF_2 in moles per liter given that $K_{sp} = 4.0 \times 10^{-11}$.
    b. Determine the F^- ion concentration in a saturated solution of CaF_2.

14. Solid barium fluoride was dissolved in pure water until a saturated solution was formed. If the fluoride concentration was 0.0150 M, determine the K_{sp} of BaF_2?

PREDICTING PRECIPITATION REACTIONS AND SEPARATION OF IONS BY PRECIPITATION (16.6 – 16.7)

STUDY OBJECTIVES

1. Predict whether a precipitate will form when two solutions of salts are mixed.
2. Describe how to selectively precipitate one of two ions in a solution.
3. Determine the completeness of precipitation.

Criteria for Precipitate Formation. When two solutions containing dissolved salts are mixed, formation of an insoluble compound is always a possibility. From a knowledge of solubility rules (Chapter 4) and solubility products (Table 16.1), you can predict whether a precipitate will form. For a dissolved salt $MX(aq)$, the **ion product** (Q) is $Q = [M^+]_0[X^-]_0$, where $[\ ]_0$ stands for the initial concentrations.

Any one of three conditions for the ion product Q may exist after two solutions are mixed.

1. $Q = K_{sp}$ For a saturated solution, the value of Q is equal to K_{sp}. No precipitate will form in this case as no net reaction occurs in a system at equilibrium.
2. $Q < K_{sp}$ In an unsaturated solution the value of Q is less than K_{sp}. No precipitate will form in this case.
3. $Q > K_{sp}$ In a supersaturated solution more of the salt is dissolved than the solubility allows. In this case, an unstable situation exists, and some solute will precipitate from solution until a saturated solution is attained. At this point $Q = K_{sp}$ and equilibrium is reestablished.

Suppose 500 mL of a solution containing 2.0×10^{-5} M Ag^+ is mixed with 500 mL of solution containing 2.0×10^{-4} M Br^-. Will a precipitate form? First determine the new concentrations of Ag^+ and Br^- in the mixture. The total volume is 1 L, and so accounting for dilution the ion concentrations in the new solution are just half of what they were in the separate solutions. The initial concentrations before any precipitate forms are:

$$[Ag^+]_0 = 1.0 \times 10^{-5} \text{ M} \qquad [Br^-]_0 = 1.0 \times 10^{-4} \text{ M}$$

The ion product for AgBr is

$$Q = [Ag^+]_0[Br^-]_0$$

To predict whether a precipitate of AgBr will form, calculate the ion product Q, and compare it to the K_{sp} value.

$$Q = [Ag^+]_0[Br^-]_0 = (1.0 \times 10^{-5})(1.0 \times 10^{-4}) = 1.0 \times 10^{-9}$$

$$K_{sp} = 7.7 \times 10^{-13}$$

Therefore, $Q > K_{sp}$, and so the mixture corresponds to a supersaturated solution of AgBr, which means that some AgBr(s) will precipitate:

$$Ag^+(aq) + Br^-(aq) \rightarrow AgBr(s)$$

until a new equilibrium is reached at which point $Q = K_{sp}$.

Criteria for Separation of Ions by Precipitation. Ions in a solution can be separated from each other on the basis of the different solubilities of their salts. For example, when slowly adding a solution containing OH^- ion to a solution that contains both Ca^{2+} ions and Mg^{2+} ions, we find that $Mg(OH)_2$ precipitates before $Ca(OH)_2$. $Mg(OH)_2$ with a K_{sp} of 1.2×10^{-11} precipitates first because it is less soluble than $Ca(OH)_2$ whose K_{sp} is 8.0×10^{-6}.

The concentrations of the ions in solution also help determine which ions will precipitate first. Even though $Mg(OH)_2$ is less soluble than $Ca(OH)_2$, $Ca(OH)_2$ would precipitate first when OH^- ion is added if the concentration of Ca^{2+} ion was much greater than the concentration of Mg^{2+} ion. It is best to determine the concentration of OH^- ion needed to cause a precipitate of both compounds. Since Ca^{2+} and Mg^{2+} ion concentrations will be given, then the concentration of OH^- that must be exceeded to initiate the precipitation of $Ca(OH)_2$ and $Mg(OH)_2$ are, respectively:

$$[OH^-] = \sqrt{\frac{K_{sp}}{[Ca^{2+}]}} \quad \text{and} \quad [OH^-] = \sqrt{\frac{K_{sp}}{[Mg^{2+}]}}$$

Examples 16.10 and 16.11 illustrate the separation of Cl^- and CrO_4^{2-} (chromate) ions.

EXAMPLE 16.9 Predicting Formation of a Precipitate

a. Predict whether or not a precipitate of PbI_2 will form when 200 mL of 0.015 M $Pb(NO_3)_2$ and 300 mL of 0.050 M NaI are mixed together. Given $K_{sp}(PbI_2) = 1.4 \times 10^{-8}$.
b. If the answer is yes, what concentrations of $Pb^{2+}(aq)$ and $I^-(aq)$ will exist when equilibrium is reestablished?

•**Method of Solution for (a)**

Recall that $Pb(NO_3)_2$ and NaI are both strong electrolytes. When the solutions are mixed, will the following reaction occur?

$$Pb^{2+}(aq) + 2\,NO_3^-(aq) + 2Na^+(aq) + 2I^-(aq) \rightarrow PbI_2(s) + 2Na^+(aq) + 2\,NO_3^-(aq)$$

A precipitate will form only if the ion product exceeds the K_{sp} value; $[Pb^{2+}]_0[I^-]_0^2 > K_{sp}$.

When the two solutions are mixed, 500 mL of new solution is formed. Immediately after mixing, the initial ion concentrations would be:

$$[Pb^{2+}]_0 = 0.015\ \text{M} \times \frac{200\ \text{mL}}{500\ \text{mL}} = 6.0 \times 10^{-3}\ \text{M}$$

$$[I^-]_0 = 0.050\ \text{M} \times \frac{300\ \text{mL}}{500\ \text{mL}} = 3.0 \times 10^{-2}\ \text{M}$$

Then $\quad Q = [Pb^{2+}]_0[I^-]_0^2 = (6.0 \times 10^{-3})(3.0 \times 10^{-2})^2 = 5.4 \times 10^{-6}$

Since $Q > K_{sp}$, then a precipitate of PbI_2 will form.

•**Method of Solution for (b)**

The concentrations of Pb^{2+} and I^- remaining in solution depend on the number of moles initially and the number of moles that precipitate. The initial number of moles of Pb^{2+} ion was 0.015 M × 0.200 L = 0.0030 mol, and that of I^- ion, 0.050 M × 0.300 L = 0.015 mol. In order to determine how many moles of Pb^{2+} and I^-

precipitated from solution, consider the net reaction

$$Pb^{2+}(aq) + 2I^-(aq) \rightarrow PbI_2(s)$$

Look to see if there is a limiting reactant. Then determine the amount of the excess reactant that remains after complete reaction of the limiting reactant. Pb^{2+} ion being in the smaller amount is the limiting reactant, and will react essentially completely. The 0.0030 mol of Pb^{2+} will react with 0.0060 mol of I^-, leaving 0.009 mol of excess I^- (0.015 − 0.0060) dissolved in solution.

•Calculation

The I^- concentration is $[I^-]$ = 0.009 mol/0.50 L = 0.018 M.

The Pb^{2+} concentration is controlled by the PbI_2 solubility equilibrium. In other words, some PbI_2 dissolves by the reverse reaction.

$$PbI_2(s) \rightleftharpoons Pb^{2+} + 2I^-$$

$$K_{sp} = [Pb^{2+}][I^-]^2 = 1.4 \times 10^{-8}$$

The Pb^{2+} ion concentration in equilibrium with 0.018 M I^- ion is:

$$[Pb^{2+}] = \frac{K_{sp}}{[I^-]^2} = \frac{1.4 \times 10^{-8}}{(0.018)^2} = 4.3 \times 10^{-5} \text{ M}$$

•Comment

The percentage of Pb^{2+} ion remaining unprecipitated is:

$$\frac{4.3 \times 10^{-5} \text{ M} (0.500 \text{ L})}{0.0030 \text{ mol}} \times 100\% = 0.72\%$$

This confirms our assumption that essentially all the Pb^{2+} ion precipitated.

EXAMPLE 16.10   Selective Precipitation of an Ion

A solution contains 0.10 M Cl^- and 0.010 M CrO_4^{2-}. If $AgNO_3$ solution is added dropwise which will precipitate first, AgCl or Ag_2CrO_4?

•Method of Solution

The solubility product is a number that the product of the ion concentrations can never exceed at equilibrium. First write the two equilibria of interest and their solubility constants.

$$AgCl(s) \rightleftharpoons Ag^+(aq) + Cl^-(aq) \qquad K_{sp} = [Ag^+][Cl^-] = 1.6 \times 10^{-10}$$

$$Ag_2CrO_4(s) \rightleftharpoons 2Ag^+ + CrO_4^{2-} \qquad K_{sp} = [Ag^+]^2[CrO_4^{2-}] = 1.1 \times 10^{-12}$$

•Calculation

The highest Ag^+ ion concentration possible in a solution of 0.10 M Cl^- is

$$[Ag^+] = \frac{1.6 \times 10^{-10}}{[Cl^-]} = \frac{1.6 \times 10^{-10}}{0.10} = 1.6 \times 10^{-9} \text{ M}$$

The highest Ag^+ ion concentration possible in a solution of 0.010 M CrO_4^{2-} is:

$$[Ag^+] = \sqrt{\frac{K_{sp}}{[CrO_4^{2-}]}} = \sqrt{\frac{1.1 \times 10^{-12}}{0.010}} = 1.0 \times 10^{-5} \text{ M}$$

Silver chloride will precipitate before silver chromate because of the lower Ag^+ ion concentration needed to produce a saturated AgCl solution.

EXAMPLE 16.11   Completeness of Precipitation

In the above example AgCl precipitates before Ag_2CrO_4 as $Ag^+(aq)$ ion is added dropwise to the solution. What percentage of the Cl^- ion in solution will have precipitated when Ag_2CrO_4 just begins to precipitate?

•**Method of Solution**

As we found above, Ag_2CrO_4 begins to precipitate when $[Ag^+] = 1.0 \times 10^{-5}$ M.

•**Calculation**

The Cl^- ion concentration in equilibrium with 1.0×10^{-5} M Ag^+ ion is:

$$[Cl^-] = \frac{1.6 \times 10^{-10}}{[Ag^+]} = \frac{1.6 \times 10^{-10}}{1.0 \times 10^{-5}} = 1.6 \times 10^{-5} \text{ M}$$

The percent of Cl^- remaining in solution *unprecipitated* is:

$$\%Cl^- = \frac{1.6 \times 10^{-5} \text{ M}}{0.10 \text{ M}} \times 100\%$$

$$\%Cl^- = 0.016\%$$

Therefore, the percentage of Cl^- ion precipitated is $100\% - 0.016\% = 99.98\%$.

EXERCISES

15. If 1.0 mL of 1.0×10^{-3} M $Ba(NO_3)_2$ is added to 99.0 mL of 1×10^{-4} M Na_2CO_3, will $BaCO_3$ precipitate from this solution? $K_{sp}(BaCO_3) = 8.1 \times 10^{-9}$.

16. Will a precipitate form when 250 mL of 2.8×10^{-3} M $MgCl_2$ solution is added to 250 mL of 5.2×10^{-3} M NaF? Identify the precipitate if any.

17. Solid $Ba(NO_3)_2$ is slowly dissolved in a solution of 2.5×10^{-4} M Na_2CO_3. At what Ba^{2+} concentration will a precipitate of $BaCO_3$ just begin to form?

18. A solution contains 1.0×10^{-4} M Cu^+ and 1.0×10^{-3} M Pb^{2+}.
    a. If NaI solution is added dropwise to this solution which compound will precipitate first, CuI or PbI_2?
    b. Give the concentrations of I^- necessary to begin precipitation of each salt.
    c. What percentage of the initial Cu^+ ion concentration will remain unprecipitated when PbI_2 just begins to precipitate?

FACTORS AFFECTING SOLUBILITY: COMMON ION EFFECT AND pH (16.8 – 16.9)

STUDY OBJECTIVES

1. Calculate the solubility of a salt in a solution containing an ion common to that salt.
2. Predict the effect of pH on the solubility of a salt.

The Common Ion Effect. The effect of adding an ion common to one already in equilibrium in a solubility reaction is to lower the solubility of the salt. In the case of AgBr solubility:

$$AgBr(s) \rightleftharpoons Ag^+(aq) + Br^-(aq)$$

The addition of either Ag^+ or Br^- ions will shift the equilibrium to the left, in accord with Le Chatelier's principle, thus decreasing the amount of AgBr dissolved. Ag^+ ions can be added by pouring in a solution of $AgNO_3$. Recall that $AgNO_3$ is very soluble, and is a strong electrolyte.

$$AgNO_3(s) \rightarrow Ag^+(aq) + NO_3^-(aq)$$

The nitrate ion will not interfere with AgBr solubility because it is not a common ion.

Additional Br^- ions could be supplied by adding NaBr, for instance. Sodium bromide is very soluble and is a strong electrolyte.

$$NaBr(s) \rightarrow Na^+(aq) + Br^-(aq)$$

The sodium ion will not affect the solubility of AgBr because it is not a common ion.

Here we see that the addition of an ion that is common to one already in the solubility equilibrium shifts the equilibrium to the left, which *decreases* the solubility. This is equivalent to the case of a weak acid, where the presence of a common ion decreases the percent ionization. Several calculations involving the common ion effect appear at the end of this section.

pH and Solubility. The pH can affect the solubility of a solute in two ways. One of these is through the common ion effect. Consider the solubility equilibrium of an insoluble hydroxide such as $Mg(OH)_2$ or $Ca(OH)_2$.

$$Ca(OH)_2(s) \rightleftharpoons Ca^{2+}(aq) + 2OH^-(aq)$$

Upon the addition of NaOH, for instance, the pH of the solution will increase. The equilibrium position will shift to the left because of the added OH^- ion (a common ion); therefore the solubility of $Ca(OH)_2$ will decrease proportionately.

The other case in which pH can affect solubility is when a salt contains a basic anion such as F^-, CH_3COO^-, or CN^-. Any basic anion will react with H^+ ions present in the solution and thereby affect the solubility of the salt. Take, for instance, the effect of adding a strong acid on the solubility of silver acetate. To explain this we need two reaction steps. In the first one silver acetate dissolves, and in the second the acetate ion combines with $H^+(aq)$ ions present in the system:

$$CH_3COOAg(s) \rightleftharpoons CH_3COO^-(aq) + Ag^+(aq)$$

$$H^+(aq) + CH_3COO^-(aq) \rightarrow CH_3COOH(aq)$$

In the presence of H^+ ions the concentration of acetate ion is lowered by the occurrence of the second reaction. The second reaction occurs essentially completely because it forms a weak acid. According to Le Chatelier's principle, the solubility equilibrium will shift to the right, causing more silver acetate to dissolve.

The solubilities of salts containing anions that do *not* hydrolyze, such as Cl^-, Br^-, I^-, and NO_3^-, are not affected by pH. These anions an conjugate bases of strong acids.

EXAMPLE 16.12   Common Ion Effect

What is the solubility of $PbCl_2$ in 0.50 M NaCl solution?

•Method of Solution

Write the solubility equilibrium reaction for $PbCl_2$, and write the K_{sp} expression. The K_{sp} value is in Table 16.1.

$$PbCl_2(s) \rightleftharpoons Pb^{2+}(aq) + 2Cl^-(aq)$$

$$K_{sp} = [Pb^{2+}][Cl^-]^2 = 2.4 \times 10^{-4}$$

All of the $[Pb^{2+}]$ ion comes from the dissolution reaction. Let s = molar solubility of $PbCl_2$. The only source of Pb^{2+} is $PbCl_2$. Therefore:

$$[Pb^{2+}] = s$$

Chlorine ions are contributed by $PbCl_2$ and by NaCl (a strong electrolyte):

$$NaCl(aq) \rightarrow Na^+(aq) + Cl^-(aq)$$

therefore $[Cl^-] = 0.50$ M + 2s

A table here will help summarize the changes in concentrations.

Concentration	$PbCl_2(s) \rightleftharpoons$	$Pb^{2+}(aq)$	+ $2Cl^-(aq)$
Initial (M)		0	0.50
Change (M)		s	2s
Equilibrium (M)		s	(0.50 + 2s)

•Calculation

Substituting into K_{sp}, we get:

$$[Pb^{2+}][Cl^-]^2 = K_{sp}$$

$$(s)(0.50 + 2s)^2 = 2.4 \times 10^{-4}$$

If we assume 0.50 >> 2s, then solving the problem is greatly simplified:

$$(s)(0.50)^2 = 2.4 \times 10^{-4}$$

$$s = 9.6 \times 10^{-4} \text{ M}$$

•Comment

Note that the assumption 0.50 >> 2s was valid; 0.50 >> 1.9×10^{-3} M.

EXAMPLE 16.13 Solubility of a Metal Hydroxide

What is the solubility of $Pb(OH)_2$ in a buffer solution of pH 8.0 and in another of pH 9.0?
Given $K_{sp} = 1.2 \times 10^{-15}$.

•Method of Solution

Write the solubility equilibrium and the K_{sp} expression for $Pb(OH)_2$.

$$Pb(OH)_2(s) \rightleftharpoons Pb^{2+}(aq) + 2OH^-(aq)$$

$$K_{sp} = [Pb^{2+}][OH^-]^2 = 1.2 \times 10^{-15}$$

Let s equal the molar solubility of $Pb(OH)_2$. The concentration of Pb^{2+} will be equal to s. The concentration of OH^- ion will not be $2[Pb^{2+}]$ because in a buffer solution, the H^+ ion and OH^- ion concentrations are maintained constant by the buffer. At pH 8.0, the pOH is 6.0, and $[OH^-] = 1.0 \times 10^{-6}$ M. In the buffer solution, the $[OH^-]$ ion concentration is *maintained constant* at 1.0×10^{-6} M. The OH^- from dissolution of $Pb(OH)_2$ is neutralized by the buffer. Therefore we can substitute as follows to obtain the solubility at pH 8.

•Calculation

$$[Pb^{2+}] = s$$

$$[Pb^{2+}] = \frac{K_{sp}}{[OH^-]^2}$$

$$s = \frac{1.2 \times 10^{-15}}{(1.0 \times 10^{-6})^2} = 1.2 \times 10^{-3} \text{ M}$$

And at pH = 9, where $[OH^-] = 1.0 \times 10^{-5}$ M,

$$s = \frac{1.2 \times 10^{-15}}{(1.0 \times 10^{-5})^2} = 1.2 \times 10^{-5} \text{ M}$$

•Comment

The lower solubility of $Pb(OH)_2$ at pH 9.0, than at pH 8.0, is due to the common ion effect.

EXAMPLE 16.14 Effect of pH on Solubility

Which of the following salts will be more soluble at acidic pH than in pure water?
a. AgBr   b. $Ba(OH)_2$ c. $MgCO_3$

•Method of Solution

First we need to identify the ions present in solutions of these compounds. Then we identify those with acid-base properties.

a. $AgBr(s) \rightleftharpoons Ag^+ + Br^-$ Neither Ag^+ nor Br^- have acid-base properties and so the solubility of AgBr is not affected by pH.

b. $Ba(OH)_2 \rightleftharpoons Ba^{2+} + 2OH^-$

Comparing equilibria in water and in acidic solution, as $[H^+]$ increases, $[OH^-]$ decreases, and the solubility equilibrium will shift to the right, causing the solubility of $Ba(OH)_2$ to increase.

c. Carbonate ion is a base, and in acid solution it combines with $H^+(aq)$.

$$MgCO_3 \rightleftharpoons Mg^{2+} + CO_3^{2-}$$

$$H^+ + CO_3^{2-} \rightarrow HCO_3^-$$

With the increase in $[H^+]$ in acid solution, $[CO_3^{2-}]$ decreases, and the solubility equilibrium shifts to the right, causing more $MgCO_3$ to dissolve.
Answer: (b) and (c)

EXERCISES

19. a. What is the molar solubility of Ag_3PO_4 in 0.20 M Na_3PO_4? $K_{sp} = 1.8 \times 10^{-18}$
    b. Write the solubility equilibrium reaction for Ag_3PO_4.

20. What is the molar solubility of Ag_2CrO_4 in 0.20 M $AgNO_3$?

21. Calculate the molar solubility AgCl in a solution containing 0.010 M $CaCl_2$.

22. Which of the following salts will be more soluble in an acidic solution than in a basic solution?

    a. AgI   b. $BaCO_3$ c. $Ca(CH_3COO)_2$ d. $Zn(OH)_2$ e. $CaCl_2$

23. What is the solubility of $Pb(OH)_2$ in a solution with a pH of 10.00? Given $K_{sp} = 1.2 \times 10^{-15}$

24. Calculate the minimum pH that will just prevent precipitation of $Mg(OH)_2$ from an aqueous solution containing 0.075 M Mg^{2+}. K_{sp} $Mg(OH)_2 = 1.2 \times 10^{-11}$

COMPLEX IONS AND SOLUBILITY (16.10)

STUDY OBJECTIVES

1. Write the formation constant expression for a given complex ion.
2. Explain the effect of complex ion formation on the solubility of a salt whose metal ion forms a complex ion in solution.
3. Calculate the solubility of a salt in a solution where the metal ion of the salt forms a complex ion.

The Formation Constant. When Ag^+ ion reacts with two CN^- ions:

$$Ag^+(aq) + 2CN^-(aq) \rightleftharpoons Ag(CN)_2^-(aq)$$

it forms a complex ion $Ag(CN)_2^-$. An ion made up of the metal ion bonded to one or more molecules or ions is called a **complex ion**. Some other examples are $Ag(NH_3)_2^+$, $Ni(H_2O)_6^{2+}$, and $Cu(CN)_4^{2-}$.

Complex ions are extremely stable and so have an important effect on certain chemical species in solution. A measure of the tendency of a metal ion to form a certain complex ion is given by its formation constant, K_f.

The **formation constant** is the equilibrium constant for the reaction that forms the complex ion. For example,

$$Ag^+ + 2CN^- \rightleftharpoons Ag(CN)_2^-$$

$$K_f = \frac{[Ag(CN)_2^-]}{[Ag^+][CN^-]^2} = 1.0 \times 10^{21}$$

Table 16.4 in the text lists K_f values for selected complex ions. The more stable the complex ion, the greater the extent of reaction, and the greater the value of K_f.

The formation of a complex ion has a strong effect on the solubility of a metal salt. For example, AgI, which has a very low solubility in water, will dissolve in an aqueous solution of NaCN. The stepwise process is:

$$AgI(s) \rightleftharpoons Ag^+(aq) + I^-(aq) \qquad K_{sp}$$
$$Ag^+(aq) + 2CN^-(aq) \rightleftharpoons Ag(CN)_2^-(aq) \qquad K_f$$

$$\text{overall} \quad AgI(s) + 2CN^-(aq) \rightleftharpoons Ag(CN)_2^-(aq) + I^-(aq) \quad K$$

The formation of the complex ion in the second step causes a decrease in $[Ag^+]$. Therefore the first equilibrium shifts to the right according to Le Chatelier's principle. This is why AgI is more soluble in CN^- solution than in pure water.

The overall reaction is the sum of the two steps. The equilibrium constant for the overall reaction is the product of the equilibrium constants of the two steps:

$$K = K_{sp} \times K_f$$

Tables 16.1 and 16.4 of the text give, respectively:

$$K_{sp} = 8.3 \times 10^{-17}$$
$$K_f = 1.0 \times 10^{21}$$

Therefore

$$K = 8.3 \times 10^4$$

The large value of K shows that AgI is very soluble in NaCN solution.

EXAMPLE 16.15   Concentration of Uncomplexed Metal Ion

Calculate the concentration of free Ag^+ ions in a solution formed by adding 0.20 mol of $AgNO_3$ to 1 L of 1.0 M NaCN.

•Method of Solution

In this solution Ag^+ ions will complex with CN^- ions, and the concentration of Ag^+ will be determined by the equation

$$Ag^+ + 2CN^- \rightleftharpoons Ag(CN)_2^- \qquad K_f = 1.0 \times 10^{21}$$

Since K_f is so large, we expect the Ag^+ to react essentially quantitatively to form 0.20 mol of $Ag(CN)_2^-$. To find the concentration of free Ag^+ at equilibrium, use the equilibrium constant expression

$$K_f = \frac{[Ag(CN)_2^-]}{[Ag^+][CN^-]^2}$$

Rearranging gives

$$[Ag^+] = \frac{[Ag(CN)_2^-]}{K_f[CN^-]^2}$$

•**Calculation**

Substitute all known equilibrium concentrations into the above equation.
At equilibrium; $[Ag(CN)_2^-] = 0.20$ M, and

$$[CN^-] = [CN^-]_0 - 2[Ag(CN)_2^-]$$

$$[CN^-] = 1.0 \text{ M} - 0.40 \text{ M} = 0.60 \text{ M}$$

Therefore

$$[Ag^+] = \frac{(0.20)}{(1.0 \times 10^{21})(0.60)^2}$$

$$[Ag^+] = 5 \times 10^{-22} \text{ M}$$

•**Comment**

This concentration corresponds to only three Ag^+ ions per 100 mL!

EXAMPLE 16.16   Effect of Complex Formation on Solubility

Calculate the molar solubility of silver bromide in 6.0 M NH_3.

•**Method of Solution**

The solubility of AgBr is determined by two equilibria. The solubility and complex ion equilibria are

$$AgBr(s) \rightleftharpoons Ag^+ + Br^- \qquad K_{sp} = 7.7 \times 10^{-13}$$
$$Ag^+ + 2NH_3 \rightleftharpoons Ag(NH_3)_2^+ \qquad K_f = 1.5 \times 10^7$$

overall $AgBr(s) + 2NH_3 \rightleftharpoons Ag(NH_3)_2^+ + Br^- \qquad K = K_{sp} \times K_f$

The equilibrium constant of this net reaction controls the solubility of AgBr.

$$K = K_{sp} \times K_f = 1.2 \times 10^{-5}$$

$$K = \frac{[Ag(NH_3)_2^+][Br^-]}{[NH_3]^2}$$

•**Calculation**

Let s equal the solubility of AgBr in 6.0 M NH_3.

Concentration	$AgBr(s) + 2NH_3 \rightleftharpoons$	$Ag(NH_3)_2^+ +$	Br^-
Initial (M)	6.0	0	0
Change (M)	−2s	+s	+s
Equilibrium (M)	(6.0 − 2s)	s	s

Substitute into the K expression for the overall reaction

$$K = \frac{(s)(s)}{(6.0 - 2s)^2} = 1.2 \times 10^{-5}$$

The left-hand side is a perfect square. Taking the square root of both sides gives

$$\sqrt{\frac{s^2}{(6.0 - 2s)^2}} = \sqrt{(1.2 \times 10^{-5})}$$

$$\frac{s}{6.0 - 2s} = 3.5 \times 10^{-3}$$

$$s = 2.1 \times 10^{-2} \text{ M}$$

•**Comment**

The solubility of AgBr in pure water is 8.8×10^{-7} M. The enhanced solubility in this example is due to the formation of the complex ion $Ag(NH_3)_2^+$.

EXERCISES

25. Write equations for the formation reaction of each of the following complex ions.
    a. $Cu(NH_3)_4^{2+}$
    b. $Cu(CN)_4^{2-}$
    c. $Ag(CN)_2^-$

26. Calculate the concentration of Cu^{2+} ions in a solution formed by adding 0.0500 mol of $CuSO_4$ to 0.500 L of 0.500 M NaCN.

27. Determine the equilibrium constant for the following reaction.
    $$AgBr(s) + 2CN^- \rightleftharpoons Ag(CN)_2^- + Br^-$$

28. Calculate the molar solubility of silver bromide in 1.0 M NH_3.

CONCEPTUAL QUESTIONS

1. Why is a solution of HBr and NaBr with approximately equal concentrations of each not a buffer?

2. What is the effect on the pH of the following solutions when:
    a. solid sodium acetate (CH_3COONa) is added to a dilute solution of acetic acid?
    b. solid NaCl is added to a dilute solution of NaOH?
    c. solid KOH is added to a dilute solution of acetic acid?

3. Describe how to choose an acid-base indicator for the titration of 50 mL of a 0.10 M HCN solution with 0.10 M NaOH.

4. Explain how changing the pH affects the solubility of $Zn(OH)_2$.

5. Why does an indicator change from its acid color to its base color over a range of pH values.

PRACTICE TEST

1. A buffer solution is prepared by mixing 500 mL of 0.60 M CH_3COOH with 500 mL of 1.00 M CH_3COONa solution. What is the pH of this solution?

2. A buffer solution is prepared by mixing 300 mL of 0.10 M HNO_2 with 200 mL of 0.40 M $NaNO_2$.
   a. Calculate the pH of the resulting solution.
   b. What is the new pH after 2.0 mL of 2.0 M HCl are added to this buffer?

3. What is the optimum pH of an HCN / CN^- buffer?

4. Which one of the following equimolar mixtures is suitable for making a buffer solution with an optimum pH of about 9.2?
   a. $NaC_2H_3O_2$ and $HC_2H_3O_2$
   b. NH_4Cl and NH_3
   c. HF and NaF
   d. HNO_2 and $NaNO_2$
   e. NaCl and HCl

5. A 20.0 mL portion of a solution of 0.0200 M HNO_3 is titrated with 0.0100 M KOH.
   a. What is the pH at the equivalence point?
   b. How many mL of KOH are required to reach the equivalence point?
   c. What will the pH be after only 10.0 mL of KOH are added?
   d. What will the pH be after 45.0 mL of KOH are added?

6. Consider the titration of 50.0 mL of 0.10 M CH_3COOH with 0.10 M NaOH. What is the pH at the equivalence point?

7. Calculate the pH of a solution prepared by mixing 25.0 mL of 0.10 M HCl and 25.0 mL of 0.25 M CH_3COONa?

8. Which of the following is the *least* soluble?
   a. SrF_2 $K_{sp} = 2.8 \times 10^{-9}$
   b. $Zn(OH)_2$ $K_{sp} = 1.8 \times 10^{-14}$
   c. PbI_2 $K_{sp} = 1.4 \times 10^{-8}$
   d. BaF_2 $K_{sp} = 1.7 \times 10^{-6}$

9. At a certain temperature the solubility of barium chromate ($BaCrO_4$) is 1.8×10^{-5} mol/L. What is the K_{sp} value at this temperature?

10. If the solubility of $Fe(OH)_2$ in water is 7.7×10^{-6} mol/L at a certain temperature, what is its K_{sp} value at that temperature?

11. What is the molar solubility of silver phosphate (Ag_3PO_4) in water? $K_{sp} = 1.8 \times 10^{-18}$

12. AgCl is most soluble in:
   a. 0.1 M $AgNO_3$
   b. 0.2 M NaCl
   c. 2.0 M HCl
   d. pure water

13. Will a precipitate of $BaSO_4$ form when 400 mL of 0.020 M Na_2SO_4 is added to 700 mL of 0.001 M $BaCl_2$? K_{sp} $(BaSO_4) = 1.1 \times 10^{-10}$

14. Some municipal water supplies contain Ca^{2+} at a concentration of 3×10^{-3} M. Will a precipitate of $CaSO_4$ form when 0.10 mol of Na_2SO_4 is dissolved in 0.50 L of this water?

15. Will a precipitate of MgF_2 form when 600 mL of solution that is 2.0×10^{-4} M in $MgCl_2$ is added to 300 mL of 1.1×10^{-2} M NaF solution? K_{sp} $(MgF_2) = 6.4 \times 10^{-9}$

16. A solution of NaOH is added dropwise to a solution that is 0.100 M Ca^{2+} and 0.010 M Mg^{2+}.
   a. Which cation precipitates first?
   b. What concentration of OH^- is necessary to begin precipitation?
   c. What is the concentration of the "first cation to precipitate" when the second cation begins to precipitate?
   d. What percentage of the initial Mg^{2+} concentration remains in solution when the Ca^{2+} begins to precipitate?

17. Which of the following salts should be more soluble in acid solution than in pure water?
   a. $CaCO_3$ b. $MgCl_2$ c. $NaNO_3$ d. LiBr   e. KI

18. Which of the following is the most stable complex ion?
   a. $Ag(S_2O_3)_2^{3-}$ $K_f = 1.0 \times 10^{13}$
   b. $Ag(NH_3)_2^+$ $K_f = 1.5 \times 10^7$
   c. $Ag(CN)_2^-$ $K_f = 1.0 \times 10^{21}$

19. Calculate the concentration of free copper ion in a solution made from 1.0×10^{-2} M $Cu(NO_3)_2$ and 1.0 M NH_3.

20. What is the molar solubility of AgCl in 3.0 M NH_3?

21. What is the molar solubility of AgBr in 0.100 M $Na_2S_2O_3$? Given:

$$Ag^+ + 2\,S_2O_3^{2-} \rightleftharpoons Ag(S_2O_3)_2^{3-} \qquad K_f = 1.0 \times 10^{13}$$

22. a. Will a precipitate form when 400 mL of 0.20 M K_2CrO_4 is added to 400 mL of 0.10 M $AgNO_3$? K_{sp} $(Ag_2CrO_4) = 1.1 \times 10^{-12}$
   b. If yes, what is the silver ion concentration left in solution?

23. What is the minimum concentration of aqueous NH_3 required to prevent AgCl(s) from precipitating from 1.0 L of solution prepared from 0.20 mol of $AgNO_3$ and 0.010 mol of NaCl?

ANSWERS

Exercises

1. 1.34% versus 0.09%
2. a. pH = 4.66   b. no
3. 6.50
4. a. initial pH is 9.39 and final is 9 45   b. $NH_4^+ \rightleftharpoons NH_3 + H^+$
   c. $OH^- + NH_4^+ \rightarrow H_2O + NH_3$
5. pH = 2.12
6. pH > 7. Usually between 8 – 11 depending on the strength of the conjugate base of the weak acid.
7. A titration of a strong acid with a weak base.
8. a. $HBr(aq) + NaOH(aq) \rightarrow H_2O(l) + NaBr(aq)$ b. pH = 1.40
9. 4.37
10. 7
11. a. $BaCO_3(s) \rightleftharpoons Ba^{2+}(aq) + CO_3^{2-}(aq)$ $K_{sp} = [Ba^{2+}][CO_3^{2-}]$
    b. $CaF_2(s) \rightleftharpoons Ca^{2+}(aq) + 2F^-(aq)$ $K_{sp} = [Ca^{2+}][F^-]^2$
    c. $Al(OH)_3(s) \rightleftharpoons Al^{3+}(aq) + 3OH^-(aq)$ $K_{sp} = [Al^{3+}][F^-]^3$
    d. $Ag_3PO_4(s) \rightleftharpoons 3Ag^+(aq) + PO_4^{3-}(aq)$ $K_{sp} = [Ag^+]^3[PO_4^{3-}]$
12. $K_{sp} = 4.9 \times 10^{-6}$
13. a. $s = 2.2 \times 10^{-4}$ M   b. $[F^-] = 4.4 \times 10^{-4}$
14. $K_{sp} = 1.7 \times 10^{-6}$
15. No
16. Yes, MgF_2
17. $[Ba^{2+}] = 3.2 \times 10^{-5}$ M
18. a. $CuI(s)$ b. For CuI to ppt $[I^-]$ must be 5.1×10^{-8} M. For PbI_2 to ppt $[I^-]$ must be 3.7×10^{-3} M.
    c. 0.0014%
19. a. $s = 2.1 \times 10^{-6}$ M   b. $Ag_3PO_4(s) \rightleftharpoons 3Ag^+ + PO_4^{3-}$
20. $s = 2.8 \times 10^{-11}$ M
21. $s = 8.0 \times 10^{-9}$ M
22. b, c, and d
23. $s = 1.2 \times 10^{-7}$ M
24. pH = 9.1
25. a. $Cu^{2+}(aq) + 4NH_3 \rightleftharpoons Cu(NH_3)_4^{2+}$
    b. $Cu^{2+}(aq) + 4CN^- \rightleftharpoons Cu(CN)_4^{2-}$
    c. $Ag^+(aq) + 2CN^- \rightleftharpoons Ag(CN)_2^-$
26. 1.0×10^{-22} M
27. $K = K_{sp} \times K_f = 7.7 \times 10^8$
28. $s = 3.5 \times 10^{-3}$ M

Conceptual Questions

1. A solution of HBr and NaBr cannot be a buffer because the HBr is a strong acid. Its conjugate base Br^- ion has no base strength. This means that the buffer cannot neutralize added strong acid. A reaction such as $H^+ + Br^- \rightarrow HBr$ cannot occur.

2. a. pH increases b. no change in pH c. pH increases

3. First determine what ions are present in the solution at the equivalence point. Here the salt formed by neutralization is NaCN. Second calculate the pH at the equivalence point. At the equivalence point we have a 0.050 M NaCN solution. Cyanide ion undergoes hydrolsis $CN^- + H_2O \rightleftharpoons HCN + OH^-$. This equilibrium determines the pH. Knowing the K_b for CN^- ion calculate the OH^- ion concentration, and then the pH as in Example 15.17. Third, look at Table 16.1 in the text, and find an indicator that changes color over a pH range that includes the pH at the equivalence point.

4. The solubility reaction of $Zn(OH)_2$ is:

$$Zn(OH)_2 \rightleftharpoons Zn^{2+}(aq) + 2OH^-(aq)$$

The pH of a solution directly affects the OH^- ion concentration. To adjust the pH upward we must add more OH^- ion to a solution. In terms of the solubility equilibrium OH^- ion is a common ion and so the solubility equilibrium is shifted to the left as pH increases. Increasing the pH lowers the solubility of $Zn(OH)_2$.

5. Indicators are weak acids that ionize according to the equation:
$$HIn \rightleftharpoons H^+ + In^-$$
When $[HIn] / [In^-] > 10$ the solution has the color of HIn. When $[HIn] / [In^-] > 0.1$ the solution has the color of HIn. The equilibrium has to shift a long way for the ratio to change from 10:1 and 1:10. You can see from the equation that to shift the equilibrium you must change the H^+ ion concentration. Therefore, the indicator can change from its acid color to its base color only over a range of pH values.

Practice Test

1. pH = 4.96
2. a. pH = 3.77 b. 3.66
3. 9.31
4. b. NH_4Cl and NH_3
5. a. pH = 7.0 b. 40.0 mL c. 2.00 d. 10.89
6. Yellow
7. pH = 4.92
8. b. $Zn(OH)_2$
9. $K_{sp} = 3.2 \times 10^{-10}$
10. $K_{sp} = 1.8 \times 10^{-15}$
11. $s = 1.6 \times 10^{-5}$ M
12. d. pure water
13. Yes
14. Yes
15. No
16. a. Mg^{2+} b. $[OH^-] = 3.5 \times 10^{-5}$ M c. $[Mg^{2+}] = 1.5 \times 10^{-7}$ M d. 1.5×10^{-3}% of Mg^{2+} remains.
17. a. $CaCO_3$
18. c. $Ag(CN)_2^-$
19. $[Cu^{2+}] = 2.35 \times 10^{-16}$ M
20. s = 0.13 M
21. s = 0.042 M
22. a. Yes b. $[Ag^+] = 3.8 \times 10^{-6}$ M
23. $[NH_3] = 0.91$ M

Chapter Seventeen
CHEMISTRY IN THE ATMOSPHERE

- Earth's Atmosphere and Its Regions (17.1 – 17.2)
- Depletion of the Ozone Layer (17.3)
- The Greenhouse Effect and Acid Rain (17.5 – 17.6)
- Photochemical Smog (17.7)
- Indoor Air Pollution (17.8)

EARTH'S ATMOSPHERE AND ITS REGIONS (17.1 – 17.2)

STUDY OBJECTIVES

1. List the major chemical components and several of the minor components of the atmosphere.
2. Describe the approximate altitude and characteristics of four regions of the Earth's atmosphere.
3. Describe the chemical processes responsible for light emission by the aurora borealis and aurora australis.

Earth's Atmosphere. Earth's atmosphere is unique in the solar system due to its high oxygen (O_2) content. Molecular oxygen makes up 21 percent of the atmosphere by volume. Table 17.1 in the text shows that N_2, O_2, and Ar make up over 99.9% of the atmosphere by volume. The other noble gas elements (besides argon) are only present in trace amounts. Associated with the high oxygen content is the formation of ozone (O_3), a molecule that absorbs ultraviolet light and filters it from sunlight. This radiation is known to be harmful to many of the life forms that inhabit Earth's surface.

Nitrogen is the major constituent of the atmosphere. Nitrogen molecules are held together by a triple bond and therefore are extremely stable. Biological and industrial *nitrogen fixation* converts the elemental form of nitrogen into nitrogen compounds. Lightning plays an important role by initiating the production of about 30 million tons of nitric acid each year. There are three steps in the mechanism of nitric acid production. Lightning provides the energy to break up the N_2 molecule and initiates the first step in the reaction sequence.

$$N_2(g) + O_2(g) \rightarrow 2NO(g)$$
$$2NO(g) + O_2(g) \rightarrow 2NO_2(g)$$
$$2NO_2(g) + H_2O(l) \rightarrow HNO_3(g) + HNO_2(g)$$

The nitrogen in nitric acid is an essential plant nutrient.

Carbon dioxide, the fourth most abundant component, makes up only 0.033% of the air. Carbon dioxide is extremely important to Earth's heat balance even though its relative concentration is low. In this chapter we will review the effects of air pollutants on the ozone layer, on the Earth's temperature, and on other phenomena related to the atmosphere.

Regions of the Atmosphere. For convenience the atmosphere is divided into four regions with respect to altitude. The lowest region is the *troposphere.* This region extends up to about 10 km. The temperature decreases with altitude in the troposphere from about 25°C at sea level to –55°C at 10 km. The troposphere contains about 80% of the mass of the atmosphere and virtually all the precipitation, clouds, and water vapor. All the weather occurs in the troposphere. Remember that the atmosphere gradually thins out as the altitude increases. The atmospheric pressure at 10 km is about 230 mm Hg.

Above 10 km we enter the *stratosphere* which extends upward to 50 km. In the stratosphere the temperature increases with altitude from about –55°C at 10 km to –10°C at 50 km. The atmospheric pressure at 50 km is only 0.5 mm Hg. For the most part there is very little mixing of the contents of the troposphere and the stratosphere.

Above the stratosphere we enter the *mesosphere*. As in the troposphere, the temperature decreases with altitude, falling to –90°C at 80 km. From 80 km to 100 km we find the *ionosphere* (also called the thermosphere). In this region, the temperature increases slightly with increasing altitude. The region gets its name from the ions that are produced there by high energy solar radiation. The temperatures associated with the different regions of the atmosphere are shown in Figure 17.3 of the text.

Auroral Displays. Chemical processes in the ionosphere produce the phenomena called "northern lights," or aurora borealis in the Northern hemisphere and aurora australis in the Southern hemisphere. In Chapter 7 you learned that excited atoms emit light when electrons jump from a higher to a lower energy state. Excited O atoms and N_2^+ ions are produced in the ionosphere when protons from solar flares are ejected from the Sun and eventually collide with molecules in the ionosphere.

$$N_2 + energy \rightarrow N_2^+{}^* + e^-$$
$$O_2 + energy \rightarrow 2O^*$$

The green and red colors seen in the auroral displays are emitted by O^* atoms, and the blue and violet colors are emitted by $N_2^+{}^*$ ions. The symbols contain an asterisk to indicate excess energy.

During the emission process excited atoms (denoted by an asterisk) drop to the ground state and emit photons (hv).

$$O^* \rightarrow O + hv$$
$$N_2^+{}^* \rightarrow O + hv$$

The auroral displays occur near the Earth's poles because solar protons are attracted there by Earth's magnetic field.

EXERCISES

1. What is nitrogen fixation. Write an equation for an example.
2. Name the regions of the atmosphere in order of increasing altitude. What region has dramatic weather changes?
3. a. What chemical species emits green and red light in auroral displays?
   b. What chemical species emits blue and violet light in auroral displays?

DEPLETION OF THE OZONE LAYER (17.3)

STUDY OBJECTIVES

1. Write the steps in the mechanism of ozone formation.
2. Write chemical equations describing the destruction of ozone in the stratosphere.

The Ozone Layer. Ozone (O_3) is a gas that is present in the atmosphere in a layer or shell that extends around the entire Earth called the "ozone layer." This layer is centered in the stratosphere about 25 km above the Earth's surface. Ozone molecules have the ability to absorb some of the ultraviolet (UV) light present in sunlight. It is in this way that the ozone layer protects life on Earth from some of the harmful effects of sunlight. Ultraviolet light causes skin cancer and eye damage in humans and mutations in plants.

The term "ozone layer" is misleading because it may imply that the layer is pure ozone. Actually the concentration of ozone is extremely low. Ozone is spread thinly at a concentration of about 10 ppm at elevations between 15 and 35 km. If all the ozone in the atmosphere was collected into a layer of pure ozone at standard temperature and pressure, it would only be 3 mm thick! The low concentration and unstable nature of ozone combine to make the ozone layer a fragile part of the atmosphere.

Formation and Destruction of Ozone. Ozone is formed in the ozone layer by a two step mechanism initiated by the action of UV light on oxygen molecules. In the first step, UV-light dissociates molecular oxygen into oxygen atoms in a process called *photodissociation*. The energy of the UV-photon is enough to break the O—O bond. In the second step, the oxygen atom combines with an oxygen molecule, forming ozone.

$$O_2(g) + h\nu \rightarrow 2O(g)$$
$$O(g) + O_2(g) \rightarrow O_3(g) \text{ (occurs twice)}$$

net $3O_2(g) \rightarrow 2O_3(g)$

Ozone, being an unstable molecule, tends to undergo reactions that convert it to the more stable O_2 molecule. Ozone absorbs UV light with wavelengths in the range 200 to 300 nm. When ultraviolet light is absorbed by ozone, it too undergoes photodissociation:

$$O_3(g) + h\nu \rightarrow O_2(g) + O(g)$$

Atomic oxygen can also destroy ozone.

$$O(g) + O_3(g) \rightarrow 2O_2(g)$$

Catalytic Depletion of Ozone. The concentration of ozone in the stratosphere depends on a balance between the rate of its production from molecular oxygen and its rate of destruction.

$3O_2(g) \rightarrow 2O_3(g)$ production of ozone
$2O_3(g) \rightarrow 3O_2(g)$ destruction of ozone

Ozone is said to be in a steady state. This refers to the situation where the rate of ozone production is approximately equal to its rate of destruction. Any process that speeds up the destruction of ozone will cause its stratospheric concentration to decrease.

The depletion of stratospheric ozone has been associated with man-made substances called chlorofluorocarbons, or CFCs for short. The CFCs were developed in the 1930s by chemists who were seeking a new refrigerant to replace toxic and corrosive ammonia and sulfur dioxide then in commercial use as refrigerants. Fluorocarbon-12 (CF_2Cl_2) was their answer, and it quickly found application in refrigerators, freezers, and air conditioners. In the 1950s, fluorocarbon-11 ($CFCl_3$) became a basic propellant for the new aerosol can industry, fostering the development of the spray can. Soon, it was also developed as a blowing agent for Styrofoam in insulation, cushions, plastic furniture, and sealants. While there are many fluorocarbon compounds, fluorocarbon-11 and fluorocarbon-12 find the widest applications. The advantage of CFCs was that they were nontoxic, noncorrosive, and nonreactive.

There are no known chemical reactions in the troposphere that destroy CFCs. Because of their inertness, CFCs accumulate in the troposphere and gradually diffuse up to the stratosphere. Since ozone filters out UV light, there is more UV light in the ozone layer than below it. In the ozone layer CFCs are bombarded by high-energy UV radiation and break apart. This photodissociation of fluorocarbon molecules breaks a C—Cl bond, and releases a chlorine atom. In the case of chlorofluorocarbon-12 the reaction is:

$$CCl_2F_2 + h\nu \rightarrow CClF_2 + Cl$$

The Cl atom attacks an ozone molecule, converting it to the more stable oxygen molecule, and forming an intermediate molecule, chlorine monoxide ClO. Compounding the problem is that once a chlorine atom has destroyed an ozone molecule, then it is regenerated when the intermediate ClO reacts with an O atom. Recall that O atoms are prevalent in the stratosphere.

$$Cl(g) + O_3(g) \rightarrow ClO(g) + O_2(g)$$
$$ClO(g) + O(g) \rightarrow O_2(g) + Cl(g)$$

net $\qquad O(g) + O_3(g) \rightarrow 2O_2(g)$

Since Cl atoms accelerate the destruction of ozone and are regenerated, they are true *catalysts*. An increased rate of destruction of ozone has the expected effect of lowering the atmospheric ozone concentration.

The net result of ozone depletion is that UV light, which is usually absorbed by ozone, is no longer sufficiently filtered from sunlight. More UV light can reach the Earth's surface. The National Research Council predicts 10,000 additional cases of skin cancer in the U.S. for each 1% depletion in the ozone layer. This problem will be around for a long time because currently the troposphere is a giant reservoir of CFCs. Estimates are that there are already enough CFCs in the troposphere to cause depletion of ozone for the next 100 years!

Currently intense efforts are underway to develop CFC substitutes. The idea is to maintain the noncorrosive properties but make a molecule that is a little more reactive. HCFC-123 ($C_2HCl_2F_3$), where the H represents hydrogen, is a promising possibility. Introduction of a hydrogen atom into CFC molecules makes them more reactive. Chemists are synthesizing a number of HCFC molecules to try as the first generation of substitutes. The goal is to find molecules that are more easily oxidized in the troposphere, before they reach the stratosphere.

Polar Ozone Holes. Since 1985 a vast hole in the ozone layer appears during the Spring months over Antarctica. It fills in during the summer and fall, but opens again each Spring. The size of this hole has increased every spring since then. Laboratory studies of ice particles, that mimic polar stratospheric clouds (PSCs) that form over Antarctica during the winter, again place the blame on chlorine species from CFCs for this dramatic loss in ozone. The concentration of the reaction intermediate chlorine monoxide in the ozone hole is large enough to account for the ozone loss.

Chlorine can be stored in chlorine nitrate, a so-called reservoir molecule. Chlorine nitrate is an unstable molecule that is formed when chlorine monoxide reacts with nitrogen dioxide.

$$ClO(g) + NO_2(g) \rightarrow ClONO_2(g)$$

To explain the observed ozone depletion during the Antarctic spring, the CFC theories where adjusted to include the reaction of chlorine nitrate with HCl on the surfaces of PSCs. These clouds contain HCl. Acting as a heterogeneous catalysis, these PSCs provide a surface for reactions converting HCl and chlorine nitrate to molecular chlorine.

$$HCl + ClONO_2 \rightarrow Cl_2 + HNO_3$$

The first sunlight of spring can trigger the release of chlorine atoms by photodissociation of chlorine molecules.

$$Cl_2 + h\nu \rightarrow Cl + Cl$$

The only sure way to save the ozone layer is a complete worldwide ban on CFC manufacture. To this end 24 nations met in Montreal in 1988. The Montreal Protocol is designed to achieve a 35% reduction in worldwide CFC production.

EXERCISES

4. a. What region of the atmosphere contains the ozone layer?
    b. Ozone is distributed over approximately what range of altitudes?
5. What is the average concentration of ozone in the ozone layer?

6. What is the overall reaction for the formation of ozone in the stratosphere?
7. Give three reactions that are examples of photodissociation.
8. What does the abbreviation CFC stand for? Write the formulas of two CFCs.
9. What is a "molecular reservoir" of atmospheric chlorine.
10. What is the Montreal Protocol?

THE GREENHOUSE EFFECT AND ACID RAIN (17.5 – 17.6)

STUDY OBJECTIVES

1. Explain how some gases produce a greenhouse effect.
2. Describe the proposed chemical events by which sulfur dioxide leads to acid rain.

The Greenhouse Effect. Each year sees another 9 gigatons (Gtons) of carbon dioxide added to the Earth's atmosphere. This is only about 1/1,000,000 of the total mass of the atmosphere. Though relatively small in amount, carbon dioxide has an important effect on the Earth's average temperature. Carbon dioxide, water, and a few other gases such as methane are called "greenhouse gases." This means that they act like the glass in a greenhouse. They let sunlight through, but do not let heat out. The main components of the air, O_2 and N_2, are not greenhouse gases.

"Greenhouse warming" occurs in the following way. Visible light from the sun passes directly through the main components of the atmosphere: N_2, O_2, CO_2, and H_2O vapor. This radiant energy is absorbed by Earth's oceans and land areas and warms the surface. All warm objects radiate infrared radiation, called IR. The oceans and land areas emit IR, which passes directly through N_2 and O_2 into outer space. The loss of this energy would mean that the Earth would be a cooler place. However, $CO_2(g)$ and $H_2O(g)$ absorb IR, and in effect slow down the loss of heat from the surface. Therefore, Earth's surface is many degrees warmer than it would be without $CO_2(g)$ and $H_2O(g)$ in the air. This is the so-called greenhouse effect. Some carbon dioxide has always been present in the atmosphere, and some warming is considered desirable. Since atmospheric carbon dioxide is increasing at the rate of 1 ppm per year, this is expected to lead to an "enhanced greenhouse effect."

Greenhouse gases reduce the heat loss to outer space by absorbing IR. Infrared radiation increases the vibrational energy level of molecules that absorb it. Then these vibrationally excited molecules reemit IR as they drop to the ground state. Statistically 1/2 of the IR is aimed downward toward Earth's surface, and 1/2 is aimed upward toward outer space. The IR aimed back toward the surface eventually causes the temperature of the air to rise.

Scientists have been trying for many years to estimate rates of global warming and its effects. The carbon dioxide content of the atmosphere is expected to double by the year 2050. Some of the predicted effects are:

1. A temperature increase of 3°C to 5°C by that year.
2. Melting of glaciers and icecaps that will cause a 2-ft rise in sea level with its accompanying flooding of sea coasts and major cities.
3. Widespread changes in climate.

Suggestions to lower carbon dioxide emissions center around less dependence on fossil fuels for energy. Carbon dioxide emissions could be reduced by increasing efficiency of automobiles, home heating, and electric power production. Additional reduction could result from replacing existing fossil fuel electric power plants with ones that utilize solar energy and nuclear energy.

Methane, nitrous oxide, and CFCs are also strong absorbers of IR. These gases are present in only trace amounts, but enhance the greenhouse effect significantly because they absorb IR at wavelengths that water and CO_2 cannot absorb. In a sense they close an open window that would allow some IR to escape the greenhouse.

Acid Rain. The term "acid rain" was coined in 1872 by an English chemist who used it to describe the increasingly acid precipitation that fell on the industrial city of Manchester. In the century since then, acid rain has grown to be an environmental problem of global proportions. Acid rain is rainwater with a pH of less than 5.5. Precipitation in the northeastern part of the U.S. has an average pH of 4.3 and in some specific storms the pH has been as low as 2.8.

The text points out that "normal" rain is slightly acid because it contains carbonic acid. As rain passes through air containing carbon dioxide, the CO_2 dissolves and forms carbonic acid. Carbon dioxide is an acid anhydride.

$$CO_2(g) + H_2O(l) \rightarrow H_2CO_3(aq)$$

Because carbonic acid is a weak acid, the pH of rain normally will not go below 5.5. The presence of several stronger acids accounts for the lower pH values of acid rain. Acid rain usually contains sulfuric acid (H_2SO_4) and nitric acid (HNO_3). We will discuss only sulfuric acid.

These acids do not start out in the atmosphere as oxoacids. The acidic oxides, SO_2 and SO_3, are the precursors of the oxoacids. Sulfur dioxide is produced by the combustion of fossil fuels such as coal and from the processing of sulfide ores at smelters.

$$2ZnS(s) + 3O_2(g) \rightarrow 2ZnO(s) + 2SO_2(g) \quad \text{smelting an ore}$$

Coal and petroleum contain between 1% and 5% S. Combustion of sulfur in fossil fuels yields SO_2.

$$S(s) + O_2(g) \rightarrow SO_2(g) \quad \text{combustion of sulfur}$$

Once in the atmosphere some of the sulfur dioxide is oxidized to sulfur trioxide by reactions that are currently being intensely studied. The effects of light and ozone on this oxidation are mentioned in the text. The net oxidation reaction is

$$2SO_2(g) + O_2(g) \rightarrow 2SO_3(g)$$

In the final step, sulfur trioxide combines with water to form sulfuric acid. Sulfur trioxide is converted to sulfuric acid by rainwater.

$$SO_3(g) + H_2O(l) \rightarrow H_2SO_4(aq)$$

Streams and lakes show the most dramatic effects of acid rain. It is known that natural waters with a low pH can kill fish eggs, salamander eggs, and frog eggs. The extent of change in acidity of a lake or stream when under the stress of acid rain is determined mainly by the buffering capacity of the surrounding soil. Watershed soils containing limestone are alkaline and can resist rapid changes in pH, making them less susceptible to harm.

Trees are also susceptible to acid rain. Acid rainfall causes damage to leaves and the growing tissues of trees. The needles of firs, spruces, and pines turn yellow and fall off. Among the possible causes are acid rain and ozone from polluted air. These pollutants damage the cell membranes of needles, allowing nutrients to escape.

Acid precipitation can acidify the soil, interfering with nutrient availability. To grow normally trees require adequate supplies of 16 elements. Several of these, notably Ca, Mg, and K are taken up by tree roots as cations from aqueous solution in the soil. When sulfuric acid is deposited in soil by precipitation, nutrient cations can be leached from the root zone. This can greatly affect the health of forests.

Sulfur dioxide emissions can be reduced by removing SO_2 after combustion, but before it leaves the stack and is released into the atmosphere. Powdered limestone ($CaCO_3$) is injected into the hot gases leaving the combustion zone. Heat causes the carbonate to decompose into quicklime (CaO). The quicklime then reacts with SO_2 to form calcium sulfite ($CaSO_3$).

$$CaCO_3(s) \rightarrow CaO(s) + CO_2(g)$$
$$CaO(s) + SO_2(g) \rightarrow CaSO_3(s)$$

About half of the SO_2 is removed by contact with the dry CaO. The remaining SO_2 must be removed by spraying the hot gases with a suspension of quicklime. This process, called "scrubbing," creates huge amounts of calcium sulfite to dispose of. Retrofitting of scrubbers onto established power plants is very expensive and significantly raises the cost of electricity .

EXERCISES

11. Give three sources of CO_2 in the atmosphere.
12. Name three "greenhouse" gases besides carbon dioxide and water.
13. Below what pH is rain considered to be "acid rain?"
14. SO_2 and SO_3 are acid anhydrides. Write reactions to show their conversion into acids.

PHOTOCHEMICAL SMOG (17.7)

STUDY OBJECTIVES

1. Distinguish between primary and secondary pollutants.
2. Describe how ozone is formed in polluted air.

Photochemical smog is formed by the reactions of automobile exhaust in the presence of sunlight. Reactions initiated by photons are called photochemical reactions. Smog begins with certain *primary pollutants*. These substances may or may not be objectionable by themselves. These are transformed by sunlight or by ordinary chemical reactions into *secondary pollutants*. Secondary pollutants are involved directly in the buildup of smog. Nitric oxide is a good example of a primary pollutant. It is formed at the high temperatures inside an internal combustion engine when nitrogen and oxygen from air react. Nitric oxide does not build up in the air because it is rapidly converted to nitrogen dioxide.

$$N_2(g) + O_2(g) \rightarrow 2NO(g)$$
$$2NO(g) + O_2(g) \rightarrow 2NO_2(g)$$

Nitrogen dioxide is involved in a chain of reactions that produce ozone. First NO_2 is photochemically decomposed by sunlight.

$$NO_2(g) + h\nu \rightarrow NO(g) + O(g)$$

Oxygen atoms initiate a number of reactions in polluted air. An important one is the formation of ozone.

$$O(g) + O_2(g) \rightarrow O_3(g)$$

Exposure to 0.1 to 1.0 ppm of ozone produces headaches, burning eyes, and irritation to the respiratory passages. Another secondary pollutant is peroxyacetyl nitrate, better known as PAN. PAN literally brings tears to your eyes. It is an example of a **lachrymator**, a compound that causes burning of the eyes and tears. Onions, as you know, contain a lachrymator.

Unburned hydrocarbons in automobile exhaust also produce secondary air pollutants. The oxidation of unburned hydrocarbons produces various alcohols and organic acids. These can condense to produce an aerosol. Aerosols are liquid droplets dispersed in air. They are objectionable because they reduce visibility and make the air look hazy.

Efforts to control smog are usually focused on reducing the source of primary pollutants. The catalytic converters on automobiles are designed to remove NO, CO, and unburned hydrocarbons from automobile exhaust. The catalyst contains platinum and palladium metals. Nitrogen monoxide, a primary pollutant, is converted back to nitrogen and oxygen by a reaction whose rate is increased by the catalyst in the converter.

$$2NO(g) \rightarrow N_2(g) + O_2(g)$$

Carbon monoxide and unburned hydrocarbons are both oxidized to carbon dioxide by the catalytic converter, as shown by the following equations (C_5H_{12} represents a typical hydrocarbon).

$$2CO(g) + O_2(g) \rightarrow 2CO_2(g)$$
$$C_5H_{12}(g) + 8O_2(g) \rightarrow 5CO_2(g) + 6H_2O(l)$$

EXERCISES

15. Name three primary pollutants.
16. Name three pollutants that are removed by catalytic converters.
17. Is ozone in a city a primary or secondary pollutant?

INDOOR AIR POLLUTION (17.8)

STUDY OBJECTIVES

1. Describe the source of radon and how it gets into buildings.
2. Describe why the main health effects of radon are to the lungs.

Radon Gas. All isotopes of radon are radioactive. Concern about radon in homes began in 1984 when a worker at a nuclear power plant in Pennsylvania found that he was setting off the plant's radiation monitor alarms in the morning upon *arriving* at the plant. The source of the radioactivity was traced to the worker's home, which measured an extremely high level of radioactivity in the basement air. The air in his home contained an isotope of radon. Radon-222 atoms undergo radioactive decay by emitting alpha particles. Radon is always associated with uranium deposits. Some uranium occurs naturally in most soils and rocks in widely varying amounts. Radon-222 has a half-life of 3.8 days. This gives it time to migrate up out of the ground and to enter buildings through cracks in foundations. It can also dissolve in ground water and enter a home via well water.

The element radon has been known since its discovery in 1900. As a member of the noble gas group of elements, radon is a colorless, odorless, and tasteless gas with very little tendency to combine with other elements. Therefore, it tends to stay in the air. Radon atoms in the air can be inhaled into the lungs. If radon atoms decay in the lungs, we get a dose of radiation. In addition, its decay products, particularly polonium-218 and polonium-214, are deposited in the lungs. These isotopes are radioactive and contribute more to the radiation dose than that received from radon itself. These alpha-emitting isotopes are of the greatest health concern because they are solids and become trapped in respiratory passages. Here their radioactivity can damage nearby cells, which over a long period of time leads to lung cancer. In order to better assess the danger to people, health workers are trying to sort out the importance of the three main causes of lung cancer: cigarette smoking, secondary smoke, and radon gas.

EXERCISES

18. What is the formula for radon gas? To what group of elements does it belong?
19. Radon decays into two extremely dangerous radioisotopes. What are they?
20. What is the principle health effect of radon gas?

CONCEPTUAL QUESTIONS

1. Ozone occurs naturally in the atmosphere; so is it correct to call it a pollutant?
2. Is it more feasible to try to reduce the effects of acid rain by reducing SO_2 emissions from power plants or by neutralizing acidic lakes, streams, and soils.

PRACTICE TEST

1. What noticable change marks the transition from the troposphere to the stratosphere?
2. a. Write out the steps in the mechanism by which ozone is formed in the stratosphere.
   b. What other naturally occurring molecule absorbs UV light in the stratosphere?
3. a. Write out the steps in the mechanism of ozone destruction by chlorine atoms.
   b. Identify the catalyst and the intermediate.
4. What does the abbreviation HCFC stand for? Write the formula of HCFC-123.
5. What is the greenhouse effect?
6. Name two ways in which deforestation contributes to the greenhouse effect that involve CO_2 levels.
7. Give two sources of methane in the atmosphere.
8. By what means does Earth lose heat?
9. Write chemical equations that show what happens when acid rain reacts with iron and with limestone.
10. a. List three primary pollutants removed from automobile exhaust by catalytic converters.
    b. Explain briefly the role of each in the formation of smog.
11. The strength of the C—Cl bond in CF_2Cl_2 is 318 kJ/mol. What is the wavelength of a photon that has enough energy to break a C—Cl bond? Is this in the UV or visible region of the spectrum?
12. List three effects of acid rain.

ANSWERS

Exercises

1. Forming compounds from the element. $N_2(g) + O_2(g) \rightarrow 2NO(g)$
2. Troposphere, stratosphere, mesosphere, and ionosphere. The troposphere has constantly changing weather.
3. a. O* atoms   b. N_2^+* ions
4. a. the stratosphere   b. 15–35 km
5. 10 ppm
6. $3O_2(g) \rightarrow 2O_3(g)$
7. $O_2(g) + h\nu \rightarrow 2O(g)$ $O_3(g) + h\nu \rightarrow O_2(g) + O(g)$ $CCl_2F_2 + h\nu \rightarrow CClF_2 + Cl$
8. CFC = chlorofluorocarbon. CF_2Cl_2 and $CFCl_3$
9. Chlorine monoxide
10. An international agreement to limit CFC production.
11. Combustion of a carbon compound. Respiration by animals and plants. Decompositon of dead animals and plants. Volcanoes
12. Methane, nitrous oxide, and CFCs.
13. 5.6
14. $SO_2(g) + H_2O(l) \rightarrow H_2SO_3(aq)$ $SO_3(g) + H_2O(l) \rightarrow H_2SO_4(aq)$
15. Carbon monoxide, nitrogen monoxide, and unburned hydrocarbons.
16. Those in #15.
17. Secondary pollutant.
18. Ra.   Noble gas elements.
19. Polonium-214 and polonium-218.
20. Lung cancer.

Conceptual Questions

1. The answer lies in just where the ozone is. Ozone occurs naturally in the stratosphere which is too far away for people to come into contact with ozone. However in the troposphere ozone concentrations are significantly increased by automobile exhaust reacting with sunlight. Humans (and other living things) can inhale too much ozone. Health effects include severe respiratory and eye irritation. Ozone in cities is a secondary pollutant.
2. Liming is the process of adding quicklime to lakes and soils to reduce their acidity. This process is very expensive due to the enormous transportation costs involved. It is much less expensive to remove SO_2 at the power plant as it is formed. Think about how much easier it is to "collect" sulfur atoms when they are concentrated in one place, rather than when they are spread out over the landscape as sulfuric acid.

Practice Test

1. In the troposphere the temperature decreases as altitude increases. In the stratosphere temperature increases with altitude.
2. a. $O_2(g) + h\nu \rightarrow 2O(g)$
    $O(g) + O_2(g) \rightarrow O_3(g)$
    b. O_2
3. a. $Cl(g) + O_3(g) \rightarrow ClO(g) + O_2(g)$
    $ClO(g) + O(g) \rightarrow O_2(g) + Cl(g)$
    b. The catalyst is atomic Cl, and the intermediate is chlorine monoxide, ClO.
4. HCFC stands for hydrochlorofluorocarbon. CF_3CHCl_2
5. The trapping of heat near the Earth's surface by atmospheric gases, particularly carbon dioxide.
6. Combustion and decomposition of trees and plants produces CO_2. Removing trees reduces photosynthesis which would normally consume CO_2.
7. Landfills and natural gas leaks.
8. Emission of IR to space.
9. $Fe(s) + 2H^+(aq) \rightarrow Fe^{2+}(aq) + H_2(g)$
    $CaCO_3(s) + 2H^+(aq) \rightarrow Ca^{2+}(aq) + CO_2(g) + H_2O(l)$
10. a. Carbon monoxide, NO, and unburned hydrocarbons.
    b. CO is toxic but may not be involved in reactions producing smog. NO is oxidized to NO_2. Photodecomposition of NO_2 leads to ozone and PAN. Oxidation of unburned hydrocarbons produces alcohols and organic acids that eventually form aerosols and haze.
11. The wavelength is 376 nm, which is in the UV region.
12. 1. Kills fish, salamnders, and frogs.   2. Damage to leaves and needles of trees. Toxic to vegetation.
    3. Damage to stone buildings and monuments.

Chapter Eighteen
ENTROPY, FREE ENERGY, AND EQUILIBRIUM

- The Second Law of Thermodynamics (18.2 – 18.3)

- Gibbs Free Energy (18.4)

- Free Energy and Equilibrium (18.5)

THE SECOND LAW OF THERMODYNAMICS (18.2 – 18.3)

STUDY OBJECTIVES

1. Explain the meaning of the term "spontaneous process."
2. Predict, for a given process, whether the entropy of the system increases or decreases.
3. State the second law of thermodynamics.
4. Calculate the standard entropy change for a given reaction using a table of standard absolute entropies.

Spontaneous Processes. An important part of experimental chemistry deals with spontaneous reactions, that is, reactions that take place without having to continually supply energy from outside the system. One goal of thermodynamics is to be able to predict whether a reaction will take place when a given set of reactants is brought together. In this chapter we discuss those properties of a system that can be used as criteria for predicting spontaneous reactions. The release of heat by a reaction was once thought to be an indication that the reaction was spontaneous. The sign of the enthalpy change, ΔH, by itself is not an adequate guide to spontaneity because while some spontaneous reactions are known to be exothermic (ΔH is $-$), many endothermic reactions (ΔH is $+$) are known to be spontaneous as well.

It is also important to remember that the term *spontaneous* doesn't necessarily mean a fast reaction rate. The rate of a reaction is controlled by several factors as we discussed in Chapter 13: reactant concentrations , catalysts, and temperature. Thermodynamics only tells us whether or not the reaction will occur, but it tells us nothing about how fast.

Entropy. In addition to the heat absorbed or released in a spontaneous process, another factor called entropy must be considered. **Entropy** is a measure of the disorder or randomness of a system. The *entropy (S)* is a state function that increases in value as the disorder or randomness of the system increases. Entropy has the units J/K·mol. Intuitively we consider a system to be "ordered" if it is arranged according to some plan or pattern. The system is "disordered" when its parts are helter-skelter and their arrangement is random.

Order and disorder in chemical systems are discernible at the molecular level. Crystalline solids are highly ordered, with molecules or ions occupying fixed lattice sites, and with the unit cell repeated identically over and over again. Liquids are less ordered than solids because the solid lattice has broken down, and molecules or ions have kinetic energy and therefore are in random motion. The molecular motion in liquids increases the disorder compared to that of solids.

Gases are more random than liquids. On vaporization, the molar volume increases almost 1000-fold, and it is much more difficult to locate the position of any one molecule. The entropy of a substance increases as its molecules are distributed over an increasingly greater volume. For a given substance, the molecular disorder increases from the solid to the gaseous state.

$$S_{solid} < S_{liquid} < S_{gas}$$

Entropy Changes. We can estimate whether the change in entropy in a process is positive or negative. For chemical reactions in which solids or liquids are converted to gases, the entropy change ΔS is positive. It is negative for the condensation of a gas or the freezing of a liquid. When a crystal of a salt dissolves in water, the disorder increases due to the increase in freedom of motion of ions in the solution compared to the highly ordered crystal. Heating also increases the entropy of a system. The higher the temperature the more molecular motion and its accompanying disorder. Figure 18.2 in the text shows several processes that lead to an increase in entropy.

Hints for estimating the sign of ΔS:

1. If a reaction produces an increase in the number of moles of *gaseous* compounds, ΔS is positive.
2. If the total number of moles of *gaseous* compounds is decreased, ΔS is negative.
3. If there is no net change in the total number of gas molecules, then ΔS is either a *small* positive or a *small* negative number.

The Second Law. The **second law of thermodynamics** is concerned with predicting the direction of spontaneous change. It states that *the entropy of the universe increases in a spontaneous change*. The term *universe* used here refers to a system and all of it's surroundings. For an isolated system:

$$\Delta S_{univ} = \Delta S_{sys} + \Delta S_{surr} \geq 0$$

where ΔS_{sys} stands for the entropy change of the system, and ΔS_{surr} is the entropy change of the surroundings. For any spontaneous change:

$$\Delta S_{sys} + \Delta S_{surr} > 0$$

and for a reaction at equilibrium (no net change):

$$\Delta S_{sys} + \Delta S_{surr} = 0$$

Entropy Changes in Chemical Reactions. Not only can the sign ΔS for a chemical reaction be estimated, but the actual value of ΔS can also be calculated. The *absolute* value of the entropy S of one mole of an element or compound can be determined by very careful experimentation. Appendix 3 lists experimental values of the absolute entropy of a number of substances in their standard states at 1 atm and 25°C. Recall that the degree superscript " ° " refers to the standard state of the substance. These values are called absolute or sometimes standard entropies.

One can confirm that the entropy of a gas is greater than that of a liquid by comparing the *absolute entropy* of H_2O in the liquid and gas phases.

S° of $H_2O(l)$ = 69.94 J/K·mol
S° of $H_2O(g)$ = 188.72 J/K·mol

For a reaction:

$$aA + bB \rightarrow cC + dD$$

the standard entropy change of the reaction is given by:

$$\Delta S^{o}_{rxn} = \Sigma nS° \text{ (products)} - \Sigma mS° \text{ (reactants)}$$

where m and n are stoichiometric coefficients. When applied to the above reaction, we get

$$\Delta S^{o}_{rxn} = [cS°(C) + dS°(D)] - [aS°(A) + bS°(B)]$$

EXAMPLE 18.1   Changes in Entropy

Predict the sign of ΔS_{sys} for each of the following reactions using just the qualitative ideas discussed in above.

a. $H_2O_2(l) \rightarrow H_2O(l) + \frac{1}{2}O_2(g)$

b. $H^+(aq) + OH^-(aq) \rightarrow H_2O(l)$

c. $CaO(s) + CO_2(g) \rightarrow CaCO_3(s)$

•Method of Solution

a.   The number of moles of gaseous compounds in the products is greater than in the reactant. Entropy increases during this reaction. *Answer:* The sign of ΔS is +.

b.   Two reactants combine into one product in this reaction. Order is increased and so entropy decreases. *Answer:* The sign of ΔS is −.

c.   The number of gas molecules is decreased. *Answer:* The sign of ΔS is −.

EXAMPLE 18.2   The Second Law

The solubility of silver chloride is so low that it precipitates spontaneously from many solutions. The entropy change of the system is negative for this process.

$$Ag^+(aq) + Cl^-(aq) \rightarrow AgCl(s) \quad \Delta H° = -65 \text{ kJ}$$

Since ΔS decreases in this spontaneous reaction, shouldn't the reaction be nonspontaneous?

•Method of Solution

For a spontaneous reaction, the second law states that $\Delta S_{univ} > 0$:

$$\Delta S_{sys} + \Delta S_{surr} > 0$$

In order to predict whether a reaction is spontaneous, both ΔS_{sys} and ΔS_{surr} must be considered, not just ΔS_{sys}. Since the entropy change of the system is negative, the only way for the inequality to be true is if ΔS_{surr} is positive and greater in amount than ΔS_{sys}. In this case, this is a reasonable assumption because the reaction is exothermic. This means that heat is released to the surroundings, which causes increased thermal motion and disorder of molecules in the surroundings. Therefore, the sum of ΔS_{sys} and ΔS_{surr} is greater than zero, even though ΔS_{sys} is negative.

EXAMPLE 18.3   Calculation of the Entropy Change for a Reaction

Use absolute entropies to calculate the standard entropy of reaction ΔS_{rxn}^o :

$$H_2(g) + \frac{1}{2}O_2(g) \rightarrow H_2O(l)$$

•Method of Solution

The standard entropy change is given by:

$$\Delta S_{rxn}^o = S°(H_2O(l)) - [S°(H_2) + \frac{1}{2}S°(O_2)]$$

Look up the standard entropy values in Appendix 3 of the text.

$$\Delta S_{rxn}^o = 1 \text{ mol} \left(\frac{69.9 \text{ J}}{K \cdot mol} \right) - \left[1\text{mol}\left(\frac{130.1 \text{ J}}{K \cdot mol} \right) + \frac{1}{2}\text{mol}\left(\frac{205.0 \text{ J}}{K \cdot mol} \right) \right]$$

$$\Delta S_{rxn}^o = 69.9 \text{ J/K} - 233.5 \text{ J/K} = -163.6 \text{ J/K}$$

•Comment

This reaction is known to be spontaneous. The value of ΔS°_{rxn} applies only to the system. We are not calculating ΔS_{univ} here.

EXERCISES

1. For each pair of substances, choose the one having the higher standard entropy value at 25°C.
   a. $CS_2(s)$ or $CS_2(l)$ b. $SO_2(g)$ or $CO_2(g)$ c. $BaSO_4(s)$ or $BaSO_4(aq)$

2. Predict, using the intuitive ideas about entropy, whether ΔS_{rxn} will be positive, negative, or essentially zero for each of the following:
   a. $CuSO_4(s) \rightarrow Cu^{2+}(aq) + SO_4^{2-}(aq)$

   b. $SO_2(g) + \frac{1}{2}O_2(g) \rightarrow SO_3(g)$

   c. $Ca(OH)_2(s) + CO_2(g) \rightarrow CaCO_3(s) + H_2O(g)$

   d. $Ag^+(aq) + 2CN^-(aq) \rightarrow Ag(CN)_2^-(aq)$

3. Using tabulated values, calculate ΔS° for the following reaction:

   $N_2(g) + 3H_2(g) \rightarrow 2NH_3(g)$

GIBBS FREE ENERGY (18.4)

STUDY OBJECTIVES

1. Calculate free energy changes for chemical reactions, given a table of standard free energies of formation.
2. Predict, using given ΔS° and ΔH° values, at what temperature a reaction will be spontaneous under standard conditions.
3. Calculate ΔS for phase transitions.

Gibbs Free Energy. Quite often it is not possible to calculate ΔS_{surr}. This makes the second law difficult to apply when it is in the form given in the previous section. The Gibbs free energy, expressed in terms of enthalpy and entropy, refers only to the system, yet can be used to predict spontaneity. The Gibbs free energy change (ΔG) for a reaction carried out at constant temperature and pressure is given by:

$$\Delta G = \Delta H - T\Delta S$$

where both ΔH and ΔS refer to the system.

The Gibbs free energy change is equal to the maximum possible work (w) that can be obtained from a process. Any process that occurs spontaneously can be utilized to perform useful work. The Gibbs free energy of the system will *decrease* ($\Delta G < 0$) in a spontaneous process, and will *increase* ($\Delta G > 0$) in a nonspontaneous process. The free energy criteria (at constant temperature and pressure) are summarized as follows:

If $\Delta G < 0$, the forward reaction is spontaneous.
If $\Delta G = 0$, the reaction is at equilibrium.
If $\Delta G > 0$, the forward reaction is nonspontaneous. The reverse reaction will have a negative ΔG and will be spontaneous.

The **standard free-energy change of reaction,** ΔG°_{rxn}, is the free-energy change for a reaction when it occurs under standard state conditions, when reactants in their standard states are converted to products in their standard states.

Calculation of ΔG°_{rxn}. The free energy change for a reaction can be calculated in two ways. When both ΔH and ΔS are known, then $\Delta G = \Delta H - T\Delta S$ will give the free energy change at the temperature T. ΔG°_{rxn} can also be calculated from standard free energies of formation in a manner analogous to the calculation of ΔH° by using enthalpies of formation of the reactants and products. The standard free energies of formation (ΔG°_f) of selected compounds are tabulated in Appendix 3 of the text. Just as for the standard enthalpies of formation, the free energies of formation of elements in their standard states are equal to zero.

For a general reaction

$$aA + bB \rightarrow cC + dD$$

The standard free energy change is given by

$$\Delta G^{\circ}_{rxn} = [c\,\Delta G^{\circ}_f(C) + d\,\Delta G^{\circ}_f(D)] - [a\,\Delta G^{\circ}_f(A) + b\,\Delta G^{\circ}_f(B)]$$

In general:

$$\Delta G^{\circ}_{rxn} = \sum n\,\Delta G^{\circ}_f(\text{products}) - \sum m\,\Delta G^{\circ}_f(\text{reactants})$$

where n and m are stoichiometric coefficients. Example 18.4 illustrates this type of calculation.

Temperature and the Free Energy Change. From the equation $\Delta G = \Delta H - T\Delta S$ we can see that temperature too will influence the spontaneity of reaction. If both ΔH and ΔS are positive then at low temperature, as long as $\Delta H > T\Delta S$, ΔG is positive and the process will be nonspontaneous. However as temperature increases, the $T\Delta S$ term increases and eventually $\Delta H = T\Delta S$. At this point ΔG is zero. With further T increase, $T\Delta S > \Delta H$, making $\Delta G < 0$, and the reaction becomes spontaneous. Table 18.1 summarizes the four possible situations affecting the ΔG of a reaction.

Table 18.1 Enthalpy and Entropy Factors that the sign of ΔG°_{rxn}

ΔH	ΔS	ΔG
+	+	ΔG is positive at low temperatures and negative at high temperatures.
+	–	ΔG is positive at all temperatures
–	+	ΔG is negative at all temperatures
–	–	ΔG is negative at low temperatures and positive at high temperatures

In the above situation, where both ΔH° and ΔS° are positive, the temperature at which $\Delta H^{\circ} = T\Delta S^{\circ}$, and also at which $\Delta G^{\circ} = 0$, can be calculated from the equation $T = \Delta H^{\circ}/\Delta S^{\circ}$. Above this temperature, the reaction favors the products at equilibrium. See Example 18.5 for an example.

Phase Transitions. For a phase transition, $\Delta G = 0$ when the two phases coexist in equilibrium. For example, at the boiling point (T_{bp}) the liquid and vapor phases are in equilibrium, and

$$\Delta H_{vap} - T_{bp}\,\Delta S_{vap} = 0$$

rearranging gives $\Delta S_{vap} = \dfrac{\Delta H_{vap}}{T_{bp}}$

This equation allows the calculation of the entropy of vaporization from knowledge of the heat of vaporization and the boiling point.

EXAMPLE 18.4   Calculation of the Free Energy Change for a Reaction

Calculate ΔG°_{rxn} at 25°C for the following reaction using Appendix 3 and given:

$$\Delta G^\circ_f (Fe_2O_3) = -741.0 \text{ kJ/mol}$$

$$2Al(s) + Fe_2O_3(s) \rightarrow Al_2O_3(s) + 2Fe(s)$$

•**Method of Solution**

$$\Delta G^\circ_{rxn} = [\Delta G^\circ_f (Al_2O_3) + 2\Delta G^\circ_f (Fe)] - [2\Delta G^\circ_f (Al) + \Delta G^\circ_f (Fe_2O_3)]$$

$$= [1 \text{ mol } (-1576.41 \text{ kJ mol}^{-1}) + 0] - [0 + 1 \text{ mol } (-741.0 \text{ kJ mol}^{-1})]$$

$$\Delta G^\circ_{rxn} = -1576.41 \text{ kJ} + 741.0 \text{ kJ} = -835.4 \text{ kJ}$$

EXAMPLE 18.5   Effect of Temperature on ΔG

Hydrated lime $Ca(OH)_2$ can be reformed into quicklime CaO by heating.

$$Ca(OH)_2(s) \rightarrow CaO(s) + H_2O(g)$$

At what temperatures is this reaction spontaneous under standard conditions (that is, where H_2O is formed at 1 atm pressure)? Given the following data on $Ca(OH)_2$ that is not in the Appendix:

$$\Delta H^\circ_f (Ca(OH)_2) = -986.2 \text{ kJ/mol}$$

$$S^\circ(Ca(OH)_2) = 83.4 \text{ J/K·mol}$$

•**Method of Solution**

This reaction is nonspontaneous at room temperature. The temperature above which the reaction becomes spontaneous under standard conditions corresponds to $\Delta G^\circ = 0$, and is given by

$$T = \frac{\Delta H^\circ}{\Delta S^\circ}$$

ΔH° and ΔS° must be calculated separately.

•**Calculation**

$$\Delta H^\circ = [\Delta H^\circ_f (CaO) + \Delta H^\circ_f (H_2O)] - [\Delta H^\circ_f (Ca(OH)_2)]$$

From Appendix 3 and the given data

$$\Delta H^\circ = 1 \text{ mol}(-635.55 \text{ kJ /mol}) + 1 \text{ mol}(-241.83 \text{ kJ/mol}) - 1 \text{ mol}(-986.2 \text{ kJ/mol})$$

$$= 108.82 \text{ kJ (or } 1.088 \times 10^5 \text{ J)}$$

$$\Delta S^\circ = S^\circ(CaO) + S^\circ(H_2O) - S^\circ(Ca(OH)_2)$$

$$\Delta S° = 1 \text{ mol}(39.8 \text{ J/K·mol}) + 1 \text{ mol}(188.7 \text{ J/K·mol}) - 1 \text{ mol}(83.4 \text{ J/K·mol})$$

$$= + 145.1 \text{ J/K}$$

The temperature at which $\Delta G°$ is equal to zero is:

$$T = \frac{\Delta H°}{\Delta S°} = \frac{1.088 \times 10^5 \text{ J}}{145.1 \text{ J/K}} = 750 \text{ K}$$

At temperatures above 750 K the reaction is spontaneous.

•Comment

Recall that this is an approximate value because of the assumption that neither $\Delta H°$ nor $\Delta S°$ change appreciably from there values calculated at 25°C.

EXAMPLE 18.6   Entropy of Fusion

The heat of fusion of water (ΔH_{fus}) at 0°C is 6.02 kJ/mol. What is ΔS_{fus} for 1 mole of H_2O at the melting point?

•Method of Solution

$$\Delta S_{fus} = \frac{\Delta H_{fus}}{T_{mp}} = \frac{6.02 \times 10^3 \text{ J/mol}}{273 \text{ K}}$$

$$= +22.1 \text{ J/K·mol}$$

•Comment

The increase in entropy upon melting of the solid corresponds to the higher degree of molecular disorder in the liquid state as compared to the solid state.

EXERCISES

4. Calculate $\Delta G°_{rxn}$ for the following reaction

$$3NO_2(g) + H_2O(l) \rightarrow 2HNO_3(l) + NO(g)$$

Given the following free energies of formation:

	$\Delta G°_f$ (kJ/mol)
H_2O	−237.2
HNO_3	−79.9
NO	86.7
NO_2	51.8

5. Calculate $\Delta G°_{rxn}$ for the following reaction at 298 K:

$$O_3(g) \rightarrow O_2(g) + O(g)$$

Given $\Delta H° = 106.5$ kJ and $\Delta S° = 127.3$ J/K.

6. The following reaction is nonspontaneous at 25°C.

$$Cu_2O(s) \rightarrow 2Cu(s) + \frac{1}{2}O_2(g) \quad \Delta G° = 141 \text{ kJ}$$

If $\Delta S° = 75.8$ J/K, above what temperature will the reaction become spontaneous?

7. The enthalpy of vaporization of mercury is 58.5 kJ/mol and the normal boiling point is 630 K. What is the entropy of vaporization of mercury?

8. What is the sign of ΔG for the melting of ice at 5°C?

FREE ENERGY AND EQUILIBRIUM (18.5)

STUDY OBJECTIVES

1. Calculate ΔG, the free energy change under nonstandard state conditions.
2. Calculate an equilibrium constant from a knowledge of $\Delta G°$, and vice versa.

ΔG and $\Delta G°$. Recall that $\Delta G°$ refers to the standard free energy change. All the values we have calculated so far relate to processes in which the reactants are present in their standard states and are converted to products in their standard states. However, in many cases neither the reactants nor the products are present at standard concentration (1 M) and standard pressure (1 atm). Under nonstandard state conditions, we use the symbol ΔG.

The relationship between ΔG and $\Delta G°$ is:

$$\Delta G = \Delta G° + RT \ln Q$$

where R is the gas constant (8.314 J/K·mole), T is the absolute temperature, and Q is the reaction quotient. For a certain reaction at a given temperature the value of $\Delta G°$ is constant, but the value of Q depends on the composition of the reacting mixture; therefore ΔG will depend on Q. To calculate ΔG, first find $\Delta G°$, then calculate Q from the given concentrations of reactants and products, and substitute into the preceding equation.

Under special conditions this equation reduces to an extremely important relationship. At equilibrium, $Q = K$, and therefore $\Delta G = 0$. The equation then becomes:

$$0 = \Delta G° + RT \ln K$$

or

$$\Delta G° = - RT \ln K$$

This equation relates the equilibrium constant of a reaction to its standard free energy change. Thus, if $\Delta G°$ can be calculated, K can be determined, and vice versa. In the equation K_p is used for gases and K_c for reactions in solution.

Three possible relationships exist between $\Delta G°$ and K, because $\Delta G°$ can be negative, positive, or zero.

1. When $\Delta G°$ is *negative*, ln K is positive and K > 1. The products are favored over reactants at equilibrium. The extent of reaction is large.
2. When $\Delta G°$ is *positive*, ln K is negative and K < 1. The reactants are favored over products at equilibrium. The extent of reaction is small.
3. When $\Delta G° = 0$, ln K is zero and K = 1. The reactants and products are equally favored at equilibrium.

When calculating ΔG and $\Delta G°$ using the two equations above, we need to express both free energy changes in units of kJ/mol. The "per mole" refers to "a mole of reaction" which is the reaction of exactly the molar amounts written in the balanced equation.

EXAMPLE 18.7   Calculating the Equilibrium Constant

The standard free energy change for the reaction

$$\frac{1}{2}N_2(g) + \frac{3}{2}H_2(g) \rightleftharpoons NH_3(g)$$

is $\Delta G^o_{rxn} = 26.9$ kJ/mol at 700 K. Calculate the equilibrium constant at this temperature.

•Method of Solution

The equilibrium constant is related to the standard free energy change by the equation:

$$\Delta G^o_{rxn} = -RT \ln K_p$$

•Calculation

Since the gas constant R has units involving joules and the free energy change has units involving kilojoules, we must be careful to use consistent units. In terms of joules, we get

$$26.9 \times 10^3 \text{ J/mol} = -(8.31 \text{ J/mol·K})(700 \text{ K}) \ln K_p$$

$$-4.62 = \ln K_p$$

Taking the antilog of both sides:

$$K_p = e^{-4.62}$$

Use of a calculator with an e^x key yields:

$$K_p = 9.8 \times 10^{-3}$$

EXAMPLE 18.8   ΔG at Nonstandard State Conditions

Using data given in the preceding example, calculate ΔG at 700 K if the reaction mixture consists of 30.0 atm of H_2, 20.0 atm of N_2, and 0.500 atm of NH_3.

•Method of Solution

Under nonstandard conditions, ΔG is related to the reaction quotient Q by the equation

$$\Delta G = \Delta G^\circ + RT \ln Q_p$$

where

$$Q_p = \frac{P_{NH_3}}{P_{N_2}^{1/2} P_{H_2}^{3/2}} = \frac{(0.500)}{(20.0)^{1/2}(30.0)^{3/2}}$$

$$Q_p = 6.80 \times 10^{-4}$$

From Example 18.7, $\Delta G^\circ = 26.9$ kJ/mol. Substitution yields:

$$\Delta G = 26.9 \text{ kJ/mol} + (8.31 \text{ J/K·mol})(700 \text{ K}) \ln (6.80 \times 10^{-4})$$

$$= 26.9 \text{ kJ/mol} - (42,400 \text{ J/mol} \times \frac{1 \text{ kJ}}{10^3 \text{ J}}) = 26.9 \text{ kJ/mol} - 42.4 \text{ kJ/mol}$$

$$= -15.5 \text{ kJ/mol}$$

•**Comment**

By making the partial pressures of N_2 and H_2 high and that of NH_3 low, the reaction is spontaneous in the forward reaction. This condition corresponds to $Q_p < K_p$, and so the reaction proceeds in the forward direction until $Q_p = K_p$.

EXERCISES

9. Explain the difference between ΔG and $\Delta G°$.

10. Hydrogen peroxide (H_2O_2) decomposes according to the equation:

$$H_2O_2(l) \rightarrow H_2O(l) + \tfrac{1}{2}O_2(g)$$

a. Is this reaction spontaneous at 25°C?
b. From the following data calculate the value of K_p for this reaction at 25°C.

$\Delta H° = -98.2$ kJ
$\Delta S° = +70.1$ J/K

11. The autoionization of water at 25°C has the equilibrium constant

$$2H_2O(l) \rightleftharpoons H_3O^+(aq) + OH^-(aq) \quad K_c = 1.0 \times 10^{-14}$$

Calculate the value of $\Delta G°$ for this reaction.

12. The equilibrium constant for the reaction:

$$AgBr(s) \rightleftharpoons Ag^+(aq) + Br^-(aq)$$

is the solubility product constant, $K_{sp} = 7.7 \times 10^{-13}$ at 25°C. Calculate ΔG for the reaction when $[Ag^+] = 1.0 \times 10^{-2}$ M and $[Br^-] = 1.0 \times 10^{-3}$ M. Is the reaction spontaneous or nonspontaneous at these concentrations?

13. Calculate ΔG for the following reaction at 25°C when the pressure of CO_2 is 0.001 atm.

$$CaCO_3(s) \rightarrow CaO(s) + CO_2(g)$$

Given $\Delta H° = 177.8$ kJ and $\Delta S° = 160.5$ J/K.

CONCEPTUAL QUESTION

1. When the environment is contaminated by a toxic chemical spill or an oil spill, the substance tends to disperse. How is this consistent with the second law of thermodynamics. In the same regard, which requires less work: cleaning the environment after a spill, or keeping the substance contained before a spill.

PRACTICE TEST

1. Which of the following processes are spontaneous?
   a. melting of ice at $-10°C$ and 1 atm pressure
   b. evaporation of water at $30°C$ when the relative humidity is less than 100 percent
   c. Water + NaCl(s) $\rightarrow$ salt solution

2. From each pair of substances, choose the one having the larger standard entropy at $25°C$.
   a. $H_2O(l)$ or $H_2O(g)$ b. $SiO_2(s)$ or $CO_2(g)$ c. $Ag^+(g)$ or $Ag^+(aq)$

   d. $F_2(g)$ or $Cl_2(g)$ e. $2Cl(g)$ or $Cl_2(g)$

3. Predict, using the intuitive ideas about entropy, whether ΔS_{sys} will be positive, negative, or essentially zero for each of the following:

   a. $Ca(OH)_2(s) + CO_2(g) \rightarrow CaCO_3(s) + H_2O(g)$
   b. $CuSO_4(s) \rightarrow Cu^{2+}(aq) + SO_4^{2-}(aq)$
   c. $2HCl(g) + Br_2(l) \rightarrow 2HBr(g) + Cl_2(g)$
   d. $SO_2(g) + \frac{1}{2}O_2(g) \rightarrow SO_3(g)$
   e. $Cu^{2+}(aq) + 4NH_3(aq) \rightarrow Cu(NH_3)_4^{2+}(aq)$

4. At the boiling point, $35°C$, the heat of vaporization of MoF_6 is 25 kJ/mol. Calculate ΔS for the vaporization of MoF_6.

5. Calculate $\Delta G°_{rxn}$ for the following reaction at 298 K:

   $$2H_2(g) + CO(g) \rightleftharpoons CH_3OH(g)$$

   given that $\Delta H° = -90.7$ kJ and $\Delta S° = -221.5$ J/K for this process.

6. For the reaction at 298 K

   $$Mg(s) + \frac{1}{2}O_2(g) \rightarrow MgO(s)$$

   $\Delta H° = -602$ kJ and $\Delta G° = -569$ kJ. Calculate $\Delta S°$.

7. Using Appendix 3 of the text calculate $\Delta G°$ values for the following reactions:

   a. $3CaO(s) + 2Al(s) \rightarrow 3Ca(s) + Al_2O_3(s)$
   b. $ZnO(s) \rightarrow Zn(s) + \frac{1}{2}O_2(g)$

8. Consider the following three reactions. Which one will have the greatest equilibrium constant?

   a. $N_2 + O_2 \rightleftharpoons 2NO$
   b. $N_2 + 2O_2 \rightleftharpoons N_2O_4$
   c. $N_2 + \frac{1}{2}O_2 \rightleftharpoons N_2O$

   Given: $\Delta G°_f(NO) = +86.7$ kJ/mol
   $\Delta G°_f(N_2O_4) = +98.3$ kJ/mol
   $\Delta G°_f(N_2O) = +103.6$ kJ/mol

9. Given the equilibrium constant at $400°C$ for the reaction:

   $$H_2(g) + I_2(g) \rightleftharpoons 2HI(g) \quad K_p = 64$$

   calculate the value of $\Delta G°_{rxn}$ at this temperature.

10. Calculate ΔG°_{rxn} and K_p at 25°C for the following reaction:

$$NO(g) + \frac{1}{2}O_2(g) \rightarrow NO_2(g)$$

11. The synthesis of $O_2(g)$ is often carried out in chemistry lab by the decomposition of $KClO_3$:

$$KClO_3(s) \rightarrow KCl(s) + \frac{1}{2}O_2(g)$$

for which $\Delta H^{\circ} = -44.7$ kJ and $\Delta S^{\circ} = +247.2$ J/K. Is this reaction spontaneous at 25°C under standard conditions?

12. For the reaction:

$$N_2 + O_2 \rightarrow 2NO$$

the following are given: $\Delta H^{\circ} = 180.7$ kJ and $\Delta S^{\circ} = 24.7$ J/K.
a. Is this reaction spontaneous at 25°C?
b. Above what temperature will this reaction become spontaneous under standard conditions?

13. For the reaction $2SO_2(g) + O_2(g) \rightleftharpoons 2SO_3(g)$, $K_p = 7.4 \times 10^4$ at 700 K. If, in a reaction vessel at 700 K, we have the following partial pressures, what is ΔG?

$$P_{SO_2} = 1.2 \text{ atm} \quad P_{O_2} = 0.5 \text{ atm} \quad P_{SO_3} = 50 \text{ atm}$$

Predict the direction of reaction.

ANSWERS

Exercises

1. a. $CS_2(l)$ b. $SO_2(g)$ c. $BaSO_4(aq)$
2. a. +   b. –   c. cannot predict the sign, ΔS will be essentially zero   d. –
3. –198.5 J/K
4. 8.7 kJ
5. $\Delta G^{\circ}_{rxn} = 68.6$ kJ
6. 2160 K
7. $\Delta S_{vap} = 92.6$ J/K·mol
8. negative
9. ΔG° refers to the free energy change when the reactants and products are both present in their standard states. Their concentrations are all 1 atm or 1 molar. ΔG refers to the free energy change when the reactants and products are present at concentrations other than those for the standard state.
10. a. yes   b. 7.7×10^{20}
11. $\Delta G^{\circ}_{rxn} = 79.8$ kJ
12. $\Delta G^{\circ} = +69.1$ kJ and $\Delta G = +40.6$ kJ. Therefore the reaction is nonspontaneous.
13. $\Delta G = 112.9$ kJ

Conceptual Question

1. Think about the system. Let it be the oil or toxic substance? When an oil spill spreads out it is becoming more disordered as its molecules have a much larger area to occupy. Any process that gives rise to a more random distribution of the particles of the substance in space gives rise to an increase in entropy of the substance. In the reverse process, that of cleaning up a dispersed substance, work must be done to bring the substance together into a more concentrated space. This reduces the disorder of the system. This is a nonspontaneous process requiring a continual work input. Keeping the substance contained in the first place takes less work.

Practice Test

1. b and c
2. a. $H_2O(g)$ b. $CO_2(g)$ c. $Ag^+(g)$ d. $Cl_2(g)$ e. $2Cl(g)$
3. a. Essentially zero   b. Positive   c. Positive   d. Negative   e. Negative
4. $\Delta S_{vap} = 81$ J/mol·K
5. $\Delta G^{\circ}_{rxn} = -24,700$ J
6. $\Delta S^{\circ} = -110$ J/K
7. a. $\Delta G^{\circ} = 236$ kJ   b. $\Delta G^{\circ} = 318.2$ kJ
8. b.
9. $\Delta G^{\circ}_{rxn} = -23.2$ kJ
10. $\Delta G^{\circ}_{rxn} = -34.85$ kJ; $K_p = 1.29 \times 10^6$
11. Yes, $\Delta G^{\circ} < 0$
12. a. No, $\Delta G^{\circ} > 0$ b. $T = 7,320$ K
13. $\Delta G = -17,800$ J. Spontaneous in the forward direction.

Chapter Nineteen
ELECTROCHEMISTRY

- Balancing Redox Equations (19.1)
- Electrochemical Cells and Standard Reduction Potentials (19.2 – 19.3)
- Spontaneity of Redox Reactions (19.4)
- Effect of Concentration on Cell EMF (19.5)
- Electrolysis and Its Quantitative Aspects (19.8)

BALANCING REDOX EQUATIONS (19.1)

STUDY OBJECTIVES

1. Review the basics of oxidation-reduction reactions.
2. Balance redox reactions using the ion-electron method.

Half-Reactions. Oxidation-reduction reactions were discussed in Chapter 4. It will be useful to review that material as you start this chapter. Recall that *oxidation is a loss of electrons* by a chemical species and *reduction is a gain of electrons*. Redox reactions are electron transfer reactions involving transfer of electrons from a *reducing agent* to an *oxidizing agent*.

For example, when copper metal is immersed in a solution containing Ag^+ ions a reaction occurs in which electrons are transferred spontaneously from Cu atoms to the Ag^+ ions, forming Ag atoms that plate out on the copper surface. The newly formed Cu^{2+} ions go into solution.

$$Cu(s) + 2Ag^+(aq) \rightarrow Cu^{2+}(aq) + 2Ag(s)$$

Cu is oxidized to Cu^{2+} ion, and Ag^+ ion is reduced to Ag.

All redox reactions can be divided into half-reactions. The half-reactions are:

oxidation	$Cu \rightarrow Cu^{2+} + 2e^-$
reduction	$2e + 2Ag^+ \rightarrow 2Ag$

Copper metal is the *reducing agent* because it supplies electrons to the silver ion. The silver ion is the *oxidizing agent* because it accepts electrons from copper. Two Ag^+ ions are required to accept the two electrons from one Cu atom. Before we introduce electrochemical cells we will discuss how to balance redox equations.

Ion-Electron Method. In this method the overall reaction is divided into two half-reactions, one involving oxidation and the other reduction. Each half-reaction is balanced according to mass and charge, and then the two half-reactions are added together to give the overall balanced equation. We will balance the equation below to illustrate the steps involved. This equation shows just the essential changes and is sometimes called a sketetal equation. To separate the equation into half-reactions, first identify which element is oxidized and which is reduced. Do this by writing oxidation numbers above each element.

$$\overset{+1\ -2}{H_2S} + \overset{+5\ -2}{NO_3^-} \rightarrow \overset{0}{S} + \overset{+2\ -2}{NO}$$

Note that S atoms and N atoms undergo oxidation number changes: S ($-2 \rightarrow 0$) and N ($+5 \rightarrow +2$). Sulfur is oxidized and nitrogen is reduced. Now we write the half-reactions.

oxidation	$H_2S \rightarrow S$
reduction	$NO_3^- \rightarrow NO$

Each half-reaction must be *balanced by mass*. In acidic solution, always add H^+ to balance H atoms, and add H_2O to balance O atoms. Here we add H^+ to the oxidation half-reaction, and H_2O to the reduction half-reaction.

oxidation $H_2S \rightarrow S + 2H^+$

reduction $NO_3^- \rightarrow NO + 2H_2O$

The oxidation half-reaction is now balanced by mass. The reduction half-reaction needs $4H^+$ on the left-hand side.

reduction $4H^+ + NO_3^- \rightarrow NO + 2H_2O$

This half-reaction is now balanced according to mass.

Next the *ionic charges must be balanced* by adding electrons. In the oxidation half-reaction there is a net charge of $+2$ on the right-hand side and zero on the left-hand side. To balance the charge, 2 electrons are added to the right.

oxidation $H_2S \rightarrow S + 2H^+ + 2e^-$

Notice that both sides of the equation have the same charge. In this case 0.

The reduction half-reaction can be charge balanced by adding three electrons to the left-hand side.

reduction $3e^- + 4H^+ + NO_3^- \rightarrow NO + 2H_2O$

Note that electrons are added to the reactant side of a reduction half-reaction and to the product side of an oxidation half-reaction.

The balanced redox equation is obtained by adding the two half-reactions. But first *the number of electrons shown in the two half-reactions must be the same*, since all the electrons lost during oxidation must be gained during reduction. Multiplying the oxidation half-reaction by 3 and the reduction half-reaction by 2 will make the number of electrons in the two half-reactions equal. That is, 6 electrons are given up during oxidation and 6 are gained during reduction.

$$3 \times (H_2S \rightarrow S + 2H^+ + 2e^-) \quad \text{gives} \quad 3H_2S \rightarrow 3S + 6H^+ + 6e^-$$

$$2 \times (3e^- + 4H^+ + NO_3^- \rightarrow NO + 2H_2O) \quad \text{gives} \quad 6e^- + 8H^+ + 2NO_3^- \rightarrow 2NO + 4H_2O$$

The sum of the half-reactions is the overall balanced redox equation.

$$3H_2S \rightarrow 3S + 6H^+ + 6e^-$$
$$\underline{6e^- + 8H^+ + 2NO_3^- \rightarrow 2NO + 4H_2O}$$
$$3H_2S + 8H^+ + 2NO_3^- \rightarrow 3S + 6H^+ + 2NO + 4H_2O$$

The $6H^+$ on the RHS will cancel 6 of the $8H^+$ on the LHS yielding

$$3H_2S + 2H^+ + 2NO_3^- \rightarrow 3S + 2NO + 4H_2O$$

The following steps are useful in balancing redox equations by the ion-electron method.

1. Write the skeletal equation containing the oxidizing and reducing agents and the products in ionic form.
2. Separate the equation into two half-reactions.
3. Balance the atoms other than O and H in each half-reaction separately. This is called a *mass balance*.
4a. For reactions in *acidic* medium, add H_2O to balance O atoms and H^+ to balance H atoms.
4b. For reactions in *basic* medium, first balance the atoms as you would for an acidic solution. Then, for each H^+ ion, add an OH^- ion to both sides of the half-reaction. Whenever H^+ and OH^- appear on the same side, combine them to make H_2O.
5. Add electrons to one side of each half-reaction to equalize the charges. Electrons are added to the reactant side of a reduction half-reaction, and to the product side of an oxidation half-reaction. The number of electrons added to one side of a half-reaction should make the total charge of that side equal to the charge on the other side. This procedure is called a *charge balance*.
6. Add the two half-reactions together. Before this can be done, the *number of electrons* shown in both half-reactions must be *the same*. The number of electrons in the two half-reactions can be *equalized* by multiplying one or both half-reactions by appropriate coefficients. Now add the two half-reactions.
7. Check the final equation by inspection. Recall that a properly balanced equation consists of a set of the smallest possible whole numbers.

EXAMPLE 19.1   Balancing A Redox Reaction

Balance the following redox reaction by the ion-electron (half-reaction) method.

$$Cu + HNO_3 \rightarrow Cu^{2+} + NO_2 \quad \text{(in acidic solution)}$$

•**Method of Solution**

Steps 1 and 2. The skeletal equation is given. Separate the given equation into half-reactions. Note that Cu is oxidized and nitrogen is reduced. The half-reactions are:

oxidation $\quad Cu \rightarrow Cu^{2+}$
reduction $\quad HNO_3 \rightarrow NO_2$

Step 3. The oxidation half-reaction is already balanced by mass.

Step 4a. Balance the reduction half-reaction by adding one H_2O to the right-hand side, and one H^+ to the left-hand side.

$$\text{reduction } \quad H^+ + HNO_3 \rightarrow NO_2 + H_2O$$

Step 5. To balance according to charge, add electrons to the product side of the oxidation half-reaction, and the reactant side of the reduction half-reaction.

oxidation $\quad Cu \rightarrow Cu^{2+} + 2e^-$
reduction $\quad e^- + H^+ + HNO_3 \rightarrow NO_2 + H_2O$

Step 6. Next equalize the number of electrons in the two half-reactions by multiplying the reduction half-reaction by 2.

$$2e^- + 2H^+ + 2HNO_3 \rightarrow 2NO_2 + 2H_2O$$

Now add the two half-reactions to obtain a balanced overall equation.

oxidation $\qquad\qquad Cu \rightarrow Cu^{2+} + 2e^-$
reduction $\quad 2e^- + 2H^+ + 2HNO_3 \rightarrow 2NO_2 + 2H_2O$

―――――――――――――――――――――――――――――

$$Cu + 2H^+ + 2HNO_3 \rightarrow Cu^{2+} + 2NO_2 + 2H_2O$$

Step 7. Check that the electrons on both sides cancel, and that the equation is balanced in terms of atoms (mass) and charge.

EXERCISES

1. Balance the following redox equations.

   a. $Cr_2O_7^{2-}(aq) + Br^-(aq) \rightarrow Cr^{3+}(aq) + Br_2(g)$ (acidic solution)
   b. $MnO_4^-(aq) + I^-(aq) \rightarrow MnO_2(s) + I_2(aq)$ (basic solution)

2. Balance the following redox equations.

   a. $H_2C_2O_4(aq) + MnO_4^-(aq) \rightarrow Mn^{2+}(aq) + CO_2(g)$ (acidic solution)
   b. $Zn(s) + ClO^-(aq) \rightarrow Zn(OH)_2(s) + Cl^-(aq)$ (basic solution)

ELECTROCHEMICAL CELLS AND STANDARD REDUCTION POTENTIALS (19.1)

STUDY OBJECTIVES

1. Diagram an electrochemical cell, labeling the anode, the cathode, the charges on the electrodes, and the directions of electron and ion flows.
2. Calculate the standard emf of an electrochemical cell.
3. Arrange given redox reagents in order of increasing strength as oxidizing and reducing agents.

Electrochemical Cells. Electrochemistry is the area of chemistry that deals with the interconversion of electrical energy and chemical energy. Electrochemical processes are redox reactions in which chemical energy is converted into electricity or in which electricity is used to cause a chemical reaction to occur.

A device, which utilizes a spontaneous redox reaction to supply a constant flow of electrons is called an **electrochemical cell or a voltaic cell.** The design of an electrochemical cell is such that the reactants are prevented from direct contact with each other. The oxidation half-reaction occurs at an electrode called the **anode**, and the reduction half-reaction occurs at an electrode called the **cathode**.

Figure 19.1 shows a copper-silver electrochemical cell. The anode is a bar of copper metal that is partially immersed in a solution of $CuSO_4$. The cathode is a small bar of silver that is partially immersed in a solution of $AgNO_3$. The reducing agent, Cu metal, does not come into direct contact with the oxidizing agent, Ag^+ ion. Cu atoms loose two electrons and become copper ions that enter the solution. Electrons travel through the outer circuit to the Ag electrode, where Ag^+ ions from the solution are reduced to silver atoms at the surface of the electrode. These atoms plate out on the electrode. As the reaction proceeds, the copper electrode loses mass and the silver electrode gains mass.

While electrons travel through the outer circuit from Cu to Ag (from the anode to the cathode), negative ions move through a porous barrier or a "salt bridge" from the $AgNO_3$ solution into the $CuSO_4$ solution. This motion resupplies negative charge to the anode compartment and maintains electrical neutrality in the solutions surrounding the electrodes. An electric current will flow until either the Cu metal or the Ag^+ ions are used up.

Figure 19.1 An electrochemical cell consisting of a copper electrode (anode) and a silver electrode (cathode).

The fact that electrons flow from the anode to the cathode means that a difference in electrical potential energy exists between the two electrodes. Electrons flow from where they are at higher potential to where they are at a lower potential. The difference in electrical potential between the anode and cathode in measured in volts, and is called the **cell potential** or **cell emf** which stands for **electromotive force**. The actual difference in electrical potential depends on the nature of the species involved in the half-reactions and their concentrations. The cell potential is represented by the symbol E_{cell}.

Cell Diagrams. Rather than always representing a cell by a sketch, we can use a *cell diagram*. For instance, the copper-silver cell discussed above and shown in Figure 19.1 is represented by:

$$Cu(s) \mid CuSO_4(aq) \parallel AgNO_3(aq) \mid Ag(s)$$

anode   cathode

In the diagram the anode is on the left, and the cathode is on the right. The single vertical lines indicate phase boundaries which in this cell are between the solid electrodes and the aqueous solutions. The sequence $Cu(s) \mid CuSO_4(aq)$ represents oxidation, and the sequence $AgNO_3(aq) \mid Ag(s)$ represents reduction. The double vertical lines indicate a salt bridge or porous barrier between the two solutions.

Standard Electrode Potentials. The emf of an electrochemical cell is dependent upon the nature of the half-reactions, the concentrations of the reactants and products, and the temperature. The **standard cell emf** is the voltage generated under standard state conditions. The standard cell emf has the symbol E_{cell}°. The superscript ° denotes that all reactants and products are in their standard states. The standard state of all solute ions and molecules is a concentration of 1 M, and for all gases it is a partial pressure of 1 atm. Unless stated otherwise, the temperature is assumed to be 25°C. When the concentrations of Cu^{2+} and Ag^+ ions in the above cell (Figure 19.1) are both 1.0 M, the cell emf is 0.46 V at 25°C.

Rather than construct a cell and measure its voltage every time a cell emf is needed, a method has been devised that allows the use of a table of half-reactions and standard electrode potentials. The **standard reduction potential,** E_{red}°, is the voltage associated with a reduction reaction when all metals, solutes, gases are in their standard states. A number of standard reduction potentials are listed in Table 19.1 of the textbook.

The following list summarizes the information contained in table of standard reduction:

1. Standard electrode potentials are reduction potentials, and measure the ease of reduction under standard conditions. The E^o_{red} values apply to the half-reactions when read in the forward direction. The more positive the value of E^o_{red}, the greater the tendency for the half-reaction to occur in the forward direction.

2. All half-cell reactions are reversible. Depending on the reducing strength of the other electrode that is chosen, any given electrode may act as an anode or as a cathode.

3. The species appearing on the left-hand sides of the half-reactions are **oxidizing agents**, and that E^o_{red} is related to the tendency of these oxidizing agents to accept electrons. The very strong oxidizing agents are at the top of the table. F_2 is the *strongest oxidizing agent* and Li^+ is the *weakest*.

4. The species shown on the right-hand sides of the half-reactions are **reducing agents**. For a reducing agent to supply electrons, the reaction must occur in the *reverse direction*. The strength of the reducing agents increases as you read down the table. Therefore, F^- ion is the *weakest reducing agent* and Li metal is the *strongest reducing agent*.

5. When predicting the direction of reaction under standard state conditions, you can use the **diagonal rule**. This rule states that any species on the *left hand side* of a given half-cell reaction (an oxidizing agent) will react spontaneously with a species that is shown on the *right* of any half-reaction (a reducing agent) located *below* it in the table.

6. The standard reduction potential is not affected when stoichiometric coefficients of a half-cell reaction are changed. For example,

$$Ag^+ + e^- \rightarrow Ag(s) \qquad E^o_{red} = 0.80 \text{ V}$$
$$2Ag^+ + 2e^- \rightarrow 2Ag(s) \qquad E^o_{red} = 0.80 \text{ V}$$

Electrode potentials are *intensive properties*, and therefore, do not depend on the size of the electrode or the manner in which the half-reaction is balanced.

Notice the 0.0 V potential assigned to hydrogen ion. A positive reduction potential means that the oxidizing agent (such as F_2) has a greater tendency to be reduced than the hydrogen ion, and is therefore a stronger oxidizing agent. A negative reduction potential for an oxidizing agent (such as Li^+) indicates that the hydrogen ion is more readily reduced than that substance.

Calculation of the Standard Cell EMF.

Calculation of the Standard Cell EMF. Standard reduction potentials are used to calculate the potential of an electrochemical cell. The standard cell emf (or standard cell potential), E^o_{cell}, is the difference between the standard reduction potential of the cathode, and the standard reduction potential of the anode.

$$E^o_{cell} = E^o(\text{cathode}) - E^o(\text{anode})$$

The standard emf of the silver-copper cell discussed above is calculated as follows. The overall reaction is:

$$Cu + 2Ag^+ \rightarrow Cu^{2+} + 2Ag$$

The half-reactions are

oxidation $Cu \rightarrow Cu^{2+} + 2e^-$
reduction $2Ag^+ + 2e^- \rightarrow 2Ag$

Therefore, the standard cell emf is given by

$$E^o_{cell} = E^o_{cathode} - E^o_{anode} = E^o_{Cu^{2+}/Cu} - E^o_{Ag^+/Ag}$$

The symbols Cu^{2+}/Cu and Ag/Ag^+ represent the reduction half-reactions that are related to the reduction potentials.

The reduction potentials needed here are found in Table 19.1 of the text.

$$E^o_{cathode} = E^o_{Ag^+/Ag} = 0.80 \text{ V}$$

$$E^o_{anode} = E^o_{Cu^{2+}/Cu} = 0.34 \text{ V}$$

$$E^o_{cell} = 0.80 \text{ V} - 0.34 \text{ V}$$

$$E^o_{cell} = 0.46 \text{ V}$$

The significance of a positive E^o_{cell} is that it indicates that the reaction is spontaneous under standard conditions.

EXAMPLE 19.2   Electrochemical Cell Reactions

Consider an electrochemical cell constructed from the following half-cells, linked by a KNO_3 salt bridge. A Cu electrode immersed in 1.0 M $Cu(NO_3)_2$, and a Sn electrode in 1.0 M $Sn(NO_3)_2$.
As the reaction proceeds, the Cu electrode gains mass and the Sn electrode loses mass.
a.   Which electrode is the anode and which is the cathode?
b.   Write a balanced chemical equation for the overall cell reaction.
c.   Sketch the half-cells and show the direction of flow of the electrons in the external circuit.
d.   Which electrodes do the positive and the negative ions diffuse toward? Include a salt bridge in your sketch.

•Method of Solution

a.   Reduction occurs at the cathode and oxidation occurs at the anode. Since the Cu electrode gains mass its half-reaction must involve the plating out of Cu^{2+} ions as Cu atoms. The copper electrode is the *cathode*.

$$Cu^{2+}(aq) + 2e^- \rightarrow Cu(s) \qquad \text{reduction}$$

Since the Sn electrode loses mass, Sn metal must be dissolving as it is oxidized, and so Sn is the *anode*. The Sn^{2+} ions go into solution.

$$Sn(s) \rightarrow Sn^{2+}(aq) + 2e^- \qquad \text{oxidation}$$

b.   Adding the two half-reactions:

$$Sn(s) \rightarrow Sn^{2+}(aq) + 2e^-$$
$$Cu^{2+}(aq) + 2e^- \rightarrow Cu(s)$$

overall $Sn + Cu^{2+}(aq) \rightarrow Sn^{2+}(aq) + Cu$

c.   Electrons flow from the anode to the cathode. See Figure 19.2.
d.   The positive ions (cations) diffuse toward the Cu electrode (the cathode) and the negative ions (anions) diffuse toward the Sn electrode (the anode).

Figure 19.2 An electrochemical cell consisting of a tin electrode (anode) and a copper electrode (cathode). Notice that in Figure 19.1 copper was an anode.

EXAMPLE 19.3   Standard Cell EMF

Calculate the standard cell emf (cell potential) for an electrochemical cell having the following overall cell reaction of tin metal and lead(II) ion:

$$Sn(s) + Pb^{2+}(aq) \rightarrow Sn^{2+}(aq) + Pb(s)$$

•Method of Solution

The standard emf of the cell is $E^{\circ}_{cell} = E^{\circ}_{cathode} - E^{\circ}_{anode}$

Separate the overall reaction into half-reactions, and identify the anode and cathode. We see that Pb^{2+} is reduced and tin is oxidized. Look up the necessary reduction potentials in Table 19.1 of the text.

•Calculation

$$2e^- + Pb^{2+}(aq) \rightarrow Pb(s) \quad \text{reduction/cathode}$$
$$Sn(s) \rightarrow Sn^{2+}(aq) + 2e^- \quad \text{oxidation/anode}$$

$$Sn(s) + Pb^{2+}(aq) \rightarrow Sn^{2+}(aq) + Pb(s) \quad \text{overall}$$

$$E^{\circ}_{cell} = E^{\circ}_{cathode} - E^{\circ}_{anode} = E^{\circ}_{Pb^{2+}/Pb} - E^{\circ}_{Sn^{2+}/Sn}$$

$$E^{\circ}_{cell} = -0.13 \text{ V} - (-0.14 \text{ V}) = 0.01 \text{ V}$$

•Comment

The positive value of the standard cell potential indicates that the reaction is spontaneous under standard conditions.

EXAMPLE 19.4   Relative Strength of Oxidizing and Reducing Agents

Consider the following species: Mg, MnO_4^-, Cl^-, and Cu^{2+}. Using Table 19.1 of the text, separate these into oxidizing and reducing agents. Then arrange the reducing agents according to increasing strength, and the oxidizing agents according to strength.

•Method of Solution

First we will take the half-reactions from Table 19.1 that contain the above species and list them here. The oxidizing agents appear on the left-hand side of the half-reactions in Table 19.1, and the reducing agents appear on the right-hand side. Examination of the potentials shows that MnO_4^- and Cu^{2+} ion are oxidizing agents, and Cl^- ion and Mg are reducing agents.

Oxidizing Agents	Reducing Agents	E^o_{red} (V)
$MnO_4^- + 8H^+ + 5e^- \rightleftharpoons 2Mn^{2+}(aq) + 2H_2O$		1.51
$Cl_2(g) + 2e^- \rightleftharpoons 2Cl^-(aq)$		1.36
$Cu^{2+}(aq) + 2e^- \rightleftharpoons Cu(s)$		0.34
$Mg^{2+}(aq) + 2e^- \rightleftharpoons Mg(s)$		−2.37

The greater the value of the reduction potential, the greater the tendency for the half-rxn to occur as written, and therefore, the greater the strength of the oxidizing agent. Comparing the oxidizing agents, we see that permanganate ion, MnO_4^- is stronger than Cu^{2+} ion.

For a substance on the right-hand side of the half-reactions in Table 19.1 to act as a reducing agent it must react in the reverse direction. The reducing agent strength increases as we read down the list. We see that Mg is a stronger reducing agent than chloride ion, Cl^-.

•Comment

All half-reactions are reversible. The product of a strong *oxidizing* agent such as MnO_4^- is a weak *reducing* agent Mn^{2+}.

EXERCISES

3. Calculate the standard emf (E^o_{cell}) of cells that have the following overall reactions:

    a. $2Co^{2+}(aq) + Zn^{2+}(aq) \rightarrow 2Co^{3+}(aq) + Zn(s)$
    b. $Cl_2(g) + 2Br^-(aq) \rightarrow 2Cl^-(aq) + Br_2(l)$

4. Consider a voltaic cell constructed from the following half cells, linked by a KCl salt bridge.

    - a Cu electrode in 1.0 M $Cu(NO_3)_2$ solution
    - a Pb electrode in 1.0 M $Pb(NO_3)_2$ solution.

    Answer the following questions:
    a. Which electrode is the cathode?
    b. Write a balanced chemical equation for the reaction occurring when the cell is operating.
    c. Which electrode gains mass?
    d. Which direction do electrons flow in the external circuit?
    e. Which electrode will K^+ of the KCl salt bridge move toward?

5. Given the following standard reduction potentials in acid solution:

	E^o_{red} (V)
$Sn^{4+} + 2e^- \rightleftharpoons Sn^{2+}$	+0.14
$Al^{3+} + 3e^- \rightleftharpoons Al(s)$	−1.66
$Fe^{3+} + e^- \rightleftharpoons Fe^{2+}$	+0.77

    a. Write the formula of the strongest reducing agent.
    b. Write the formula of the strongest oxidizing agent.

SPONTANEITY OF REDOX REACTIONS (19.4)

STUDY OBJECTIVES

1. Predict whether a redox reaction will be spontaneous or nonspontaneous.
2. Calculate $\Delta G°$ and K for a redox reaction, given $E°_{cell}$.

Criterion for Spontaneity. For a process carried out at constant temperature and pressure, the Gibbs free energy change is equal to the maximum amount of work (w_{max}) that can be done by the process.

$$\Delta G = w_{max}$$

Any reaction that occurs spontaneously can be utilized to perform useful work. Combustion of fuels and the oxidation of food components both provide energy to do work. The difference in free energy between the reactants and the products is the energy available to do work.

Electrical work is equal to the product of the cell emf (E_{cell}) and the total charge in coulombs carried through the circuit. The total charge in coulombs is nF, where n is *the number of moles of electrons transferred from the reducing agent to the oxidizing agent according to the balanced equation*, and F is the **Faraday constant**. The Faraday constant equals the charge in coulombs carried by one mole of electrons which is 96,500 C/mol of electrons. In most thermodynamic calculations, F is expressed in energy units: 96,500 J/V·mol e⁻.

The electrical work is given by:

$$w_{elect} = -nFE_{cell}$$

The maximum work from an electrochemical process is

$$w_{max} = w_{elect} = -nFE_{cell}$$

The negative sign is in accord with the sign convention that when work is done on the surroundings, the system loses energy.

The Gibbs free energy change for a redox reaction is:

$$\Delta G = -nFE_{cell}$$

For a spontaneous redox reaction, ΔG will be negative, and E must be positive. For reactions in which reactants and products are in their standard states:

$$\Delta G° = -nF\,E°_{cell}$$

The Equilibrium Constant. In Chapter 18 you learned that $\Delta G°$ is related to the equilibrium constant K for the reaction.

$$\Delta G° = -\,RT \ln K$$

Therefore, the equilibrium constant of a redox reaction is related to the standard cell emf (E°) by the equation.

$$-nFE° = -\,RT \ln K$$

and $\qquad E° = \dfrac{RT}{nF} \ln K$

Substituting for R and F at 25°C gives the term RT/F which equals 0.0257 V.

$$E° = \dfrac{0.0257 \text{ V}}{n} \ln K$$

Therefore reactions with large equilibrium constants generate higher standard cell emf's. When $E° > 1$, $K > 1$. Rearranging,

$$\ln K = \frac{nE°}{0.0257 \text{ V}}$$

If any one of the three quantities $\Delta G°$, K, or $E°$ is known, both of the others can be calculated. Table 19.1 summarizes the criteria for spontaneous redox reactions.

Table 19.1 Criteria for Spontaneous Redox Reactions

$\Delta G°$	K	$E°_{cell}$	Reaction under Standard Conditions
Negative	> 1	Positive	Spontaneous
0	$= 1$	0	At equilibrium
Positive	< 1	Negative	Nonspontaneous Reaction is spontaneous in the reverse direction

---

EXAMPLE 19.5   Predicting Spontaneous Redox Reactions

Predict whether the following reactions are spontaneous under standard conditions.

a. $2Fe^{3+}(aq) + 2I^-(aq) \rightarrow 2Fe^{2+}(aq) + I_2(s)$
b. $Cu(s) + 2H^+(aq) \rightarrow Cu^{2+}(aq) + H_2(g)$

•Method of Solution

For a redox reaction to be spontaneous, it must have a positive cell emf. For the reaction in part (a) we will apply the *diagonal rule*. In part (b) we will calculate the standard emf of the cell.

a. According to the diagonal rule when applied to Table 19.1 in the text; any *oxidizing agent* will react spontaneously with a reducing agent of any half-reaction located below *it* in the table. In reaction a. we ask, can Fe^{3+} oxidize I^- ion? In the table, I^- ion appears *below* and to the right of Fe^{3+}. We conclude that Fe^{3+} will oxidize I^- ion.

b. To calculate the cell emf separate the reaction into half-reactions:

reduction/cathode	$2H^+ + 2e^- \rightarrow H_2(g)$
oxidation/anode	$Cu(s) \rightarrow Cu^{2+}(aq) + 2e^-$

$$Cu(s) + 2H^+(aq) \rightarrow Cu^{2+}(aq) + H_2(g)$$

$$E°_{cell} = E°_{cathode} - E°_{anode} = E°_{H^+/H_2} - E°_{Cu^{2+}/Cu}$$

$$E°_{cell} = 0.00 \text{ V} - (0.34 \text{ V}) = -0.34 \text{ V}$$

The negative standard cell emf indicates that this reaction is not spontaneous at standard conditions. Cu will not oxidize in 1 M HCl.

•Comment

Since the cell emf changes sign when the reaction is reversed, the reaction $Cu^{2+} + H_2 \rightarrow Cu + 2H^+$ will have a positive cell emf, and the reduction of Cu^{2+} ion by H_2 will be spontaneous under standard conditions.

EXAMPLE 19.6   Free Energy Change and the Standard Cell EMF

Calculate $\Delta G°$ and the equilibrium constant at 25°C for the reaction given the standard cell potential:

$$2Br^-(aq) + I_2(s) \rightarrow Br_2(l) + 2I^-(aq) \qquad E°_{cell} = -0.54 \text{ V}$$

•**Method of Solution**

The standard Gibbs free energy change is related to the standard cell potential by the following equation in which F is the Faraday constant and n is the number of moles of electrons are transferred:

$$\Delta G° = -nFE°$$

The equilibrium constant at 25°C is also related to $E°_{cell}$:

$$\ln K = \frac{n E°}{0.0257 \text{ V}}$$

•**Calculation**

Examination of the half-reactions clearly indicates that 2 mol of electrons are transferred in the balanced equation. n = 2.

The half-reactions are:

$$
\begin{array}{ll}
\text{reduction} & 2e^- + I_2(s) \rightarrow 2I^-(aq) \\
\text{oxidation} & 2Br^-(aq) \rightarrow Br_2(l) + 2e^-
\end{array}
$$

$$\text{overall} \quad 2Br^-(aq) + I_2(s) \rightarrow Br_2(l) + 2I^-(aq)$$

Substituting into $\Delta G° = -nF E°_{cell}$

$$\Delta G° = (-2 \text{ mol})(96{,}500 \text{ J/V·mol})(-0.54 \text{ V})$$

The units of the Faraday constant must be expressed in J/(V·mol) in order to cancel the units of volts from $E°_{cell}$. Continuing with our calculation:

$$\Delta G° = 1.04 \times 10^5 \text{ J} = 104 \text{ kJ}$$

The positive value of $\Delta G°$ indicates the reaction is not spontaneous under standard conditions. Next we calculate the equilibrium constant.

$$\ln K = \frac{n E°}{0.0257 \text{ V}}$$

$$= \frac{2(-0.54 \text{ V})}{0.0257 \text{ V}}$$

$$= -18.3$$

Taking the antilog of both sides we get

$$
\begin{aligned}
K &= 10^{-18.3} \\
&= 5 \times 10^{-19}
\end{aligned}
$$

EXERCISES

6. Predict whether Br⁻ ion can reduce I_2 under standard state conditions.

    $2Br^-(aq) + I_2(s) \rightarrow Br_2(l) + 2I^-(aq)$

7. Predict whether the following reactions will occur spontaneously in aqueous solution at 25°C under standard state conditions.

    a. $Ag(s) + Fe^{3+}(aq) \rightarrow Ag^+(aq) + Fe^{2+}(aq)$
    b. $Cu^{2+}(aq) + H_2(g) \rightarrow Cu(s) + 2H^+(aq)$

8. Calculate the standard Gibbs free energy changes for the two reactions in problem 7.

9. Calculate the equilibrium constant for the following redox reaction at 25°C.
    $2Fe^{2+}(aq) + Ni^{2+}(aq) \rightarrow 2Fe^{3+}(aq) + Ni(s)$

EFFECT OF CONCENTRATION ON CELL EMF (19.5)

STUDY OBJECTIVES

1. Calculate the emf of an electrochemical cell in which the reactants and products are present at nonstandard concentrations.
2. From a knowledge of E and E°, calculate the concentrations of a given ion in a half- cell.

The Nernst Equation. We mentioned earlier that the emf of a cell depends on the nature of the reactants and products, and on their concentrations. Since the cell emf is a measure of the spontaneity of the cell reaction, we might reasonably expect the voltage to fall as reactants are consumed and products accumulate.

The equation that relates the cell emf to the concentrations of reactants and products is named after Walter Nernst. At 298 K, for a redox reaction of the type:

$$aA + bB \rightarrow cC + dD$$

The Nernst equation is

$$E = E° - \frac{0.0257 \text{ V}}{n} \ln Q$$

where E° is the standard cell emf, E is the nonstandard cell emf, 0.0257 V is a constant at 298 K, and n is the number of moles of electrons transferred according to the balanced equation. The reaction quotient Q contains the concentrations of reactants and products.

$$Q = \frac{[C]^c[D]^d}{[A]^a[B]^b}$$

The Nernst equation predicts that E will decrease as reactant concentrations decrease, or as product concentrations increase. Both changes cause Q to increase. As Q increases, ln Q increases. Therefore, an increasingly larger number is *subtracted* from E°. Therefore, *E is not a constant*, but decreases as a reaction proceeds. E° is a constant for a reaction and is characteristic of the reaction.

At equilibrium, this equation reduces to one we have seen before. When the reaction is at equilibrium no net reaction occurs and no net transfer of electrons occurs, and E = 0. Also, at equilibrium Q = K, and the Nernst equation becomes:

$$0 = E° - \frac{0.0257 \text{ V}}{n} \ln K$$

$$E° = \frac{0.0257 \text{ V}}{n} \ln K$$

Therefore, the standard cell emf is related to the equilibrium constant.

When both E_{cell} and $E°_{cell}$ are known, the Nernst equation can be used to calculate an unknown concentration. This will be illustrated in Example 19.8.

EXAMPLE 19.7 EMF for a Nonstandard Cell

Calculate the cell emf at 25°C for an electrochemical cell in which the following overall reaction occurs at the concentrations given:

$$Zn(s) + 2H^+(aq, 1.0 \times 10^{-4} \text{ M}) \rightarrow Zn^{2+}(aq, 1.5 \text{ M}) + H_2(g, 1 \text{ atm})$$

•Method of Solution

The cell emf can be calculated by use of the Nernst equation.

$$E = E° - \frac{0.0257 \text{ V}}{n} \ln \frac{[Zn^{2+}]P_{H_2}}{[H^+]^2}$$

Where

$$E°_{cell} = E°_{cathode} - E°_{anode} = E°_{H^+/H_2} - E°_{Zn^{2+}/Zn} = 0.0 \text{ V} - (-0.76 \text{ V}) = 0.76 \text{ V}$$

$n = 2$ (the number of moles of electrons transferred according to the balanced equation)
$P_{H_2} = 1$, because hydrogen is in its standard state, 1 atm.
$[Zn^{2+}] = 1.5 \text{ M}$ and $[H^+] = 1.0 \times 10^{-4} \text{ M}$

•Calculation

Substitution into the Nernst equation yields:

$$E = 0.76 \text{ V} - \frac{0.0257 \text{ V}}{2} \ln \frac{1.5}{(1.0 \times 10^{-4})^2}$$

$$= 0.76 \text{ V} - 0.24 \text{ V}$$

$$E = 0.52 \text{ V}$$

•Comment

The low concentration of H^+ (1.0×10^{-4} M) compared to its standard state value (1 M) means that the driving force for reaction will be less than in the standard state, and as we see $E < E°$.

EXAMPLE 19.8   Determining Ion Concentrations

An electrochemical cell is constructed from a silver half-cell and a copper half-cell. The copper half-cell contains 0.10 M $Cu(NO_3)_2$ and the concentration of silver ions in the other half-cell is unknown. If the Ag electrode is the cathode, and the cell emf is measured and found to be 0.10 V at 25°C, what is the Ag^+ ion concentration?

•Method of Solution

Since E_{cell} is known and $E°_{cell}$ be calculate, the Nernst equation can be used to calculate an unknown concentration. To write the reaction quotient Q we need the cell reaction. Since Ag is the cathode, the reduction half-reaction is:

reduction/cathode $\qquad Ag^+ + e^- \rightarrow Ag$

Then Cu must be oxidized:

oxidation/anode $\qquad$ $Cu \rightarrow Cu^{2+} + 2e^-$

The overall reaction is $Cu + 2Ag^+ \rightarrow Cu^{2+} + 2Ag$. Therefore $Q = [Cu^{2+}]/[Ag^+]^2$.

•**Calculation**

The standard cell potential when calculated as we have show several times is 0.46 V. Also n = 2. The Nernst equation is:

$$E = E° - \frac{0.0257 \text{ V}}{n} \ln \frac{[Cu^{2+}]}{[Ag^+]^2}$$

Substitution gives:

$$0.10 \text{ V} = 0.46 \text{ V} - \frac{0.0257 \text{ V}}{2} \ln \frac{(0.10)}{[Ag^+]^2}$$

This equation must be solved for the unknown concentration $[Ag^+]$.

$$\frac{2(0.10 \text{ V} - 0.46 \text{ V})}{-0.0257 \text{ V}} = \ln \frac{(0.10)}{[Ag^+]^2}$$

$$12.2 = \ln \frac{(0.10)}{[Ag^+]^2}$$

Take the antilog of both sides:

$$10^{12.2} = \frac{(0.10)}{[Ag^+]^2}$$

$$1.5 \times 10^{12} = \frac{(0.10)}{[Ag^+]^2}$$

and

$$[Ag^+] = \sqrt{\frac{0.10}{1.5 \times 10^{12}}}$$

$$[Ag^+] = 2.6 \times 10^{-7} \text{ M}$$

EXERCISES

10. Determine the cell potential (E_{cell}) for the following reaction where the concentrations are as shown.

    $2Ag^+(1 \text{ M}) + H_2(1 \text{ atm}) \rightarrow 2Ag(s) + 2H^+ \text{ (pH = 7.12)}$

11. The cell emf for the following reaction is 0.98 V.

    $2Ag^+(aq, 1 \text{ M}) + H_2(g, 1 \text{ atm}) \rightarrow 2Ag(s) + 2H^+(aq, ? \text{ M})$
    Calculate the concentration of hydrogen ions.

12. Calculate $E°$, E_{cell}^o, and ΔG for the following cell reaction.

    $3Zn(s) + 2Cr^{3+}(0.0010 \text{ M}) \rightarrow 3Zn^{2+}(0.010 \text{ M}) + 2Cr(s)$

ELECTROLYSIS AND ITS QUANTITATIVE ASPECTS (19.8)

STUDY OBJECTIVES

1. Diagram an electrolytic cell, showing the reactions that occur at the anode and cathode.
2. Choose the most likely oxidation and reduction processes to be involved in the electrolysis of a given aqueous solution.
3. Calculate the amount of a substance produced by the flow a given electrical current for a given time, or calculate the time required for a given current to produce a given amount of product.

Electrolysis of Molten Salts. Electrolysis is the process in which electrical energy is used to cause nonspontaneous redox reactions to occur. An electrolytic cell is the apparatus for carrying out electrolysis. A schematic diagram of an electrolytic cell is shown in Figure 19.3. A battery or other DC power supply serves as an "electron pump" that supplies electrons to the cathode, where chemical species are reduced. Electrons resulting from the oxidation of chemical species are withdrawn from the anode and return to the battery. In the electrolytic cell the cathode is negative and the anode is positive. This is the opposite of an electrochemical cell.

Molten calcium chloride can be decomposed into the elements calcium and chlorine in an electrolytic cell. Let's examine the redox processes and the movement of ions and electrons in the apparatus. First, no current can flow through the cell unless it is filled with a liquid because ions in a solid crystal do not move from place to place. $CaCl_2$ melts at 780°C. The electrode in the electrolysis cell that is attached to the battery's negative terminal is the cathode. Electrons flow from the battery onto the cathode. Ca^{2+} ions are attracted to the cathode and are reduced to Ca metal. The anode of the electrolysis cell is attached to the positive terminal of the battery. Chloride ions drift toward the anode where they are oxidized. The electrons travel from the anode to the battery. For every electron leaving the battery, one must return. The reactions at the electrodes are:

cathode/reduction $\qquad Ca^{2+}(l) + 2e^- \rightarrow Ca(s)$

anode/oxidation $\qquad 2Cl^-(l) \rightarrow Cl_2(g) + 2e^-$

Overall $\qquad Ca^{2+}(l) + 2Cl^-(l) \rightarrow Ca(s) + Cl_2(g)$

Molten $CaCl_2$ consists of Ca^{2+} and Cl^- ions

Figure 19.3 Schematic diagram of an electrolytic cell. During electrolysis the cations (M^+) migrate to the cathode where they are reduced. Anions (X^-) migrate to the anode, where they are oxidized.

Voltage Requirement. The minimum voltage required to bring about electrolysis can be estimated from the standard reduction potentials. For the above reaction:

$$E^o_{cell} = E^o_{cathode} - E^o_{anode} = E^o_{Ca^{2+}/Ca} - E^o_{Cl_2/Cl^-}$$

$$E^o_{cell} = -2.87 \text{ V} - 1.36 \text{ V} = -4.23 \text{ V}$$

The negative cell potential means that this is a nonspontaneous reaction.

This reaction can be forced to occur by connecting the electrodes to an external source of electrical energy such as a battery or voltage supply, as in Figure 19.3. The electrolysis will occur when the external voltage is greater than the negative voltage of the nonspontaneous reaction. The external source of electrical energy must provide more than 4.23 V.

Competing Electrode Reactions in Aqueous Solution. A complicating factor in the electrolysis of aqueous solutions containing a dissolved salt is that water molecules can be reduced to H_2 or oxidized to O_2 in preference to the solute species.

Reduction of water: $2H_2O(l) + 2e^- \rightarrow H_2(g) + 2OH^-$ $E^o_{red} = -0.83 \text{ V}$

Oxidation of water: $H_2O(l) \rightarrow \frac{1}{2}O_2(g) + 2H^+ + 2e^-$ $-E^o_{red} = -1.23 \text{ V}$

Because oxidation is the reverse of reduction, the potentials measuring the ease of oxidation are $-E^o_{red}$.

We must consider whether water molecules are oxidized or reduced instead of ions of the salt. In the electrolysis of aqueous KI solution, for example, $H_2(g)$ is formed at the *cathode*, and I_2 at the *anode*. $H_2(g)$ is coming from the reduction of water. Water has a higher (more positive) standard reduction potential than K^+, which indicates that it is reduced more easily than potassium ion. K^+ ions drift to the cathode, but no K metal is formed.

$2H_2O(l) + 2e^- \rightarrow H_2(g) + 2OH^-(aq)$ $E^o_{red} = -0.83 \text{ V}$

$K^+(aq) + e^- \rightarrow K(s)$ $E^o_{red} = -2.93 \text{ V}$

Molecular iodine is formed at the anode during electrolysis of aqueous KI. This means the anode reaction is:

$$2I^-(aq) \rightarrow I_2 + 2e^-$$

rather than

$$H_2O(l) \rightarrow \frac{1}{2}O_2(g) + 2H^+(aq) + 2e^-$$

This preference reflects that I^- is easier to oxidize than H_2O.

$2I^- \rightarrow I_2(s) + 2e^-$ $-E^o_{red} = -0.53 \text{ V}$

$H_2O \rightarrow \frac{1}{2}O_2(g) + 2H^+ + 2e^-$ $-E^o_{red} = -1.23 \text{ V}$

The higher (more positive) the value of $-E^o_{red}$, the more favorable is the oxidation. Iodide ion has a higher oxidation potential than H_2O, and so iodide ion oxidized preferentially.

In summary. *If there is more than one reducible species in solution, the species with the greater reduction potential will be preferentially reduced. If there is more than one oxidizable species in solution, the species with the higher* $-E^o_{red}$ *will be preferentially oxidized.*

Overvoltage. The higher standard oxidation potential of I^- indicates that it should be more readily oxidized than H_2O. However, the existence of an overvoltage is a complicating factor. **Overvoltage** is the additional voltage over and above the standard potential required to cause electrolysis. Overvoltage affects H_2O oxidation more than any other process by lowering the ease of oxidation by about 0.6 V. According to the oxidation potentials $(-E^o_{red})$, I^- ion and Br^- ion, will be oxidized in the presence of H_2O, and Cl^- ion would not be. Due to overvoltage effects that reduce the tendency of H_2O to be oxidized, Cl^- ion can also be oxidized in aqueous solution. Keep in mind that *standard oxidation potentials do not always make the correct prediction of which species will be oxidized when one of the oxidizable species is* H_2O.

Quantitative Aspects of Electrolysis. Our knowledge of the quantitative relationships in electrolysis is due mostly to the work of Michael Faraday. He observed that the *quantity of a substance* undergoing chemical change during electrolysis is proportional to the *quantity of electrical charge* that passes through the cell. The quantity of electrical charge is expressed in coulombs. The electrical charge of 1 mole of electrons is called 1 Faraday which is equal to 96,500 coulombs. 1 mole $e^- = 1$ F $= 96,500$ C.

The number of electrons shown in the half-reaction is the link between the number of moles of substance reacted and the number of faradays of electricity required. We obtain the following conversion factors for the two half-reactions shown below:

$$K^+ + e^- \rightarrow K \qquad\qquad\qquad Ca^{2+} + 2e^- \rightarrow Ca$$

$$\frac{1 \text{ mol K}}{1 \text{ mol } e^-} \qquad\qquad\qquad \frac{1 \text{ mol Ca}}{2 \text{ mol } e^-}$$

The number of coulombs passing through an electrolytic cell in a given period of time is related to the electrical current in amperes (A) where 1 ampere is equal to 1 coulomb per second.

$$1 \text{ A} = 1 \text{ C/s}$$

In an electrolysis experiment, the measured quantities are current and the time that the current flows. The number of coulombs passing through the cell is

$$\text{Total charge in coulombs (C)} = \text{current(A)} \times \text{time (t)}$$

$$\text{Charge in coulombs} = \frac{C}{s} \times s$$

When an electrical current of known amperes flows for a known time the amount of chemical change can be calculated according to the following road map.

$$\text{current} \times \text{time} \rightarrow \text{coulombs} \rightarrow [\text{MOLES } e^- \text{ Transferred}] \rightarrow \underset{\text{product}}{\text{moles of}} \rightarrow \underset{\text{product}}{\text{grams of}}$$

The number of moles of electrons is the key term. It links the experimental variables to the theoretical yield. Reversing the road map allows one to calculate the time required to produce a specific amount of chemical change with a given electrical current.

EXAMPLE 19.9 Electrolysis of Aqueous Solutions

Predict the products of the electrolysis of aqueous $MgCl_2$ solution.

•Method of Solution

In aqueous solution the metal ion is not the only species the can be reduced. H_2O can be reduced as well. The two possible reduction reactions are:

$$
\begin{array}{ll}
Mg^{2+} + 2e^- \rightarrow Mg & E^o_{red} = -2.37 \text{ V} \\
2H_2O + 2e^- \rightarrow H_2 + 2OH^- & E^o_{red} = -0.83 \text{ V}
\end{array}
$$

Of the two, water has a higher standard reduction potential, and therefore has a greater tendency to be reduced than Mg^{2+}. The two oxidizable species are H_2O and Cl^- ion. The possible oxidation reactions and their $-E^o_{red}$ potentials are:

$$
\begin{array}{ll}
H_2O \rightarrow \frac{1}{2}O_2 + 2H^+ + 2e^- & -E^o_{red} = -1.23 \text{ V} \\
2Cl^- \rightarrow Cl_2 + 2e^- & -E^o_{red} = -1.36 \text{ V}
\end{array}
$$

The oxidation potentials indicate that H_2O is more readily oxidized than chloride ion. However, in this particular case, as discussed in Section 19.7 of the text, Cl^- ion is actually oxidized. The large overvoltage for O_2 formation prevents its production when Cl^- ion is there to compete. The overall reaction is:

$$2H_2O + 2Cl^- \rightarrow H_2 + 2OH^- + Cl_2$$

•**Comment**

The only example of overvoltage that you will need to be concerned with is in the oxidation of H_2O. It lowers the $-E^o_{red}$ by about 0.6 V to –1.83 V.

EXAMPLE 19.10   Quantitative Aspects of Electrolysis

How many moles of electrons are transferred in an electrolytic cell when a current of 12 amps flows for 16 hours?

•**Method of Solution**

First find the number of coulombs passing through the cell in 16 hours, and then convert to moles of electrons. Our road map is:

current × time → coulombs → moles of electrons
Recall 12 A = 12 C/s

•**Calculation**

The number of coulombs in 16 hours is given by multiplying the current times the time (in seconds).

$$\frac{12\ C}{1\ s} \times 16\ h \times \frac{3600\ s}{1\ h} = 691{,}000\ C$$

The Faraday constant is the charge of one mole of electrons. To three significant figures $F = 9.65 \times 10^4$ C. The number of moles of electrons is:

$$691{,}000\ C \times \frac{1\ mol\ e^-}{9.65 \times 10^4\ C} = 7.2\ mol\ e^-$$

EXAMPLE 19.11   Quantitative Aspects of Electrolysis

How many grams of copper metal will be deposited from a solution of $CuSO_4$ by the passage of 3.0 A of electrical current through an electrolytic cell for 2.0 hours?

•**Method of Solution**

Cu^{2+} is reduced according to the half-equation

$$Cu^{2+} + 2e^- \rightarrow Cu$$

This half-equation tells us that 2 moles of electrons are required to produce 1 mol of Cu.

$$\frac{1\ mol\ Cu}{2\ mol\ e^-}$$

The number of coulombs passing through a cell is given by the current times the time.

$$3.0\ A \times 2.0\ h \times \frac{1\ C\ s^{-1}}{1\ A} \times \frac{3600\ s}{1\ h} = 2.16 \times 10^4\ C$$

From this number of coulombs we can find the number of moles of electrons passing through the cell. The road map will be:

current × time → coulombs → [MOLES e^-] → mol Cu → g Cu

•Calculation

The number of moles of electrons is the link from the current measurements to the molar amount of Cu formed. The number of moles of electrons is:

$$2.16 \times 10^4 \, C \times \frac{1 \text{ mol } e^-}{9.65 \times 10^4 \, C} = 0.224 \text{ mol } e^-$$

The number of moles of Cu is:

$$0.224 \text{ mol } e^- \times \frac{1 \text{ mol Cu}}{2 \text{ mol } e^-} = 0.112 \text{ mol Cu}$$

The number of grams of Cu is

$$0.112 \text{ mol Cu} \times \frac{63.5 \text{ g}}{1 \text{ mol Cu}} = 7.11 \text{ g Cu}$$

Of course, this calculation could be carried out using the factor label method.

$$3.0 \, A \times \frac{1 \, C \, s^{-1}}{1 \, A} \times 2.0 \, h \times \frac{3600 \, s}{1 \, h} \times \frac{1 \text{ mol } e^-}{9.65 \times 10^4 \, C} \times \frac{1 \text{ mol Cu}}{2 \text{ mol } e^-} \times \frac{63.5 \text{ g}}{1 \text{ mol Cu}} = 7.11 \text{ g Cu}$$

EXERCISES

13. a. Write the half-reaction that occurs at the anode during the electrolysis of aqueous lithium bromide.
    b. Write the half-reaction that occurs at the cathode during the electrolysis of aqueous lithium bromide?

14. How many coulombs are required to cause reduction of 0.20 mol of Cr^{3+} to Cr?

15. How many moles of electrons are transferred in an electrolytic cell when a current of 5.0 amps flows for 10.0 hours?

16. How many moles of electrons are required to electroplate 29.4 g of nickel from a solution containing Ni^{2+} ion?

17. How many minutes are required to electroplate 25.0 g of metallic Cu using a constant current of 20.0 A?

18. A current of 0.80 A was applied to an electrolytic cell containing molten $CdCl_2$ for 2.5 hours. Decide on the appropriate half-reaction, and calculate the mass of cadmium metal deposited.

CONCEPTUAL QUESTIONS

1. a. Consider the solution in the anode half-cell of Figure 19.1. How is the concentration of Cu^{2+} ions related to the concentration of SO_4^{2-} ions before any reaction occurs? How are the concentrations of these two ions related after reaction occurs? b. How does the solution stay electrically neutral?

2. Electrolysis experiments provide one of the most accurate ways to determine Avogadro's number. Electrolysis of a $NiSO_4$ solution by a current of 1.02 A for 252 s yielded a deposit of 0.1531 g Ni. Use this information to calculate a value for Avogadro's number. Do not use the Faraday constant, but you will need the electron charge: 1.602×10^{-19} C.

3. Why are different products obtained when molten $AlBr_3$ and aqueous $AlBr_3$ are electrolyzed using inert electrodes?

PRACTICE TEST

1. Draw the cell diagrams for each of the redox reactions given below. You may use platinum as an inert electrode.
   a. $2Al(s) + 3H_2SO_4(aq) \rightarrow Al_2(SO_4)_3(aq) + 3H_2(g)$
   b. $Fe_2(SO_4)_3 + 3Pb(s) \rightarrow 3PbSO_4(s) + 2Fe(s)$
   c. $CuSO_4(aq) + H_2(g) \rightarrow Cu(s) + H_2SO_4(aq)$
   d. $2NaBr(aq) + I_2(g) \rightarrow Br_2(l) + 2NaI(aq)$
   e. $2FeCl_2(aq) + SnCl_2(aq) \rightarrow 2FeCl_3(aq) + Sn(s)$
   f. $2FeCl_2(aq) + SnCl_4(aq) \rightarrow 2FeCl_3(aq) + SnCl_2(aq)$

2. Consider an electrochemical cell constructed from the following half-cells that are linked by a porous membrane. (1) an Au electrode dipped into 1.0 M $Au(NO_3)_3$ and (2) an Fe electrode dipped into 1.0 M $FeSO_4$. Answer the following questions:
   a. Which electrode is the cathode?
   b. Write a balanced equation for the reaction occurring while the cell is discharging.
   c. What emf should the cell generate?
   d. In what direction will electrons flow in the outer circuit?
   e. Toward which electrode will positive ions migrate?

3. Arrange the following species in order of increasing strength as oxidizing agents:
   Ce^{4+}, O_2, H_2O_2, SO_4^{2-}

4. Arrange the following species in order of increasing strength as reducing agents:
   Zn, Ni, H_2, F_2.

5. a. Will $O_2(g)$ oxidize I^- ion in acid solution under standard conditions?
   b. Will $O_2(g)$ oxidize Br^-?
   c. Will $O_2(g)$ oxidize Cl_2?

6. Calculate $\Delta G°$ and the equilibrium constant (K) for the following reaction.

   $Fe^{3+}(aq) + Ag(s) \rightarrow Fe^{2+}(aq) + Ag^+(aq)$

7. At what ratio of $[Fe^{2+}]/[Fe^{3+}]$ would the reaction given in problem 6 become spontaneous if $[Ag^+] = 1.0$ M?
8. Calculate the cell emf for the following reaction:

   $2Ag^+(0.10 \text{ M}) + H_2(1 \text{ atm}) \rightarrow 2Ag(s) + 2H^+(pH = 8)$

9. What is the minimum voltage required to bring about electrolysis of a solution of Cu^{2+} and Br^- at standard concentrations?

10. What change in the emf of a hydrogen-copper cell will occur when $NaOH(aq)$ is added to the solution in the hydrogen half-cell?

11. Consider a uranium-bromine cell in which U is oxidized and Br_2 is reduced. The half-reactions are

    $U^{3+}(aq) + 3e^- \rightarrow U(s)$ $E°_{U^{3+}/U} = ?$
    $Br_2(l) + 2e^- \rightarrow 2Br^-(aq)$ $E°_{Br_2/Br^-} = 1.07$ V
    If the standard cell emf is 2.91 V, what is the standard reduction potential for uranium?

12. When the concentration of Zn^{2+} ion is 0.15 M, the measured voltage of the Zn-Cu cell is 0.40 V. What is the Cu^{2+} ion concentration?

13. A hydrogen electrode is immersed in an acetic acid solution. This electrode is connected to another consisting of an iron nail dipping into 0.10 M $FeCl_2$. If E_{cell} is found to be 0.24 V, what is the pH of the acetic acid solution?

14. How many moles of electrons are transferred in an electrolytic cell when a current of 2.0 amps flows for 6 hours?

15. How many moles of electrons are required to electroplate 6.0 g of chromium from a solution containing Cr^{3+}?

16. a. How many grams of nickel can be electroplated by passing a constant current of 5.2 A through a solution of $NiSO_4$ for 60.0 min?
    b. How many grams of cobalt can be electroplated by passing a constant current of 5.2 A through a solution of $CoCl_3$ for 60.0 min?

17. How long will it take to produce 54 kg of Al metal by the reduction of Al^{3+} in an electrolytic cell using a current of 500 amps?

18. Balance the following redox reactions by the ion-electron method.
    a. $H_2O_2 + I^- \rightarrow I_2$ (acidic solution)
    b. $Cr_2O_7^{2-} + H_3AsO_3 \rightarrow Cr^{3+} + H_3AsO_4$ (acidic solution)
    c. $Cl_2 \rightarrow ClO_4^- + Cl^-$ (basic solution)
    d. $Ag(s) + NO_3^-(aq) \rightarrow NO_2(g) + Ag^+(aq)$ (acidic solution)

General Problems

19. Given the reduction potential of the standard hydrogen electrode, and the half-equation:

    $2H_2O + 2e^- \rightarrow H_2(g) + 2OH^-(aq)$ $E° = -0.83$ V

    calculate the value of the ionization constant for water.

    $H_2O \rightleftharpoons H^+(aq) + OH^-(aq)$

20. A cell was constructed using the standard hydrogen electrode as one half-cell, and a lead electrode in a 0.10 M K_2CrO_4 solution in contact with undissolved $PbCrO_4$ as the other half-cell. The potential of the cell was measured to be 0.50 V with the Pb electrode as the anode. From these data determine the K_{sp} of $PbCrO_4$.

21. A 40 watt light bulb is powered by a lead storage battery that has 20.0 g of Pb available for the anode? For how many hours will the bulb provide illumination? Assume the voltage is constant at 1.5 V. Recall: 1 watt = 1 J/s.

ANSWERS

Exercises

1. a. $14H^+ + Cr_2O_7^{2-}(aq) + 6Br^-(aq) \rightarrow 2Cr^{3+}(aq) + 3Br_2(g) + 7H_2O$

   b. $2MnO_4^-(aq) + 6I^-(aq) + 4H_2O \rightarrow 2MnO_2(s) + 3I_2(aq) + 8OH^-(aq)$

2. a. $5H_2C_2O_4(aq) + 2MnO_4^-(aq) + 6H^+(aq) \rightarrow 2Mn^{2+}(aq) + 10CO_2(g) + 8H_2O(l)$

   b. $Zn(s) + ClO^-(aq) + H_2O(l) \rightarrow Zn(OH)_2(s) + Cl^-(aq)$

3. a. $E^{\circ}_{cell} = -2.58$ V   b. $E^{\circ}_{cell} = 0.29$ V

4. a. Cu   b. $Cu^{2+}(aq) + Pb(s) \rightarrow Cu(s) + Pb^{2+}(aq)$ c. Cu

   d. From Pb to Cu.   e. Cu

5. a. Al(s)   b. Fe^{3+}

6. No

7. a. No   b. Yes

8. a. $\Delta G^{\circ} = 2.90$ kJ   b. $\Delta G^{\circ} = -65.6$ kJ

9. $K_c = 3.0 \times 10^{-35}$

10. $E = 1.22$ V

11. 9.0×10^{-4} M

12. $E^{\circ}_{cell} = 0.02$ V $E_{cell} = 0.01$ V $\Delta G = -5.79$ kJ

13. a. $2Br^- \rightarrow Br_2 + 2e^-$

    b. $2H_2O + 2e^- \rightarrow H_2(g) + 2OH^-(aq)$

14. 57,900 C

15. 1.86 F

16. 0.808 F

17. 63.3 min

18. 4.19 g Cd

Conceptual Questions

1. a. After: $[Cu^{2+}] > [SO_4^{2-}]$ b. Cl^- ions migrate into the solution from the salt bridge. Two chloride ions for each excess Cu^{2+} ion.

2. Avogadro's number is a ratio. Its units are "atoms per mole." We are given the mass of Ni so we can use it to find the number of moles. Now to count the atoms. Actually we will count the electrons and then divide by 2 to get the number of Ni atoms. As usual the total charge in coulombs flowing through the cell is equal to the current (A) times the time (t). Dividing the total charge by the charge on one electron gives the total number of electrons passing through the cell. The number of Ni atoms produced by the reduction of Ni^{2+} ions is just half the number of electrons. Divide the number of Ni atoms by its corresponding number of moles in order to get Avogadro's number.

3. Molten $AlBr_3$ consists of Al^{3+} ions and Br^- ions. When it undergoes electrolysis the Al^{3+} ion is reduced to Al metal, and the Br^- ion is oxidized to Br_2. An aqueous solution of $AlBr_3$ consists of H_2O molecules, as well as Al^{3+} ions and Br^- ions. As you can tell from reduction potentials, Al^{3+} ion will not be reduced in aqueous solution. Rather the H_2O molecules are and the products are H_2 and OH^- ion. In aqueous solution Br^- ion will be oxidized just as it was in the molten state.

Practice Test

1. a. $Al(s) \mid Al^{3+}(aq) \parallel H^+(aq) \mid H_2(g) \mid Pt(s)$

    b. $Pb(s) \mid Pb^{2+}(aq) \parallel Fe^{3+}(aq) \mid Fe(s)$

    c. $Pt(s) \mid H_2(g) \mid H^+(aq) \parallel Cu^{2+}(aq) \mid Cu(s)$

    d. $Pt(s) \mid Br^-(aq) \mid Br_2(l) \parallel I_2(s) \mid I^-(aq) \mid Pt(s)$

    e. $Pt(s) \mid Fe^{2+}(aq), Fe^{3+}(aq) \parallel Sn^{2+}(aq) \mid Sn(s)$

    f. $Pt(s) \mid Fe^{2+}(aq), Fe^{3+}(aq) \parallel Sn^{4+}(aq), Sn^{2+}(aq) \mid Pt(s)$

2. a. Au   b. $2Au^{3+}(aq) + 3Fe(s) \rightarrow 2Au(s) + 3Fe^{2+}(aq)$

    c. 1.94 V   d. from Fe to Au   e. Au

3. $SO_4^{2-} < O_2 < Ce^{4+} < H_2O_2$

4. $F_2 < H_2 < Ni < Zn$

5. a. Yes   b. Yes   c. No

6. $\Delta G° = 2.9$ kJ; K = 0.31

7. $[Fe^{2+}]/[Fe^{3+}] = 0.31$

8. E = 1.21 V

9. 0.73 V

10. The cell emf will increase due to a decrease in $[H^+]$. The cell reaction is $Cu^{2+} + H_2 \rightarrow Cu + 2H^+$

11. −1.84 V

12. $[Cu^{2+}] = 3.1 \times 10^{-25}$ M

13. pH = 3.88

14. 0.45 F

15. 0.35 F

16. a. 5.69 g Ni   b. 3.81 g Co

17. 322 h

18. a. $H_2O_2 + 2I^- + 2H^+ \rightarrow I_2 + 2H_2O$

    b. $Cr_2O_7^{2-} + 3H_3AsO_3 + 8H^+ \rightarrow 2Cr^{3+} + 3H_3AsO_4 + 4H_2O$

    c. $4Cl_2 + 8OH^- \rightarrow ClO_4^- + 7Cl^- + 4H_2O$

    d. $Ag(s) + 2H^+(aq) + NO_3^-(aq) \rightarrow NO_2(g) + Ag^+(aq) + H_2O$

19. $K_w = 9.0 \times 10^{-15}$

20. $[Pb^{2+}] = 3.0 \times 10^{-13}$ M; $K_{sp} = 3.0 \times 10^{-14}$

21. 0.19 h

Chapter Twenty
METALLURGY AND THE CHEMISTRY OF METALS

- Metallurgical Processes (20.1 – 20.2)
- The Band Theory of Conductivity (20.3)
- The Alkali Metals (20.5)
- The Alkaline Earth Metals (20.6)
- Aluminum (20.7)

METALLURGICAL PROCESSES (20.1 – 20.2)

STUDY OBJECTIVES

1. List the procedures used to separate minerals from gangue.
2. Describe briefly the means used to produce and purify a metal.
3. Describe the steps in the production of iron.

Preliminary Treatment. Metallurgy is the science and technology of extracting metals from their ores and preparing these metals for use. Most metals are found in nature in the combined state as minerals. A *mineral* is a naturally occurring substance with a characteristic range of composition. An **ore** is a mineral deposit of high enough concentration to make an economical recoverery of a desired metal. Ores generally must be cleaned before the metal is extracted. The waste materials, called *gangue*, are removed by *flotation, ferromagnetic separation,* or *distillation of mercury amalgams.*

Reduction of Metals. The process of producing a free metal is always one of reduction. Reduction of metal oxides is usually less complex than reduction of sulfides. **Roasting** is a preliminary operation used to convert a metal sulfide or carbonate into an oxide. Two types of reductions are ordinarily employed:

1. *Chemical reduction* is the use of a reducing agent to prepare the elemental form of a metal from a compound. Calcium, magnesium, aluminum, hydrogen, and carbon are often used in chemical reductions.
2. *Electrolytic reduction* is necessary to produce the very electropositive metals such as sodium, magnesium and aluminum.

Iron. Iron is prepared by a chemical reduction process. The raw materials for making iron are (1) iron ore, either hematite Fe_2O_3 or magnetite Fe_3O_4, (2) coke, and (3) limestone. These three materials are mixed and fed into a huge blast furnace (Figure 20.3 text). A strong blast of air preheated to 1500°C is blown in at the bottom, where oxygen in the air reacts with the coke. This reaction supplies heat for the furnace, and carbon monoxide which is the reducing agent in the reaction of the iron oxides.

$$2C(g) + O_2(g) \rightarrow 2CO(g)$$
$$Fe_2O_3(s) + 3CO(g) \rightarrow 2Fe(l) + 3CO_2(g)$$

A temperature gradient is set up in the furnace. The temperature at the bottom is 1500°C, and it decreases with height; falling to about 250°C at the top. Much of the carbon dioxide produced is reduced by reaction with excess coke as follows:

$$CO_2(g) + C(s) \rightarrow 2CO(g)$$

thereby forming more of the principal reducing agent. The gas that escapes at the top of the furnace is mostly nitrogen and carbon monoxide. The molten iron runs to the bottom where it is withdrawn periodically.

Limestone serves as a flux in the removal of impurities such as silica (SiO_2) and alumina (Al_2O_3). The limestone decomposes at temperatures above 900°C, forming calcium oxide and carbon dioxide.

$$CaCO_3(s) \rightarrow CaO(s) + CO_2(g)$$

Calcium oxide unites with silica and alumina to form a glassy, molten substance called *slag* which is composed mainly of $CaSiO_3$ and some calcium aluminate. Slag is less dense than iron, and so it collects as a pool on top of the metal. It is drawn off, leaving the molten iron. Iron prepared in this manner contains many impurities and is called pig iron, or cast iron. Cast iron is made into steel by further treatment in a basic oxygen furnace.

Purification of Metals. Once a metal is prepared it may need to be further purified. Distillation, electrolysis, and zone refining are three common procedures. Metals with low boiling points, such as mercury (357°C), cadmium (767°C), and zinc (907°C), are referred to as volatile. They can be purified by *fractional distillation*. On heating, these metals vaporize, leaving behind any nonvolatile impurities. Condensation of the vapor yields the purified metal.

Metals can also be purified by *electrolysis*. The metals that plate out on a cathode can be controlled by the voltage at which the electrolysis is carried out. In the purification of copper, Cu^{2+} is reduced much more readily than iron or zinc ions, which are common impurities. At just the right voltage Cu is plated out and the metal impurities remain in solution.

Zone refining takes advantage of the fact that when liquids begin to freeze, the impurities tend to remain in the liquid phase. See Figure 20.8 of the textbook. Extremely pure metals can be obtained by repeating this process a number of times.

EXAMPLE 20.1   Chemical Reduction

Chemical reduction is the use of a reducing agent to prepare the elemental form of a metal from a compound. Look up the reduction potentials of calcium, magnesium, and hydrogen listed in Table 19.1 in the text and suggest five metals that calcium, and magnesium can reduce, but that hydrogen cannot reduce.

•Method of Solution

Ca and Mg are strong reducing agents. Their reduction potentials are quite low which means that they have a strong tendency to transfer electrons to other metal cations.

$$Ca^{2+}(aq) + 2e^- \rightarrow Ca(s) \qquad E^o_{red} = -2.87$$
$$Mg^{2+}(aq) + 2e^- \rightarrow Mg(s) \qquad E^o_{red} = -2.37 \text{ V}$$
$$2H^+ + 2e^- \rightarrow H_2(g) \qquad E^o_{red} = 0 \text{ V}$$

H_2 is not as strong a reducing agent as the others.

Ca(s) and Mg(s) can be used to reduce ions of metals having higher reduction potentials (the diagonal rule) such as Zn^{2+}, Cr^{3+}, Co^{2+}, Ni^{2+}, and Sn^{2+}, but not Li^+ and K^+. H_2 cannot reduce any of these ions, but is used to reduce more noble metals such as Cu^{2+}.

EXAMPLE 20.2   Steel Production

How are the impurities present in iron ore, such as silica and sulfur, removed during steel production?

•**Method of Solution**

Acidic impurities such as silica and sulfides combine with CaO, a base, to form a molten slag.

$$SiO_2 + CaO \rightarrow CaSiO_3$$
$$FeS + CaO \rightarrow CaS + FeO$$

The slag has a lower density than molten iron. Therefore it collects as a pool on top of the metal. This slag is drawn off and used as a component in making cement.

EXERCISES

1. Distinguish between an ore and a mineral.
2. Why must electrolytic reduction be used in the production of some metals such as lithium and sodium?
3. What is zone refining used for?
4. How is limestone used in iron production?
5. Write an equation for the chemical reduction of Cr_2O_3 with Al metal yielding chromium.

THE BAND THEORY OF CONDUCTIVITY (20.3)

STUDY OBJECTIVES

1. Compare the energies of the conduction bands and valence bands in metals, insulators, and semiconductors.
2. Use the band theory to describe conduction in metals.

The Conduction Band.   The *band theory* is the result of the application of molecular orbital theory to metals. In metals, the atoms lie in a three-dimensional array and take part in bonding that spreads over the entire crystal. Take, for example, an alkali metal atom that carries a single valence electron in an s orbital. Recall that the atomic orbitals of two atoms will overlap when the atoms are close together. This results in the formation of two molecular orbitals: a bonding orbital and an antibonding orbital. The total number of molecular orbitals produced always equals the number of atomic orbitals that overlap to produce them. The overlap of four atomic orbitals from four atoms, for instance, will form four molecular orbitals: two bonding and two antibonding orbitals. A crystal made up of N atoms with overlapping atomic orbitals will have a total of N molecular orbitals. These bonding and antibonding orbitals are so closely spaced in terms of energy that they are called a "band." The formation of a band is shown in Figure 20.1.

The band containing the valence shell electrons (3s for Na) is called the **valence band**. For alkali metals there are N valence electrons and N orbitals in the band. These N orbitals have a capacity for 2N electrons. Any band that is either vacant or partially filled is called a **conduction band**.

Figure 20.1 The formation of an energy band by the successive overlap of atomic orbitals. a. When the orbitals of two atoms overlap, one bonding orbital and one antibonding orbital of significantly different energies are formed. b. When the orbitals of four atoms overlap, four molecular orbitals are formed. c. When the orbitals of N atoms overlap where N is a large number, N orbitals are formed. With so many orbitals, the orbital energies differ only infinitesimally from each other and form a virtually continuous band.

Electrical Conduction. The band theory explains conduction in the alkali metals in the following way. When a voltage is applied across a piece of sodium metal, conduction occurs. The current is the result of electrons in the 3s band being free to jump from atom to atom. The free movement of electrons is possible for two reasons. In alkali metals the conduction band and the valence band are the same. The orbitals within the band are so similar in energy that an electron does not need to gain appreciable energy to reach the conduction band. Also, as discussed above, the conduction band is only half filled to capacity which means that an electron is free to move through the entire metal.

In an **insulator**, such as glass or plastic, the valence band is filled. Thus the next vacant higher-energy band becomes the conduction band. An energy gap exists between the valence band and the conduction band, as shown in Figure 20.10 of the text. This large separation prevents electrons in insulators from entering the conduction band.

A **semiconductor**, such as Si and Ge, has a filled valence band and an empty conduction band, but in contrast to an insulator, a relatively small gap exists between these bands. A relatively small amount of thermal energy will promote an electron into the conduction band. Thus as temperature increases, the conductivity of semiconductors increases correspondingly.

EXAMPLE 20.3 Band Theory

Show for sodium that the valence band is half-filled.

•Method of Solution

Sodium atoms have filled 1s, 2s, and 2p orbitals; therefore the corresponding bands in the solid are filled. The valence band for Na is the 3s. In the metal the 3s orbitals of N sodium atoms overlap to form a total of N molecular orbitals that make up the valence band. The capacity of this band is 2N electrons. Since each Na atom has one 3s electron, the N Na atoms have N electrons in the valence band. This means that the valence band is half-filled. The 3s band of Na is the conduction band.

•Comment

The highest occupied band is called the *valence band.* This unfilled valence band is called the *conduction band* of sodium metal.

EXERCISES

6. Define the terms valence band and conduction band.
7. Define the terms conductor, insulator, and semiconductor.

THE ALKALI METALS (20.5)

STUDY OBJECTIVES

1. Describe the chemical properties of the alkali metal elements.
2. Describe the sources and preparation of sodium and potassium metals.

Properties of the Alkali Metals. As a group, the alkali metals are extremely reactive and never occur naturally in the elemental form. They are the least electronegative group of elements and exist as +1 cations combined with halides, sulfate, carbonate, and silicate ions. A number of properties of these elements are listed in Table 20.4 of the text. Compared with other metals, they are very soft and have low melting points and low densities. Lithium, sodium and potassium will float on water. The low density is a result of their large atomic volume and the fact that they all possess a body-centered crystal structure which has a low packing efficiency.

Sources and Preparations. The preparation of Group 1A metals requires large amounts of energy because their positive ions are difficult to reduce. Lithium, being the most active metal, is prepared by electrolysis of molten LiCl.

Sodium and potassium occur in a wide variety of minerals, and are the sixth and seventh most abundant elements in Earth's crust. Sodium compounds are so abundant and wide-spread that it is difficult to find matter free of this element. Sodium chloride makes up about two-thirds of the solid matter dissolved in sea water. The minerals *carnallite* ($KMgCl_3 \cdot 6H_2O$), and *sylvite* (KCl) are found in ancient lake and seabeds and serve as commercial sources of potassium and its compounds.

Both metals, sodium and potassium, were first prepared in 1807 by the English chemist Humphry Davy. He used the electrolysis of the corresponding moist hydroxides. As their discoverer, Davy was allowed to name the elements giving them the above names. However, you will notice that neither has an element symbol consistent with its name. At the time both potash and soda were recognized as carbonates of these unisolated metal elements. Potassium obtained its name because it was the metallic element in potash, potassium carbonate. The metallic element of soda (sodium carbonate) Davy called sodium.

Sodium and Potassium. Metallic sodium is prepared commercially by the electrolysis of molten sodium chloride. Most of the sodium made in the United States is produced in the Downs cell (Figure 19.17 text) by the electrolysis of a mixture of sodium chloride and calcium chloride. This electrolyte mixture melts at 505°C, whereas pure NaCl melts at 801°C. The lower temperature reduces the cost of production.

A Downs cell consists of a carbon anode and an iron cathode. The chloride ions are oxidized at the anode, and the sodium ions are reduced at the cathode. The half-reactions are:

anode	$2Cl^- \rightarrow Cl_2(g) + 2e^-$
cathode	$2Na^+ + 2e^- \rightarrow 2Na(l)$

The liquid sodium is drawn off and kept from contact with chlorine and oxygen. Sodium can also be obtained by the electrolysis of molten sodium hydroxide.

Potassium is made by reaction of potassium chloride with sodium vapor in the absence of air at 900°C.

$$KCl(l) + Na(g) \rightarrow K(g) + NaCl(l)$$

This reaction should occur to only a small extent. However, removal of potassium vapor as it is formed drives the reaction to completion.

Reactions of Alkali Metals. The alkali metals are extremely reactive. The reactivity increases with atomic mass. Sodium forms two oxides on reaction with oxygen: sodium oxide (Na_2O) and sodium peroxide (Na_2O_2).

$$2Na(s) + \frac{1}{2}O_2(g) \rightarrow Na_2O(s)$$
$$2Na(s) + O_2(g) \rightarrow Na_2O_2(s)$$

The reaction of potassium with oxygen forms the peroxide (K_2O_2), but in addition when K burns in air it forms potassium superoxide

$$K(s) + O_2(g) \rightarrow KO_2(s)$$

Potassium superoxide is a source of O_2 in breathing equipment. KO_2 reacts with water:

$$2KO_2(s) + 2H_2O(l) \rightarrow 2KOH(aq) + O_2(g) + H_2O_2(aq)$$

Sodium chloride, sodium carbonate, sodium bicarbonate, sodium hydroxide, sodium nitrate, potassium hydroxide, and potassium nitrate are important compounds of these elements.

EXERCISES

8. List four physical properties of the alkali metals.
9. List three chemical properties of the alkali metals.

THE ALKALINE EARTH METALS (20.6)

STUDY OBJECTIVES

1. Write chemical equations to show the preparations of magnesium and calcium.
2. Describe the chemical properties and give the major uses of these elements and their compounds.

Properties of the Alkaline Earth Metals. The alkaline earth metals are somewhat less reactive than the alkali metals. Except for beryllium the alkaline earth metals have similar chemical properties. The oxidation number of the Group 2A metals in almost all compounds is +2. The electronegativity of these elements is low enough that they form predominately ionic compounds with nonmetals. Table 20.5 in the text lists some common properties of these metals. As with the other groups of elements in the periodic table, metallic character increases as you read down the group. The properties of magnesium and calcium are reviewed below. The formulas of the important ores of magnesium and calcium are given in Table 20.1.

Table 20.1 Ores Containing Magnesium and Calcium

Element	Rank (Earth's crust)	Ore	Formula
Magnesium	6th	brucite	$Mg(OH)_2$
		Dolomite	$MgCO_3 \cdot CaCO_3$
		epsomite	$MgSO_4 \cdot 7H_2O$
Calcium	5th	limestone	$CaCO_3$
		gypsum	$CaSO_4 \cdot 2H_2O$
		fluorite	CaF_2

Preparation of Magnesium and Calcium. Seawater is an important source of magnesium. It contains 0.13 percent Mg^{2+} by weight. $Mg(OH)_2$ being an insoluble compound is precipitated from seawater by adding $Ca(OH)_2$.

$$Mg^{2+}(aq) + Ca^{2+}(aq) + 2OH^-(aq) \rightarrow Mg(OH)_2(s) + Ca^{2+}(aq)$$

Magnesium hydroxide is then converted to magnesium chloride by neutralization with hydrochloric acid. Metallic magnesium is obtained by electrolysis of molten magnesium chloride.

$$Mg^{2+} + 2Cl^- \xrightarrow{\text{electrolysis}} Mg(l) + Cl_2(g)$$

Calcium is obtained from limestone. Heating limestone causes it to decompose into calcium oxide and carbon dioxide. Calcium metal is prepared by a thermal reduction process, rather than electrolysis. Calcium oxide is reduced with aluminum at high temperature (1200°C).

$$CaCO_3(s) \rightarrow CaO(s) + CO_2(g)$$
$$6CaO(s) + 2Al(l) \rightarrow 3Ca(l) + Ca_3Al_2O_6(s)$$

Properties of Magnesium and Calcium. Magnesium is less reactive than calcium. Magnesium does not react with cold water, but it reacts slowly with steam. Calcium, on the other hand, reacts with cold water.

$$Mg(s) + H_2O(g) \rightarrow MgO(s) + H_2(g)$$
$$Ca(s) + 2H_2O(l) \rightarrow Ca(OH)_2(aq) + H_2(g)$$

Both MgO and CaO react with water to give hydroxides.

$$MgO(s) + H_2O(l) \rightarrow Mg(OH)_2(s) \quad \text{(reacts with steam)}$$
$$CaO(s) + H_2O(l) \rightarrow Ca(OH)_2(s) \quad \text{(reacts with cold water)}$$

When magnesium burns in air, considerable nitride is formed along with the oxide.

$$2Mg(s) + O_2(g) \rightarrow 2MgO(s)$$
$$3Mg(s) + N_2(g) \rightarrow Mg_3N_2(s)$$

Uses of the Alkaline Earth Metals. Beryllium has excellent alloying qualities and is used to make alloys that are corrosion resistant. Beryllium is used as a "window" in X-ray tubes.

The major uses of magnesium are also in alloys. Because of its low atomic mass, it is a good lightweight structural metal. It is also used in batteries, and for cathodic protection of buried metal pipelines and storage tanks.

Metallic calcium finds use mainly in alloys. Calcium salts are used as dehydrating agents; anhydrous calcium chloride, for example, has a strong affinity for water. Quicklime (CaO) is used is steel production, and in the removal of SO_2 from coal-fired power plants. Slaked lime ($Ca(OH)_2$) is used in water treatment.

Strontium nitrate and carbonate are used in fireworks and highway flares to provide their brilliant red color. Barium metal is the most active of the alkaline earth metals and so it has very few uses.

EXAMPLE 20.4   Properties of Alkaline Earth Metals

Identify the element among the alkaline earth metals (excluding radium) that will have the following properties.
a. Most reactive oxide towards water.
b. Lowest electronegativity
c. Smallest atomic radius
d. Lowest first ionization energy.

•**Method of Solution**

Applying peroidic trends in these properties.
a. Ba because reactivity increases as you read down in the group.
b. Ba because electronegativity decreases as you read down in the group.
c. Be because atomic radius increases as you read down in the group.
d. Ba because ionization energy decreases as you read down in the group.

EXAMPLE 20.5   Electrolytic Reduction of Calcium

Starting with Ca^{2+} ions in limestone, write equations to show how you would obtain pure calcium, using electrolysis, rather than thermal reduction.

•**Method of Solution**

Electrolysis of molten metal chlorides is often used to prepare pure metals. First limestone, chalk, or sea shells, all of which contain $CaCO_3$, must be decomposed at 900°C.

$$CaCO_3(s) \rightarrow CaO(s) + CO_2(g)$$

The CaO(s) is slaked to yield calcium hydroxide.

$$CaO(s) + H_2O(l) \rightarrow Ca(OH)_2(s)$$

This base is then neutralized with hydrochloric acid to give the desired salt $CaCl_2$.

$$Ca(OH)_2(aq) + 2HCl(aq) \rightarrow CaCl_2(aq) + 2H_2O(l)$$

After drying, the $CaCl_2$ can be melted and then electrolyzed.

$$CaCl_2(l) \xrightarrow{\text{electrolysis}} Ca(l) + Cl_2(g)$$

EXERCISES

10. Give the sources of magnesium and calcium.
11. Write chemical formulas for limestone, quicklime, and slaked lime.

ALUMINUM (20.7)

STUDY OBJECTIVES

1. Write chemical equations to show the reactions of aluminum with acids, bases, oxygen, and other metal oxides.
2. Write chemical equations that show the amphoteric nature of $Al(OH)_3$.
3. List several uses of aluminum and its compounds.

Source and Preparation. Aluminum is the third most abundant element in the earth's crust, making up 7.5 percent by mass. It is too active chemically to occur free in nature, but is found in compounds in over 200 different minerals. The most important ore of Al is bauxite, which contains hydrated aluminum oxide ($Al_2O_3 \cdot 2H_2O$), along with silica and hydrated iron oxide ($Fe_2O_3 \cdot 2H_2O$). Another important mineral is cryolite (Na_3AlF_6), which is used in the metallurgy of aluminum. The Hall process for the production of aluminum metal is described in Figure 20.19 of the text.

Aluminum is obtained from bauxite ore ($Al_2O_3 \cdot 2H_2O$). The first step in its preparation is to separate pure aluminum oxide from the silica and iron impurities. First, bauxite is pulverized and digested with sodium hydroxide solution. This converts the silica into soluble silicates, and aluminum oxide is converted to the aluminate ion AlO_2^- which remains in solution. However this digestion treatment has no effect on the iron which remains as insoluble $Fe_2O_3(s)$. It is removed by filtration. Aluminum hydroxide is then precipitated by acidification (with carbonic acid) to about pH 6. The precipitate is heated strongly to produce pure Al_2O_3. The chemical changes involving aluminum are:

$$Al_2O_3(s) + 2OH^-(aq) \rightarrow 2\,AlO_2^-(aq) + H_2O(l)$$
$$AlO_2^-(aq) + H^+(aq) + H_2O(l) \rightarrow Al(OH)_3(s)$$
$$2Al(OH)_3(s) \rightarrow Al_2O_3(s) + 3H_2O(g)$$

The aluminum ions in Al_2O_3 can be reduced to metallic aluminum efficiently only by electrolysis. The melting point of Al_2O_3 is 2050°C, which makes electrolysis of pure molten Al_2O_3 extremely expensive owing to the need to maintain the high temperature. In the Hall process, Al_2O_3 is dissolved in molten cryolite (Na_3AlF_6) which melts at 1000°C. The use of cryolite makes it possible to lower the temperature of electrolysis by 1050°C! The mixture is electrolyzed to produce aluminum and oxygen.

cathode $2Al^{3+} + 6e^- \rightarrow 2Al(l)$

anode $3O^{2-} \rightarrow \frac{3}{2}O_2 + 6e^-$

_____

overall $2Al^{3+} + 3O^{2-} \rightarrow 2Al(l) + \frac{3}{2}O_2$

Properties. Pure aluminum is a silvery-white metal with low density (2.7 g/cm^3) and high tensile strength. It is malleable and can be rolled into thin foils. Its electrical conductivity is about 65 percent that of copper.

Aluminum is an *amphoteric* element, reacting with both acids and bases.

$$2Al(s) + 6HCl(aq) \rightarrow AlCl_3(aq) + 3H_2(g)$$
$$2Al(s) + 2NaOH(aq) + 2H_2O(l) \rightarrow 2NaAlO_2(aq) + 3H_2(g)$$

Aluminum has a strong affinity for oxygen. Indeed the metal is usually covered by an oxide film.

$$4Al(s) + 3O_2(g) \rightarrow 2Al_2O_3(s)$$

This layer of Al_2O_3 forms a compact, adherent, protective, surface coating on the metal and is responsible for preventing further oxidation and corrosion. Because of this surface oxide layer, Al is practically insoluble in weak or dilute acids.

The large enthalpy of formation of aluminum oxide ($\Delta H_f^o = -1670$ kJ/mol) makes the metal an excellent reducing agent. Thus a variety of metals can be produced in a series of similar reactions involving Al powder with the corresponding metal oxides. So much heat is liberated in these reactions, called *aluminothermic reactions,* that the metal is usually obtained in the molten state. The *thermite reaction,* used in welding steel and iron, is one example.

$$2Al(s) + Fe_2O_3(s) \rightarrow Al_2O_3(l) + 2Fe(l)$$

Aluminum hydroxide is an amphoteric hydroxide, dissolving in both acid and base.

$$Al(OH)_3(s) + 3H^+(aq) \rightarrow Al^{3+}(aq) + 3H_2O(l)$$
$$Al(OH)_3(s) + OH^-(aq) \rightarrow Al(OH)_4^-(aq)$$

Uses. Aluminum metal is used in high-voltage transmission lines because it is cheaper and lighter than copper. Its chief use is in aircraft construction. Aluminum is employed also as a solid propellant in rockets. This is another example of the great affinity that aluminum has for oxygen. Ammonium perchlorate, NH_4ClO_4, is the oxidizer in the reaction.

The formation of aluminum hydroxide precipitate is used in water treatment plants. The process requires large amounts of aluminum sulfate.

$$Al_2(SO_4)_3(aq) + 3Ca(OH)_2(aq) \rightarrow 2Al(OH)_3(s) + 3CaSO_4(aq)$$

Aluminum hydroxide is a gelatinous substance. As it settles in treatment pools it coprecipitates suspended matter such as bacteria and colloidal sized particles. This process clarifies drinking water.

Alums are compounds which have the general formula:

$$M^+M^{3+}(SO_4)_2 \cdot 12H_2O$$

where $M^+ = K^+$, Na^+, NH_4^+, and $M^{3+} = Al^{3+}$, Cr^{3+}, Fe^{3+}. They are used in the dying industry where the formation of the gelatinous aluminum hydroxide plays in "fixing" the dye to the cloth.

EXAMPLE 20.6   Amphoterism

Explain the meaning of the term amphoterism. Write equations for the amphoteric behavior of aluminum.

•Method of Solution

An amphoteric substance can react either as an acid or as a base. Aluminum reacts with hydrochloric acid and with strong bases as follows:

$$2Al(s) + 6HCl(aq) \rightarrow AlCl_3(aq) + 3H_2(g)$$
$$2Al(s) + 2NaOH(aq) + 2H_2O(l) \rightarrow 2NaAlO_2(aq) + 3H_2(g)$$

EXAMPLE 20.7   Oxidation of Aluminum

Aluminum is a good reducing agent, and its reduction potential is quite negative ($E_{red}^o = -1.66$ V) which means that aluminum should react with water and liberate hydrogen. But we know that airplanes do not dissolve in thunderstorms. Explain.

•**Method of Solution**

Aluminum readily forms the oxide Al_2O_3 when exposed to air. The oxide forms a tenacious surface film that protects the aluminum metal from further corrosion due to water. Therefore it is the presence of this oxide layer that makes aluminum practically insoluble even in dilute acids.

EXERCISES

12. What is the role of cryolite, Na_3AlF_6, in aluminum production?
13. Write chemical equations that show the amphoteric nature of $Al(OH)_3$.
14. The enthalpy of formation of Al_2O_3 is large and exothermic. Why doesn't aluminum metal just completely oxidize away.

PRACTICE TEST

1. Distinguish between:
   a. Leaching and flotation
   b. Roasting and reduction

2. Can carbon be used to reduce Al_2O_3 to $Al(s)$? *Hint:* Is the following reaction spontaneous?
   $$2Al_2O_3(s) + 2C(s) \rightarrow 4Al(s) + 3CO_2(g)$$

3. When a solution of NaOH is added dropwise to a test tube containing a solution of Al^{3+}, a white gelatinous precipitate is formed. Upon continued addition of NaOH, the precipitate disappears. Write the chemical equations to explain this observation.

4. In the production of sodium, the metal is prevented from coming into contact with Cl_2 and O_2. Write chemical equations for the reactions of these elements with sodium.

5. What is the agent that actually reduces Fe in the blast furnace? Write a chemical equation to show the reaction.

6. Distinguish between an alloy and an amalgam.

7. According to the band theory, why is copper a conductor?

8. Considering that electrons in metals have random thermal motion like molecules of gas, explain why electrical conductivity of conductors *decreases* with increasing temperature?

9. Why does the electrical conductivity of semiconductors *increase* with increasing temperature?

10. What are the sources of aluminum and potassium?

11. What is quicklime? How is it made?

12. Aluminum metal was first prepared in 1825 by the action of potassium metal on aluminum chloride. Write a balance equation for this reaction.

13. What is the chemical form of aluminum in bauxite?

14. How is aluminum separated from Fe_2O_3?

15. What is an alum? Which alum is used in baking powder? *Hint:* Read a label. What is its function in baking powder?

16. What is the role of aluminum sulfate in water purification?

17. What is the pH of a 0.10 M aluminum chloride solution if it is hydrolyzed to the extent of 10 percent at room temperature.

$$[Al(H_2O)_6]^{3+} \rightarrow H^+ + [Al(H_2O)_5OH]^{2+}$$

ANSWERS

Exercises

1. A mineral is a naturally occurring substance with a characteristic range of chemical composition. An ore is a mineral, or a mixture of minerals, from which a particular metal can be profitably extracted.

2. The electropositive metals can be reduced by a more electropositive metal such as Li, but there is no metal electropositive enough to reduce Li^+ to Li. Electrolytic reduction is the only way to prepare Li. It is possible to reduce other metals with Li. Rather than prepare Li by electrolysis and then use it to reduce aluminum, magnesium, or sodium, it is more convenient simply to prepare these metals in one step by electrolytic reduction.

3. Zone refining is a technique for purifying metals. Figure 20.8 in the text explains how it works.

4. Limestone is used to remove impurities from iron during production. Limestone decomposes to lime (CaO). CaO reacts with SiO_2 and Al_2O_3 impurities from the iron ore forming calcium silicate and calcium aluminate. The mixture of calcium silicate and calcium aluminate is known as slag.

5. $Cr_2O_3(s) + 2Al(s) \rightarrow 2Cr(l) + Al_2O_3(s)$

6. A band is a large number of molecular orbitals that are closely spaced in energy. The valence band is a set of closely spaced MOs that are filled with electrons. The conduction band is a set of closely spaced empty orbitals. An electron can travel freely through the metal since the conduction band is void of electrons.

7. A conductor is capable of conducting an electrical current. In a conductor there is essentially no energy gap between the valence band and the conduction band. Insulators are materials that do not conduct. In an insulator the energy gap between the valence band and the conduction band is much greater than the gap in a conductor. Semiconductors are normally not conductors, but will conduct electricity at higher temperatures, or when combined with small amounts of certain elements.

8. Good conductors of heat and electricity. Low density. Soft enough to cut with a knife. Low melting points.

9. Extremely reactive. React with water to form hydroxides. React with oxygen to form a varity of oxides, peroxides and superoxides.

10. Magnesiun is precipitated from seawater. Calcium is obtained from limestone.

11. $CaCO_3$, CaO, and $Ca(OH)_2$ respectively.

12. Molten cryolite is used as a solvent for alumina, Al_2O_3, in aluminum production. Cryolite melts at 1000°C as compared to 2050°C for alumina and therefore it lowers the energy consumption.

13. Reaction with acid: $Al(OH)_3(s) + 3H^+(aq) \rightarrow Al^{3+}(aq) + 3H_2O(l)$
    Reaction with base: $Al(OH)_3(s) + OH^-(aq) \rightarrow Al(OH)_4^-(aq)$

14. A fresh surface of aluminum does react readily with oxygen and forms a surface coating of Al_2O_3. This layer is very impenetrable and does not allow oxygen and water through to continue a reaction with the rest of the aluminum.

Practice Test

1. a. *Leaching* is the selective dissolution of a metal from an ore. *Flotation* is a technique used to separate mineral particles from waste clays and silicates called gangue.

   b. *Roasting* involves heating an ore in the presence of air. The idea is to convert metal carbonates and sulfides to oxides, which can be more conveniently reduced to yield pure metals. *Reduction* is the gain of electrons by a cation to yield the pure metal.

2. No. Use Appendix 3 of the text to calculate $\Delta G°$. The sign of $\Delta G°$ indicates whether the reaction will go or not.

3. $Al^{3+} + 3OH^- \rightarrow Al(OH)_3(s)$
   $Al(OH)_3(s) + OH^- \rightarrow Al(OH)_4^-(aq)$

4. $2Na + Cl_2 \rightarrow 2NaCl$
   $2Na + O_2 \rightarrow Na_2O_2$
   $2Na + \frac{1}{2}O_2 \rightarrow Na_2O$

5. Carbon monoxide. $Fe_2O_3 + 3CO \rightarrow 2Fe + 3CO_2$

6. An *alloy* is a mixture of two or more metals, or of metals and nonmetal elements that have metallic properties. An *amalgam* is an alloy containing mercury.

7. Copper atoms have a half-filled 4s orbital. This means that copper metal has a half-filled valence band. A partially filled valence band is a conduction band.

8. The increasingly random motion of electrons due to a temperature rise opposes motion in one unified direction.

9. As temperature increases more electrons acquire enough energy to "jump" the energy gap between the valence band and the conduction band.

10. The source of Al is bauxite. Potassium is obtained from sylvite.

11. Quicklime is CaO. It is made the thermal decomposition of limestone.

12. $3K + AlCl_3 \rightarrow Al + 3KCl$

13. $Al_2O_3 \cdot 2H_2O$

14. The bauxite is pulverized and digested with sodium hydroxide solution. This converts the aluminum oxide to the aluminate ion AlO_2^- which remains in solution. However treatment with base has no effect on the iron oxide which remains as insoluble $Fe_2O_3(s)$, and is removed by filtration. Aluminum hydroxide is then precipitated by acidification to about pH 6.

15. Alums are sulfates with the general formula $M^+M^{3+}(SO_4)_2 \cdot 12H_2O$, where $M^+ = K^+$, Na^+, NH_4^+, and $M^{3+} = Al^{3+}$, Cr^{3+}, Fe^{3+}. Sodium aluminum sulfate is used in baking powder as an anticaking agent.

16. $Al_2(SO_4)_3$ is used to clarify water through its reaction with $Ca(OH)_2$. The precipitation of $Al(OH)_3$ traps dirt and dust particles.

17. pH = 2

Chapter Twenty-One
NONMETALLIC ELEMENTS AND THEIR COMPOUNDS

- Hydrogen (21.1 – 21.2)
- Carbon (21.3)
- Nitrogen and Phosphorus (21.4)
- Oxygen and Sulfur (21.5)
- The Halogens (21.6)

HYDROGEN (21.1 – 21.2)

STUDY OBJECTIVES

1. Write chemical equations to show the reactions used to prepare hydrogen.
2. Describe the chemical properties and major uses of hydrogen.

General Properties of Nonmetals. The nonmetal elements with the exception of hydrogen are located in the upper right hand corner of the periodic table. The properties of nonmetals are more varied than those of metals. Several nonmetal elements are gases, and nonmetals exhibit both positive and negative oxidation numbers. In this chapter we will discuss the chemistry of a number of important nonmetallic elements: hydrogen, carbon, nitrogen, phosphorus, oxygen, sulfur, fluorine, chlorine, bromine, and iodine.

Hydrogen. Hydrogen is the most abundant element in the universe, and the tenth most abundant element in Earth's crust where it is found in combination with other elements. At ordinary temperatures elemental hydrogen is a diatomic molecule. Commercially, the most important large-scale preparation is the reaction between propane and steam in the presence of a catalyst at 900°C:

$$C_3H_8(g) + 3H_2O(g) \rightarrow 3CO(g) + 7H_2(g)$$

In another process steam is passed over a bed of red-hot coke (the water gas reaction):

$$C(s) + H_2O(g) \xrightarrow{1000°C} CO(g) + H_2(g)$$

Small quantities of hydrogen are prepared in the laboratory by the reaction of zinc with hydrochloric acid.

$$Zn(s) + 2HCl(aq) \rightarrow ZnCl_2(aq) + H_2(g)$$

Properties of Hydrogen. Hydrogen can be oxidized to the H^+ ion and can be reduced to the H^- ion (hydride ion) as well. The H^+ ion has the same charge as the metal alkali metal ions, and the H^- ion is similar to the halide ions. It also forms diatomic molecules as do the halogens elements. As a result of these properties, there is no totally satisfactory position for hydrogen in the periodic table.

Hydrogen forms binary compounds called *hydrides* with a large number of elements; it reacts directly with most active metals to form *ionic hydrides* in which hydrogen is an H^- ion.

$$2Na(s) + H_2(g) \rightarrow 2NaH(s)$$
$$2Al(s) + 3H_2(g) \rightarrow 2AlH_3(s)$$

The ionic character of these metal hydrides increases as we move down a group of the periodic table. Metal hydrides react with water to form hydrogen gas and the metal hydroxide.

$$NaH(s) + H_2O(l) \rightarrow NaOH(aq) + H_2(g)$$

Hydrogen in compounds with nonmetals such as CH_4, NH_3, and H_2S has an oxidation state of +1. These are *covalent hydrides*.

Uses of Hydrogen. Hydrogen is used in the synthesis of ammonia, an important fertilizer. The Haber process for synthesizing ammonia was discussed in Chapter 14. Ammonia which results from the reaction of hydrogen with nitrogen is an example of a *covalent hydride*.

$$N_2(g) + 3H_2(g) \rightarrow 2NH_3(g)$$

Hydrogen is used in the production of margarine and cooking oil to convert polyunsaturated fats which contain $C = C$ double bonds into partially hydrogenated products.

Hydrogen may play an important role in our energy future in terms of the "hydrogen economy." As fossil fuel supplies dwindle, hydrogen gas could take on more of a role as a fuel. Nuclear energy could be used to generate electricity which in turn could be used to electrolyze water to obtain hydrogen.

$$\text{electrical energy} + H_2O(l) \rightarrow H_2(g) + \frac{1}{2}O_2(g)$$

The combustion of hydrogen as a fuel would release energy.

$$H_2(g) + \frac{1}{2}O_2(g) \rightarrow H_2O(l) \qquad \Delta H° = -285.8 \text{ kJ}$$

EXAMPLE 21.1   Hydrogen Compounds

Give an example of an ionic hydride and an example of a covalent hydride. What is the oxidation number of hydrogen in each?

•Method of Solution

In ionic compounds, hydrogen has an oxidation number of -1. The hydrides of the alkali metals and some of the alkaline earth metals are ionic hydrides. CaH_2 is an example of an ionic hydride. Hydrogen has an electronegativity of 2.1 and that of calcium is 1.0. Thus hydrogen has an oxidation number of -1 in CaH_2. H_2S is a covalent hydride. The electronegativity of sulfur is 2.5. The lower polarity of the H—S bond is indicated by a Δ value of only 0.4. The oxidation number of hydrogen is $+1$ in H_2S.

EXAMPLE 21.2   Comparison of Ionic and Covalent Hydrids

Compare the reactions of water with covalent hydrides and ionic hydrides.

•Method of Solution

Covalent hydrides either do not react with water at all, which is the case with CH_4, or are *acids* in water. One of them, ammonia, is a weak base.

$$H_2S(aq) + H_2O(l) \rightleftharpoons HS^-(aq) + H_3O^+(aq)$$

Ionic hydrides are *strong Brønsted bases*; they readily accept a proton from water.

$$H^-(aq) + H_2O(l) \rightarrow H_2(g) + OH^-(aq)$$

Lithium hydride is an ionic hydride.

$$LiH(s) + H_2O(l) \rightarrow LiOH(aq) + H_2(g)$$

EXERCISES

1. Write an equation to show how aluminum can be used to prepare hydrogen in the laboratory.
2. What is the oxidation number of hydrogen in AlH_3, in H_2S?
3. Give an example of an ionic hydride.
4. Describe the hydride ion as a Brønsted acid or a base? Write an equation for the reaction of the hydride ion with water.

CARBON (21.3)

STUDY OBJECTIVES

1. Describe the chemical and physical properties of carbon.
2. Describe some of the inorganic compounds of carbon.

Properties of Carbon. Carbon makes up only 0.09% of Earth's crust by mass. Carbon occurs in nature as the free element in the allotropic forms graphite and diamond. Carbon occurs as CO_2 in the atmosphere and as the carbonate in limestone and chalk. Fossil fuels contain a large percentage of carbon. Carbon is the essential element in living organisms.

Pure carbon in the form of diamond is the hardest substance known. The microscopic structure of diamond is such that each carbon atom is covalently linked to four other carbon atoms (see Figure 11.28 in the text). Therefore, a diamond is a single giant molecule. The melting point of a diamond is extremely high, 3550°C. Graphite is the "lead" in lead pencils.

On heating carbon and silicon at 1500°C a compound called silicon carbide or *carborundum* (SiC) is formed. Silicon carbide is almost as hard as diamond and has the diamond structure. Its melting point is about 2700°C. The hardness of diamond and carborundum are put to important uses. Synthetic diamonds are not of gem quality, but are used as abrasives in cutting concrete and many other hard substances. Carborundum is used for cutting, grinding, and polishing metals and glasses.

Carbon forms two important classes of compounds called *carbides* and *cyanides*. CaC_2 and Be_2C contain carbon in the form of C_2^{2-} and C^{4-} ions, respectively. The carbide ion contains a triple bond, $^-C \equiv C^-$. These ions are strong Brønsted bases and react with water as follows.

$$C_2^{2-}(aq) + 2H_2O(l) \rightarrow 2OH^-(aq) + C_2H_2(g)$$
$$C^{4-}(aq) + 4H_2O(l) \rightarrow 4OH^-(aq) + CH_4(g)$$

Cyanides contain the cyanide anion, $C \equiv N^-$. Hydrogen cyanide, sodium cyanide, and potassium cyanide are well-known cyanide compounds.

Oxides of Carbon. Carbon forms two important oxides, CO and CO_2. Carbon monoxide is a colorless, odorless, and poisonous gas. It is formed by incomplete combustion of carbon compounds. Carbon monoxide burns readily in air.

Carbon dioxide is a colorless and odorless gas. It 's concentration in air is 350 ppm. At this concentration, it is not toxic. In enclosed low-lying places such as abandoned mines, CO_2 concentrations can build up to suffocating levels. Carbon dioxide is formed by the complete combustion of carbon-containing compounds and fuels. If oxygen is restricted from reaching the combustion zone, carbon monoxide is formed instead. Since CO_2 cannot be oxidized further, it finds use in fire extinguishers. Dry ice (solid carbon dioxide) is used as a refrigerant.

EXERCISES

5. Diamond is the highest melting substance. Why is the melting point of diamond so high?
6. Write formulas for the cyanide ion and two kinds of carbide ions.
7. Which oxide of carbon is flammable?

NITROGEN AND PHOSPHORUS (21.4)

STUDY OBJECTIVES

1. Describe the chemical properties and uses of nitrogen and some of its compounds.
2. Describe the allotropic forms of phosphorus, and list the main phosphorus compounds.

Nitrogen (N_2). The elemental form of nitrogen is extremely stable owing to its triple bond, $N \equiv N$. Molecular nitrogen is obtained from air, of which nitrogen makes up 78 percent by volume. When air under a pressure of 30 atm is cooled to $-190°C$, it liquefies. Fractional distillation of liquid air yields pure N_2. Nitrogen is used in the laboratory to create atmospheres devoid of oxygen. This is desirable when working with chemicals that are attacked and degraded by oxygen, or that burn spontaneously in it. Liquid N_2 is used as a coolant, providing a temperature of $-196°C$.

Nitrides (N^{3-}). Most nitrogen compounds are covalent, however nitrogen forms ionic compounds with certain metals on heating. Li_3N, Mg_3N_2, and Ca_3N_2 are examples. The nitride ion is a very strong Brønsted base and readily accepts protons from water, forming ammonia:

$$Mg_3N_2(s) + 6H_2O \rightarrow 3Mg(OH)_2 + 2NH_3$$

Ammonia (NH_3). The quantity of ammonia produced each year is second in tonnage only to sulfuric acid production. Ammonia is synthesized by the Haber process, which uses nitrogen gas and hydrogen gas (from hydrocarbon cracking).

$$3H_2 + N_2 \rightarrow 2NH_3$$

The main use of ammonia is in the production of ammonium and nitrate fertilizers. Ammonia is an important ingredient in such diverse substances as explosives and nylon.

Liquid ammonia (b.p. $-33.4°C$) is a solvent resembling water. It undergoes autoionization

$$2NH_3(l) \rightleftharpoons NH_4^+ + NH_2^-$$

The value of the equilibrium constant at $-50°C$ is $K = [NH_4^+][NH_2^-] = 1 \times 10^{-33}$. Alkali metals dissolve in liquid ammonia, producing beautiful blue solutions containing a positive ion and an electron.

$$Li(s) \rightarrow Li^+ + e^-$$

Both the cation and the electron are highly solvated by ammonia. The solvated electron e⁻ is responsible for the characteristic color of these solutions.

Nitrogen Oxides (N_2O, NO, & NO_2). "Fixed nitrogen" refers to nitrogen present in chemical compounds. Nitrogen is fixed, that is incorporated into compounds, by the action of lightning in the atmosphere, by high temperature, by the Haber process, and by certain types of soil bacteria. An appreciable amount of nitric oxide (NO) is formed in electrical storms and in internal combustion engines.

$$N_2 + O_2 \rightarrow 2NO$$

Nitric oxide reacts very rapidly with oxygen to yield nitrogen dioxide.

$$2NO + O_2 \rightarrow 2NO_2$$

Nitrogen dioxide is a key molecule in the formation of photochemical smog. NO_2 is prepared in the laboratory by the action of concentrated nitric acid on copper.

$$Cu(s) + 4HNO_3(aq) \rightarrow Cu(NO_3)_2(aq) + 2H_2O(l) + 2NO_2(g)$$

Nitric acid is a major inorganic acid. Over 8.0 million tons were produced in the United States in 1985. Nitric acid is synthesized by the Ostwald process, which uses ammonia and oxygen as starting materials. Reaction occurs at 1000°C in the presence of a platinum-rhodium catalyst.

$$4NH_3(g) + 5O_2(g) \rightarrow 4NO(g) + 6H_2O(g)$$

Then oxygen is introduced.

$$2NO(g) + O_2(s) \rightarrow 2NO_2(s)$$

The resulting NO_2 readily dissolves in water, yielding a mixture of nitrous acid and nitric acid.

$$2NO_2(g) + H_2O(l) \rightarrow HNO_2(aq) + HNO_3(aq)$$

Then nitrous acid is eliminated by heating.

$$3HNO_2(aq) \rightarrow HNO_3(aq) + H_2O(l) + 2NO(g)$$

Nitric acid is a strong acid, and an oxidizing acid. Aqua regia, the only acid that can dissolve gold, is 1 part nitric acid and 3 parts hydrochloric acid. Nitric acid is used to manufacture fertilizers and explosives such as nitroglycerin and TNT.

Nitrous oxide (N_2O) is a colorless gas with a pleasing odor and a sweet taste. It is used as an anesthetic in dental and minor surgery.

Phosphorus. There are several *allotropic forms* of phosphorus. The most important ones are white and red phosphorus. Their structures are shown in Figure 21.9 in the text. White phosphorus consists of discrete P_4 molecules. The white solid melts at 44°C and is insoluble in water. It is usually stored under water to prevent contact with air. White phosphorus is extremely reactive and bursts into flame when exposed to air.

$$P_4(s) + 5O_2(g) \rightarrow P_4O_{10}(s)$$

When white phosphorus is heated to 300°C in the absence of air it is slowly converted to red phosphorus. Red phosphorus has a polymeric structure. It is much more stable than white phosphorus. It melts at 600°C and is stable toward oxygen at room temperature. Red phosphorus must be heated to about 400°C before it will ignite in air.

Oxides of Phosphorus. Tetraphosphorus hexaoxide (P_4O_6) and tetraphosphorus decaoxide (P_4O_{10}) are two important oxides of phosphorus (Figure 21.10 text). Both oxides are acidic.

$$P_4O_6(s) + 6H_2O(l) \rightarrow 4H_3PO_3(aq)$$
$$P_4O_{10}(s) + 6H_2O(l) \rightarrow 4H_3PO_4(aq)$$

Phosphoric Acid. Industrially, phosphoric acid is prepared by the reaction of sulfuric acid with calcium phosphate:

$$Ca_3(PO_4)_2(s) + 3H_2SO_4(aq) \rightarrow 2H_3PO_4(aq) + 3CaSO_4(aq)$$

Phosphoric acid is not a particularly strong acid, as indicated by its first ionization constant.

$$H_3PO_4 \rightleftharpoons H^+ + H_2PO_4^- \qquad K_{a1} = 7.5 \times 10^{-3}$$

Large quantities of phosphoric acid are produced commercially every year. Its main uses are in the production of phosphate fertilizers, cattle feed additives, and water softeners (builders) in household detergents.

EXAMPLE 21.3 Oxidation Numbers of Nitrogen

List the known oxidation numbers of nitrogen and write the formula of a compound or ion that exemplifies each oxidation number.

•Method of Solution

Oxidation Number	Compound	Formula
+ 5 (maximum)	nitrate ion	NO_3^-
+4	nitrogen dioxide	NO_2
+3	nitrite ion	NO_2^-
+2	nitric oxide	NO
+1	nitrous oxide	N_2O
0	nitrogen gas(not a compound)	N_2
−1	chloramine	NH_2Cl
−2	hydrazine	N_2H_4
−3	ammonia and nitride	NH_3 and N^{3-}

EXAMPLE 21.4 Allotropic Forms

Starting with white phosphorus, write equations for the steps needed to prepare phosphoric acid.

•Method of Solution

First, burn the white phosphorus in excess oxygen to make tetraphosphorus decaoxide.

$$P_4(s) + 5O_2(g) \rightarrow P_4O_{10}(s)$$

The oxide is an acidic oxide and is converted to phosphoric acid in water.

$$P_4O_{10}(s) + 6H_2O(l) \rightarrow 4H_3PO_4(aq)$$

EXERCISES

8. What are the oxidation numbers of nitrogen in nitrate ion and nitrite ion?
9. Write three equations that show the steps in the production of nitric acid from ammonia.
10. Compare the stability of white and red phosphorus.

OXYGEN AND SULFUR (21.5)

STUDY OBJECTIVES

1. Describe the allotropic forms of oxygen and sulfur.
2. Distinguish between oxides, peroxides, and superoxides.
3. List some of the compounds of oxygen and sulfur and describe their chemical properties.

Allotropic Forms of Oxygen. Oxygen is the most abundant element in Earth's crust, making up about 46% by mass. As the free element oxygen exists in two allotropic forms, diatomic oxygen (O_2) and ozone (O_3). Diatomic oxygen makes up 21 percent of Earth's atmosphere by volume. Industrially, oxygen is obtained by liquefaction of air followed by fractional distillation. In the laboratory, oxygen is obtained by heating potassium chlorate in the presence of MnO_2 catalyst.

$$KClO_3(s) \rightarrow KCl(s) + \frac{3}{2}O_2(g)$$

Ozone is a less stable form of oxygen. It is produced by the action of an electrical discharge or ultraviolet (UV) light on diatomic oxygen.

$$O_2(g) + \text{UV-light} \rightarrow O(g) + O(g)$$
$$O(g) + O_2(g) \rightarrow O_3(g)$$

Ozone's presence in the atmosphere prevents the sun's UV light from reaching Earth's surface. Ozone absorbs the light in the following reaction:

$$O_3(g) + \text{UV-light} \rightarrow O_2(g) + O(g)$$

Both ozone and diatomic oxygen are strong oxidizing agents.

$$O_2(g) + 4H^+ + 4e^- \rightarrow 2H_2O(l) \qquad E^o_{red} = 1.23 \text{ V}$$
$$O_3(g) + 2H^+ + 2e^- \rightarrow O_2(g) + H_2O(l) \qquad E^o_{red} = 2.02 \text{ V}$$

Ozone is used mainly to purify drinking water, to deodorize air and sewage gases, and to bleach waxes, oils, and textiles. About 65 percent of all the diatomic oxygen produced in the United States is used in the *basic oxygen process* for making steel. Pure oxygen gas is blown over the surface of molten iron to lower the carbon content by burning it to CO_2. At the same time silicon and phosphorus impurities are oxidized and later removed as slag.

Hydrogen Peroxide. Besides forming water, oxygen forms another important binary compound with hydrogen: hydrogen peroxide (H_2O_2). Hydrogen peroxide is never used as a pure substance because of its tendency to explode.

Metal peroxides contain the peroxide ion, O_2^{2-}. These ions are strong Brønsted bases. All the metal peroxides hydrolyze in acidic solutions to form hydrogen peroxide. Hydrogen peroxide can be prepared in the laboratory from barium hydroxide octahydrate and sulfuric acid.

$$BaO_2 \cdot 8H_2O(s) + 2H_2SO_4(aq) \rightarrow BaSO_4(aq) + H_2O_2(aq) + 8H_2O(l)$$

In aqueous solutions, H_2O_2 can act both as an oxidizing agent and a reducing agent. For example, H_2O_2 oxidizes iron(II) ion to iron(III) ion.

$$H_2O_2(aq) + 2Fe^{2+}(aq) + 2H^+(aq) \rightarrow 2Fe^{3+}(aq) + 2H_2O(l)$$

It reduces silver oxide to metallic silver:

$$H_2O_2(aq) + Ag_2O(s) \rightarrow 2Ag(s) + H_2O(l) + O_2(g)$$

Hydrogen peroxide readily decomposes on heating and in the presence of various catalysts.

$$2H_2O_2(aq) \rightarrow 2H_2O(l) + O_2(g)$$

This is a disproportionation reaction in which one O atom in H_2O_2 is reduced and the other one is oxidized.

Sulfur. Although it is not very abundant, sulfur is readily available because of its occurrence in large deposits of the free element. Sulfur is extracted from these underground deposits by the *Frasch process*. There are several allotropic forms of sulfur. The most important are rhombic sulfur and monoclinic sulfur. Rhombic sulfur is the more stable form at room temperature. In this form, sulfur atoms are linked to each other as puckered, eight-membered rings, S_8. Above 96°C, monoclinic sulfur is more stable. This form melts at 119°C. Above 150°C, the pale yellow liquid becomes more viscous as the rings break open and the newly formed molecular chains entangle. Sulfur has a wide range of oxidation numbers in its compounds, as shown in Table 21.3 of the text. The major end uses of sulfur are in fertilizers (60%) and in chemicals (20%).

Hydrogen Sulfide. H_2S (oxidation number –2) is a colorless, highly toxic gas, with an offensive odor. When dissolved in water the resulting solution is called hydrosulfuric acid. $H_2S(aq)$ is a weak acid. In the laboratory, $H_2S(aq)$ is used to separate groups of metal ions for their qualitative analysis. It is often prepared in the laboratory by the hydrolysis of thioacetamide.

$$CH_3CSNH_2(aq) + 2H_2O + H^+(aq) \rightarrow CH_3COOH(aq) + NH_4^+(aq) + H_2S(aq)$$

Oxides of Sulfur. The major oxides of sulfur are sulfur dioxide and sulfur trioxide, SO_2 and SO_3. Sulfur dioxide is a colorless gas with a pungent odor and is quite toxic. It dissolves in water forming sulfurous acid.

$$SO_2(g) + H_2O(l) \rightarrow H_2SO_3(aq)$$

Sulfur dioxide is slowly oxidized in air to sulfur trioxide.

$$2SO_2(g) + O_2(g) \rightarrow 2SO_3(g)$$

Sulfur trioxide dissolves in water to form sulfuric acid.

$$SO_3(g) + H_2O(l) \rightarrow H_2SO_4(aq)$$

Sulfur dioxide is a major air pollutant. It is formed during combustion of sulfur-containing oil and coal. Smelters and coal-fired electric power plants are the primary sources of sulfur oxides.

Sulfuric Acid. More sulfuric acid is produced each year than any other industrial chemical. It is formed from SO_3 and H_2O as shown above. First, sulfur is burned in air to make SO_2. Second, a vanadium oxide catalyst is used to convert SO_2 to SO_3 in the presence of oxygen. Third, SO_3 and water react as above.

Hot, concentrated sulfuric acid is a strong oxidizing agent. In such a solution, the oxidizing agent is actually the sulfate ion. Copper is oxidized by sulfuric acid by the following reaction:

$$Cu(s) + 2H_2SO_4(aq) \rightarrow CuSO_4(aq) + SO_2(g) + 2H_2O(l)$$

EXAMPLE 21.5 Oxygen Compounds

Distinguish between the three types of oxides: the normal oxide, the peroxide, and the superoxide.

•Method of Solution

Normal oxides contain the O^{2-} ion in which oxygen has an oxidation state of –2. *Peroxides* contain an oxygen-oxygen bond. Metal peroxides are ionic and contain the O_2^{2-} ion. Hydrogen peroxide is a covalent peroxide (H—O—O—H) with oxygen in the –1 oxidation state. The superoxide ion O_2^- has oxygen in the –1/2 oxidation state.

EXAMPLE 21.6 Oxidation Numbers of Sulfur

List the known oxidation numbers of sulfur and write the formula of a compound or ion that exemplifies each oxidation number.

•Method of Solution

Oxidation Number	Compound	Formula
+6	sulfate ion	H_2SO_4
+4	sulfite ion	$CaSO_3$
+2	sulfur dichloride	SCl_2
+1	disulfur dichloride	S_2Cl_2
0	sulfur (the element)	S_8
–2	sulfide ion	Na_2S

EXERCISES

11. Which is a stronger oxidizing agent, diatomic oxygen or ozone?
12. What is the principle use of O_2 in the United States?
13. What is the oxidation number of O in H_2O_2, H_2O and O_2?
14. Hydrogen peroxide can decompose by a disproportionation reaction. Write the equation for this reaction.
15. Write formulas of oxide, peroxide, and superoxide compounds that include a metal.
16. Starting with elemental sulfur, $S(s)$, show the steps needed to prepare sulfuric acid.

THE HALOGENS (21.6)

STUDY OBJECTIVES

1. Write chemical equations to show the preparations of the halogens.
2. Describe the chemical properties and give the major uses of these elements and their compounds.

Occurrence and Preparation. The elements in Group 7 are called the *halogens*. In the elemental state they form diatomic molecules, X_2. Because of their high reactivity the halogens are always found combined with other elements. Chlorine, bromine, and iodine occur as halides in seawater, and fluorine occurs in the minerals fluorspar (CaF_2) and cryolite (Na_3AlF_6).

Fluorine is prepared by the electrolysis of liquid HF at 70°C.

$$2HF(l) \xrightarrow{\text{electrolysis}} H_2(g) + F_2(g)$$

Chlorine is prepared industrially by the electrolysis of a concentrated aqueous NaCl solution (brine). This is called the *chlor-alkali process* (NaOH is the alkali).

$$2NaCl(aq) + 2H_2O(l) \xrightarrow{\text{electrolysis}} 2NaOH(aq) + H_2(g) + Cl_2(g)$$

Because chlorine is a stronger oxidizing agent than bromine or iodine, free Br_2 and I_2 can be generated by bubbling Cl_2 gas through solutions containing bromide and iodide ions.

$$Cl_2(g) + Br^-(aq) \rightarrow 2Cl^-(aq) + Br_2(l)$$
$$Cl_2(g) + 2I^-(aq) \rightarrow 2Cl^-(aq) + I_2(s)$$

Chlorine, bromine, and iodine can be prepared in the laboratory by heating the corresponding alkali halides in concentrated sulfuric acid in the presence of MnO_2.

$$2NaCl(aq) + 2H_2SO_4(aq) + MnO_2(s) \rightarrow Cl_2(g) + Na_2SO_4(aq) + MnSO_4(aq) + 2H_2O(l)$$

Oxoacids of the Halogens. Most of the halides (halogens with an oxidation number of –1 as in chloride) can be classified into two categories: ionic and covalent. The fluorides and chlorides of many metallic elements, especially those belonging to the Group 1A and 2A metals, are ionic compounds. The halides of nonmetals are covalent compounds.

Because fluorine is the most electronegative element and because of its unusually weak
F—F bond, fluorine has a chemistry somewhat different from the rest of the halogens. Five differences are pointed out in Section 21.6 of the text.

- Fluorine is the most reactive of all of the halogens.
- Hydrogen fluoride exhibits hydrogen bonding. None of the other hydrogen halides do.
- Hydrofluoric acid is a weak acid. All other HX acids are strong.
- Fluorine reacts with cold NaOH solution.
- Silver fluoride is soluble. All other silver halides are insoluble.

The halogens form a series of oxoacids with the following general formulas:

HOX	HXO_2	HXO_3	HXO_4
hypohalous acid	halous acid	halic acid	perhalic acid

Table 21.5 in the text lists the oxoacids of the halogens.

Hypochlorous, hypobromous, and hypoiodous acids can be prepared by the reaction of the elemental halogen with water.

$$X_2 + H_2O \rightleftharpoons HX + HOX$$

hydrohalic   hypohalous

acid   acid

The smaller the radius of the halogen atom, the farther the equilibrium lies to the right. Mixtures of these halogen acids react with bases to produce *hypohalite salts* (NaOX)

$$HX + HOX + 2NaOH \rightarrow NaX + NaOX + 2H_2O$$

When the elemental form of a halogen, X_2, is dissolved in a base the net reaction is

$$X_2 + 2NaOH \rightarrow NaX + NaOX + H_2O$$

This is another example of a type of redox reaction called a *disproportionation* reaction. In this type of reaction the same element (the halogen) is both oxidized and reduced. One halogen atom in X—X is the reducing agent and the other is the oxidizing agent.

Interhalogens are compounds formed when molecules of two different halogens react. The interhalogen molecules consist of an atom of the larger halogen bound to an odd number of atoms of the smaller halogen. They have the following formulas: XX', XX'_3, XX'_5, and XX'_7, where X and X' are different halogens and X is the larger atom of the two.

Uses of the Halogens.
The halogens and their compounds find many uses in industry, the home, and in the field of health. *Fluorine* is used in fluoride toothpaste, in uranium isotope separation as UF_6, and in the manufacture of Teflon for electrical insulators, plastics, and cooking utensils.

Chlorine is widely used as a bleaching agent and as a disinfectant for water purification. Household bleach is 5.25 percent NaOCl by mass. The major end uses of chlorine are in manufacturing polymers such as polyvinyl chloride, in chemical solvents, in the pulp and paper industry, and in water treatment.

Bromine is used to prepare dibromoethane ($BrCH_2CH_2Br$) which is a gasoline additive, and silver bromide (AgBr), used in photographic film. *Iodine* finds uses as a medicinal antiseptic. NaI is added to table salt to prevent goiter, and silver iodide is used in cloud seeding.

EXAMPLE 21.7 Preparation of F_2

The chemistry of fluorine differs in many ways from that of the rest of the halogens. For instance, it cannot be prepared by the electrolysis of an aqueous solution of its sodium salt, as Cl_2 is prepared. Write chemical equations to show the steps in the industrial process of making F_2. Give the source of the fluorine, and mention why it cannot be prepared in aqueous solution.

•Method of Solution

The industrial process starts with the mineral fluorspar, CaF_2. This is treated with sulfuric acid to make hydrogen fluoride.

$$CaF_2(s) + H_2SO_4(l) \rightarrow 2HF(l) + CaSO_4(s)$$

The hydrogen fluoride is mixed with KF and electrolyzed

$$2HF(l) \rightarrow H_2(g) + F_2(g)$$

The F_2 molecule has such a strong attraction for electrons that it reacts violently with water, hot glass, and most metals. If F_2 were formed in an aqueous solution, it would immediately oxidize water to oxygen gas.

EXAMPLE 21.8   Reaction of Chlorine with Sodium Hydroxide

When chlorine is dissolved in base, a solution of aqueous sodium hypochlorite is formed. A 5.25% solution is sold as liquid laundry bleach. Write the chemical equation for the reaction. Assign oxidation numbers and identify the oxidizing agent and the reducing agent.

•**Method of Solution**

$$Cl_2(g) + 2NaOH(aq) \rightarrow NaCl(aq) + NaOCl(aq) + H_2O(l)$$

If we assign oxidation numbers to all the chlorine atoms, we can see that chlorine is both oxidized and reduced. None of the other elements in the equation changes oxidation state. Thus chlorine is the oxidizing agent and the reducing agent.

$$\begin{array}{c} \overset{\displaystyle \overbrace{}^{\text{reduction}}}{} \\ \overset{0 \quad\; 0 \qquad\qquad\quad\; -1 \qquad +1}{Cl-Cl + 2NaOH \rightarrow NaCl + NaOCl + H_2O} \\ \underbrace{}_{\text{oxidation}} \end{array}$$

This is an example of a disproportionation reaction.

EXERCISES

17. Which halogens are prepared by electrochemical oxidation?
18. Write a chemical equation for the preparation of elemental bromine from seawater.
19. Which of the hydrogen halides are strong acids in water?
20. Which halogen is the strongest oxidizing agent among all the elements.
21. Show the similarity of chlorine and bromine by writing equations for their reactions:
    a. with I^- ion.
    b. with NaOH(aq).

CONCEPTUAL QUESTIONS

1. Explain why hydrogen does not have a unique position in the periodic table.
2. Explain how Br_2 can act as both an oxidizing agent and a reducing agent.

PRACTICE TEST

1. Write chemical equations to show the two commercial methods of preparation of $H_2(g)$.

2. What are two important uses of hydrogen?

3. Write a chemical equation to show the synthesis of boric acid from kernite.

4. Coke is used to reduce Fe_2O_3 in iron production. What is it? How can it be converted to graphite?

5. List the major uses of: a. N_2 b. NH_3 c. HNO_3

6. What is *aqua regia?*

7. What are the acid anhydrides of HNO_3, HNO_2, H_3PO_4, and H_3PO_3?

8. Write a balanced overall equation that shows the synthesis of nitric acid from ammonia.

9. Why is calcium phosphate not used directly as a fertilizer?

10. What chemical compounds contain phosphate in commercial fertilizers?

11. Give the formulas and names of the allotropic forms of oxygen and sulfur.

12. List several uses of O_2.

13. Distinguish between the peroxide ion and the superoxide ion.

14. Write chemical equations to show the preparations of the elemental forms of the halogens.

15. Write the chemical formulas of the halic acids.

16. Describe the colors and the standard physical states at 25°C of the diatomic halogens.

17. Write equations for the preparation of three interhalogens: ClF_3, BrF_5, and IF_7.

ANSWERS

Exercises

1. $2Al(s) + 3HCl(aq) \rightarrow AlCl_3(aq) + \dfrac{3}{2}H_2(g)$

2. $-1, +1$

3. NaH

4. Brønsted base. $H^-(aq) + H_2O(l) \rightarrow OH^-(aq) + H_2(g)$

5. The melting point of diamond is related to the energy required to break C—C bonds. In diamond each carbon atom is covalently linked to four other carbon atoms. To turn diamond into a liquid about half of these bonds must be broken. This requires extremely high temperatures.

6. CN^-, a C_2^{2-} and a C^{4+} ions

7. CO

8. -5 and -3

9. $4NH_3(g) + 5O_2(g) \rightarrow 4NO(g) + 6H_2O(g)$
   $2NO(g) + O_2(s) \rightarrow 2NO_2(s)$
   $2NO_2(g) + H_2O(l) \rightarrow HNO_2(aq) + HNO_3(aq)$

10. White phosphorus is very unstable compared to red phosphorus.

11. ozone

12. steelmaking

13. $-1, -2, 0$

14. $2H_2O_2(aq) \rightarrow 2H_2O(l) + O_2(g)$

15. K_2O, K_2O_2, and KO_2

16. $S(s) + O_2(g) \rightarrow SO_2(g)$
    $2SO_2(g) + O_2(g) \rightarrow 2SO_3(g)$
    $SO_3(g) + H_2O(l) \rightarrow H_2SO_4(aq)$

17. F_2 and Cl_2

18. $Cl_2(g) + 2Br^-(aq) \rightarrow 2Cl^-(aq) + Br_2(l)$

19. HCl, HBr, and HI
20. F_2
21. a. $Cl_2(g) + 2I^-(aq) \rightarrow 2Cl^-(aq) + I_2(aq)$
    $Br_2(g) + 2I^-(aq) \rightarrow 2Br^-(aq) + I_2(aq)$
    b. $Cl_2(g) + 2NaOH(aq) \rightarrow NaOCl(aq) + NaCl(aq) + H_2O(l)$
    $Br_2(g) + 2NaOH(aq) \rightarrow NaOBr(aq) + NaBr(aq) + H_2O(l)$

Conceptual Questions

1. Because hydrogen can be oxidized to the H^+ ion, it resembles the alkali metals. And because it can be reduced to the H^- ion it resembles the halogen elements. Hydrogen has an electron configuration of $1s^1$ making it similar to the alkali metals. It also forms diatomic molecules just as the halogens do. H_2 is a gas at room temperature. As a result of these properties, hydrogen resembles the alkali metals and the halogens.
2. Br_2 can act as both an oxidizing agent and a reducing agent because Br has three oxidation states, namely -1 in Br^- ion, 0 in Br_2, and $+1$ in OBr^-. Therefore, Br_2 can be an oxidizing agent by accepting electrons and being reduced to Br^- ion. And Br_2 can also be a reducing agent by donating electrons and being oxidized to the OBr^- ion.

Practice Test

1. $C_3H_8(g) + 3H_2O(g) \rightarrow 3CO(g) + 7H_2(g)$
    $C(s) + H_2O(g) \rightarrow CO(g) + H_2(g)$
2. To make ammonia and partially hydrogenated food products.
3. $Na_2B_4O_7 \cdot 4H_2O + H_2SO_4 + H_2O \rightarrow 4H_3BO_3 + Na_2SO_4$
4. When coal is distilled in the absence of oxygen to release volatile hydrocarbons, the high carbon residue is called coke. Coke is made into graphite by the Acheson process, in which an electric current is passed through coke for several days.
5. a. Create atmospheres devoid of O_2; a coolant.
    b. Ammonia fertilizer
    c. Nitrate fertilizer and explosives
6. A mixture of 3 volumes of concentrated HCl to 1 volume of concentrated HNO_3.
7. NO_2, NO, P_2O_5, P_2O_3
8. $12NH_3 + 21O_2 \rightarrow 8HNO_3 + 14H_2O + 4NO$
9. It is insoluble.
10. Diammonium phosphate, $(NH_4)_2HPO_4$; superphosphate $Ca(HPO_4)_2 \cdot H_2O$
11. O, O_2, O_3; rhombic S_8, monoclinic S_8
12. In the blast furnace for steel making, medical uses, welding.
13. Peroxide ion O_2^{2-}; superoxide O_2^-
14. $CaF_2 + H_2SO_4 \rightarrow 2HF + CaSO_4$
    $2HF \rightarrow H_2 + F_2$
    $2NaCl + 2H_2O \rightarrow 2NaOH + H_2 + Cl_2$
    $Cl_2 + 2I^- \rightarrow 2Cl^- + I_2$
15. HIO_3, $HBrO_3$, $HClO_3$. The comparable fluorine acid does not exist.
16. $I_2(s)$, black crystals; $Br_2(l)$, dark red liquid; $Cl_2(g)$, a yellow-green gas; $F_2(g)$, pale yellow gas.
17. $Cl_2 + 3F_2 \rightarrow 2ClF_3$
    $KBr + 3F_2 \rightarrow KF + BrF_5$
    $KI + 4F_2 \rightarrow KF + IF_7$

Chapter Twenty-Two
TRANSITION METAL CHEMISTRY AND COORDINATION COMPOUNDS

- Properties of the Transition Metals (22.1)
- Coordination Compounds (22.3)
- The Structure of Coordination Compounds (22.4)
- Bonding in Coordination Compounds (22.5)

PROPERTIES OF THE TRANSITION METALS (22.1)

STUDY OBJECTIVES

1. Describe several general characteristics of the transition metal elements.
2. Explain why there is only a gradual decrease in atomic radius when moving across a series of transition elements.

Electron Configurations of Atoms and Ions. In this chapter, we will focus on the first-row transition elements from scandium to copper. Recall that the transition metals have atoms or ions with *incompletely filled d subshells*.

The electron configurations of the first-row transition elements were discussed in Section 7.9 of the text. For the representative elements potassium and calcium, the 4s orbital is lower in energy than the 3d orbital. The 3d orbitals begin to fill only after the 4s orbital is complete. For elements after Ca $[Ar]4s^2$, electrons are added one at a time to the 3d subshell beginning with scandium, the first transition element.

The electron configurations are given in Table 22.1 below. The configurations of chromium $[Ar]4s^13d^5$ and copper $[Ar]4s^13d^{10}$ are exceptions to the order of filling. The basis for these exceptions is that the energies of a 3d orbital and a 4s orbital are almost the same for these elements. For chromium, the 4s and 3d orbitals are almost identical in energy, and so the electrons enter all *six* of the available orbitals singly in accordance with Hund's rule. In the case of copper, the exception shows the importance of extra stability associated with a completely filled 3d subshell.

Table 22.1 Electron Configurations of Atoms and Ions

	Atom	+2 ion	+3 ion
Sc	$4s^23d^1$	$3d^1$	[Ne]
Ti	$4s^23d^2$	$3d^2$	$3d^1$
V	$4s^23d^3$	$3d^3$	$3d^2$
Cr	$4s^13d^5$	$3d^4$	$3d^3$
Mn	$4s^23d^5$	$3d^5$	$3d^4$
Fe	$4s^23d^6$	$3d^6$	$3d^5$
Co	$4s^23d^7$	$3d^7$	$3d^6$
Ni	$4s^23d^8$	$3d^8$	
Cu	$4s^13d^{10}$	$3d^9$	——

To understand the electron configurations of the ions shown in Table 22.1, it helps to recall that electrons are *removed* from the 4s orbital before they are taken out of the 3d. The energies of the 3d and 4s orbitals are not as close together in the *ions* of the transition metals as they are in the neutral *atoms*. In fact, for the ions of transition elements the 3d orbitals are lower in energy than the 4s orbitals. Therefore, the electrons most easily lost are those in the outermost principal energy level, the ns. Additional electrons may then be lost from the $(n - 1)d$ orbital.

Properties of Transition Elements. For the representative elements, properties such as the atomic radius, ionization energy, and electronegativity vary markedly from element to element as the atomic number increases across any period. In contrast, the chemical and physical properties of the transition metal elements vary only slightly as we read across a period. Table 22.1 in the text gives a number of properties of the transition elements for comparison. The characteristics of transition metals are summarized below.

1. *General Physical Properties*. Transition metals have relatively high densities, high melting and boiling points, and high heats of fusion and vaporization. See Table 22.2 in the text.
2. *Atomic Radius*. Recall that the atomic radii of representative elements decrease markedly as we read across a period of elements. In contrast, the atomic radii of transition metals decrease only slightly as we read across a series of these elements.

The decrease in radii is consistent with an increase in effective nuclear charge as atomic number increases. All the first-row transition elements have 4s orbitals as the outermost occupied orbitals. The atomic radius will depend on the strength of the nuclear charge felt by the 4s electrons. The greater the charge, the smaller the atom. The effective nuclear charge experienced by the outermost electrons depends on the shielding provided by inner electrons.

For transition metals the nuclear charge increases from scandium to copper, and electrons are being added to an inner 3d subshell. These 3d electrons shield the 4s electrons from the increasing nuclear charge for the most part. Consequently the 4s electrons of the first-row transition elements feel only a slightly increasing effective nuclear charge as the atomic number increases, and the atomic radii makes only a gradual decrease.

3. *Ionization Energy*. The first ionization energies of the first transition metal series are remarkably similar, increasing very gradually from left to right. There is a slight increase over the first five elements then the ionization energy barely changes from iron to copper. See Table 22.1 in the text.
3. *Variable Oxidation States*. The common oxidation states for the transition elements from Sc to Cu are +2 and +3 (Table 22.1 text). Some of the elements exhibit the +4 state, and Mn even shows +5, +6, and +7. Transition metals usually exhibit their highest oxidation states in compounds with oxygen, fluorine, or chlorine. $KMnO_4$ and $K_2Cr_2O_7$ are examples. The variability of oxidation states for transition metal ions results from the fact that the 4s and 3d subshells are similar in energy. Therefore, an atom can form ions of roughly the same stability by losing different numbers of electrons. The +3 oxidation states are more stable at the beginning of the series, but toward the end the +2 oxidation states are more stable.

EXAMPLE 22.1   Transition Metal Elements

What distinguishes a transition element from a representative element?

•Method of Solution

The *representative elements* are those in which all the inner subshells are filled and the last electron enters an outer s or p subshell. The *transition elements* are those in which an inner d subshell is partially filled. For example, the outer electron configuration of Sc is $4s^2 3d^1$ and for Co it is $4s^2 3d^7$.

EXAMPLE 22.2   Properties of Transition Elements

Why do the transition metals have higher densities than the Group 1A and 2A metals?

•Method of Solution

Densities of solids are related to atomic and ionic radii. As the atoms get smaller, the solids tend to get more dense. The effective nuclear charge experienced by the outermost electrons in transition metal atoms is greater than that experienced in Group 1A and 2A metal atoms.
Answer: The atomic radii of the transition elements, in general, are smaller than those of the Group 1A and 2A elements of the same period.

EXERCISES

1. Write the electron configurations of: a. a Mn atom   b. a Mn^{2+} ion   c. a Mn^{3+} ion.
2. Which atom in each pair has the larger atomic radius? a. V or Fe   b. V or Nb
3. Which has the higher ionization energy, Ti or Fe?

COORDINATION COMPOUNDS (22.3)

STUDY OBJECTIVES

1. Define the terms: complex ion, ligand, donor atom, coordination number, and coordination compound.
2. Determine the oxidation number and coordination number of metal atoms in coordination compounds.
3. Name coordination compounds when given their formulas, and write their formulas when given their names.

   Terminology. A **complex ion** consists of a central metal cation to which several anions or molecules are bonded. The complex ion may be positively or negatively charged. The metal cation is called the **central atom**, and the attached anions and molecules are called **ligands**. The free ligands each have at least one unshared pair of electrons which can be donated to the electron-deficient metal ions. The **donor atom** is the atom in the ligand that is directly bonded to the metal. Some typical ligands are Cl^-, CN^-, NH_3, H_2O, and $H_2NCH_2CH_2NH_2$.
   A neutral species containing a complex ion is called a **coordination compound**. These compounds usually have complicated formulas. Two examples are :

$$[Ag(NH_3)_2]Cl \quad \text{and} \quad K_3[Fe(CN)_6]$$

The complex ion is shown enclosed in brackets. In the silver compound, Cl^- is a free (meaning uncomplexed) chloride ion, and in the iron compound each K^+ is a free potassium ion. K^+ and Cl^- ions in the above formulas are examples of *counter ions*. They serve to balance or neutralize the charge of the complex ion. The **coordination number** is defined as the number of donor atoms surrounding the central metal atom. Thus the coordination number of Pt^{2+} in $[Pt(NH_3)_4]^{2+}$ is 4, and that of Co^{2+} in $[Co(NH_3)_6]^{2+}$ is 6.
   Some ligands contain more than one donor atom. Ligands that coordinate through two bonds are called **bidentate ligands**. Those with more than two donor atoms are referred as **polydentate ligands**. Ethylenediamine (en) and oxalate ion (ox) are bidentate ligands:

$$H_2N-CH_2-CH_2-NH_2$$

Ethylenediamine (en)

Oxalate ion (ox), $C_2O_4^{2-}$

Bidentate and polydentate ligands are often called **chelating ligands** (pronounced key-lateing). The name derives from the Greek word *chele* meaning "claw." Chelate complexes are extra stable because two bonds must be broken to separate a metal from a ligand. $EDTA^{4-}$ is an excellent chelating ligand because it has six donar atoms. Its structure is shown in Figure 22.8 in the text.

Oxidation Number. Each complex ion carries a net charge that is the sum of the charges on the central atom (or ion) and on each of the ligands. In $Fe(CN)_6^{3-}$, for example, each cyanide ion (CN^-) contributes a -1 charge, and the iron ion a $+3$ charge. The charge of the complex ion then is; $(+3) + 6(-1) = -3$. The *oxidation number* of iron is $+3$ which is the same as its ionic charge. See Example 22.3 for a calculation.

Naming Coordination Compounds. Thousands of coordination compounds are known. The rules that have been developed for naming then are summarized below:

1. The cation is named before the anion.
2. Within the complex ion, the ligands are named first, followed by the metal ion. Ligands are listed in alphabetical order.
3. The names of anionic ligands end with the letter o, whereas neutral ligands names are usually the same as the names of the molecules. The exceptions are H_2O (aquo), CO (carbonyl), and NH_3 (ammine). Table 22.3 in the textbook lists the names of some common ligands.
4. When several ligands of a particular kind are present, use the Greek prefixes di-, tri-, tetra-, penta-, and hexa. Thus the ligands in $[Co(NH_3)_4Cl_2]^+$ are "tetraamminedichloro." If the ligand itself contains a Greek prefix, use the prefixes bis (2), tris (3), and tetrakis (4) to indicate the number of ligands present. The ligand ethylenediamine already contains the term di; therefore bis(ethylenediamine) is used to indicate two ethylenediamine ligands.
5. The oxidation number of the metal is written in Roman numerals following the name of the metal.
6. If the complex is an anion, attach the ending -ate to the name of the metal. $Fe(CN)_6^{3-}$ is named hexacyanoferrate (III) ion.

See Examples 22.3, 22.4, and 22.5 for the application of terminology and nomenclature rules to coordination compounds.

EXAMPLE 22.3   Terminology of Coordination Compounds

A certain coordination compound has the formula $[Co(NH_3)_4Cl_2]Cl$.
a. Which atom is the central atom?
b. Name the ligands and point out the donor atoms.
c. What is the charge on the complex ion?
d. What is the oxidation number (O.N.) of the central atom?

•**Method of Solution**

a. Inside the brackets the metal atom is written first and this is the central atom to which all ligands are bonded. Cobalt is the central atom.
b. The ligands are written next. These are ammonia and chloride ion. The donor atoms are N and Cl, respectively.
c. Since only one chloride ion is needed to balance the charge of the complex ion, the complex ion must have a charge of $+1$.
d. The charge of the central ion plus the sum of charges of ligands = the charge of the complex ion. Ammonia is a neutral ligand and will not affect the complex ion charge. Since two chloride ions are -2, in order for the complex ion to have a $+1$ charge, then Co must be a $+3$ ion.

O.N.(Co) + 2 O.N.(Cl) = complex ion charge
O.N.(Co) + 2(−1) = + 1
O.N.(Co) = +3

EXAMPLE 22.4   Naming Coordination Compounds

Give a systematic name for each of the following compounds:
a. $[Ag(NH_3)_2]Br$
b. $Ni(CO)_4$
c. $K_2[Cd(CN)_4]$

•Method of Solution

a.   The two ammonia ligands are represented by the term "diammine." The oxidation state of the Ag ion is +1. The bromide ion is not complexed with silver. It is a counter ion. The systematic name is diamminesilver(I) bromide.

b.   The term tetracarbonyl signifies that there are four carbon monoxide ligands. The oxidation state of the nickel atom is zero. Systematic name: tetracarbonylnickel(0).

c.   In this case the complex ion is an anion, thus cadmium will be referred to as cadmate. Since K is +1 and each cyanide ligand is –1, Cd must have a +2 charge. The name is potassium tetracyanocadmate(II).

EXAMPLE 22.5   Writing Formulas from the Names

Write formulas for the following coordination compounds:
a.   diaquodicyanocopper(II)
b.   potassium hexachloropalladate(IV)
c.   dioxalatocuprate(II) ion

•Method of Solution

a.   *diaquo* refers to two water molecules, and *dicyano* to two cyanide ions. Since the copper ion has a +2 charge, this coordination compound is neutral. *Answer:* $[Cu(H_2O)_2(CN)_2]$

b.   The ending *-ate* indicates that the complex ion must be an anion. *Answer:* $K_2[Pd(Cl)_6]$.

c.   The complex ion is an anion. *Answer:* $[Cu(C_2O_4)_2]^{2-}$

EXERCISES

4.   What is the charge on the complex ion in $K_2[PtCl_4]$?
5.   What is the coordination number of Hg in $[Hg(en)_2]^{2-}$?
6.   What is the oxidation number of Fe in $[Fe(CN)_6]^{4-}$?
7.   What is the coordination number and oxidation number of the central atom in $K[Cr(NH_3)_3Cl_3]$?
8.   Give the systematic name for $[CoCl_3Br_3]^{4-}$.
9.   Give the systematic name for $[Ni(en)_2(NH_3)_2]^{2+}$.
10. Write formulas for the following coordination compounds:
     a. diammineoxolatocopper(II)
     b. potassium amminetrichloroplatinate(II)

THE STRUCTURE OF COORDINATION COMPOUNDS (22.4)

STUDY OBJECTIVES

1. Describe geometric and optical isomers.
2. Define the terms chiral and enantiomers.

Common Structures. The geometry of a coordination complex is defined by the arrangement of the donar atoms of the ligands around the central metal atom. Complexes with coordination numbers of two, four, and six are the most common. Complexes with a general formula ML_2^{n+}, where M is the metal atom or ion and L is a donar atom, are linear with ligands on opposite sides of the metal atom. Other geometries are as follows:

General formula	Coodination number	Structure
ML_2^{n+}	2	Linear
ML_4^{n+}	4	Tetrahedral or square planar
ML_6^{n+}	6	Octahedral

Geometric Isomers. When two or more compounds have the same composition but a different arrangement of atoms, the compounds are called **isomers**. Isomerism is a characteristic feature of coordination compounds. **Stereoisomers** are compounds that have the same types and numbers of atoms bonded together in the same sequence, but with different spatial arrangements. There are two types of stereoisomers: geometric isomers and optical isomers.

Geometric, or *cis-trans*, isomers are distinguished by the position of like ligands or groups. The isomer with the like-groups in adjacent positions is called the the *cis* isomer, and the one with like-groups across from each other is called the *trans* isomer. *Cis* and *trans* isomers of coordination compounds generally have different physical and chemical properties.

Two forms of the complex $Pt(NH_3)_2Cl_2$ have been prepared. Both have square planar structures. The trans form has no dipole moment, whereas the cis form has an appreciable dipole moment.

The simplest type of geometric isomerism of octahedral complexes occurs in cases where four of the six ligands of the complex are the same. Then the cis and trans forms correspond to those shown below. The two X ligands are closest to each other in the cis isomer. In the trans isomer they are across the complex ion from each other.

Optical Isomers. **Optical isomers** are nonsuperimposable mirror images of one another. They are also called *enantiomers*. Enantiomers have the same relationship to one another as do your right and left hands. If you place your left hand parallel to your right hand facing each other, you get the same effect that you would get if you placed one hand in front of a mirror. Your left hand is the mirror image of your right hand. However, your left hand is not superimposable upon your right hand. Your hands are enantiomers, that is, nonsuperimposable mirror images of one another.

The enantiomers of *cis*-dichlorobis(ethylenediamine)cobalt(III) ion are shown in Figure 22.13 in the text. The mirror image of the trans isomer can be superimposed on the original after rotating it by 90°. Therefore, there is only one form of the trans isomer. The mirror image of the cis isomer *cannot be superimposed* on the original no matter how it is rotated. The cis isomer exists as enantiomers.

Not all objects are enantiomers. An ordinary chair, for example, looks the same in the mirror as when viewed directly. A chair is superimposable on its mirror image. An object which is not identical with its mirror image is said to be *chiral,* from the Greek word for hand (see Figure 22.12 in the text). Example 23. 7 illiustrates chiral complexes.

Enantiomers are called optical isomers because they differ with respect to the direction in which they rotate the plane of polarization of plane-polarized light when light is passed through the substance. The isomer that rotates the plane of polarized light to the right is said to be *dextrorotatory*. Its mirror image will rotate the plane to the left, *levorotatory*. A **racemic** mixture is an equimolar mixture of the two optical isomers. Such a mixture produces no net rotation of plane polarized light.

EXAMPLE 22.6   Geometric Isomers

Sketch the geometric isomers of $[Cr(en)_2Br_2]^+$.

•Method of Solution

First identify the ligands. Recall that "en" stands for ethylenediamine, a neutral *bidentate* ligand. Bromide ion is the other ligand. The, the central ion is Cr^{3+} and its coordination number is 6 because it bonds to six donar atoms. The complex will be octahedral. N — N represents ethylenediamine..

For an octahedral complex, ligands at adjacent corners are referred to as cis to each other. The ligands at opposite corners are trans to each other. Two geometric isomers are possible. One has two Br^- ions in cis positions, and in the other isomer the Br^- ligands are trans.

cis   trans

EXAMPLE 22.7   Optical Isomers

For the complex ion in the preceding example, indicate if either of the isomers exhibit optical isomerism.

•Method of Solution

Sketch the mirror images of the cis and trans isomers, and look for nonsuperimposable mirror images, the enantiomers.

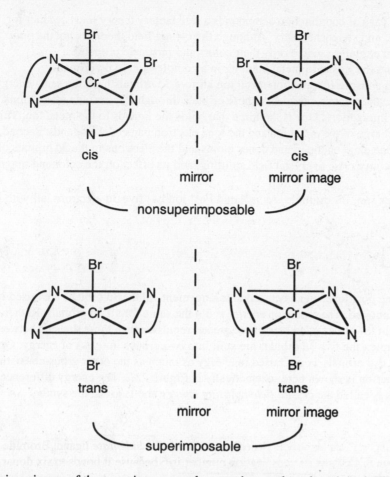

The mirror image of the trans isomer can be superimposed on the original. Therefore, there is only one form of the trans isomer. The mirror image of the cis isomer *cannot be superimposed* on the original no matter how it is rotated. The cis isomer exists in two forms (enantiomers). The cis isomer is chiral. The trans isomer is not.

EXERCISES

11. Distinguish between geometric isomers and optical isomers.
12. Descirbe the term chiral.

BONDING IN COORDINATION COMPOUNDS (22.5)

STUDY OBJECTIVES

1. Explain how the interaction between the ligands and the d orbitals of the central metal atom in a coordination compound produces crystal-field splitting.
2. Explain the relationship of the color of a complex ion to the wavelength of light absorbed.
3. Predict the number of unpaired electrons in a given complex ion.

Crystal Field Theory. In the case of coordination compounds a satisfactory theory must account for properties such as color, magnetism, and stereochemistry. Although the crystal field theory is not the only theory of coordination compounds, it explains satisfactorily their colors and magnetic properties.

The *crystal field model* considers the ligand to metal bonding in a complex ion to be primarily electrostatic, rather than covalent. In an isolated transition metal ion all five 3d orbitals have the same energy. The effect of the ligands is to change the relative energies of these orbitals through electrostatic interactions. There are two types of electrostatic interactions. One is the force that holds the ligands to the metal ion. This is the attraction between the positive charge of the metal ion and the lone electron pairs of the ligands. Second, there is the repulsion between the lone pairs on the ligand donor atoms and the electrons in the 3d orbitals of the metal. It is this latter interaction that gives rise to crystal field splitting, and its effect on the color and magnetic properties of the complex ion.

Consider the $Fe(CN)_6^{3-}$ complex ion, for example. An isolated Fe^{3+} ion has five 3d electrons, all with the same energy.

When the six cyanide ligands become positioned in an octahedral arrangement, all the 3d orbitals are raised in energy due to the repulsion just mentioned. The repulsion energy is not the same for all 3d orbitals. Rather, those orbitals that are directed straight toward a ligand experience greater repulsion than those directed between ligands. Thus, in an octahedral complex the five 3d orbitals are split into two groups in terms of energy. One group, comprising the d_{xy}, d_{yz} and d_{xz} orbitals, is not raised in energy as much as the other group consisting of the d_{z^2} and $d_{x^2-y^2}$ orbitals. The situation is shown here schematically in Figure 22.2. The energy difference between these two sets of d orbitals is called the *crystal field splitting energy* and is given the symbol Δ.

Figure 22.1 Crystal field splitting. On the left are the energy levels of the five 3d orbitals in a free Fe^{3+} ion. On the right are the same d orbitals in the same ion when it is a part of the octahedral complex ion $Fe(CN)_6^{3-}$.

Color. The color of a coordination compound results from electron transitions from the lower-energy set of orbitals to the higher-energy orbitals. Absorption of light occurs when the energy of an incoming photon is equal to the difference in energy Δ, and an electron transition occurs. The requirement for light absorption is

$$\Delta = E_{photon}$$

Recall from Chapter 7 that the energy of a photon is $E_{photon} = h\nu$, where h is Planck's constant, and ν is the frequency of the radiation. Therefore, $\Delta = h\nu$.

A substance has a color because it absorbs light at certain wavelengths in the visible part of the electromagnetic spectrum (from 400 to 700 nm) and it reflects the other wavelengths. Light that is a combination of all visible wavelengths appears white. When white light impinges on a coordination compound, and light of a certain wavelength is absorbed, the reflected light is missing that component and no longer appears white to the eye.

If the complex absorbs all colors except orange, the complex appears orange. The complex will also appear orange if it absorbs only light of the color blue. In a complementary manner, if the complex ion absorbed only orange, it would appear blue. Blue and orange are referred to as *complementary colors*.

The following "color wheel" allows you to estimate the color of reflected light when light of a given color (or wavelength) is absorbed. The colors that appear across from each other are complementary colors. For example, the hexacyanoferrate(II) ion absorbs light in the visible region of the spectrum at about 410 nm. The absorbed light is violet. The color of the complex is the color directly across the wheel from violet. The complex will appear yellow, the complementary color of violet.

The extent of crystal field splitting, that is, the magnitude of Δ, determines the wavelength of light absorbed, and thus the color of the complex. Each ligand has a different strength of interaction with the d orbitals of the metal ion. Thus each splits the orbital energies by a different amount. The ligands can be arranged according to increasing values of Δ. This establishes a *spectrochemical series*.

$$I^- < Br^- < Cl^- < OH^- < F^- < H_2O < NH_3 < en < CN^- < CO$$

CO and CN^- are called *strong-field* ligands because they produce a relatively large splitting, whereas I^- and Br^- are *weak-fields* ligands because they split d orbital energies to a lesser extent.

Magnetic Properties. In Chapter 7 you learned that paramagnetic substances have at least one unpaired electron and diamagnetic substances have all their electrons paired. The Fe^{3+} ion has a $3d^5$ configuration. Two of its octahedral complexes, $[FeF_6]^{3-}$ and $[Fe(CN)_6]^{3-}$ are both paramagnetic, and yet they have different magnetic properties. $[FeF_6]^{3-}$ has five unpaired electrons, but $[Fe(CN)_6]^{3-}$ has only one. The former is called a *high-spin complex*, and the latter a *low-spin complex*.

The reason for this difference lies in the spectrochemical series and the value of Δ. Remember that according to Hund's rule, electrons in orbitals with similar energies prefer to be unpaired. Energy is required to make electrons pair up. Therefore, if Δ is small, as in the case of the weak-field ligand F^-, the electrons enter all five orbitals one at a time without pairing in accord with Hund's rule. For small Δ values, pairing of electrons would not occur until the sixth electron is added.

However, when Δ is large enough, Hund's rule no longer applies to the entire set of d orbitals. The lowest energy configuration corresponds to placing all of the electrons in the three lowest-energy 3d orbitals first, before any enter the two higher-energy orbitals. This is the case for $[Fe(CN)_6]^{3-}$ because CN^- is a strong-field ligand and produces large Δ values. Figure 22.2 shows the energy level diagrams for hexafluoroferrate (III) ion and hexacyanoferrate (III) ion.

So far we have focused on octahedral complexes. Crystal field splitting occurs for tetrahedral and square planar complexes as well. See Figures 22.23 and 22.24 in the text for the splitting diagrams of these complexes.

$[FeF_6]^{3-}$
5 unpaired electrons
(high spin)

$[Fe(CN)_6]^{3-}$
1 unpaired electrons
(low spin)

Figure 22.2 Effect of crystal field splitting on pairing of d electrons in two Fe(III) complex ions. When Δ is small, Hund's rule applies to all five orbitals. When Δ is large, the three lower-energy orbitals fill before the two higher orbitals.

EXAMPLE 22.8   Crystal Field Theory

$[Ti(H_2O)_6]^{3+}$ ion absorbs light at a wavelength of 498 nm.
a.   Calculate the crystal field splitting energy Δ.
b.   Determine the color of the complex.

•Method of Solution for (a)

a.   Absorption of light occurs when the energy of an incoming photon is equal to the difference in energy Δ between the lower energy 3d orbitals, and the higher energy 3d orbitals of the titanium ion.
The requirement for light absorption is

$$\Delta = E_{photon}$$

$$\Delta = h\nu = \frac{hc}{\lambda}$$

where h is Planck's constant, c is the velocity of light, and λ is the wavelength of light, 498 nm.

•Calculation for (a)

$$\Delta = \frac{(6.63 \times 10^{-34} \text{ J s})(3.00 \times 10^8 \text{ m/s})}{498 \text{ nm} \times \left(\frac{10^{-9} \text{ m}}{1 \text{ nm}}\right)}$$

$$\Delta = 3.99 \times 10^{-19} \text{ J}$$

•Method of Solution for (b)

The complex absorbs light with a wavelength of 498 nm. On the color wheel this is green light. A solution that absorbs green light will reflect all the colors except green. According to the color wheel the solution will be red.

EXAMPLE 22.9   Crystal Field Theory

$[CoF_6]^{4-}$ is a high-spin paramagnetic complex, and $[Co(en)_3]^{2+}$ is a low-spin paramagnetic complex. Draw the crystal field splitting diagrams for these two complex ions and show the proper relationship of the two Δ values.

•Method of Solution

The crystal field splitting diagram for a Co^{2+} ion ($3d^7$) must show 7 electrons in the 3d orbitals. In a high-spin complex Δ is small enough so that electrons can enter 3d orbitals one at a time according to Hund's rule. Pairing of electrons will not occur until after the fifth electron has been added. In the low-spin complex, Δ is much larger and the electrons enter the lower-energy orbitals first. Only after 6 electrons are positioned in the lower-energy set of orbitals, can electrons then be placed in the upper two orbitals.

$[CoF_6]^{4-}$
3 unpaired electrons
(high-spin)

$[Co(en)_3]^{2+}$
1 unpaired electrons
(low-spin)

EXERCISES

13. If a complex ion absorbs blue light, what color is the complex ion?
14. Which ligand is the stronger-field ligand, ethylenediamine or bromide ion?
15. Distinguish between a high-spin complex and a low-spin complex.
16. In the complex ion ML_6^{n+}, M^{n+} has six d electrons and L is a weak field ligand. How many unpaired electrons are there in this complex?

CONCEPTUAL QUESTION

1. The color of oxyhemoglobin is due to complex ions involving Fe(II) ions. Oxyhemoglobin (with O_2 bound to iron) is a low-spin, Fe(II) complex. Five of the donor atoms are N atoms and do not change. The sixth donor atom is O in the O_2 molecule. Deoxyhemoglobin (without the O_2) has a water ligand in place of O_2 and is a high-spin complex. Both complexes are octahedral Fe^{2+} complexes.
   a. How many unpaired electrons occupy the 3d orbitals of the iron(II) ion in each case?
   b. Oxyhemoglobin is red and deoxyhemoglobin is blue. Explain in terms of the crystal field theory why the two forms of hemoglobin have different colors.

PRACTICE TEST

1. Write the electron configuration of an Fe atom; an Fe^{2+} ion; an Fe^{3+} ion.
2. Write the formula of the coordination compound containing a Pt^{4+} central atom, three chlorides, three ammonia ligands, and one uncomplexed nitrate ion.
3. Find the oxidation number of the central atom in $Na_2[Co(H_2O)_2I_4]$.
4. For the coordination compound $[Co(NH_3)_6]Cl_3$ give the oxidation state and coordination number of the central atom.

5. For the coordination compound $[Ni(en)_2(NH_3)_2]Cl_2$ give the oxidation state and coordination number of the central atom.

6. Name the following coordination compounds and complex ions.
   a. $[Pt(NH_3)_4Cl_2]Cl_2$
   b. $[Ni(NH_3)_6]^{2+}$
   c. $[Cr(OH)_4]^-$
   d. $[Co(CN)_6]^{3-}$
   e. $K_2[Cu(CN)_4]$

7. Write the formula for each of the following coordination compounds and ions.
   a. tetrahydroxoaluminate(III) ion
   b. tetraiodomercurate(II) ion
   c. potassium dichlorobis(oxalato)cobaltate(III)
   d. tris(ethylenediamine)nickel(II) sulfate

8. Sketch two geometrical isomers for $[Ni(NH_3)_2(CN)_4]^{2-}$.

9. Sketch the structures of the square planar complex *trans*-$[Pt(Cl)_2(NH_3)_2]$, and its mirror image. Are the two structures enantiomers?

10. Why are chiral substances said to be optically active?

11. Illustrate each of the following by giving formulas, geometry, or energy diagrams of:
    a. octahedral complex of $NiCl_6^{4-}$.
    b. cis and trans isomers of $Ni(CO)_2(CN)_2$.
    c. $[Fe(H_2O)_6]^{3+}$ is a high-spin complex.
    d. a bidentate ligand.
    e. a coordination number of 2

12. If green light is absorbed by a complex in solution, what is the color of the solution?

13. Sketch the crystal field splitting diagram for the low-spin $[Fe(CN)_6]^{4-}$ ion.

14. Sketch the crystal field splitting diagram for the tetrahedral complex $[FeCl_4]^-$.

15. Distinguish between inert and labile complexes.

16. Complexes of zinc are never paramagnetic. Explain.

17. In the complex ion $[ML_6]^{n+}$, M^{n+} has four d electrons, and L is a weak-field ligand. How many unpaired electrons are there in this complex?

18. $[Ni(H_2O)_6]^{2+}$ is green and $[Ni(en)_3]^{2+}$ is violet. Which has the larger value of Δ?

ANSWERS

Exercises

1. a. $3d^5 4s^2$ b. $3d^5$ c. $3d^4$
2. a. V   b. Nb
3. Fe
4. −2
5. 4
6. +2
7. coordination number = 6,   oxidation number = +2
8. tribromotrichlorocobalt(II) ion.
9. diamminebis(ethylenediamine)nickel(II) ion
10. a. $[Cu(NH_3)_2(C_2O_4)]$ b. $K[PtNH_3Cl_3]$
11. Geometric isomers have the same numbers and kinds of atoms but the arrangement of atoms is different even though the same bonds are present. Optical isomers (enantiomers) are mirror images that cannot be superimposed on each other.

12. Chiral objects are mirror images that cannot be superimposed on one another no matter how they are rotated.
13. orange
14. ethylenediamine
15. The terms low-spin and high-spin refer to magnetic properties of complex ions. In a high-spin complex the electrons are arranged so that they are unpaired as much as possible. In a low-spin complex there are more electrons that are paired. High-spin complexes are more paramagnetic than low-spin complexes.
16. 4

Conceptual Question

1. a. Six d-electrons total. In the low-spin O_2 containing complex there are no unpaired electrons. In the high-spin complex with H_2O there are four unpaired electrons.
    b. The two forms of hemoglobin have different colors due to the differing amount of crystal field splitting (Δ) in the iron complexes. O_2 must be a stronger-field ligand than H_2O. The splitting is much greater in the low-spin O_2-containing complex than in the high-spin H_2O-containing complex. Therefore the color of light absorbed will be different for the two complexes, leading to different observed colors. From the complementary colors you can tell that the O_2 complex absorbs green light and the H_2O complex absorbs orange light.

Practice Test

1. Fe $[Ar]4s^23d^6$
    Fe $[Ar]3d^6$
    Fe $[Ar]3d^5$
2. $[Pt(NH_3)_3Cl_3]NO_3$
3. +2
4. The coordination number is 6; the oxidation state is +3.
5. The coordination number is 6; the oxidation state is +2.
6. a. tetraamminedichloroplatinum(IV) chloride
    b. hexaamminenickel(II) ion
    c. tetrahydroxochromate(III) ion
    d. hexacyanocobaltate(III) ion
    e. potassium tetracyanocuprate(II)
7. a. $[Al(OH)_4]^-$
    b. $[HgI_4]^{2-}$
    c. $K_3[Co(C_2O_4)_2Cl_2]$
    d. $[Ni(en)_3]SO_4$
8.

9. The two structures are not enantiomers.

10. Because when plane-polarized light passes through a solution of a chiral substance, the plane of the polarized light is rotated.

11. a.

b.

cis   trans

c.

d.

e. $[H_3N—Ag—NH_3]^+$

12. Red

13.

14.

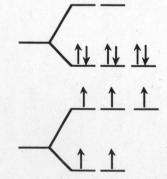

15. The terms inert and labile are kinetic terms, that is, they refer to relative rates of ligand substitution reaction. Complexes that undergo rapid ligand exchange are said to be labile complexes. In inert complexes the bonds between the central atom and the ligands are broken and re-formed relatively infrequently.

16. Zinc has ten d electrons that completely fill the five d orbitals.

17. 4

18. $[Ni(en)_3]^{2+}$

Chapter Twenty-Three
NUCLEAR CHEMISTRY

- The Nature of Nuclear Reactions (23.1)
- The Stability of Nuclei (23.2)
- Natural Radioactivity (23.3)
- Nuclear Transmutation (23.4)
- Nuclear Fission and Fusion (23.5 – 23.6)
- Biological Effects of Radiation (23.8)

THE NATURE OF NUCLEAR REACTIONS (23.1)

STUDY OBJECTIVES

1. Write balanced nuclear equations for radioactive decay and nuclear transmutation processes.
2. Compare chemical reactions and nuclear reactions.

Nuclear Reactions and Nuclear Equations. This chapter emphasizes changes that occur within the nucleus of an atom. To begin, we will discuss two types of nuclear reactions: radioactive decay and nuclear transmutation. **Radioactivity** is described as the spontaneous emission of particles and/or radiation by unstable atomic nuclei. These processes often result in the formation of a new element. The kinds of particles emitted from various nuclei are shown in Table 23.1.

Table 23.1 Particles From Radioactive Decay

Particle	Mass (amu)	Charge	Symbol
alpha	4.0	+2	$_2^4\alpha$ or $_2^4\text{He}$
beta	0.0005	−1	$_{-1}^0\beta$ or $_{-1}^0\text{e}$
positron	0.0005	+1	$_{+1}^0\beta$ or $_{+1}^0\text{e}$
gamma	0	0	$_0^0\gamma$

A nucleus can also undergo change by nuclear **transmutation**. In this process a nucleus reacts with another nucleus, an elementary particle, or a photon (gamma particle) to produce one or more new nuclei.

Radioactive decay and nuclear transmutation processes are described by nuclear equations. These equations use isotopic and elementary particle symbols to represent the reactants and products of nuclear reactions. For example, in the first nuclear transmutation ever observed (in 1919) alpha particles were used to bombard nitrogen-14 nuclei. The observed products were atoms of oxygen-17 and protons. The nuclear equation is

$$_7^{14}\text{N} + {}_2^4\text{He} \rightarrow {}_8^{17}\text{O} + {}_1^1\text{H}$$

The balancing rules for nuclear equations are given below and are applied to the above equation:

1. The sum of the mass numbers of the reactants must equal the sum of the mass numbers of the products (conservation of mass number). $14 + 4 = 18 = 17 + 1$
2. The sum of the nuclear charges of the reactants must equal the sum of the nuclear charges of the products (conservation of atomic number). $7 + 2 = 9 = 8 + 1$

These rules will be illustrated further in the example problems.

Comparison of Chemical and Nuclear Reactions. Table 23.1 of the text lists a number of comparisons between chemical and nuclear processes. Keep in mind that in chemical reactions the number of atoms of each element is conserved. Only changes in chemical bonding occur. However, in nuclear reactions the composition of the atomic nucleus is altered, and so elements are converted from one to another. Since a nucleus is so extremely small, and in some cases contains large numbers of positively charged protons, the energy changes associated with nuclear changes are much greater than energy changes in chemical reactions.

EXAMPLE 23.1 Nuclear Equations

Complete the following nuclear equations.

a. $^{14}_{7}N + ^{1}_{0}n \rightarrow ^{14}_{6}C +$ _____

b. $^{226}Ra \rightarrow ^{4}_{2}\alpha +$ _____

•**Method of Solution**

a. According to rule 1, the sum of the mass numbers of the reactants must equal the sum of the mass numbers of the products. $14 + 1 = 15 = 14 + A$. Therefore, the unknown product will have a mass number of 1. According to rule 2, the sums of the nuclear charges on both sides of the equation must be the same. $7 + 0 = 7 = 6 + Z$. Therefore the nuclear charge of the unknown product must be 1, making it a proton.

$$^{14}_{7}N + ^{1}_{0}n \rightarrow ^{14}_{6}C + ^{1}_{1}P$$

b. Note that the atomic number of radium is missing. The periodic table indicates that Ra is element number 88. Balancing the mass numbers first, we find that the unknown product must have a mass number of 222. Balancing the nuclear charges next, we find that the atomic number of the unknown must be 86. Element number 86 is radon.

$$^{226}_{88}Ra \rightarrow ^{4}_{2}\alpha + ^{222}_{86}Rn$$

EXERCISES

1. Write symbols for alpha particles, beta particles, and gamma rays.
2. How many protons (p), neutrons (n), and electrons (e) does an atom of the radioisotope phosphorus-32 contain?
3. Complete the following nuclear reactions:

    a. $^{14}_{7}N + ^{1}_{0}n \rightarrow ^{14}_{6}C +$ _____

    b. $^{210}Po \rightarrow ^{4}_{2}\alpha +$ _____

    c. $^{4}_{2}He + ^{9}_{4}Be \rightarrow ^{12}_{6}C +$ _____

    d. $^{15}_{7}N \rightarrow$ _____ $+ ^{0}_{-1}\beta$

4. When a nucleus decays by positron emission, the atomic number of the decay product will be (higher or lower) than the original radioisotope.

THE STABILITY OF NUCLEI (23.2)

STUDY OBJECTIVES

1. Compare relative stabilities of given nuclei by applying stability rules.
2. Predict, using the n : p ratio, whether a given isotope will decay by beta decay or positron decay.
3. Calculate the nuclear binding energies of given nuclei.

Nuclear Stability. Little is known about the forces that hold the nucleus together. However, some interesting facts emerge if we examine the numbers of protons and neutrons found in those nuclei that are stable. Nuclei can be classified according to whether they contain even or odd numbers of protons and neutrons. The number of stable isotopes of each of the four types of nuclei classified in this way are shown in Table 23.2.

Table 23.2 The Number of Stable Isotopes

Number of protons	even	even	odd	odd
Number of neutrons	even	odd	even	odd
Number of stable isotopes	157	52	50	8

The following rules are useful in predicting nuclear stability:

1. Nuclei with even numbers of both protons and neutrons are generally more stable than those with odd numbers of these particles. Nuclei that contain certain specific numbers of protons and neutrons within a nucleus ensure an extra degree of stability. These so-called *magic numbers* for protons and for neutrons are 2, 8, 20, 28, 50, 82, and 126.
2. Nuclei with even numbers of protons or neutrons are generally more stable than those with odd numbers of these particles.
3. All isotopes of elements after bismuth (Z = 83) are radioactive.

The Belt of Stability. The principal factor for determining whether a nucleus is stable is the neutron to proton ratio. Figure 23.1 shows a plot of the number of neutrons versus the number of protons in various isotopes. Each dot represents a *stable* isotope. The stable nuclei are located in an area of the graph known as the *belt of stability*. In this figure we see that at low atomic numbers stable nuclei possess a neutron to proton ratio of about 1.0. Above Z = 20 the number of neutrons always exceeds the number of protons in stable isotopes. The n : p ratio increases to about 1.5 at the upper end of the belt of stability.

If you were given the symbol of a radioisotope, without any experience, it would be impossible to tell its mode of decay. But with knowledge of the belt of stability, you can make accurate predictions of the expected mode of decay.

Isotopes with too many neutrons lie above the belt of stability. The nuclei of these isotopes decay in such a way as to lower their n : p ratio. Thus one of its neutrons may decay into a proton and a beta particle.

$$\,^1_0\text{n} \rightarrow \,^1_1\text{p} + \,^0_{-1}\beta$$

The proton remains in the nucleus, and the beta particle is emitted from the atom. The loss of a neutron and the gain of a proton produces a new isotope with two important properties. It has a *lower n : p ratio,* and thus is more likely to be stable. Also, the daughter product has an atomic number that is *one greater* than the decaying isotope due to the additional proton. Consider the decay of carbon-14 for example. Carbon-14 is continually produced in the upper atmosphere by the interaction of cosmic rays with nitrogen. Carbon-14 has a higher n : p ratio than either of carbon's stable isotopes (C-12 and C-13), and decays by beta decay.

$$\,^{14}_6\text{C} \rightarrow \,^{14}_7\text{N} + \,^0_{-1}\beta$$

Note that the product isotope $^{14}_{7}N$ is one atomic number greater than carbon. Also, it is stable. Its n : p ratio is 1.0.

Figure 23.1 A graph showing all the stable isotopes when plotted as neutron number versus proton number. The shaded area represents the belt of stability. In order for isotopes with Z = 20 to be stable, the ratio of n : p must be about 1.1. For isotopes with Z = 40 to be stable, the ratio of n : p must be about 1.3 : 1.0. For isotopes with Z = 80 to be stable, the ratio of n : p must be about 1.5 : 1.0.

Isotopes with too many protons have a low n : p ratio and lie below the belt of stability. These isotopes tend to decay by positron emission because this process produces a new isotope with a higher n : p ratio. During *positron emission* a proton emits a positron, $^{0}_{+1}\beta$, and becomes a neutron.

$$^{1}_{1}p \rightarrow {}^{1}_{0}n + {}^{0}_{+1}\beta$$

The neutron remains in the nucleus, and the positron is ejected from the atom. Thus, a product nucleus will contain one less proton and one more neutron than the parent nucleus. The n : p ratio increases due to positron decay.

Electron capture accomplishes the same end, that is, a higher n : p ratio. Some nuclei decay by capturing an orbital electron of the atom.

$$^{1}_{1}p + {}^{0}_{-1}e \rightarrow {}^{1}_{0}n$$

Lanthanum-138, a naturally occurring isotope with an abundance of 0.089 percent, decays by electron capture.

$$^{138}_{57}La + {}^{0}_{-1}e \rightarrow {}^{138}_{56}Ba$$

Electron capture is accompanied by X-ray emission.

Nuclear Binding Energy. One of the important consequences of Einstein's theory of relativity was the discovery of the equivalence of mass and energy. The total energy content (E) of a system of mass m is given by Einstein's theory:

$$E = mc^2$$

where c is the velocity of light $(3.0 \times 10^8$ m/s). Therefore, *the mass of a nucleus is a direct measure of its energy content*. It was discovered in the 1930s that the measured mass of a nucleus is always *smaller* than the sum of the separate masses of its constituent nucleons. This difference in mass is called the *mass defect*. When the mass defect is expressed as energy by applying Einstein's equation, it is called the binding energy of the nucleus. The **binding energy** is the energy required to break up a nucleus into its component protons and neutrons. The binding energy provides a quantitative measure of nuclear stability. The greater the binding energy the more stable the nucleus is toward decomposition.

In terms of the $^{17}_{8}O$ nucleus for instance, the binding energy (BE) is the energy required for the process

$$^{17}_{8}O + BE \rightarrow 8\,^{1}_{1}p + 9\,^{1}_{0}n$$

The mass defect Δm is equal to the total mass of the products minus the total mass of the reactants.

$$\Delta m = [8(\text{proton mass}) + 9(\text{neutron mass})] - (^{17}_{8}O \text{ nuclear mass})$$

Here we can substitute atomic masses, which include the electrons, for the nuclear masses.

$$\Delta m = [8(^{1}_{1}H \text{ atomic mass}) + 9(\text{neutron mass})] - (^{17}_{8}O \text{ atomic mass})$$

This works because the mass of 8 electrons in the 8 hydrogen atoms is canceled by the mass of the 8 electrons in the oxygen atom.

$$8(^{1}_{1}H \text{ atomic mass}) = 8(\text{proton mass}) + 8(\text{electron mass})$$

$$(^{17}_{8}O \text{ atom mass}) = (^{17}_{8}O \text{ nuclear mass}) + 8(\text{electron mass})]$$

Continuing by using atomic masses given in the text (Section 23.2) and in Table 23.3 below.

$$\Delta m = [8(^{1}_{1}H \text{ atomic mass}) + 9(\text{neutron mass})] - (^{17}_{8}O \text{ atomic mass})$$

$$\Delta m = [8(1.007825 \text{ amu}) + 9(1.008665 \text{ amu})] - (16.999131 \text{ amu}) = 0.141454 \text{ amu}$$

The eight protons and nine neutrons have more mass than the oxygen-17 nucleus. The binding energy is:

$$\Delta E = \Delta mc^2$$

The calculation of binding energy will be illustrated in Example 23.4.

In comparing the stability of any two different nuclei, we must account for the different numbers of nucleons per nucleus. A satisfactory comparison of nuclear stabilities can be made by using the **binding energy per nucleon**, that is, *the binding energy of each nucleus divided by the total number of nucleons* in the nucleus. This is one of the most important properties of a nucleus. When the BE per nucleon is plotted as a function of the atomic mass we get the curve of binding energy as shown in Figure 23.2 of the text. Note that at first it rises rapidly with increasing atomic mass, and then it reaches a maximum at mass 56. Above mass 56, the binding energy drops slowly as atomic mass increases. Table 23.1 compares the total binding energy and the binding energy per nucleon for several isotopes.

Table 23.3 Binding Energies of Selected Isotopes

| | Mass (amu) | Binding Energy (J) | |
		Total	Per Nucleon
$^2_1$H	2.01410	3.57×10^{-13}	1.78×10^{-13}
$^3_2$He	3.01603	1.24×10^{-12}	4.13×10^{-13}
$^4_2$He	4.00260	4.52×10^{-12}	1.13×10^{-12}
$^{16}_8$O	15.99491	2.04×10^{-11}	1.28×10^{-12}
$^{17}_8$O	16.999131	2.10×10^{-11}	1.24×10^{-12}
$^{56}_{26}$Fe	55.934939	7.90×10^{-11}	1.41×10^{-12}
$^{238}_{92}$U	238.0508	2.89×10^{-10}	1.22×10^{-12}

EXAMPLE 23.2 Nuclear Stability

Using the stability rules, rank the following isotopes in order of increasing nuclear stability.

$$^{39}_{20}\text{Ca} \qquad ^{40}_{20}\text{Ca} \qquad ^{30}_{15}\text{P}$$

•Method of Solution

Phosphorus-30 should be the least stable because it has odd numbers of both protons and neutrons. Calcium-39 has an even number of protons (20) and an odd number of neutrons. With an even number of protons and a "magic number" at that, it should be more stable than P-30. Of the three isotopes Calcium-40 should be the most stable. It has an even number of protons and of neutrons. Both numbers are "magic numbers."

$$^{30}_{15}\text{P} < {}^{39}_{20}\text{Ca} < {}^{40}_{20}\text{Ca}$$

EXAMPLE 23.3 Types of Radioactive Decay

The only stable isotope of sodium is sodium-23. What type of radioactivity would you expect from sodium-25?

•Method of Solution

Sodium-23 has 11 protons and 12 neutrons and is in the belt of stability. Sodium-25 must have two more neutrons, and so has a higher n : p ratio than the stable isotope. Sodium-25 will decay by $^0_{-1}\beta$ emission.

EXAMPLE 23.4 Nuclear Binding Energy

Calculate the nuclear binding energy of the light isotope of helium, helium-3. The atomic mass of $^3_2$He is 3.01603 amu.

•Method of Solution

The binding energy is the energy required for the process

$$^3_2\text{He} \rightarrow 2\,^1_1\text{p} + {}^1_0\text{n}$$

where $\Delta m = [2(\text{proton mass}) + (\text{neutron mass})] - {}^{3}_{2}\text{He}$ (nuclear mass)

The mass difference is calculated using atomic masses:

$$\Delta m = [2({}^{1}_{1}\text{H atomic mass}) + (\text{neutron mass})] - ({}^{3}_{2}\text{He atomic mass})$$

$$\Delta m = [2(1.007825 \text{ amu}) + 1.008665 \text{ amu}] - 3.01603 \text{ amu}$$
$$\Delta m = 8.29 \times 10^{-3} \text{ amu}$$

Using Einstein's equation:

$$\Delta E = (\Delta m)c^2 = (8.29 \times 10^{-3} \text{ amu}) \times (3.00 \times 10^8 \text{ m s}^{-1})^2$$

$$= 7.46 \times 10^{14} \text{ amu m}^2 \text{ s}^{-2}$$

$$= 7.46 \times 10^{14} \text{ amu m}^2 \text{ s}^{-2} \times \frac{1.00 \text{ kg}}{6.022 \times 10^{26} \text{ amu}} \times \frac{1 \text{ J}}{1 \text{ kg m}^2 \text{ s}^{-2}}$$

The binding energy is:

$$\Delta E = 1.24 \times 10^{-12} \text{ J/atom}$$

Each ${}^{3}_{2}\text{He}$ atom contains three nucleons. The binding energy per nucleon is:

$$\text{BE per nucleon} = \frac{1.24 \times 10^{-12} \text{ J/atom}}{3 \text{ nucleons/atom}}$$

The binding energy per nucleon is:

$$\Delta E = 4.13 \times 10^{-13} \text{ J/nucleon}$$

•Comment

By combining the above conversion factors into one constant, the number of steps in future calculations can be lessened.

$$? \text{ J/amu} = (3.00 \times 10^8 \text{ m s}^{-1})^2 \times \frac{1.00 \text{ kg}}{6.022 \times 10^{26} \text{ amu}} \times \frac{1 \text{ J}}{1 \text{ kg m}^2 \text{ s}^{-2}}$$

$$= 1.49 \times 10^{-10} \text{ J/amu}$$

This constant is a useful factor relating energy to mass in amu. Applying this to the mass defect (0.14145 amu) calculated earlier in the discussion for ${}^{17}_{8}\text{O}$, the binding energy is 2.10×10^{-11} J/atom. See Table 23.3.

EXERCISES

5. For each pair of nuclei, predict which one is the more stable.

   a. ${}^{3}_{2}\text{He}$ or ${}^{4}_{2}\text{He}$.

   b. ${}^{26}_{13}\text{Al}$ or ${}^{27}_{13}\text{Al}$

6. Fluorine has only one stable isotope, fluorine-19.
   a. The nucleus of fluorine-18 lies below the belt of stability. Write an equation for the decay of fluorine-18.
   b. The nucleus of fluorine-21 lies above the belt of stability. Write an equation for the decay of fluorine-21.

7. How many atomic mass units are in one kilogram?

8. Which has more mass, the nucleus of aluminum–27 or 13 protons and 14 neutrons?

9. What is the binding energy of $^{27}_{13}Al$? Calculate its binding energy per nucleon. The atomic mass of aluminum-27 is 26.981541 amu.

10. How much energy is released when one Po-214 atom decays by alpha emission?

$$^{214}_{84}Po \rightarrow {}^{4}_{2}\alpha + {}^{210}_{82}Pb$$

Given the atomic masses: Pb-214 = 213.99519 amu, Po-210 = 209.98286 amu, and He-4 = 4.00260 amu.

NATURAL RADIOACTIVITY (23.3)

STUDY OBJECTIVES

1. Describe general features of the uranium series.
2. Use the first-order rate equation to determine the amount of a radioisotope remaining after a given time.
3. Calculate the age of a rock sample given information about the amount of a particular radioisotope present in the rock.

Natural Radioactivity. A number of isotopes exist in nature that have an n : p ratio that places them outside the belt of stability. Radioisotopes that occur naturally on Earth give rise to *natural radioactivity.* Uranium, thorium, radon, potassium-40, carbon-12, and tritium (H-3) are naturally occurring radioisotopes. Uranium, which is fairly abundant in Earth's crust, is the start of a radioactive decay series. The series is a sequence of decay reactions that change U-238 ultimately to a stable isotope of lead. The uranium series is shown in Table 23.3 of the textbook. The uranium decay series includes all the elements between lead and uranium.

All of the radioisotopes in this series decay by alpha or beta emission. In the first step U-238 decays by alpha emission.

$$^{238}_{92}U \rightarrow {}^{234}_{90}Pu + {}^{4}_{2}\alpha$$

The decay product thorium-234 is also radioactive and decays by beta emission.

$$^{234}_{90}Pu \rightarrow {}^{234}_{91}Pa + {}^{0}_{-1}\beta$$

The product is also radioactive and the series continues by a number of steps to end at Pb-206.

Rates of Decay. Radioactive decay rates obey first-order kinetics.

decay rate = number of atoms disintegrating per unit time = λN

where λ is the first-order rate constant, called the decay constant, and N is the number of atoms of the particular radioisotope present in the sample being studied. Recall that the half-life is related to the rate constant by

$$t_{1/2} = 0.693/\lambda$$

The integrated first-order equation is

$$\ln \frac{N}{N_0} = -\lambda t$$

Where N is the number of atoms of the radioisotope present in the sample after time t has elapsed, and N_0 is the number of atoms of the radioisotope present initially. If N_0, N, and t are known, then we can calculate the rate constant, λ.

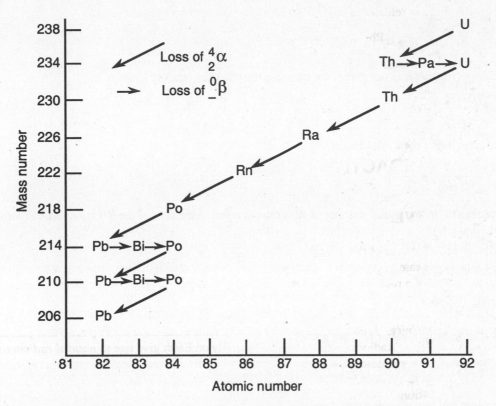

Figure 23.2. The uranium decay series involves 14 steps as uranium decays eventually into lead.

The object of radioactive dating is to determine the age of geological and archaeological samples and specimens. The age (t in the calculation) of certain rocks, for instance, can be estimated from analysis of the number of atoms of a particular radioisotope present now (N), as compared to the number present originally when the rock was formed (N_0).

Rearranging the above equation, the age t is given by:

$$t = -\frac{1}{\lambda} \ln \frac{N}{N_0}$$

The value of the initial number of atoms N_0 is the sum, N + D, where D is the number of daughter nuclei resulting from the decay of atoms of the radioisotope. The original number of atoms of a radioisotope present in a rock sample is equal to the number N remaining at time t, plus the number of daughter atoms (D).

EXAMPLE 23.5 Radioactive Dating

The rubidium-87/strontium-87 method of dating rocks was used to analyze lunar samples. The half-life of Rb-87 is 4.9×10^{10} yr.

$$^{87}_{37}\text{Rb} \rightarrow {}^{87}_{38}\text{Sr} + {}^{0}_{-1}\beta$$

Estimate the age of moon rocks in which the mole ratio of Rb-87 to Sr-87 is 40.

•**Method of Solution**

The age of the rock can be calculated from first-order equation relating time to the number of nuclei remaining:

$$t = -\frac{1}{\lambda} \ln \frac{N}{N_0}$$

Since the decay constant k is not given, we must use the equation relating k and $t_{1/2}$:

$$\lambda = \frac{0.693}{t_{1/2}}$$

This can be substituted for k, yielding:

$$t = -\frac{t_{1/2}}{0.693} \ln \frac{N}{N_0}$$

Since Rb decays to Sr, the initial number of Rb atoms is equal to the sum of the Rb atoms and Sr atoms present.

$$N_0^{Rb} = N^{Rb} + N^{Sr}$$

Given that $N^{Rb}/N^{Sr} = 40$, then:

$$N^{Sr} = \tfrac{1}{40} N^{Rb}$$

Therefore, after substitution for N^{Sr}, we get:

$$N_0^{Rb} = N^{Rb} + \tfrac{1}{40} N^{Rb} = 1.025\, N^{Rb}$$

Substituting into the rate equation, we get:

$$t = -\frac{4.9 \times 10^{10} \text{ yr}}{0.693} \ln \frac{N^{Rb}}{1.025\, N^{Rb}} = -(7.07 \times 10^{10} \text{ yr}) \ln 0.976$$

$$t = 1.7 \times 10^9 \text{ yr}$$

EXERCISES

11. The uranium decay series starts with uranium-238 and ends with lead-206. Each step in the series involves the loss of either an alpha or a beta particle. In the entire series how many alpha particles and how many beta particles are emitted?

12. The rates of decay of all radioisotopes follow the same type of rate equation. What type of equation is it?

13. Cobalt-60 has a half-life of 5.26 years. a. Calculate the decay constant for this isotope. b. How much cobalt-60 will remain from a 20.0 mg sample after 8.75 years?

14. A 2.52 mg sample of pure uranium-238 has a decay rate (activity) of 31.2 dps due to U–238. a. What is the decay constant for U-238? b. What is the half-life of U–238?

15. The C–14 activity of some ancient corn was found to be 10 disintegrations per minute per gram of C. If present day plant life gives 15.3 dpm/g C, how old is the corn? The half-life of C–14 is 5730 y.

16. Estimate the age of rocks in which the mole ratio of U-238 to Pb-206 is 0.75. The half-life of U-238 is 4.5×10^9 yr.

17. Analysis of a sample of uranite ore yields 3.2 g of U-238 and 1.5 g Pb-206. Assuming there was no Pb-206 present initially, how old is the rock? The half-life of U-238 is 4.5×10^9 yr.

NUCLEAR TRANSMUTATION (23.4)

STUDY OBJECTIVES

1. Determine the products of nuclear transformations and use the abbreviated notation for transformation reactions.
2. Predict the product resulting from neutron capture by a given isotope, which is then followed by beta decay.

Artificial Radioactivity. Nuclear transmutation is the process of converting one element into another. Artificial radioactivity results when an unstable nucleus is produced by transmutation. Irene Curie and Frederic Joliot discovered this phenomenon in 1933 while bombarding light elements with α particles from radioactive sources. For example, when aluminum atoms are bombarded with alpha particles the product is phosphorus-30 which is radioactive. Phosphorus-30 decays by positron emission and has a half-life of 2.5 min. It does not occur naturally in phorphorus compounds.

$$^{27}_{13}\text{Al} + ^{4}_{2}\text{He} \rightarrow ^{30}_{15}\text{P} + ^{1}_{0}\text{n}$$
$$^{30}_{15}\text{P} \rightarrow ^{30}_{14}\text{Si} + ^{0}_{+1}\beta$$

Neutrons readily produce "artificial" radioactivity because they are easily captured by stable nuclei, with the result that a new nucleus is formed that has a higher n : p ratio. This leads to a product that decays by beta decay. Neutron capture by chlorine-37 yields chlorine-38.

$$^{1}_{0}\text{n} + ^{37}_{17}\text{Cl} \rightarrow ^{38}_{17}\text{Cl}$$
$$^{38}_{17}\text{Cl} \rightarrow ^{38}_{18}\text{Ar} + ^{0}_{-1}\beta$$

Chlorine-38 has a short half-life, and is not found naturally on Earth. Note that capture of a neutron, followed by beta decay, yields a new element (Ar) with an atomic number one greater than the original element. This procedure of neutron capture followed by beta decay of the product nucleus has been used to synthesize elements that were "missing" from the periodic table, such as Tc and Pm, elements number 43 and 61, respectively. All isotopes of these elements are radioactive.

Using neutrons as bombarding particles is convenient because neutrons have no charge and therefore are not repelled by target nuclei. When projectiles are positively charged such as protons and alpha particles, they must have considerable kinetic energy in order to overcome the electrostatic repulsion as these particles approach and collide with the target nucleus.

Nuclear transmutation processes can be abbreviated according to the following format.

Target nucleus (bombarding particle, ejected particle) Product nucleus

The alpha bombardment reaction above can be written:

$$^{27}_{13}\text{Al} \, (\alpha,\text{n}) \, ^{30}_{15}\text{P}$$

The symbols n, p, d, α, e, and γ are used in this notation to represent neutron, proton, deuteron, alpha particle, electron, and gamma ray.

EXAMPLE 23.6 Synthesizing a Transuranium Element

The first transuranium element to be synthesized by scientists was neptunium, atomic number 93. Devise a means to produce neptunium starting with U-238 and neutrons.

•Method of Solution

Neutron capture by a nucleus followed by beta decay produces a new nucleus of one atomic number higher than the original. Neptunium is one atomic number beyond uranium. So start with uranium and bombard it with neutrons.

$$_{92}^{238}U + {}_0^1n \rightarrow {}_{92}^{239}U$$

$$_{92}^{239}U \rightarrow {}_{93}^{239}Np + {}_{-1}^0\beta$$

•Comment

Uranium's other naturally occuring isotope, U-235, will not produce neptunium in the same way because it undergoes neutron-induced fission.

EXERCISES

18. Predict the product of the following bombardment reactions.

    a. $_{13}^{27}Al + {}_0^1n \rightarrow X + {}_1^1H$

    b. $_5^{11}B + {}_2^4He \rightarrow X + {}_1^1H$

19. Predict the product of the following bombardment reactions.

    a. $_5^{11}B\,(p, \alpha)X$

    b. $_5^{10}B\,(n,p)X$

20. Devise a scheme by which bombardment of molybdenum with neutrons and subsequent β decay of the product could be used to prepare an isotope of the missing element technetium, $_{43}^{99}Tc$.

NUCLEAR FISSION AND FUSION (23.5 – 23.6)

STUDY OBJECTIVES

1. Describe both fission and fusion, and relate these processes to the curve of binding energy per nucleon versus mass number.
2. Compare the energy released by fission and fusion with that of an ordinary chemical reaction.

Nuclear Fission. The curve of binding energy per nucleon versus mass number (Figure 23.2) shows that the most stable nuclei are those with masses close to iron-56, which is the most stable nucleus. During **fission** a heavy nucleus of mass greater than about 230 amu splits into two lighter nuclei whose masses are usually between 80 and 160 amu. Since the two smaller nuclei are more stable than the larger nucleus, energy is released in the process.

Although many nuclei heavier than uranium can undergo fission, the most important ones are U-233, U-235, and Pu-239. These isotopes undergo fission upon capture of a neutron. It is important to realize that naturally occurring uranium consists of two isotopes, U-235 and U-238, but that only U-235 is fissionable with thermal neutrons. The term thermal neutrons refers to those existing at temperatures around 25°C. Since the nucleus does not repel a neutron, as it does an alpha particle, for instance, neutron-induced fission will occur at ordinary temperatures.

The reaction is quite complex because the same two products are not formed by all fissioning nuclei. The two reactions below show just two out of many possibilities.

$$^{235}_{92}U + ^{1}_{0}n \begin{cases} \longrightarrow ^{141}_{56}Ba + ^{92}_{36}Kr + 3\,^{1}_{0}n \\ \\ \longrightarrow ^{137}_{52}Te + ^{96}_{40}Zr + 2\,^{1}_{0}n \end{cases}$$

The actual distribution of product yields is shown in Figure 23.7 of the text.

A significant feature of fission is that on the average 2 to 3 neutrons are released per fission event. Since neutrons are required to initiate fission, and because neutrons are also products of fission, a nuclear chain reaction is possible.

The energy released during nuclear fission depends somewhat on just what products are formed. The energy released from the fission of one mole of U-235 atoms can be calculated from the equation $\Delta E = \Delta mc^2$. The calculation shows that about 2.0×10^{10} kJ are released per mole of uranium. This amount of energy is *70 million times* the amount released in the exothermic chemical reaction in which 1 mol of H_2 reacts with 1/2 mol of O_2 to form 1 mol of water!

Nuclear Fusion. Radioactivity and nuclear fission are processes in which matter "comes apart." The energy releasing processes that occur on the sun are ones in which matter is fused. **Nuclear fusion** is the combining of small nuclei, such as hydrogen, to form a larger, more stable nucleus. Such a nucleus will have a higher average binding energy per nucleon, and fusion reactions will be exothermic. Because all nuclei are positively charged they must collide with enormous force in order to combine (fusion). This means that the atoms that will undergo fusion must be heated to millions of degrees. Fusion reactions are called *thermonuclear* reactions because they occur only at very high temperatures, such as those in the sun.

The reaction that accounts for the tremendous release of energy by the sun is believed to be the stepwise fusion of four hydrogen nuclei to produce one helium nucleus. The net process is

$$4\,^{1}_{1}H \rightarrow ^{4}_{2}He + 2\,^{0}_{+1}e \qquad \Delta E = -4.3 \times 10^{-12} \text{ J}$$

One gram of hydrogen upon fusion releases the energy equivalent to the combustion of 20 tons of coal. The fusion of four moles of H atoms by the preceding equation releases 2.6×10^{9} kJ of energy.

Our sun is made up of mostly hydrogen (90 percent) and helium (9 percent). In its interior the temperatures are estimated to reach 15 million °C. At these temperatures hydrogen will fuse to form helium; but helium, with its greater nuclear charge, will not fuse to form the heavier elements. The heavier elements up to iron are formed by fusion reactions that occur in exploding stars, called nova and supernova, where the temperatures can reach 2 billion °C.

EXAMPLE 23.7 Relationship of Mass and Energy

Calculate the mass of hydrogen that must undergo nuclear fusion each day in order to provide just the fraction of the daily energy output of the sun that reaches the earth, which is 1.5×10^{22} J.

•Method of Solution

Using the following equation:

$$4\,^{1}_{1}H \rightarrow ^{4}_{2}He + 2\,^{0}_{+1}e \qquad \Delta E = -4.3 \times 10^{-12} \text{ J}$$

we see that fusion of 4 H atoms yields 4.3×10^{-12} J. We can set up the calculation:

$$\text{g H atoms} = \frac{1.5 \times 10^{22} \text{ J}}{1 \text{ day}}$$

$$\text{g H atoms} = \frac{1.5 \times 10^{22} \text{ J}}{1 \text{ day}} \times \frac{4 \text{ H atoms}}{4.3 \times 10^{-12} \text{ J}} \times \frac{1.66 \times 10^{-24} \text{ g}}{\text{H atom}}$$

$$\text{g H atoms} = 2.3 \times 10^{10} \text{ g}$$

Expressing the answer in tons, we get:

$$2.3 \times 10^{10} \text{ g} = 25,500 \text{ tons of hydrogen}$$

EXAMPLE 23.8 Comparing Nuclear Fission and Fusion

Compare fission and fusion with respect to the temperatures required and the nature of the by-products of these processes.

•Method of Solution

Neutron-induced fission occurs at ordinary temperatures, whereas nuclear fusion requires temperatures in the millions of degrees. Fission of heavy elements yields hundreds of isotopes of elements with intermediate atomic numbers. These isotopes, by and large, have an excess of neutrons and are beta emitters. On the other hand, fusion of "light" nuclei yields stable isotopes of low-to-medium atomic mass.

EXERCISES

21. Which process, nuclear fusion or nuclear fission, requires temperatures of millions of degrees.
22. Complete the following fission reaction of plutonium.

$$^{239}_{94}\text{Pu} + ^{1}_{0}\text{n} \rightarrow ^{144}_{58}\text{Ce} + X + 3^{1}_{0}\text{n}$$

23. Complete the following nuclear fission reaction.

$$^{2}_{1}\text{H} + ^{3}_{1}\text{H} \rightarrow X + ^{1}_{0}\text{n}$$

BIOLOGICAL EFFECTS OF RADIATION (23.8)

STUDY OBJECTIVES

1. Describe how radiation interacts with matter.
2. Define radiation dose in units of rads and rems.

Interaction of Radiation with Matter. In passing through matter alpha, beta, and gamma rays lose energy chiefly by interaction with electrons which make up most of the volume of matter. Alpha and beta particles, on colliding with electrons, forcefully eject these electrons from atoms and molecules, and thereby produce ions. These particles lose only a small fraction of their energy in a single collision with an electron. Because alpha and beta particles are extremely energetic, thousands of collisions are required to bring them to rest. These particles produce "tracks" of ionization. Alpha, beta, and gamma rays are known as *ionizing radiation*.

Most devices for detecting radioactivity depend on the formation of ions. The best known instrument for detecting radiation is the Geiger counter in which ions produced by a particle trigger a pulse of electricity that is counted electronically. Darkening of photographic plates, discharging of electroscopes, and damage to biological tissue all involve ionization.

Units of Radiation Dose. Two units used to measure radiation dose are the rad and the rem. The *rad* (radiation absorbed dose) is defined as 0.01 joule of energy absorbed per kilogram of any absorbing material. The millirad is one-thousandth of a rad. Because beams of different radiations cause very different biological damage even when the body absorbs the same amount of energy from each type, it is necessary to define a unit specifically for biological tissue. The unit of biologically effective dose is the *rem* (<u>r</u>adiation <u>e</u>quivalent in <u>m</u>an), which is the absorbed dose in rads multiplied by the <u>r</u>elative <u>b</u>iological <u>e</u>ffectiveness factor, *RBE*. The millirem is one thousandth of a rem. For beta and gamma rays $RBE = 1.0$; for fast neutrons and alpha particles $RBE = 10$. Thus, a dose of one rad of alpha radiation is equivalent to 10 rem.

$$\text{dose (in rem)} = RBE \times \text{dose (in rad)}$$

EXERCISE

24. How does a rem differ from a rad?

PRACTICE TEST

1. What similarities and differences exist between beta particles and positrons?
2. Complete and balance the following nuclear equations.

   a. $^{239}_{94}\text{Pu} \rightarrow {}^{4}_{2}\alpha + \underline{\hspace{1cm}}$

   b. $\underline{\hspace{1cm}} + {}^{6}_{3}\text{Li} \rightarrow 2\ {}^{4}_{2}\text{He}$

   c. $^{90}_{38}\text{Sr} \rightarrow \underline{\hspace{1cm}} + {}^{0}_{-1}\beta$

   d. $^{10}_{5}\text{B} + {}^{4}_{2}\text{He} \rightarrow \underline{\hspace{1cm}} + {}^{1}_{0}\text{n}$

   e. $^{56}_{26}\text{Fe} + {}^{1}_{0}\text{n} \rightarrow \underline{\hspace{1cm}}$

3. Rank the following nuclides in order of increasing nuclear stability:

   $^{40}_{20}\text{Ca} \qquad {}^{39}_{20}\text{Ca} \qquad {}^{11}_{5}\text{B}$

4. With reference to the belt of stability, state the modes of decay you would expect for the following:

   a. $^{13}_{7}\text{N}$ b. $^{26}_{13}\text{Al}$ c. $^{28}_{13}\text{Al}$

5. a. Calculate the binding energy and the binding energy per nucleon of $^{27}_{13}\text{Al}$. The atomic mass of Al-27 is 26.98154 amu.

   b. Compare this result to the binding energy of $^{28}_{14}\text{Si}$ which has an even number of protons and neutrons. The atomic mass of Si-28 is 27.976928 amu.

6. Cobalt-60 is used in radiation therapy. It has a half-life of 5.26 years.

   a. Calculate the rate constant for radioactive decay.

   b. What fraction of a certain sample will remain after 12 years?

7. Radioactive decay follows first-order kinetics. If 20 percent of a certain radioisotope decays in 4.0 years, what is the half-life of this isotope?

8. Estimate the age of a bottled wine that has a tritium (^3_1H) content that is 3/4 that of environmental water obtained from the area where the grapes were grown. $t_{1/2} = 12.3$ yr.

9. Analysis of a sample of uranite ore shows that the ratio of U-238 atoms to Pb-206 atoms is 3.8. Assuming there was no Pb-206 present initially, how old is the rock? $t_{1/2}$ (U-238) $= 4.5 \times 10^9$ yr.

10. Consider the following fusion reactions:

    a. $^1_1H + ^2_1H \rightarrow ^3_2He$

    b. $^3_2He + ^3_2He \rightarrow ^4_2He + 2^1_1H$

    c. $^2_1H + ^3_1H \rightarrow ^4_2He + ^1_0n$

    Given the atomic masses: $^3_2He = 3.016029$ amu; $^4_2He = 4.002603$ amu; $^1_1H = 3.017005$ amu. Which of the above has the largest change in energy as indicated by its change in mass Δm?

11. One atom of element 109 was prepared by bombardment of a target of bismuth-209 with accelerated nuclei of iron-58. Write a balance nuclear equation to show the formation of the isotope of element 109 with a mass number of 266.

12. Write nuclear equations that show how Pu-239 is formed in a "breeder" reactor.

13. $^{137}_{55}Cs$ is a fission product of $^{235}_{92}U$. If it is formed along with three neutrons, what is the other isotope formed?

14. What is the mode of decay expected for "light" nuclei which are unstable because of low n : p ratio?

15. One natural radioactive series begins with $^{238}_{92}U$ and ends with $^{206}_{82}Pb$. All steps in the series are either alpha or beta decay. How many α particles and β particles are emitted?

16. The β particles emitted by carbon-14 atoms have a maximum energy of 2.5×10^{-14} J per particle. What is the dose in rads when 8.0×10^{10} carbon-14 atoms decay, and all the energy from their decay is absorbed by 2.0 kg of matter?

17. A sample of biological tissue absorbs a dose of 1.0 rad of alpha radiation. How many rems is this?

General Problem

18. Radium-226 is an α emitter with a half-life of 1600 years. If 2.0 g of radium were allowed to undergo decay for 10 years, and all of the α particles were collected over the that time as helium gas, what would be the mass and the volume of the He at STP?

ANSWERS

Exercises

1. $^4_2\alpha$ $^0_{-1}\beta$ $^0_0\gamma$
2. 15 protons, 17 neutrons, and 15 electrons
3. a. ^1_1H b. $^{206}_{82}Pb$ c. ^1_0n d. $^{15}_8O$

4. lower

5. a. $_2^4\text{He}$ b. $_{13}^{27}\text{Al}$

6. a. $_9^{18}\text{F} \rightarrow\ _8^{18}\text{O} +\ _{-1}^{0}\beta$

   b. $_9^{21}\text{F} \rightarrow\ _{10}^{21}\text{Ne} +\ _{-1}^{0}\beta$

7. 6.022×10^{26} amu = 1.00 kg

8. 13 protons and 14 neutrons

9. 1.34×10^{-12} J/nucleon

10. 1.45×10^{-12} J/nucleon

11. 5 alpha particles and 6 beta particles

12. the integrated first-order rate equation

13. a. 0.132 yr^{-1} b. 6.3 mg

14. a. 4.89×10^{-18} s^{-1} b. 4.49×10^{9} y

15. 3520 y

16. 3.6×10^{9} y

17. 2.8×10^{9} y

18. a. $_{12}^{27}\text{Mg}$ b. $_6^{14}\text{C}$

19. a. $_4^{8}\text{Be}$ b. $_4^{10}\text{Be}$

20. $_{42}^{98}\text{Mo} +\ _0^1\text{n} \rightarrow\ _{42}^{99}\text{Mo}$

    $_{42}^{99}\text{Mo} \rightarrow\ _{43}^{99}\text{Tc} +\ _{-1}^{0}\beta$

21. fusion

22. $_{36}^{93}\text{Kr}$

23. $_2^4\text{He}$

24. Both a rad and a rem refer to the absorption of the same amount of energy per kilogram of an object. However, for living matter the actual damage done varies for each type of particle absorbed. The rem is the dosage unit used for biological material. It is the absorbed dose in rads multiplied by the relative biological effectiveness factor, *RBE*. For beta and gamma rays *RBE* = 1.0; for fast neutrons and alpha particles *RBE* = 10. dose (rem) = *RBE* × dose (rad)

Practice Test

1. Beta particles and positrons have the same mass, 0.00055 amu. Beta particles have one unit of negative charge while positrons have one unit of positive charge.

2. a. $_{94}^{239}\text{Pu} \rightarrow\ _2^4\alpha +\ _{92}^{235}\text{U}$

   b. $_1^2\text{H} +\ _3^6\text{Li} \rightarrow 2\ _2^4\text{He}$

   c. $_{38}^{90}\text{Sr} \rightarrow\ _{39}^{90}\text{Y} +\ _{-1}^{0}\beta$

   d. $_5^{10}\text{B} +\ _2^4\text{He} \rightarrow\ _7^{13}\text{N} +\ _0^1\text{n}$

   e. $_{26}^{56}\text{Fe} +\ _0^1\text{n} \rightarrow\ _{26}^{57}\text{Fe}$

3. $^{11}_{5}B < ^{39}_{20}Ca < ^{40}_{20}Ca$. B-11 has an odd number of protons and an odd number of neutrons (rule 2). Ca-39 has an even number of protons and an odd number of neutrons (rule 1). Ca-40 has an even number of protons and an even number of neutrons. It also has a "magic number" of protons and neutrons (rule 2).

4. a. positron emission b. positron emission c. beta decay

5. a. 3.6×10^{-11} J/atoms; 1.3×10^{-12} J/nucleon
   b. 3.8×10^{-11} J/atoms; 1.4×10^{-12} J/nucleon

6. a. $\lambda = 0.132$ /yr   b. $N/N_0 = 0.205$

7. $t_{1/2} = 12.4$ yr

8. $t = 5.1$ yr

9. $t = 1.5 \times 10^9$ yr

10. c. $\Delta m = -1.98 \times 10^{-2}$ amu

11. $^{209}_{83}Bi + ^{58}_{26}Fe \rightarrow ^{266}_{109}X + ^{1}_{0}n$

12. $^{238}_{92}U + ^{1}_{0}n \rightarrow ^{239}_{92}U$

   $^{239}_{92}U \rightarrow ^{239}_{93}Np + ^{0}_{-1}\beta$

   $^{239}_{93}Np \rightarrow ^{239}_{94}Pu + ^{0}_{-1}\beta$

13. $^{235}_{92}U + ^{1}_{0}n \rightarrow ^{137}_{55}Cs + ^{96}_{37}Rb + 3\,^{1}_{0}n$

14. positron emission

15. 8 alpha particles account for the mass change; 8 alpha and 6 beta particles account for the change in positive charge.

16. 0.10 rad

17. 10 rem

18. 2.3×10^{19} Ra atoms decayed in 10 years; 2.3×10^{19} He atoms have a mass of 1.6×10^{-4} g and occupy a volume of 0.87 cm^3 at STP.

Chapter Twenty-Four
ORGANIC CHEMISTRY

- Aliphatic Hydrocarbons (24.1 – 24.2)
- Aromatic Hydrocarbons (24.3)
- Functional Groups (24.4)

ALIPHATIC HYDROCARBONS (24.1 – 24.2)

STUDY OBJECTIVES

1. Name four different types of aliphatic hydrocarbons and draw a structural formula of a typical example of each.
2. Name hydrocarbons of these four groups given a structural formula. Given a systematic name, draw a structural formula.
3. Write equations for addition reactions to alkenes and alkynes.
4. Apply Markovnikov's rule concerning the addition of unsymmetrical reagents to alkenes.

Organic Chemistry. The heart of organic chemistry is the carbon atom. Carbon is a key ingredient of about seven million chemical compounds primarily because of its ability to form long chains of self-linked atoms. The existence of structural and geometric isomers contributes strongly to the number of organic or carbon-containing compounds.

The **hydrocarbons** are an important class of organic compound that consist only of the elements carbon and hydrogen. Hydrocarbons are divided into two classes: aliphatic and aromatic. **Aromatic hydrocarbons** contain one or more benzene rings. **Aliphatic hydrocarbons** do not contain benzene rings. Four groups of aliphatic hydrocarbons are known. These are alkanes, alkenes, alkynes, and cycloalkanes.

Alkanes. The general formula for an alkane is C_nH_{2n+2}, where n is the number of carbon atoms in the molecule, n = 1, 2, 3···. When n = 1, we have the simplest member of the alkane family methane (CH_4). The alkanes make up a *homologous series;* a series of compounds differing in the number of carbon atoms. As n increases one at a time, we can generate the formulas of the entire series of alkanes. The names and formulas of the first ten straight-chain *alkanes* are given in Table 24.1. The first part of each name represents the number of C atoms in the molecule. The ending '-ane' is common to all alkanes. If you learn these names, they will prove to be very useful in naming other organic compounds.

Table 24.1 Names of the First Ten Alkanes.

Formula	Name
CH_4	Methane
C_2H_6	Ethane
C_3H_8	Propane
C_4H_{10}	Butane
C_5H_{12}	Pentane
C_6H_{14}	Hexane
C_7H_{16}	Heptane
C_8H_{18}	Octane
C_9H_{20}	Nonane
$C_{10}H_{22}$	Decane

Alkenes. Alkenes contain $C = C$ double bonds, and members of this homologous series have the general formula C_nH_{2n}. Alkenes are named with the same root word as alkanes to indicate the number of carbon atoms, but all names end in "ene."

C_2H_4	$CH_2 = CH_2$	ethene (ethylene)
C_3H_6	$CH_3CH = CH_2$	propene
C_4H_8	$CH_3CH_2CH = CH_2$	butene

Alkynes. Alkynes contain $C \equiv C$ triple bonds, and have the general formula C_nH_{2n-2}. The names of alkynes end with "yne."

C_2H_2	$CH \equiv CH$	ethyne (acetylene)
C_3H_4	$CH_3C \equiv CH$	propyne
C_4H_6	$CH_3CH_2C \equiv CH$	butyne

Cycloalkanes. There is also a type of *alkane* that has atoms bonded into ring configurations. These are the *cycloalkanes*. They have the same general formula as the alkenes, C_nH_{2n}, but do not have double bonds. Neither are these aromatics. Two cycloalkanes are shown below.

```
H2C—CH2              H2C—CH2
  \ /                 |   |
  CH2                H2C—CH2
cyclopropane         cyclobutane
```

Structural Isomers. Until now, we have considered only aliphatic hydrocarbons that have straight chains of carbon atoms. Branching of hydrocarbon chains is very common. Butane (C_4H_{10}) can be straight chained, and branched chained.

```
CH3—CH2—CH2—CH3   and   CH3—CH—CH3
                              |
                             CH3
       n-butane           2-methylpropane
```

Note that both molecules have the same molecular formula, but they have different arrangements of atoms (different structures). These are actually two distinguishable compounds with their different structures producing slightly different chemical and physical properties. Molecules that have the same molecular formula, but a different structure are called **structural isomers**. Straight chain hydrocarbons are called *normal* and use the symbol n. The straight chain form of C_4H_{10} has the name n-butane. Naming branched chain hydrocarbons is covered in the next section. In the alkane series, as the number of C atoms increases, the number of possible structural isomers increases dramatically. For example, C_4H_{10} has 2 isomers, C_6H_{14} has 5, and $C_{10}H_{22}$ has 75 isomers.

Nomenclature. The rules for naming hydrocarbons according to the IUPAC system are briefly summarized as follows:

1. The systematic name of an alkane is based on the number of carbon atoms in the longest carbon chain. For alkanes, the longest carbon chain is given a name corresponding to the alkane with the same number of C atoms. The parent name of a compound with 5 carbon atoms in the longest carbon chain is pentane.

2. Groups attached to the main chain are called **substituent groups**. Substituent groups that contain only hydrogen and carbon atoms are called **alkyl groups**. When an H atom is removed from an alkane the fragment is called an alkyl group. This group can be attached to the longest chain. Alkyl groups are named by dropping the ending *-ane* and adding *yl* to the alkane name. Therefore CH_3- is a methyl group, C_2H_5- is an ethyl group, and C_3H_7 is a propyl group. Table 24.2 in the text lists the names of six common alkyl groups.

3. The locations of alkyl groups that are attached to the main chain must also be included in the name. Number the carbon atoms in the longest chain. The numbering should start at the end of the chain such that the side groups will have the smallest numbers.
4. Prefixes such as *di*, *tri*, and *tetra* are used when more than one substituent of the same kind is present. Combine the names of substituent groups, their locations, and the parent chain into the hydrocarbon name. Arrange the substituent groups alphabetically followed by the parent name of the main chain.
5. Alkanes can have many different types of substituents besides alkyl groups. Halogenated hydrocarbons contain fluorine, chloroine, bromine and iodine atoms which are named fluoro, chloro, bromo, and iodo, respectively. NH_2-, NO_2-, and $CH_2=CH-$ are called amino, nitro, and vinyl, respectively.
6. When naming alkenes we indicate the positions of the carbon-carbon double bonds, and in alkynes the position of the triple bond is numbered. For alkenes and alkynes, the parent name is derived from the longest chain that contains the double or triple bond. Then parent name will end in *ene* for alkenes and *yne* for alkynes. Number the C atoms in the main chain by starting at the end nearer to the double or triple bond, and use the lower number of the carbon atom in the $C=C$ bond.

These rules are applied in Examples 24.1 and 24.2.

Addition Reactions. Alkenes, alkynes, and aromatics are called *unsaturated hydrocarbons*. This means that they can acquire more hydrogen atoms in an addition reaction called *hydrogenation*.

$$CH_2=CH_2 + H_2 \rightarrow CH_3-CH_3$$
$$CH\equiv CH + 2H_2 \rightarrow CH_3-CH_3$$

Alkanes are called *saturated hydrocarbons* because they cannot acquire additional hydrogen atoms. The carbon atoms in a saturated hydrocarbon are already bonded to the maximum number of H-atoms.

$$CH_3-CH_3 + H_2 \rightarrow \text{no reaction}$$

In an *addition reaction* a small molecule such as H_2 is added to an unsaturated hydrocarbon. The addition reaction occurs at the $C=C$ double bond. One atom of the small molecule links to one of the carbon atoms of the double bond, while the other atom attaches to the other carbon atom. Other examples are:

$$CH_2=CH_2 + Cl_2 \rightarrow CH_2Cl-CH_2Cl$$
$$CH_2=CH_2 + H_2O \rightarrow CH_3-CH_2OH$$
$$CH_2=CH_2 + HCl \rightarrow CH_3-CH_2Cl$$

Markovnikov's Rule. The two carbons of a double bond suffer different fates during addition reactions with unsymmetrical reagents such as HCl, HBr, and H—OH (water). For example, two different compounds might possibly be formed by reaction of 1-butene with HCl.

$$CH_3CH_2CH=CH_2 + HCl \rightarrow CH_3CH_2CHCl-CH_3 \text{ or } CH_3CH_2CH_2-CH_2Cl$$
$$\qquad\qquad\qquad\qquad\qquad\qquad \text{observed product} \qquad\qquad \text{not found}$$

The rule that predicts which of the two products is formed is called **Markovnikov's rule**. This rule states: *In the addition of unsymmetrical reagents to alkenes, the positive group of the reagent adds to the carbon that already has the most hydrogen atoms.* In HCl, HBr, and H_2O the hydrogen atoms have a partially positive charge because they are bonded to more electronegative atoms. Thus, in the above reaction with $CH_3CH_2CH=CH_2$, the H atom from HCl bonds to the terminal C atom. The Cl atom bonds to the CH group.

Geometric Isomers of Alkenes. $C\!=\!C$ double bonds are completely rigid. Consider the structures of *cis*- and *trans*-2-butene.

$$
\begin{array}{ccc}
CH_3 & & H \\
\diagdown & & \diagup \\
& C\!=\!C & \\
\diagup & & \diagdown \\
H & & CH_3
\end{array}
\qquad\qquad
\begin{array}{ccc}
CH_3 & & CH_3 \\
\diagdown & & \diagup \\
& C\!=\!C & \\
\diagup & & \diagdown \\
H & & H
\end{array}
$$

trans-2-butene   *cis*-2-butene

Neither structure can rotate around the $C\!=\!C$ bond to become the other structure. These two molecules have different physical properties and can be separated from one another. Therefore, these two structures represent two different compounds and are isomers. Many pairs of such isomers have been isolated, and they have no tendency to undergo interconversion.

The rigidity of the double bond gives rise to a new kind of isomerism: **geometric isomerism**, which is sometimes called cis-trans isomerism. The relative positions of atoms in space is known as a *configuration*. The isomers are named cis or trans depending on their configuration. The *cis* configuration has like groups on the same side of the double bond.

$$
\begin{array}{ccc}
CH_3 & & CH_3 \\
\diagdown & & \diagup \\
---C & \!=\! & C--- \\
\diagup & & \diagdown \\
H & & H
\end{array}
\qquad cis
$$

The trans configuration has like groups on the opposite sides of the double bond. *Trans* means across.

$$
\begin{array}{ccc}
CH_3 & & H \\
\diagdown & & \diagup \\
---C & \!=\! & C--- \\
\diagup & & \diagdown \\
H & & CH_3
\end{array}
\qquad trans
$$

EXAMPLE 24.1   Naming Hydrocarbons

Name the molecule that has the following structure:

$$
\begin{array}{l}
\quad\quad H_3C \quad CH_2\!-\!CH_3 \\
\quad\quad\quad | \quad\quad\; | \\
CH_3\!-\!CH_2\!-\!C\!-\!CH\!-\!CH_2\!-\!CH_3 \\
\quad\quad\quad\;\, 4\,| \quad\; 3 \quad 2 \quad\;\; 1 \\
\quad\quad\quad\;\, 5\;CH_2 \\
\quad\quad\quad\quad\;\; | \\
\quad\quad\quad 6\;CH_2\!-\!CH_3 \\
\quad\quad\quad\quad\quad\quad 7
\end{array}
$$

•**Method of Solution**

First identify the longest continuous carbon chain. Two equivalent chains containing seven carbon atoms are noticeable. Number the carbon atoms as shown. There are two ethyl groups, one at carbon 3 and one at carbon 4, and one methyl group also bonded to carbon 4. The parent name is heptane. Placing the names of the substituent groups (alphabetically ethyl is before methyl) and their position numbers in front of the parent alkane name we get:
3,4-diethyl-4-methylheptane.

•Comment

If the chain had been numbered in reverse order the name would be 4,5-diethyl-5-methylheptane. Since 3,4 is smaller than 4,5 the first name is the correct one.

EXAMPLE 24.2   Naming Hydrocarbons

Name the following hydrocarbon:

$$
\begin{array}{cccc}
6 & 5 & 4 & \\
CH_3\!-\!CH_2\!-\!CH\!-\!CH_3 & \\
& & | & \\
& & CH\!=\!CH\!-\!CH_3 & \\
& & 3\quad\ 2\quad\ 1 &
\end{array}
$$

•Method of Solution

Locate the longest chain that contains the double bond. This chain contains six C atoms so the root name will be hexene. Number the chain so that the double bond has the smallest number, as shown. Note the methyl group on carbon 4. The name is 4-methyl-2-hexene.

•Comment

The 2 indicates the position of the $C\!=\!C$ double bond, which connects carbons numbered 2 and 3.

EXAMPLE 24.3   Addition Reactions

Give the structure of the product of the following reaction.

$$
\begin{array}{c}
CH_3CH\!=\!C\!-\!CH_3 + HCl \\
| \\
CH_3
\end{array}
$$

•Method of Solution

This solution involves the addition of an unsymmetrical reagent to a double bond. According to Markovnikov's rule the positive portion of the reagent (in this case an H atom) adds to the carbon atom in the double bond that already has the most hydrogen atoms. The Cl atom adds to the other C atom. The product will be

$$
\begin{array}{c}
Cl \\
| \\
CH_3CH_2\!-\!C\!-\!CH_3 \\
| \\
CH_3
\end{array}
$$

EXERCISES

1. Write the general formulas of alkanes, alkenes, and alkynes.
2. Which of the classes of hydrocarbons in number 1 are unsaturated hydrocarbons?
3. Give the IUPAC name for the compound with the following structure.

4. Give the IUPAC name for the following compound.

$$CH_3—CH=CH—\overset{\overset{\displaystyle CH_3}{|}}{CH}—CH_3$$

5. Name the following compounds.
    a. CH_3CHFCH_3 b. $(CH_3)_3CCH_2CHClCH_3$

6. Write the structure for 3,3-dimethyl pentane.

7. What is the product of the following addition reaction: $Cl_2 + CH_3CH=CH_2$?

8. What is the product of the addition reaction $HBr + CH_3CH=CH_2$?

9. Namethe following: a. $CH_3(CH_2)_3CH=CHCH_3$ b. $CH_3(CH_2)_3C\equiv CH_3$

10. Draw all the isomers of C_4H_8.

AROMATIC HYDROCARBONS (24.3)

STUDY OBJECTIVES

1. Describe the structure of aromatic hydrocarbons.
2. Determine the product of a substitution reaction.

   Benzene. Benzene is the parent compound of a class of hydrocarbons called aromatic hydrocarbons. The hydrogen to carbon ratio in the molecular formula of benzene C_6H_6 suggests that it is highly unsaturated. In 1865 Kekule suggested that benzene had a cyclic structure with three double bonds.

Benzene can be represented as two resonance structures.

Alternatively, benzene can be represented in terms of delocalized molecular orbitals.

Regardless of which symbol is used, the hydrogen atoms are usually not explicitly written. Rather we must remember that there is one attached to each carbon.

Aromatic hydrocarbons contain benzene rings and do not have a simple general formula. Benzene is the simplest compound in this group.

Nomenclature of Aromatic Hydrocarbons. The naming of monosubstitited benzenes is straightforward.

bromobenzene   nitrobenzene

If more than one substituent is present, we must indicate the position of the two groups relative to each other. The prefixes ortho- (o), meta- (m), and para- (p) are used to indicate their relative positions.

o-dichlorobenzene   m-dichlorobenzene   p-dichlorobenzene

Reactions of Aromatic Compounds. Halogens react with benzene by a **substitution reaction** rather than an addition reaction. In this reaction an atom or group of atoms *replaces* an H atom in the benzene ring. Chlorine atoms can be substituted for hydrogen atoms by reacting benzene and chlorine in the presence of an $FeCl_3$ catalyst.

EXERCISES

11. Name the following compounds.

a.   b.   c.

12. How does benzene differ in reactivity from unsaturated aliphatic hydrocarbons?

FUNCTIONAL GROUPS (24.4)

STUDY OBJECTIVES

1. Define the term functional group. Sketch the general structural formulas of alcohol, ether, ketone, aldehyde, carboxyl, ester, and amine groups.
2. Write structural formulas for products of the oxidation of alcohols and aldehydes, for products of the reactions of carboxylic acids with alcohols, and for products of the reactions of esters with water and sodium hydroxide.

Alcohols. Alkanes are quite inert toward most substances, the main exception being oxygen. The alkane portion of organic compounds is combustable. Certain groups of atoms within carbon compounds always react in the same way regardless of the chain length of the alkane portion. The group of atoms that is largely responsible for the chemical behavior of a molecule is called a **functional group**. For instance, *alcohols* contain an alkane group and a hydroxyl group, OH. Methanol (CH_3OH), ethanol (C_2H_5OH), and n-propanol (C_3H_7OH) are the first three alcohols of a homologous series. In each of the three, chemical reactions occur at the OH group, rather than at the less reactive alkane group. The hydroxyl group is the functional group in each, and the three alcohols react toward other reagents in the same way.

Reactions of alcohols discussed in the textbook are oxidation by oxidizing agents, displacement of hydrogen by alkali metals, and esterification. An oxidizing agent removes two H atoms, one from the hydroxyl group and one from the adjacent carbon atom, forming an aldehyde (shown below). Permanganate ion is an oxidizing agent.

$$CH_3CH_2CH_2{-}OH \xrightarrow{MnO_4^-} CH_3CH_2CHO$$

Quite often, in organic reactions, we do not write completely balanced equations. Here the emphasis is on the chemical change of the alcohol functional group.

Alkali metals displace $H_2(g)$ from alcohols. Potassium is a reducing agent.

$$2CH_3CH_2OH + 2K \rightarrow 2CH_3CH_2OK + H_2$$

Esterification will be discussed under carboxylic acids.

Ethers. The functional group in ethers is C—O—C or, more generally, R—O—R, where R stands for any alkyl group such as those listed in Table 24.2 of the textbook. Ethers and alcohols are structural isomers. As illustrated in the following diagram, both dimethyl ether and ethanol have the same chemical formula.

$$CH_3OCH_3 \qquad \leftarrow [C_2H_6O] \rightarrow \qquad CH_3CH_2OH$$
dimethyl ether   ethanol
T_b –25°C T_b 78°C

The boiling points given show that ethers are much more volatile than the corresponding alcohols. This is because of the absence of hydrogen bonding in ethers. Alcohols contain the polar hydroxyl group, $O^{\delta-}{-}H^{\delta+}$, which can participate in hydrogen bonding to neighboring alcohol molecules. For many years diethyl ether, $CH_3CH_2OCH_2CH_3$, was used as an anesthetic.

Aldehydes and Ketones. The funtional group in these compounds is the carbonyl group, C=O. In aldehydes the carbonyl group is at the end of the alkane chain. In ketones it is not a terminal group.

Aldehydes are easily oxidized to carboxylic acids.

$$CH_3CH_2\overset{\overset{\displaystyle O}{\|}}{C}H \;\xrightarrow{MnO_4^-}\; CH_3CH_2\overset{\overset{\displaystyle O}{\|}}{C}-OH$$

Ketones generally are less reactive than aldehydes.

$$CH_3CH_2\overset{\overset{\displaystyle O}{\|}}{C}CH_3 \;\xrightarrow{MnO_4^-}\; \text{no reaction}$$

Carboxylic Acids. The carboxyl group is the functional group in organic acids.

$$\overset{\overset{\displaystyle O}{\|}}{-C}-OH \quad \text{abbreviated as} \quad -COOH$$

This group is weakly acidic. The ionizable hydrogen atom accounts for the acidic properties of carboxylic acids.

$$CH_3COOH(aq) \;\rightleftharpoons\; CH_3COO^-(aq) + H^+(aq)$$

A homologous series of organic acids can be generated starting with formic acid.

HCOOH	methanoic acid (formic acid)
CH_3COOH	ethanoic acid (acetic acid)
CH_3CH_2COOH	propanoic acid
$CH_3CH_2CH_2COOH$	butanoic acid

Fatty acids are carboxylic acids that contain more than four carbon atoms. Two fatty acids are:

$CH_3(CH_2)_{14}COOH$	palmitic acid
$CH_3(CH_2)_{16}COOH$	stearic acid

Esters. The ester functional group is

$$\overset{\overset{\displaystyle O}{\|}}{-C}-OR \quad \text{or} \quad -COOR$$

where R stands for any alkyl group. This functional group resembles the carboxylic acid group except that the R group replaces the H atom. Carboxylic acids react with alcohols to form the compounds called esters. The reaction that changes an acid to an ester is called an **esterification** reaction.

$$\underset{\text{acetic acid}}{CH_3\overset{\overset{\displaystyle O}{\|}}{C}-OH} + \underset{\text{methanol}}{H-OCH_3} \;\rightarrow\; \underset{\text{methyl acetate}}{CH_3\overset{\overset{\displaystyle O}{\|}}{C}-OCH_3} + H_2O$$

Esterification is an example of a *condensation* reaction. A "condensation" is a reaction in which parts of two molecules become joined by formation of a new covalent bond to make a new, larger molecule. Esterification reactions are reversible, and at equilibrium a mixture is formed that contains all four substances.

Saponification is the alkaline hydrolysis of an ester. In this reaction the base reacts directly with an ester to split it into *a salt of an organic acid* and *an alcohol*.

$$C_2H_5COOCH_3 + NaOH \rightarrow C_2H_5COONa + CH_3OH$$

Amines. Organic bases contain the —NH_2 functional group, which is called an amino group. Molecules containing an amino group are called amines. The amines may be considered as derivatives of ammonia (NH_3). Amines have the general formula R_3N, where R may be H, an alkyl group, or an aromatic group. Molecules of the type RNH_2 are called primary amines. Those with R_2NH are secondary amines, and R_3N is a tertiary amine.

EXAMPLE 24.4   Identifying Functional Groups

Indicate the functional groups in the following formulas:

a. $C_5H_{11}OH$
b. CH_3CHO
c. $C_3H_7OCH_3$
d. $CH_3COC_2H_5$
e. CH_3COOCH_3

•Method of Solution

Learning to recognize functional groups requires memorization of their structural formulas. Table 24.5 (textbook) shows a number of the important functional groups.

a. $C_5H_{11}OH$ contains a *hydroxyl* group. It is an alcohol.

b. CH_3CHO is a way to represent $CH_3\overset{\displaystyle O}{\overset{\|}{C}}$—H on a single line of type.
   C=O is a carbonyl group. CH_3CHO is an aldehyde.

c. $C_3H_7OCH_3$ contains a C—O—C group and is an ether.

d. $CH_3COC_2H_5$ is a way to represent a ketone on a single line. C=O is a carbonyl group.

e. CH_3COOCH_3 is a condensed structural formula for:

$$CH_3-\overset{\displaystyle O}{\overset{\|}{C}}-O-CH_3$$

   CH_3COOCH_3 is an ester functional group.

EXAMPLE 24.5   Predicting Reaction Products

Predict the product or products of the following reactions:

a. $CH_3CH(OH)CH_3 + Cr_2O_7^{2-} \rightarrow$
b. $CH_3CHO + MnO_4^- \rightarrow$
c. $HCOOH + CH_3OH \rightarrow$
d. $CH_3CH_2COOCH_3 + NaOH \rightarrow$

•Method of Solution

a. This is the reaction of a secondary alcohol with an oxidizing agent. The product is a ketone:

$$CH_3-\overset{\displaystyle O}{\overset{\|}{C}}-CH_3$$

b. This is the reaction of an aldehyde with an oxidizing agent. The product is a carboxylic acid: CH_3COOH.

c. The reaction of an alcohol with an acid yields an ester and water.

$$\overset{\displaystyle O}{\overset{\displaystyle \|}{HCOCH_3}} + H_2O$$

d. This is a saponificaton of an ester to form the sodium salt of a carboxylic acid and an alcohol.

$$\overset{\displaystyle O}{\overset{\displaystyle \|}{CH_3CH_2CONa}} + CH_3OH$$

EXERCISES

13. Draw structures for the following functional groups: aldehyde, amine, carboxylic acid, ester.
14. Give two chemical properties of alcohols.
15. Oxidation of what type of compound yields a ketone?
16. What elements are present in an amine group?
17. Which of the following functional groups exhibit hydrogen bonding? carboxylic acids, alcohols, ethers.
18. What are the reactants in a saponificaton reaction that yields $C_6H_{13}COONa$ and C_2H_5OH?
19. Identify the functional group or type of molecule in the following.
    a. CH_3OH
    b. CH_3CHO
    c. CH_3COCH_3
    d. CH_3COOCH_3
    e. $CH_3CH_2CH_3$

PRACTICE TEST

1. Give the systematic name for each of the following structural formulas.
    a. $CH_3(CH_2)_7CH_3$

    b.
$$\begin{array}{c} \quad\quad\quad\quad CH_3 \\ \quad\quad\quad\quad | \\ CH_3-CH_2-CH-CH=CH_2 \end{array}$$

    c.
$$\begin{array}{c} CH_3-CH_2-CH-CH_2-OH \\ | \\ CH_2-CH_3 \end{array}$$

    d.
$$\begin{array}{c} CH_3-CH_2-C\equiv C-CH_2-CH-CH_3 \\ \quad\quad\quad\quad\quad\quad\quad\quad | \\ \quad\quad\quad\quad\quad\quad\quad\quad C_3H_7 \end{array}$$

    e.
$$\begin{array}{c} \quad\quad\quad\quad\quad CH_2-CH_3 \\ \quad\quad\quad\quad\quad | \\ CH_3-CH_2-CH-CH-CH-CH_3 \\ \quad\quad\quad | \quad\quad | \\ \quad\quad\quad CH_2-CH_2\ CH_3 \\ \quad\quad\quad\quad\quad | \\ \quad\quad\quad\quad\quad CH_3 \end{array}$$

f.

g.

$$
\begin{array}{ccc}
\text{H} & & \text{CH}_2\text{CH}_3 \\
\backslash & & / \\
& \text{C} = \text{C} & \\
/ & & \backslash \\
\text{CH}_3 & & \text{H}
\end{array}
$$

h. $CH_3-CH-CH_2-CH_3$
$\qquad\quad |$
$\qquad\quad OH$

i. CH_3-OH

2. Draw structural formulas for:
   a. 3-ethyl-4-methyl-4-isopropylheptane
   b. 3-bromo-2,5-dimethyl-*trans*-3-hexene
   c. 3-methyl-3-hexanol
   d. trichloroacetic acid

3. Draw structural formulas of all isomers of C_5H_{10}. Include both structural and geometric isomers.

4. List the functional groups by name that are shown in the following molecules.
   a. $CH_3-CH = CH-CH-CH_2-NH_2$
      $\qquad\qquad\qquad\quad |$
      $\qquad\qquad\qquad\quad OH$

   b.
   $$
   \begin{array}{cccc}
   & O & O & O \\
   & \| & \| & \| \\
   HO-C-CH_2-C-CH_2-CH_2-CH
   \end{array}
   $$

5. Give the structure of the organic product of each of the following reactions:
   a.
   $$
   \begin{array}{c}
   OH \\
   | \\
   CH_3CH_2CHCH_3
   \end{array}
   \xrightarrow[\text{agent}]{\text{oxidizing}}
   $$

   b.
   $$
   \begin{array}{c}
   O \\
   \| \\
   CH_3CH
   \end{array}
   \xrightarrow[\text{agent}]{\text{oxidizing}}
   $$

   c. $CH_3C \equiv CH + Cl_2 \rightarrow$

   d. $C_3H_8 + Cl_2 \xrightarrow{\text{light}}$

   e.
   $$
   \bigcirc\!\!=\; + H_2 \xrightarrow[\text{catalyst}]{\text{Pt}}
   $$

   f. $CH_3CH_2CH = CHCH_2CH_3 + H_2O \xrightarrow[\text{dil}]{H_2SO_4}$

   g. $CH_3CHCH_3 + Cr_2O_7^{2-} \rightarrow$
      $\qquad\quad |$
      $\qquad\quad OH$

h.

$$CH_3CH_2CCH_3 + Cr_2O_7^{2-} \rightarrow$$
(with C=O above the second C)

i. $+ H_2O \xrightarrow[\text{dil}]{H_2SO_4}$

j.

$$CH_3COH + CH_3OH \xrightarrow{H+}$$
(with C=O above COH)

k. $CH_3CH_2CH = CHCH_3 + H_2 \xrightarrow{Pt}$

6. a. Sketch structural formulas of the hydrolysis products of the following ester:

$$CH_3-CH-C-O-CH_3 + H_2O \xrightarrow{H+}$$
(with CH₃ above the CH, and O above the C)

b. Sketch structures of the products of the following saponification reaction:

$$CH_3(CH_2)_{12}COO(CH_2)_4CH_3 + NaOH \rightarrow$$

7. Define the following terms:
   a. an alkene   b. homologous series   c. an addition reaction   d. functional group.

8. How many structural isomers are there of $C_4H_{10}O$?

9. Show how the following chemical change can be carried out in two steps.

$$CH_3CH_2CH_2OH \rightarrow CH_3CHCH_3$$
(with OH below the second C)

ANSWERS

Exercises

1. C_nH_{2n+2}, C_nH_{2n}, and C_nH_{2n-2}
2. alkenes and alkynes
3. 3-methylhexane
4. 4-methyl-2-pentene
5. a. 2-fluoropropane   b. 2-chloro-4,4-dimethylpentane
6.

$$CH_3-CH_2-C-CH_2-CH_3$$
(with CH₃ above and CH₃ below the central C)

7. CH_3CHCl—CH_2Cl

8. CH_3CHBr—CH_3

9. a. 2-heptene   b. 1-hexyne

10. Six isomers. 1–Butene, *cis*- and *trans*-2-butene, 2-methyl propene and 2 cycloalkanes.

11. a. o-dinitrobenzene   b. p-dibromobenzene   c. m-bromonitrobenzene

12. Benzene is much less reactive than alkenes. Rather than addition reactions, benzene undergoes substitution reactions.

13.

$$
\begin{array}{cccc}
O & H\ \ O & & O \\
\| & |\ \ \ \ \| & & \| \\
\text{—C—H} & \text{—N—H—C—O—H} & & \text{—C—O—R}
\end{array}
$$

14. Alcohols react with carboxylic acids to form esters. Also alcohols are easily oxidized.

15. A secondary alcohol

16. Nitrogen and hydrogen

17. Carboxylic acids and alcohols

18. An ester; $C_6H_{13}COOC_2H_5$ and NaOH.

19. a. alcohol   b. aldehyde   c. ketone   d. ester   e. alkane

Practice Test

1. a. *n*-nonane
   b. 3-methyl-1-pentene
   c. 2-ethyl-1-butanol
   d. 6-methyl-3-nonyne
   e. 3,4-diethyl-2-methylheptane
   f. 4-ethyl-3-methylheptane
   g. *trans*-2-pentene
   h. 2-butanol
   i. methanol

2. a.

$$
\begin{array}{c}
\qquad\qquad\ C_2H_5\ \ \ CH_3 \\
\qquad\qquad\ |\qquad\ | \\
CH_3\!-\!CH_2\!-\!CH\!-\!C\!-\!CH_2\!-\!CH_2\!-\!CH_3 \\
\qquad\qquad\qquad | \\
\qquad\qquad\ CH_3\!-\!CH\!-\!CH_3
\end{array}
$$

   b.

$$
\begin{array}{c}
\qquad\ CH_3\qquad\ H \\
\qquad\ |\qquad\quad | \\
CH_3\!-\!CH\!-\!C\!=\!C\!-\!CH\!-\!CH_3 \\
\qquad\quad |\qquad\qquad | \\
\qquad\quad Br\qquad\quad\ CH_3
\end{array}
$$

   c.

$$
\begin{array}{c}
\qquad\qquad\ CH_3 \\
\qquad\qquad\ | \\
CH_3\!-\!CH_2\!-\!C\!-\!CH_2\!-\!CH_2\!-\!CH_3 \\
\qquad\qquad\ | \\
\qquad\qquad\ OH
\end{array}
$$

   d. CCl_3–$COOH$

3. There are five alkenes: 1-pentene, 3-methyl-1-butene, 2-methyl-1-butene, and *cis*- and *trans*--2-pentene. There are three cycloalkanes: cyclopentane, methylcyclobutane, and dimethylcyclopropane.

4. a. Carbon-carbon double bond, hydroxyl, amine
   b. Carboxyl, carbonyl (ketone), carbonyl (aldehyde)

5. a.
$$CH_3CH_2\overset{\overset{\displaystyle O}{\|}}{C}CH_3$$

b. CH_3COOH
c. $CH_3CCl_2CHCl_2$
d. C_3H_7Cl, $C_3H_6Cl_2$, $C_3H_5Cl_3$, etc.

e.

f. $CH_3CH_2CH_2\overset{\overset{\displaystyle }{}}{C}HCH_2CH_3$
$\qquad\qquad\qquad |$
$\qquad\qquad\qquad OH$

g.
$$CH_3\overset{\overset{\displaystyle O}{\|}}{C}CH_3$$

h. no reaction

i.

j.
$$CH_3\overset{\overset{\displaystyle O}{\|}}{C}OCH_3$$

k. $CH_3CH_2CH_2CH_2CH_3$

6. a.
$$CH_3\overset{\overset{\displaystyle CH_3}{|}}{C}H\overset{\overset{\displaystyle O}{\|}}{C}\text{—}OH + CH_3OH$$

b. $CH_3(CH_2)_{12}COONa + CH_3(CH_2)_4OH$

7. a. A straight chain or branch chain hydrocarbon with one or more double bonds
    b. Organic compounds with the same functional group that differ only in the number of carbon atoms.
    c. A reaction in which a small molecule is added to an organic compound. The organic molecule must contain a $C = C$ bond or $C = O$ (carbonyl) group.
    d. A characteristic grouping of atoms that imparts certain chemical and physical properties when incorporate into organic molecules

8. Seven (four alcohols and three ethers)

9.
$$CH_3CH_2CH_2OH \xrightarrow[\text{conc}]{H_2SO_4} CH_3CH = CH_2 + H_2O$$

$$CH_3CH = CH_2 + H_2O \xrightarrow[\text{dil}]{H_2SO_4} CH_3\overset{\overset{\displaystyle }{}}{C}HCH_3$$
$$\qquad\qquad\qquad\qquad\qquad\qquad\qquad\qquad |$$
$$\qquad\qquad\qquad\qquad\qquad\qquad\qquad\qquad OH$$

Chapter Twenty-Five
SYNTHETIC AND NATURAL ORGANIC POLYMERS

- Synthetic Organic Polymers (25.1 – 25.2)
- Proteins (25.3)
- Nucleic Acids (25.4)

SYNTHETIC ORGANIC POLYMERS (25.1 – 25.2)

STUDY OBJECTIVES

1. Define monomer and give several examples of addition polymers.
2. Define copolymer and give several examples of condensation polymers.
3. Describe the steps in the addition polymerization process.

Polymers. The word polymer means "many parts." A **polymer** is a compound with an unusually high molecular mass, consisting of a large number of small molecular units that are linked together. The small unit that is repeated many times is called a **monomer**. A typical polymer molecule contains a chain of monomers several thousand units long. Polymers are often called macromolecules.

Proteins, nucleic acids, carbohydrates, and rubber are natural polymers. Synthetic polymers such as nylon, polyester, and polyethylene are organic compounds.

Addition Polymers. Addition polymers are made by adding monomer to monomer until a long chain is produced. Ethylene and its derivatives are excellent monomers for addition polymers. In an addition reaction, the polymerization process is initiated by a radical. When ethylene is heated to 250°C under high pressure (1000–3000 atm) in the presence of a little oxygen or benzoyl peroxide (the initiator), addition polymers with molecular masses of about 30,000 amu are obtained. This reaction is represented by:

ethylene a segment of polyethylene

The general equation for addition polymerization is:

monomer   repeating unit

Polyethylene is an example of a **homopolymer**, which is a polymer made up of only one type of monomer. Substitution of one or more hydrogen atoms in ethylene with Cl atoms, phenyl groups, acetate,

478

cyano groups, and F atoms provides a wide selection of monomers from which to make various homopolymers. For example, substitution of a Cl atom for an H atom in ethylene gives the monomer called vinyl chloride, $CH_2=CHCl$. Polymerization of vinyl chloride yields the polymer, polyvinyl chloride.

vinyl chloride   polyvinyl chloride
monomer   repeating unit

Table 25.1 in the text gives the names, structures, and uses of a number of monomers and addition polymers.

Addition Polymerization. The addition polymerization process occurs by several steps which are summarized below. The first step is called *initiation*. This is followed by a *chain growth* process, and finally a *termination* reaction.

1. *Initiaition.* Polymerization is initiated by a radical. A radical (also called a free radical) is a species which has an unpaired electron. The symbol for a radical contains the usual letters for the elements followed by a dot to represent the unpaired electron. $CH_3\cdot$ is the symbol for a methyl radical. The letter R is the general symbol for an alkyl group and $R\cdot$ for an alkyl radical.

In a polymerization process, first the initiator molecule is dissociated by heating to yield two radicals. Then the radical adds to an ethylene molecule which generates a new radical.

$$R\text{—}R \rightarrow 2R\cdot$$
$$R\cdot + CH_2=CH_2 \rightarrow R\text{—}CH_2\text{—}CH_2\cdot$$

2. *Chain Growth.* Next, the free radical formed above can add to another molecule of ethylene.

$$R\text{—}CH_2\text{—}CH_2\cdot + CH_2=CH_2 \rightarrow R\text{—}CH_2\text{—}CH_2\text{—}CH_2\text{—}CH_2\cdot$$

The length of the carbon chain grows rapidly as the last free radical formed reacts with yet another ethylene molecule, and so on.

$$R\text{—}(CH_2\text{—}CH_2)_n\text{—}CH_2\text{—}CH_2\cdot$$

3. *Termination.* The polymerization process continues until a termination reaction occurs. When the radical ends of two chains meet they may combine. When this happens there is no new radical formed and the chain lengthening process ceases. Of course the result of this termination is the formation of a molecule of polyethylene which may contain up to 800 carbon atoms.

$$R\text{—}(CH_2\text{—}CH_2)_n\text{—}CH_2\text{—}CH_2\cdot + \cdot CH_2\text{—}CH_2\text{—}(CH_2\text{—}CH_2)_n\text{—}R \rightarrow$$
$$R\text{—}(CH_2\text{—}CH_2)_n\text{—}CH_2\text{—}CH_2\text{—}CH_2\text{—}CH_2\text{—}(CH_2\text{—}CH_2)_n\text{—}R$$

Condensation Polymers. **Copolymers** are polymers that contain two or more different monomers. When monomers A and B are linked by condensation reactions a uniform copolymer, —ABABAB— can be formed. Nylon and polyesters, such as the well-known Dacron, are copolymers.

Polyester is made by an esterification reaction. When one monomer is an alcohol and the other is a carboxylic acid, they can be joined by an esterification reaction. The alcohol and the acid both must contain two functional groups. The monomers in polyester are the dicarboxylic acid called phthalic acid, and the dialcohol, ethylene glycol.

phthalic acid   ethylene glycol

Condensation reactions differ from addition reactions in that the former always result in the formation of a small molecule such as water. Polyesters are produced from the esterification reaction between an alcohol and an acid. When phthalic acid and ethylene glycol react to form an ester, the first products are:

This molecule can react further from both ends. When this product reacts with another molecule of the diacid, the polymer chain grows longer.

Segment of a condensation polymer chain

The general formula for this polyester is:

EXAMPLE 25.1   Monomers and Polymers

Write the formulas of the monomers used to prepare the following polymers:
a. Teflon   b. Polystyrene   c. PVC

•Method of Solution

Refer to Table 25.1 of the text.
a.   Teflon is an addition polymer with the formula $-(CF_2-CF_2)_n$. It is prepared from the monomer tetrafluoroethylene ($CF_2=CF_2$).
b.   The monomer used to prepare polystyrene is styrene.

polystyrene   styrene

c.   Polyvinylchloride is prepared by the successive addition of vinyl chloride molecules, $CH_2=CHCl$.

EXERCISES

1. List three steps occurring during addition polymerization.
2. What monomer is used to make Teflon $\left(CF_2-CF_2\right)_n$?
3. Is Teflon a homopolymer or a copolymer?

PROTEINS (25.3)

STUDY OBJECTIVES

1. Write a general formula for an amino acid.
2. Describe the condensation reaction between two amino acids that result in the formation of a dipeptide.
3. Discuss the terms used to describe the four levels of protein structure.

Proteins. Proteins are truly giant molecules having molecular masses that range from about 10,000 to several million amu. Proteins play many roles in living organisms where they function as catalysts (enzymes), transport molecules (hemoglobin), contractile fibers (muscle), protective agents (blood clots), hormones (chemical messengers), and structural members (feathers, horn, nails). The word *protein* comes from the Greek word *proteios,* meaning "first." From this partial list of functions it is easy to see why proteins occupy "first place" among biomolecules in their importance to life.

Amino Acids. Even though each protein is unique, all proteins are built from the same set of amino acids. An amino acid consists of an amino group, a carboxylic acid group, a hydrogen atom, and a distinctive R-group, all bonded to the same carbon atom.

$$H_2N-\underset{\underset{R}{|}}{\overset{\overset{H}{|}}{C}}-\overset{\overset{O}{\|}}{C}-OH$$

All amino acids in proteins have a common structural feature. This is the attachment of the amino group to the carbon atom adjacent to the carboxylic acid group. The R-group is different in each amino acid, and some 20 different R-groups are found in the proteins from natural sources. The structural formulas of the amino acids are shown in Table 25.2 (text).

Amino acids in solution at near neutral pH exist as dipolar ions. This form results from the transfer of a proton from the carboxylic acid group to the basic amine group

$$^+H_3N-\underset{\underset{R}{|}}{\overset{\overset{H}{|}}{C}}-\overset{\overset{O}{\|}}{C}-O^-$$

In proteins the amino acid units are hooked together to form a polypeptide chain. The carboxyl group of one amino acid is joined to the amino group of another amino acid by the formation of a peptide bond.

$$\begin{array}{ccc}
\overset{\text{H}}{\underset{\text{R}_1}{\overset{|}{\underset{|}{\text{N}}}}}\ \overset{\text{O}}{\overset{\|}{}} & & \\
{}^{+}\text{H}_3\text{N}-\text{C}-\text{C}-\text{O}^{-} + {}^{+}\text{H}_3\text{N}-\text{C}-\text{C}-\text{O}^{-} \rightarrow {}^{+}\text{H}_3\text{N}-\text{C}-\text{C}-\text{NH}-\text{C}-\text{C}-\text{O}^{-} + \text{H}_2\text{O}
\end{array}$$

This type of reaction is another example of a condensation reaction. The new C—N covalent bond is called a *peptide bond* or an *amide bond*. The amide functional group is present in all proteins:

O
‖ ⟵ peptide bond
—C — NH —

amide group

The molecule above in which two amino acids are joined is called a *dipeptide*. Peptides are structures intermediate in size between amino acids and proteins. The term *polypeptide* refers to long molecular chains containing many amino acid units. An amino acid unit in a polypeptide chain is called a *residue*.

Structure of Proteins. Proteins are so complex that four levels of structural features have been identified. The structure of proteins is extremely important in determining just how efficiently and effectively a protein will function. A stretched-out or unfolded polypeptide chain does not exhibit biological activity, and is said to be *denatured*. The four levels of protein structure are summarized below. See Figure 25.13 of the text.

1. *The Primary Structure.* Each protein has a unique amino acid sequence of its polypeptide chain. It is the amino acid sequence that distinguishes one protein from another. Proteins differ in the numbers and kinds of amino acids, but especially in the sequence of amino acid residues.

2. *The Secondary Structure.* This refers to the spatial relationship of amino acid units that are close to one another in the sequence. A configuration that appears in many proteins is the α-helix shown in Figure 25.11 (textbook). In this configuration the polypeptide is coiled much like the arrangement of stairs in a spiral staircase. The tightly coiled polypeptide chain forms the inner part of the spiral stairway, and the R groups extend outward forming a helical pattern. The figure shows that the α-helix is stabilized by the presence of hydrogen bonds (dashed lines) between the C=O and NH groups in the peptide chains. The CO group of each amino acid is hydrogen bonded to the NH group of the amino acid that is located four amino acids ahead in the sequence. Thus amino acids spaced four apart in the sequence are actually close to one another due to the coiled or spiral arrangement.

   The α-helix is the main structural feature of the oxygen-storage protein, myoglobin, and of many other proteins. The presence of this feature in a polypeptide chain is shown in Figure 25.11 (text). The β-pleated sheet is another common protein structure. In this structure a polypeptide chain interacts strongly with adjacent chains by forming many hydrogen bonds (Figure 25.12 text).

3. *The Tertiary Structure.* Myoglobin like many other proteins is a globular protein. The polypeptide chain is folded into a compact globular shape. The folding of the chain results in some amino acid units being in very close proximity to each other even though they are widely separated in the amino acid sequence. The term *tertiary structure* refers to the spatial relationship of amino acid units that are far apart in the sequence, or more simply, to just its three-dimensional structure. In myoglobin, for example, the tertiary structure is a unique three-dimensional shape resulting from the folding of the polypeptide chain. The folding takes place quite naturally in aqueous medium due to the interactions of hydrophobic and hydrophilic R groups with water. The tertiary structure is stabilized by hydrogen bonding, dispersion forces, and ionic forces.

4. *The Quaternary Structure.* Proteins that consist of more than one polypeptide chain exhibit an additional level of structural organization called the *quaternary structure*. This structural feature involves the way in which separate chains fit together. Hemoglobin, the oxygen-carrying molecule of the blood, exhibits a quaternary structure (Chemistry in Action). It consists of four separate polypeptide chains or subunits (labeled α_1, α_2, β_1, β_2). The quaternary structure results from interaction between chains. Ionic forces and hydrogen bonds are important in holding the subunits together.

EXAMPLE 25.2   Amino Acids

Find and name five amino acids in Table 25.2 (textbook) that have polar R groups.

•Method of Solution

Recall that polar groups are those that have relatively large electronegativity differences between bonded atoms. Thus, —COOH, —OH, and —SH are polar R groups. *Answer:* The R groups in *aspartic acid* and *glutamic acid* contain carboxylic acid groups. The R groups in *serine* and *threonine* contain hydroxyl groups. And the R group in cysteine contains a polar —SH group.

EXAMPLE 25.3   A Polypeptide Chain

Sketch a portion of a polypeptide chain consisting of the amino acids cysteine, glycine, valine, and phenylalanine. Point out the peptide bonds and amide groups.

•Method of Solution

The main backbone of a polypeptide chain is made up of carbon atoms, and the amide group repeating alternately along the chain. The abbreviation for the chain is cys—gly—val—phe.

Table 25.2 (text) gives the structures of the amino acids. Substitute for each R group shown above, the distinctive R groups of the four amino acids. The chain structure will be:

There are four peptide bonds. They are the C—N bonds. Each –CO—NH– is an amide group.

EXAMPLE 25.4   Tertiary Structure of Proteins

What kinds of forces stabilize the tertiary structures of proteins?

•Method of Solution

The term tertiary structure is given to the folded three-dimensional structure. The polypeptide chain folds into a specific three-dimensional structure under the influence of hydrophobic and hydrophilic interactions. The hydrophilic R groups (polar or ionic groups are attracted to water) are spread uniformly over the outside surface of the folded polypeptide chain where they can interact with water via hydrogen bonding and ion-dipolar interactions. The chain folds in such a way that the hydrophobic R groups (nonpolar groups are repelled by water) appear on the "inside" of the molecule where water molecules cannot reach.

EXERCISES

4. What is an amino acid?
5. Sketch the dipolar ion of alanine.
6. What is an amide group and a peptide bond?
7. What is the primary structure of a protein?
8. What is a denatured protein?

NUCLEIC ACIDS (25.4)

STUDY OBJECTIVES

1. Describe the composition of nucleic acids.
2. Distinguish chemical and structural differences between DNA and RNA.
3. Draw structures of nucleotides and indicate where they link together to form strands of nucleic acids.

Components of DNA and RNA. The chemical composition of the cell nucleus was first studied in the 1860s by Friedrich Miescher. He found the major components to be protein and a new material not previously isolated. This material was found to be acidic, and so it was referred to as *nucleic acid.*

Nucleic acids are now known to be giant molecules with molecular masses in the range 1 to 10 billion amu. These molecules carry information in the form of the genetic code. Enough information is stored in nucleic acid molecules to allow the complete assembly of an entire organism. The nucleic acid molecule called *DNA* carries this information from generation to generation.

Hydrolysis of nucleic acid showed that it is composed of one part phosphoric acid, one part sugar, and one part nitrogen base. One of two sugars are present—ribose or deoxyribose. Five different nitrogen bases are found in nucleic acids. Their names are adenine, thymine, guanine, cytosine, and uracil (Figure 25.17 text).

Two types of nucleic acids are recognized. These are called DNA and RNA. The compositions of DNA and RNA are quite similar, but they differ in two significant ways. DNA contains the sugar deoxyribose and is called deoxyribonucleic acid. RNA contains the sugar ribose, hence the name ribonucleic acid. The second difference is that DNA contains the four bases adenine, thymine, guanine, and cytosine, while RNA contains adenine, uracil, guanine, and cytosine. DNA contains thymine but no uracil, whereas RNA contains uracil but no thymine. Table 25.1 summarizes the building blocks of DNA and RNA and gives useful abbreviations.

Ribose

Deoxyribose

Table 25.1 The Component Parts of DNA and RNA.

	DNA	RNA
Acid:	phosphoric acid (P)	phosphoric acid (P)
Sugar:	deoxyribose (D)	ribose (R)
N bases:	adenine (A)	adenine (A)
	thymine (T)	uracil (U)
	guanine (G)	guanine (G)
	cytosine (C)	cytosine (C)

Nucleotides. The repeating unit in nucleic acids is called a nucleotide. It is a combination of a phosphate group, a five-carbon sugar, and a nitrogen base. See Figure 25.18 (text) for the full structure of the nucleotide, deoxyadenosine monophosphate (dAMP), which contains the sugar deoxyribose, and the base adenine. Figure 25.1 below shows the structure of the nucleotide containing the base cytosine, the sugar ribose and the phosphate group. It is named cytosine monophosphate (CMP).

Figure 25.1 The molecular structure of the nucleotide cytosine monophosphate (CMP). The phosphoric acid groups of nucleotides are quite acidic, and at pH 7 these groups exist mainly as the dianion as shown.

Nucleic Acids. A strand of DNA or RNA is constructed by linking the nucleotides together with a bond from the sugar unit of one nucleotide to the phosphate unit of another nucleotide. Such a strand of DNA is represented in abbreviated form below. To simplify DNA and RNA structures we will use the abbreviations in Table 25.1: P = phosphate, D = deoxyribose, and A, G, T, and C for the bases.

In the 1940s, Edwin Chargaff studied the composition of DNA. His analysis of the base composition of DNA showed that the amount of adenine (A) always equaled that of thymine (T), and that the amount of guanine (G) equaled that of cytosine (C). These relations became known as Chargaff's rules. Watson and Crick in 1953 proposed a two stranded structure for DNA which provided an explanation of Chargaff's rules. The two-stranded structure can be shown as follows:

```
      /                 \
      P                 P
      \                 /
       D—C · · · G—D
      /                 \
      P                 P
      \                 /
       D—A · · · T—D
      /                 \
      P                 P
      \                 /
       D—G · · · C—D
      /                 \
      P                 P
      \                 /
       D—T · · · A—D
      /                 \
      P                 P
      \                 /
```

Adenine in one strand is always paired with thymine in the other strand. Guanine is always paired with cytosine. The two strands are not identical, rather they are *complementary*. Base pairing and the resulting association of the two strands are the result of hydrogen bonding. Hydrogen atoms in a base in one strand form hydrogen bonds to oxygen and nitrogen atoms of a base attached to the other strand (see Figure 25.19(a) in textbook). X-ray data suggested that DNA had the helical structure as shown in Figure 25.19(b) of the text.

RNA, on the other hand, does not follow the base-pairing rules. X-ray data and other evidence ruled out a double-helical structure for RNA. RNA is single stranded.

EXAMPLE 25.5   Nucleotide Structures

Sketch the structure of the nucleotide containing uracil that appears in RNA.

•Method of Solution

Nucleotides from RNA contain ribose as the sugar and and a phosphate dianion. The structure of the base uracil is given in Figure 25.17 in the text. The phosphate group in bonded to the sugar which in turn is bonded to the nitrogen base, uracil. The structure is:

Phosphate unit   Ribose unit   Uracil unit

EXAMPLE 25.6 Base Pairing

What types of forces cause base pairing in the double-stranded helical DNA molecule?

•Method of Solution

The N-bases thymine (T) and cytosine (C) in one strand of DNA form hydrogen bonds to the N-bases adenine (A) and guanine (G), respectively, in the other DNA strand. Hydrogen atoms that are covalently bonded to nitrogen carry a partial positive charge. These H atoms are attracted to lone electron pairs on oxygen and nitrogen atoms of another base. Since two complementary bases are attached to different strands, the hydrogen bonds hold the strands together.

adenine   thymine

EXERCISES

9. What is the 5-carbon sugar in DNA? In RNA?
10. Name four nitrogen bases found in DNA.
11. What type of bond is responsible for base pairing in DNA?
12. Only DNA has base-pairing rules. Why doesn't RNA have base-pairing rules?
13. What is the repeating unit in nucleic acids?

PRACTICE TEST

1. Sketch the structures of the monomers from which these polymers are formed.

    a. $C\;\;H\;\;Cl\;\;H\;\;Cl\;\;H$

    ... (structure) ...

    $$-\overset{\displaystyle |}{\underset{\displaystyle |}{C}}-\overset{\displaystyle |}{\underset{\displaystyle |}{C}}-\overset{\displaystyle |}{\underset{\displaystyle |}{C}}-\overset{\displaystyle |}{\underset{\displaystyle |}{C}}-\overset{\displaystyle |}{\underset{\displaystyle |}{C}}-\overset{\displaystyle |}{\underset{\displaystyle |}{C}}$$

    $H\;\;CH_3\;\;H\;\;CH_3\;\;H\;\;\;\;CH_3$

    b. $-CH_2-CCl_2-CH_2-CCl_2-CH_2-CCl_2-$

2. Draw structures for the monomers used to make the following polyester.

    $$\left[\overset{\displaystyle O}{\overset{\displaystyle \|}{C}}-(CH_2)_4-\overset{\displaystyle O}{\overset{\displaystyle \|}{C}}-O-(CH_2)_3-O\right]_n$$

3. Sketch the structure of the monomer of natural rubber.

4. What chemical elements are found in proteins?

5. List several functions of proteins in living systems.

6. What is the role of enzymes in biochemical systems?

7. Describe the structure of an amino acid.

8. Describe a dipeptide. A polypeptide.

9. Draw the structural formula of the tripeptide formed from the sequence of amino acids below. (See Table 25.2 of the textbook.) Label the peptide bonds and R groups.
   Ser—Cys—Lys

10. Consider two polypeptide chains
    1. Lue—Phe—Pro—Gly—Ala
    2. Gly—Ser—Lys—Asp—Tyr
    a. Which would be more soluble in water?
    b. Which would be more soluble in a nonpolar solvent?

11. Explain how the primary structure of a protein differs from the tertiary structure.

12. List four differences between DNA and RNA.

13. Sketch the structure of the nucleotide consisting of phosphate, deoxyribose, and adenine.

14. What is the biological role of DNA?

15. If the base sequence in one strand of DNA is A, T, G, C, T, then the base sequence in the complementary strand is __, __, __, __, __.

ANSWERS

Exercises

1. Initiation, chain growth, and termination.
2. $CF_2 = CF_2$
3. a homopolymer
4. An amino acid is a carboxylic acid that contains an amino group ($-NH_2$) attached to the carbon atom adjacent to the carboxylic acid functional group ($-COOH$).

5.
$$
{}^+H_3N-\underset{\underset{CH_3}{|}}{\overset{\overset{H}{|}}{C}}-\overset{\overset{O}{\parallel}}{C}-O^-
$$

6. An amide group is a $-CO-NH-$ group. The peptide bond is the C—N bond that links amino acids in a polypeptide or protein.
7. The sequence of amino acids in a peptide chain.
8. Denaturation is the loss of structure in a protien. The structure change may be a loss of secondary, tertiary, or quaternary structure. Sometimes the primary structure is lost.
9. Deoxyribose. Ribose.
10. Adenine, guanine, cytosine, and thymine.
11. A hydrogen bond.
12. RNA is single stranded, not double stranded.
13. A nucleotide.

Practice Test

1. a. $CH_3CH{=}CHCl$ b. $CH_2{=}CCl_2$

2.

$$\underset{}{HO{-}\overset{\displaystyle O}{\overset{\|}{C}}{-}(CH_2)_4{-}\overset{\displaystyle O}{\overset{\|}{C}}{-}OH} \qquad HOCH_2CH_2CH_2OH$$

3.

$$CH_2{=}\overset{\displaystyle CH_3}{\overset{|}{C}}{-}CH{=}CH_2$$

4. C, H, O, N, and S
5. Catalysts, transport, contractile, protection, hormones, structural elements.
6. Enzymes are catalysts.
7. Amino acids consist of an amino group, a carboxylic acid group, a hydrogen atom, and R group all bonded to the same carbon atom.
8. A dipeptide is two amino acids linked by a peptide bond. A polypeptide consists of many amino acid residues each linked to the next by peptide bonds.

9.

10. a. 2   b. 1
11. The primary structure of a protein refers to the number and the sequence of amino acids within the polypeptide chain. The chain twists and folds so that some amino acids far apart in sequence come into proximity with each other. The tertiary structure is the 3D structure that shows the folding of the polypeptide chain.
12. (1) RNA contains uracil but no thymine; DNA contains thymine but no uracil. (2) RNA contains ribose as the 5-carbon sugar; DNA contains deoxyribose. (3) RNA is single stranded; DNA is double helical. (4) DNA is found only in the nucleus; RNA is found outside the nucleus.

13.

14. DNA is the molecule of heredity; it is responsible for passing genetic information from one generation to the next.
15. T, A, C, G, A